3704634960

D1766934

HANDBOOK OF DIGITAL FORENSICS OF MULTIMEDIA DATA AND DEVICES

HANDBOOK OF DIGITAL FORENSICS OF MULTIMEDIA DATA AND DEVICES

Edited by

Anthony T.S. Ho and Shujun Li

Department of Computing and Surrey Centre for Cyber Security (SCCS)
University of Surrey, Guildford, UK

IEEE Press

Library of Congress Cataloging-in-Publication Data

Ho, Anthony T. S., 1958-
Handbook of digital forensics of multimedia data and devices / Anthony T. S. Ho and Shujun Li, Department of Computing and Surrey Centre for Cyber Security (SCCS), University of Surrey, UK.
        pages   cm
Summary: "Focuses on the interface between digital forensics and multimedia forensics, bringing two closely related fields of forensic expertise together to identify and understand the current state-of-the-art in digital forensic investigation"– Provided by publisher.
   Includes bibliographical references and index.
   ISBN 978-1-118-64050-0 (hardback)
1. Criminal investigation–Technological innovations.   2. Electronics in criminal investigation.   I. Title.
   HV8073.H596 2015
   363.25′62–dc23
                                        2015011187

A catalogue record for this book is available from the British Library.

Set in 11/13pt Times by SPi Global, Pondicherry, India

Printed in Singapore by C.O.S. Printers Pte Ltd

1   2015

Anthony T.S. Ho dedicates this book to his wife Jeanette and his sons, Shaun and Nick, for their encouragement and loving support throughout the preparation and completion of this book.

Shujun Li dedicates this book to his wife Shanny and his son Xiaocan (Luke) for their continuous support in the process of editing this book which lasted for over a year.

Contents

PART TWO   DIGITAL EVIDENCE EXTRACTION

PART FOUR   MULTIMEDIA CONTENT FORENSICS

13 Digital Image Forensics with Statistical Analysis   **483**

List of Contributors

Below is a list of all contributors to this book with their affiliations when they were working on this book.

- Guy Adams, HP Labs Bristol, UK
- Ivo Alberink, Netherlands Forensic Institute, The Hague, the Netherlands
- Banafshe Arbab-Zavar, University of Southampton, Southampton, UK
- Tiziano Bianchi, Politecnico di Torino, Torino, Italy
- John D. Bustard, University of Southampton, Southampton, UK
- Mandeep K. Dhami, Department of Psychology, Middlesex University, London, UK
- Jana Dittmann, Department of Computer Science, Otto-von-Guericke-University, Magdeburg, Germany
- Thomas Gloe, dence GmbH, Dresden, Germany
- Hagit Hel-Or, Department of Computer Science, University of Haifa, Haifa, Israel
- Anthony T.S. Ho, Department of Computing Surrey Centre for Cyber Society (SCCS), University of Surrey, Guildford, UK
- Bart Hoogeboom, Netherlands Forensic Institute, The Hague, the Netherlands
- Giuseppe F. Italiano, University of Tor Vergata, Rome, Italy
- Matthias Kirchner, University of Münster, Münster, Germany
- Bruce E. Koenig, BEK TEK LLC, Stafford, VA, USA
- Xiangwei Kong, Dalian University of Technology, Dalian, China
- Christian Kraetzer, Department of Computer Science, Otto-von-Guericke-University, Magdeburg, Germany
- Douglas S. Lacey, BEK TEK LLC, Stafford, VA, USA
- Chang-Tsun Li, University of Warwick, Coventry, UK
- Shujun Li, Department of Computing and Surrey Centre for Cyber Security (SCCS), University of Surrey, Guildford, UK
- Fabio Marturana, University of Tor Vergata, Rome, Italy

- Mark S. Nixon, University of Southampton, Southampton, UK
- Xunyu Pan, Frostburg State University, Frostburg, MD, USA
- Alessandro Piva, University of Florence, Florence, Italy
- Stephen Pollard, HP Labs Bristol, UK
- Shize Shang, Dalian University of Technology, Dalian, China
- Steven Simske, HP Printing and Content Delivery Lab (PCDL), Ft. Collins, USA
- Martin Steinebach, Fraunhofer Institute for Secure Information Technology SIT, Darmstadt, Germany
- Simone Tacconi, Ministry of the Interior, Rome, Italy
- Marco Tagliasacchi, DEI, Politecnico di Milano, Milano, Italy
- Stefano Tubaro, DEI, Politecnico di Milano, Milano, Italy
- Giuseppe Valenzise, CNRS LICI, Telecom ParisTech, Paris, France
- Derk Vrijdag, Netherlands Forensic Institute, The Hague, the Netherlands
- Xingjie Wei, University of Warwick, Coventry, UK
- Christian Winter, Fraunhofer Institute for Secure Information Technology SIT, Darmstadt, Germany
- Marcel Worring, University of Amsterdam, Amsterdam, the Netherlands
- York Yannikos, Fraunhofer Institute for Secure Information Technology SIT, Darmstadt, Germany
- Ido Yerushalmy, Department of Computer Science, University of Haifa, Haifa, Israel
- Sascha Zmudzinski, Fraunhofer Institute for Secure Information Technology SIT, Darmstadt, Germany

Foreword

Like many researchers and colleagues in multimedia security, I started working on multimedia forensics around 15 years ago, by focusing on the signal processing aspects of such a discipline. In the end, all we had to do was to apply a decades-old signal processing know-how, or slightly adapt it, to detect the subtle traces left in images, videos and audio files by the acquisition device, look for the artefacts introduced by common operators like interpolation, lossy compression, filtering, and so on, or analyze the content of images and videos to highlight inconsistencies of shadows, reflections and other geometric elements. Eventually, we could resort to machine learning tools to handle the most difficult problems for which precise analytical models were not available. After many years, I am now aware that there is much more to multimedia forensics than a bunch of tools to process multimedia signals. If the information extracted from the digital signals has to be exploited in a forensic scenario to produce digital evidence suitable for being used in a court, and for serving law enforcement agencies, the information extraction process must follow precise procedures matching national and international regulations, and the adopted tools must be designed in such a way to cope with the very specific problems and needs typical of the forensic arena. In a few words, a researcher or a practitioner wishing to enter this new and exciting field must be equipped with an interdisciplinary background involving disciplines such as criminology, law, psychology, economics, in addition to being an expert in multimedia signal processing.

For this reason, I welcome the effort made by Anthony T.S. Ho and Shujun Li in this handbook, to put together classical multimedia processing aspects, such as those covered in Parts III and IV of the handbook, together with the more technical and forensics-oriented issues discussed in Parts I and II. Thanks to their experience in the field and to the involvement of authors accustomed to apply the results of their research to real forensic cases; this handbook provides an excellent mix of the various themes broadly covered by multimedia forensics. As a newcomer in the field of law enforcement and digital forensics, I particularly appreciated the content of the first part of the handbook, providing a general view of some aspects that are usually neglected by researchers belonging to the multimedia signal processing community. On the other hand, as a signal processing expert, I enjoyed reading the second part of the handbook, since it provides an updated, complete and rigorous treatment of the most mature techniques developed by researchers in the last two decades. I am sure that this handbook has the potential to become a landmark for all those involved with multimedia forensics, either as researchers or as users.

I know by personal experience how challenging it is to assemble a handbook such as this one, and how difficult it is to engage the many valuable researchers and scientists in such an effort as Ho and Li have done. For this reason, I thank the two editors of the handbook for providing our community and all the future readers with such an excellent piece of work.

Mauro Barni
Department of Information
Engineering and Mathematics
University of Siena, Italy

Preface

Digital forensics is a rapidly growing field in both industry and academia due to the increasing number of criminal cases involving electronic evidence, which are limited not only to e-crime (a.k.a. cyber crime) but also to many traditional forms of crime since the use of digital computing devices and the internet has been so ubiquitous in today's highly digitized and networked society. While technically digital forensics is more about tools and techniques in computer science and electronic engineering, the need for digital forensics to serve law enforcement requires an inter-disciplinary approach involving many other subjects such as criminology, policing, law, psychology, economics, business and management. As a consequence of the increasing demands and the intensified research in digital forensics fields, a large number of books have been published on this topic. Searching the keyword 'digital forensics' on Amazon.com, we can find more than 5000 books published as of the time of this writing (28 October 2014).

One important area of digital forensics is forensic analysis of digital multimedia data and devices, which is commonly called 'multimedia forensics' by researchers working in this space. Research on multimedia forensics started in the late 1990s, largely grown from activities of the multimedia security community which conducts research on digital watermarking, steganography and steganalysis, especially on multimedia data. Many techniques developed in this research area have become mature and found their applications in real world, for example a number of commercial software tools have been developed and been used by law enforcement for recovering deleted and segmented digital images, identifying source devices of digital images and video sequences, detecting forged digital images, among other purposes.

While multimedia forensics as a 'visible' research area is relatively new, handling of multimedia data and devices in digital forensic practice has a much longer history. One typical example is closed-circuit television (CCTV) systems which first appeared in the 1940s and was used widely for video surveillance purposes since then. Nowadays

CCTV systems are largely digitized and often controlled over the internet, and they have been an important source of electronic evidence in many criminal cases. Another example is fingerprint and face recognition systems which have been widely used for many years, first using in analogue domain and later in digital domain. Clearly, forensic analysis and processing of electronic evidence generated by digital CCTV systems and digital fingerprint/face recognition systems require handling of multimedia data and devices, and often the techniques used are very similar to what researchers in multimedia forensics field are using. To some extent, forensic analysis of multimedia evidence has been an essential part of forensic sciences from the beginning, and it gradually moved into the digital domain when audio and imaging devices become digitized in the 1980s and 1990s. Handling multimedia data and devices in digital domain has become such an important area for law enforcement that nowadays many (if not most) law enforcement agencies with a digital forensic team also has a sub-team or a separate team focusing on multimedia evidence (see Chapter 1 of this book for examples). This can also be seen from the fact that two important Scientific Working Groups (SWGs) in digital forensics field in the United States, Scientific Working Group on Digital Evidence (SWGDE) and Scientific Working Group on Imaging Technology (SWGIT), were formed around the same time (1997–1998). Similarly, American Academy of Forensic Sciences (AAFS) has a Digital & Multimedia Sciences section whose name puts digital forensics and multimedia forensics together. In addition, audio evidence is traditionally covered as part of digital forensics and is also considered as an essential part of multimedia forensics.

As a sub-field of digital forensics, multimedia forensics or digital forensic of multimedia data and devices is growing, but so far has received less attention from digital forensics practitioners in law enforcement and industry. Searching 'multimedia forensics' in Amazon.com, we can find only less than 200 books. Most books on multimedia forensics have a narrower focus on 'multimedia forensics' research or on even narrower topics like digital image forensics only. Very few books on multimedia forensics look at the bigger picture of handling multimedia data and devices in the general context of digital forensics and electronic evidence. Multimedia forensics books are often written by multimedia forensics researchers for multimedia forensics researchers, who are mostly people in multimedia signal processing research community. One of the consequences of this practice is that practitioners in digital forensics laboratories are not well informed about recent progresses in the multimedia forensics research field and potentials of new techniques developed. While many digital forensics practitioners are indeed aware of multimedia forensics to some extent, they often found it hard to read books written for multimedia forensics researchers.

On the other hand, although there are a large number of books on digital forensics published over the years, very few of them cover multimedia forensics. For many digital forensics practitioners, multimedia data and devices especially digital images and video evidence are out of the scope of digital forensics, which are in practice often handled by a separate team who often see themselves as audio and video analysis experts rather than digital forensics experts. A consequence of this practice is that multimedia forensics researchers are less interested in digital forensics books and thus less exposed to important aspects in the general context. For instance, researchers working on multimedia forensics are generally not well informed about standard procedures of evidence handling which often define requirements of digital forensic techniques.

The lack of interaction between multimedia forensics and digital forensics communities is obviously harmful and unnecessary for the long-term healthy development of both communities and the whole society's efforts of fighting against crime as a whole. Many researchers and practitioners in both communities have noticed this problem, and in recent years we have started seeing more multimedia forensics researchers attending digital forensics conferences, and vice versa. A few books have been recently published in the hope to bridge the two relatively isolated communities. One example is *Digital Image Forensics: There Is More to a Picture than Meets the Eye* (edited by Husrev Taha Sencar and Nasir Memon, published by Springer in 2012), which has two chapters going beyond 'mainstream' multimedia forensics research to searching/extracting digital image evidence and courtroom considerations.

Although we are considering ourselves more as multimedia forensics researchers, since 2012 we have been involved in a number of research projects on digital forensics of non-multimedia data, working closely with UK law enforcement. The experience of working in both fields let us think about how the two communities could be brought closer than what they are now. This finally led to the decision of a book focusing on links between digital forensics and multimedia forensics, which should be designed not only for one community but also for both. We decided to make the book an edited one rather than a monograph authored by us because of the diverse topics we want to cover and our ambition of involving not only researchers in both fields but also contributors from digital forensics industry and law enforcement agencies. Since there have been a large number of books on digital forensics of non-multimedia evidence, for this book we decided to focus on multimedia data and devices and look at them from perspectives of both digital forensics and multimedia forensics researchers. We also decided to get digital forensics and multimedia forensics practitioners involved so that the book can benefit not only researchers but also people who are end users of research results in both fields.

To ensure the quality of the book, we first designed the structure of the book and the topics we wanted to cover, and then for each topic we invited known researchers who are conducting active research on that topic to write a self-contained chapter. All the invited chapters were reviewed by us and the contributors were guided to improve their chapters by following our comments. We encouraged all chapter contributors to include some examples or case studies to illustrate how the techniques introduced in their chapters work, in order to make the book more accessible to digital forensics practitioners. In addition to being editors of the book, we also authored the first two chapters of the book to provide necessary background information about how multimedia data and devices are currently handled in law enforcement laboratories (Chapter 1) and the state of the art of digital and multimedia forensics standards and best practices (Chapter 2). For Chapter 1, we worked closely with technical staff of three selected digital forensics laboratories (two in the United Kingdom and one in China) of different sizes and structures to produce the contents. Chapter 2 is based on a report we produced for a past project on digital forensics standards funded by the UK Home Office in 2012, but it covers more recent standards and best practice guides published after 2012 and also more standards and best practice guides on multimedia evidence handling.

As a whole, the contents of this book are organized into four parts:

1. **Multimedia Evidence Handling**: This part covers how digital forensics laboratories handle multimedia data and devices (Chapter 1), digital forensics standards and best practices (Chapter 2), machine learning-based digital triage (child pornography as a case study) (Chapter 3) and forensic authentication of digital audio and video files (Chapter 4).
2. **Digital Evidence Extraction**: This part covers different approaches of extracting evidence from multimedia data and devices that are more in the digital forensics domain, including photogrammetric methods (Chapter 5), multimedia file carving (Chapter 6), biometric forensics (Chapter 7) and analytics for image collection forensics (Chapter 8).
3. **Multimedia Device and Source Forensics**: This part covers forensic analysis of different types of multimedia devices or media such as digital cameras (Chapter 9), printers and scanners (Chapter 10), microphones (Chapter 11) and also printed documents (Chapter 12). In multimedia forensics field, these techniques are normally called source identification.
4. **Multimedia Content Forensics**: This part covers multimedia content analysis techniques for forensic purposes including statistical approaches (Chapter 13), camera based forgery detection (Chapter 14), processing history recovery (Chapter 15), anti-forensics and countermeasures (Chapter 16).

The four parts are ordered in such a way that they form a smooth information flow from digital forensics to multimedia forensics by incrementally adding more multimedia forensics-specific contents. Roughly speaking, the first two parts are more in digital forensics domain focusing on multimedia evidence handling and extraction, and the last two parts are mainly in multimedia forensics domain covering two main sub-areas – source analysis and content analysis. Each part contains four chapters focusing on different selected topics falling into the topical category of each part. We believe the 16 chapters in the book are able to cover most important topics in digital forensics of multimedia data and devices, but we would also point out that it is technically impossible to cover all topics in this space especially those emerging ones. We leave our ambition for a more complete coverage for future editions of the book.

The organization of the book allows each part to be read as a self-contained set of chapters. Chapter contributors were guided to make their chapters largely self-contained as well so that each chapter can also be read without too much dependencies on other chapters. We also worked with chapter contributors to minimize unnecessary use of mathematical material to enhance the accessibility of the book especially for readers who have less technical background. We expect the book will be a useful reference for a broader audience including researchers, practitioners and students who are studying digital forensics or multimedia forensics courses. We plan to use part of the book as further readings of our MSc course 'Multimedia Security and Digital Forensics' to be taught in the Spring semester of 2014–2015 academic year at the University of Surrey as part of our MSc in Information Security programme.

In order to cover more dynamic contents (i.e. multimedia files, more examples and important datasets) and to inform our readers about new development in a more timely manner, we also set up a web site www.wiley.com/go/digitalforensics for this book and will be updating that web site frequently in the future to add more contents that are not included in the published book. Our current plan is to include at least the following items on the book's web site: one page for each chapter which will allow any supplementary material (including multimedia files, corrections and updates) to be added; one web page on digital and multimedia forensics glossary; one web page on a list of digital forensics standards and best practice guides (as supplemental material to Chapter 2); one web page on existing datasets, free software and source code; and one web page for related books, journals, magazines, conferences and events. We will also respond to our readers' requests to add other useful materials that will help them to make full use of the book.

We hope you enjoy reading this book and find it useful for your research, work or study. We also hope the publication of this book will indeed help the digital forensics and multimedia forensics communities look at each other's work more often and work together more closely to create more useful techniques and tools that can help law enforcement investigate crimes more effectively and the whole society have a safer world to live in.

Anthony T.S. Ho and Shujun Li
Department of Computing
Surrey Centre for Cyber Security (SCCS)
University of Surrey, Guildford, UK

Acknowledgements

The editors would like to thank all contributors of this book for accepting our invitations to write a chapter for this book. We also thank all staff of John Wiley & Sons, Ltd who were involved in the creation of the book. Our special thanks go to Ms. Alexandra Jackson, Associate Commissioning Editor for Electrical Engineering of John Wiley & Sons, Ltd, who initiated the book from the publisher's side and guided the editors through the whole process until the last stage.

About the Website

This book's companion website www.wiley.com/go/digitalforensics provides you with additional resources to further your understanding, including:

- Selected multimedia files and other examples
- Source code and software tools
- Links to many useful web resources

Part One

Multimedia Evidence Handling

1

Digital Forensics Laboratories in Operation: How Are Multimedia Data and Devices Handled?

Shujun Li and Anthony T.S. Ho
Department of Computing and Surrey Centre for Cyber Security (SCCS),
University of Surrey, Guildford, UK

1.1 Introduction

This chapter looks at the operational side of digital forensics of multimedia data and devices in real-world digital forensics laboratories, especially those run by law enforcement, in order to prepare readers with a proper background for different technical aspects covered in other chapters of this book. The chapter can also be read alone to gain insights about the operational aspect of digital forensic services and practices on multimedia data and devices.

While most digital forensics laboratories handle multimedia data and devices in their everyday operation, the forensic procedures, techniques and tools used often differ from each other due to many factors such as different legislations regulating forensic practices, different digital forensic standards and best practices followed, different structures of the digital forensics laboratories and their parent bodies, etc. We realized that it is difficult to cover too many digital forensics laboratories in different countries in such a short book chapter, so we decide to focus on the following three representative laboratories (two in the United Kingdom and one in China):

Handbook of Digital Forensics of Multimedia Data and Devices, First Edition.
Edited by Anthony T.S. Ho and Shujun Li.
Companion Website: www.wiley.com/go/digitalforensics

- Digital and Electronics Forensic Service (DEFS), Metropolitan Police Service (known as "the Met" or "the Met Police", the largest territorial police force in the United Kingdom responsible for policing Greater London excluding the City of London Police area),
- Digital Forensics Team (including an affiliated Audio-Video Team, AV Team), Surrey Police (a medium-sized territorial police force responsible for policing the county of Surrey in the United Kingdom),
- Shanghai Stars Digital Forensic Center (SSDFC), Third Research Institute, Ministry of Public Security, China.

Among the three digital forensics laboratories listed, the Met's DEFS is the largest one with around 70 technical staff members. The other two are smaller in size but still have a considerably large team (with 14 and 32 technical staff members, respectively). Covering digital forensics laboratories of different sizes highlights how forensic practices of large and smaller law enforcement agencies differ from each other. The third digital forensics laboratory in the list is not an in-house team of a police force, but part of a research institute of the central police authority (Ministry of Public Security) in China. It was selected to cover a different type of digital forensic service providers and also digital forensic practices in a major legislation significantly different from the UK system and the Western systems as a whole. We have not opted to cover any private sector digital forensic service providers because the quality and scope of their services are much more diverse and may not be directly comparable with those provided by law enforcement bodies.

The rest of this chapter is organized as follows. Section 1.2 describes how multimedia data and devices are handled at the Met's DEFS, with a particular focus on the DEFS's AV Laboratory (AV Lab) which is the main forensic laboratory handling multimedia data and devices. Sections 1.3 and 1.4 describe different aspects of digital forensic practices on multimedia data and devices at the Digital Forensics Team (including the affiliated AV Team) of Surrey Police (UK) and at the Shanghai Stars Digital Forensic Center (China), respectively. After describing the three digital forensics laboratories, Section 1.5 compares digital forensics practices of the three laboratories to show common features and major differences, which lead to some key observations for future research and possible extension of this chapter. The last section briefly summarizes this chapter. An appendix is given to cover the questions that we asked during our interviews with Surrey Police and the Shanghai Stars Digital Forensic Center in China.

1.2 Digital and Electronics Forensic Service, Metropolitan Police Service, UK

The information about the Digital and Electronics Forensic Service (DEFS) of the Metropolitan Police Service was obtained through a written document provided by staff members of the DEFS. The document was not guided by the authors of the chapter, but is more a systematic summary from DEFS staff's point of view on different aspects of their digital forensic practices with a special focus on multimedia data and devices. Further e-mail communications took place after the authors received the written document to clarify unclear issues. The published content of this section was reviewed by the informants.

1.2.1 Background: Metropolitan Police Service

The Metropolitan Police Service ("the Met" hereinafter) is the largest police force in the United Kingdom with around 31 000 police officers, 13 000 police staff, 2 600 Police Community Support Officers (PCSOs) and 5 100 special constables (volunteer police officers) as of May 2014 (Metropolitan Police Service 2014a). It was established in 1829 after the first Metropolitan Police Act was passed (UK Parliament 1829). It is responsible for law enforcement in Greater London (excluding the small region called 'City of London' which is the responsibility of the City of London Police), covering a population of 7.2 million (Metropolitan Police Service 2014a). According to the statistics published on its website (Metropolitan Police Service 2014b), there were in total 771 566 notifiable offences in 2012–2013 fiscal year.

1.2.2 Digital and Electronics Forensic Service

The DEFS hereinafter provides the Met with in-house facilities and outsourced services to examine today's high-tech electronic data and devices. It provides a comprehensive range of services at one location, in its AV, computer, and telephone laboratories. Each of the three laboratories has around 20 members of technical staff. There is also a smaller laboratory for cell site analysis, which has five members of staff. In 2013 the DEFS handled 2 780 cases, leading to an average work load of 3–4 cases per technical staff member per month.

The forensic services provided by the DEFS are guided mainly by in-house standards, the standard operating procedures laid out by the Met and the Association of Chief Police Officers (ACPO) Good Practice Guide for Digital Evidence (2011).[1] All laboratories of the DEFS are also actively working towards ISO/IEC 17025 (ISO/IEC 2005) accreditation as required by the UK Forensic Science Regulator in its 2011 'Codes of Practice and Conduct for forensic science providers and practitioners in the Criminal Justice System'. A more detailed and systematic description on the ACPO guide and other standards and best practice guides can be found in Chapter 2.

Multimedia related forensic services in DEFS are mainly conducted by its AV laboratory ('AV Lab' hereinafter), so in the following we will focus on the AV Lab only. Note that some descriptions on the AV Lab also apply to the other three

---

[1] Previous editions of the ACPO guide are titled 'ACPO Good Practice Guide for Computer-Based Electronic Evidence'.

laboratories, for example the training and performance evaluation practices and the working environment.

1.2.3 AV Lab: Operational and Procedural Aspects

The AV Lab of the Met's DEFS is one of the best equipped digital forensics laboratories in the world in terms of both the range and specification of its enhancement tools, and the expertise and experience of its staff. The team is made up of forensic audio and forensic video practitioners, forensic technicians and a number of specialists in areas such as image comparison, CCTV data recovery, audio authenticity and voice comparison.

The AV Lab works with customers across the Met to retrieve, process and analyze all types of audio and video evidence. For complex cases with a lot of digital evidence, the Met's FOCUS Court Presentation System (Thomas 2010) is sometimes used to assure the maximum impact and accuracy of electronic forensic evidence within the courtroom.

An approved lab form is required for all submissions and the following services are currently available:

- Audio Services
  - Copying – format conversion and decryption, optimal replay
  - Audio enhancement – improving intelligibility and listening comfort
  - Editing – removing redacted passages for disclosure/court, for example ABE interviews
  - Processing – for example voice disguise, level optimization, speed correction and section compilations
  - Audio authenticity analysis – identifying whether a recording has been copied, edited, tampered with or deleted; assessing the provenance of a recording
  - Identification of the time and date of an audio recording through electric network frequency (ENF) analysis
  - Acoustic crime scene evaluation – could person A have heard event B from location C?
  - Voice comparison
  - Repair of broken media
- Video Services
  - Assistance with viewing CCTV/video footage
  - Decoding and conversion of CCTV/video images from either analogue or digital CCTV recorders for court purposes
  - Downloading CCTV from CCTV recorders (DVRs)
  - Court compilations of CCTV/video material
  - Data recovery from corrupted CCTV recordings
  - Production of still images
  - Image comparison

- Image enhancement
- Height estimation from images
- Speed estimation of vehicles from CCTV
- Linking an image to a source camera (PRNU analysis)
- Bespoke CCTV playback solutions

In the following, we will look at different operational/procedural aspects of digital forensic practices of the AV Lab.

1.2.3.1 Job Description

Depending on the entry grade, all employees of the AV Lab hold a minimum BTEC[2] Diploma in Electronic Engineering, but some roles require a degree and/or significant relevant experience. Insistence on electronic/technical engineering qualifications underpins the ability to give evidence in court about the 'how's and why's' of audio and video recordings. Understanding the fundamental causes of noise, intelligibility degradations and recording-chain impulse response, along with other audio and video engineering issues, enables the specialist to competently explain to a jury what has occurred and the limitations of what can be done to address those problems.

The role requires a broad knowledge of, and interest in, all modern communications and computing technologies with an emphasis on the forensic opportunities offered by their examination. Excellent communication skills and the ability to convey ideas and technical knowledge to police officers and colleagues of all ranks are essential, together with a strong technical understanding of the fundamental principles behind digital devices and the way in which they interoperate with each other and other systems. Staff work on a call out rota, attend crime scenes and give evidence in court.

1.2.3.2 Training

The AV Lab organizes its training program by defining a core skill sets for all lab roles and each staff member's current status for that knowledge or skill is assessed. This gives both in-house and external training focus for the individual. Elements include externally run foundation courses, internal training with specific software packages such as Avid (formerly known as Digidesign) Pro Tools (Avid Technology, Inc. 2014b), Apple Final Cut Pro (Apple Inc. 2014), and then detail training on the modules within them such as compression/level control, spectral subtraction, etc.

The AV Lab also has internally run courses on more general forensics-related matters such as statement writing, court skills, etc. Internal training is achieved by a mix of group training (e.g. court skills), formal one-to-one training sessions on a particular

---

[2] BTEC refers to UK's Business and Technology Education Council, the British body existing between 1984 and 1996, which awarded the vocational qualifications commonly referred to as 'BTECs'. In 1996, BTEC was merged with the University of London Examination and Assessment Council (ULEAC) to form a new awarding organization Edexcel (Pearson Education Limited) (Edexcel 2014).

topic (e.g. compression) and on-the-job training or specialist shadowing. The new entrant will undergo a full range of teaching from peers and will be carefully shadowed on their early work. The AV Lab fosters a culture of peer review, checking, comparison of techniques and discussion. Even the most experienced specialists will often ask for a second opinion or ideas for alternative approaches. This is never seen as a sign of incompetence but rather aids the dissemination of good practice and innovation.

The Met operates a Performance Development Review system for staff appraisal and development. This is designed to give 6 monthly reviews of an individual's training requirements, based on performance and his/her future work profile. Objectives are set and reported on for each 6-month period and are based on the needs of the department (e.g. 'procure and set up a data recovery PC will all relevant tools installed, by March') and those of the individual (e.g. 'complete the Met's diversity strategy training module by December'). Staff's performance is assessed in a wide range of behaviours, based on the UK's National Occupational Standards (NOS) (UK Commission for Employment and Skills 2014). Their salaries are however not performance related.

Some equipment vendors and manufacturers run training courses specifically on how to use their products. To date the AV Lab has made little use of these courses as they have emerged relatively recently and staff turnover has been low for many years. However, the AV Lab is not opposed to courses like these as long as they cover the fundamental theory behind the controls being manipulated.

1.2.3.3 Working Environment

It is vital that the working environment is free of any electromagnetic interference especially mobile phone (GSM) interference. A recommended safe distance for mobile phone activity is 10 m. Because the handset handshakes at full power at turn on and turn off, all visitors are expected to turn off their mobile phones or activate 'flight' mode before they enter the examination area, and if they are already near the examination area then they will be requested to leave before they change the mode of their phones.

1.2.3.4 Copy Quantities and Formats

When an exhibit is booked in, the officer is informed that he/she will receive back three discs in the format of Compact Disc Digital Audio (CD-DA) (International Electrotechnical Commission 1999) and/or DVD-Video (DVD FLLC 2014) where appropriate. Any agreed variations to this will be noted at the booking. Three enhanced, evidential copies will be produced. One will be a sealed copy and two will be unsealed. If there are a large number of exhibits to be copied or very long running times, then it will also be considered to produce WAV-type files and issue DVD data discs or a hard drive. Included on the copy is a READ-ME document detailing any instructions, copy format types and possible issues.

It is sometimes the case that even following an enhancement the quality of the copied recordings remains poor. In this instance, an unfiltered 'flat' copy of the exhibit is produced in addition to the filtered versions and labelled as 'un-enhanced'.

The exhibit reference numbers assigned to copies are usually in the form of the initials of the forensic audio/video engineer producing the copies, plus an incrementing number that relates to discrete sets of originals, followed by the copy format and then the copy serial number. Each copy reference number should be unique for all submitted recordings relating to a particular operation.

1.2.3.5 Triage

The Police Forensic Liaison Unit (PFLU) acts as a gate keeper for the DEFS service. This PFLU assesses each application for submission before any exhibits arrive in order to winnow out any submissions which are not clearly focused on a specific question, which would cost a disproportionate amount of time to process or which can be reduced or even eliminated in scope through alternative methods of analysis or investigation. For instance, a mobile phone and a laptop computer may be submitted in order to prove the association between two suspects. If the two suspects are found to be linked through mobile phone contact and text records, there is no need to submit the laptop. For audio it might be that an enhancement is requested to aid transcription of a surveillance probe, but an undercover officer may provide a much better quality version of the same conversation.

One major area of work which requires robust insistence on perfectly completed submission documents is the authorized editing of significant witness interview recordings. Known as ABE (Achieving Best Evidence) interviews (UK Ministry of Justice 2011), these video interviews often require editing to remove references to specific names, addresses and other identifying information. The discussions and negotiations between prosecuting and defence counsel over what will and will not be removed are usually held immediately prior to, or indeed in the middle of, a court hearing. This means that the laboratory is regularly expected to perform edits at very short notice or within tight (and sometimes impossible) timescales, which needs to be carefully handled in the triage process. The ignorance about the time taken for the technical process of capturing, editing and producing these edited versions can lead to frustrations and conflicts so the laboratory has to ensure that its workflow is as efficient as it can be, particularly in this key area.

1.2.3.6 Exhibit Security

Due to the need to protect electronic evidence from being damaged or unintentionally changed, it is crucial for forensic analysts to pay special attention to exhibit security. In the AV Lab, it is required that DVDs and CDs must be played using read-only DVD or CD ROM drives. Writable media such as compact flash cards, hard disks or USB memory sticks must be write protected using a hardware write blocker. In circumstances where hardware write blocking cannot be used, then software write blocking should be utilized. Details of the write blocking procedure should be recorded in the analyst's examination notes. In most cases it is not a requirement before an examination to create a forensic copy or duplicate of any source material using a specialized digital forensics software package such as FTK (AccessData Group, Inc. 2014a) or EnCase (Guidance Software, Inc. 2014a), although this may not apply to some types of solid-state storage media.

1.2.3.7   Contemporaneous Notes

In the AV Lab, it is required that all forensic engineers should keep notes of important information about any case such as records of movements and storage of exhibits, instructions from phone calls and emails, etc. The contemporaneous notes should be signed and dated by the engineer conducting the corresponding forensic examination.

Working notes should be made on an examination form with a record of the continuity of the processes and procedures used and include screen shots showing all relevant settings for computer-based signal processing algorithms and where appropriate software versions of plug-in, etc. In addition, the signal chain should be recorded through the whole signal path.

The notes should also include a brief understanding of the instructions and who supplied them from the submission documents so that both the objective(s) and the origin of the forensic work are clear.

1.2.3.8   Evidential Evaluation

Throughout the period working with exhibits, an evidential evaluation of the contents of the recording takes place. The purpose of this evaluation is to inform the investigating police officers of any anomalies found on the recordings and to be in a position to give advice relating to the evidential credibility of the material and offer potential solutions to its shortcomings. Any negative issues arising from the assessment of the submitted evidence regarding evidential quality or authenticity are fed back to the officer in the case (OIC) and technical program unit (TPU).

1.2.3.9   Authentication

The authentication process seeks to establish if a recording is an original and genuine record of the event(s) that took place. This normally means it should be a continuous recording, and should not have been altered or tampered with (i.e. that no information had been added, modified or deleted). This can be approached from a physical level if it is disc/tape based, wrapper level, that is header/file structure and signal level. Ultimately, detection of deliberate tampering may not be possible, however detecting a purported 'original' recording is actually some form of copy casts significant doubt on its evidential integrity.

To undertake the procedure one would normally need access to both the original recording and the original recording machine and any equipment used. In addition, a statement from the individual producing the recording should be provided and the statement should cover the equipment used, its means of deployment, and also establish the recording's evidential chain of continuity.

The authentication process then proceeds in stages as follows:

- Examination and analysis of the original recording and test results.
- Examination and analysis of the original recorder and test results.
- Comparative analysis of the original and test results.
- Production of report and conclusion from the analysis findings.

To explain the aforementioned process more clearly, let us look at how authentication is done for audio recordings.[3] Authenticity examination of digital audio material has been used in many cases and has been an ongoing research topic for the AV Lab since around 2000. Various software-based tools have been developed in house to assist with this task.

The initial phase of the examination subjects the recording to critical listening which seeks to establish agreement between the acoustic qualities of the recording and the alleged recording environment and equipment. Critical listening will also identify any key 'events' worthy of further analysis such as apparent discontinuities, inconsistencies in the background noise, or acoustic signatures which may indicate manipulation of the recording.

Following critical listening the whole recording and in particular the key events are subjected to waveform analysis which provides a visual display of the audio waveform by means of time/amplitude, frequency/amplitude and dynamic range graphs.

The next stage covers the production of test recordings on the alleged original recording apparatus, made under the same (if possible) or similar conditions and testing of all the various control functions of the recording apparatus. The resulting test recording is then subjected to the same waveform analysis as the original recording above so that a comparative analysis between the original recording results can take place.

Based on the results of the comparative analysis a set of conclusions are derived which aim to either establish the integrity and authenticity of the original recording or question its evidential reliability.

If the recording is in a digital format (e.g. a computer sound file), then further analysis of the digital data can often take place, and this would include consideration of the structure of the file and file header (metadata) present. Examination of the metadata may reveal the recording date/time and duration plus the model of recording used,

---

[3] Video recordings can be handled in a similar way although the techniques and tools used will often differ.

similarly absence of the metadata may indicate the file has been produced by digital editing software.

The original recordings are needed because they can be edited with relative ease in digital format nowadays if one has access to a general purpose personal computer (PC) and readily available editing software (many of which are freely available in the public domain).

Even a purportedly original analogue recording may have been copied to a computer, edited or manipulated then recopied back to an analogue recorder. Because of the high quality of many computer systems when dealing with sound material analogue recordings can be presented as the original with in many cases little degradation in quality.

Once material has been copied some of the key audio features of the original can be lost making establishment of its authenticity much harder. This can also be true with regard to the metadata embedded in a digital audio file which could be manipulated by an individual with sufficient knowledge making changes difficult to detect.

The aforementioned authentication process does have limitations. If the original recording is not available to analyze and there is no access to the original recording device (or any knowledge relating to its make, model and means of deployment), one cannot exclude the possibility that the recording has been altered. Examination of a copy of the original recording can be compared to looking at a facsimile of the evidence making it difficult to establish if it is an authentic reproduction of the sound heard, whereas access to the original may tell us a lot more about the evidence gathered.

Where events noted appear to challenge the authenticity of the material, the access to the original recording and an explanation of how it was used to make the recording may help settle questions regarding its integrity. For instance, if there are signal dropouts and noise present on the recording this may be explained by use of a radio link in the signal chain, however if the original equipment only included a cable-connected microphone then another explanation for these artefacts must be sought to 'prove' the recording.

1.2.3.10   Audio Enhancement and Processing

Prior to any enhancement/processing taking place, an auditory assessment coupled with FFT (Fast Fourier Transform) analysis of the recording will be undertaken wherever possible. From this evaluation an enhancement strategy will be decided upon. The primary aim of enhancement is to improve the intelligibility of speech or to make the quality of speech more acceptable to the listener.

Degradation can be caused by a number of effects including acoustic noise from background speech, reverberation caused by multiple reflections from a hard surface environment (e.g. in a corridor), electrical interference (e.g. GSM tone injection from a mobile phone), clipping where non-linear distortion is added to the signal through over-driving the signal path, distortion effect from the recording apparatus (i.e. poor quality microphone or over compression by the codec).

The AV Lab is a co-sponsor of the Centre for Law Enforcement Audio Research (CLEAR 2012), a UK Home Office-led 5-year research program which seeks to better understand the effects of filtering on speech intelligibility, transcribers' performance and developing state-of-the-art algorithms for speech cleaning. The motivation for this sponsorship is to better understand the limitations and opportunities in forensic grade speech enhancement and thus optimize filtering processes and workflows to achieve the ultimate goal of gaining better evidence/intelligence from recordings. Audio filtering has been undertaken for many years by the AV Lab and other forensic labs with the assumption that a corresponding increase in intelligibility must have been attained if a filter setup makes the recording sound 'better' to the listener. The AV Lab is questioning this assumption, and working with the other CLEAR consortium members and the academic team to better understand the effects of filtering techniques on intelligibility. The idea is to put enhancement on a more scientific footing–to know more about what the AV Lab is doing with corrupted audio and to avoid unseen pitfalls.

At the heart of the problem is assessment of the enhanced product: how do we know or how do we show that our processes have not degraded the intelligibility of the original signal. Just because we may have applied recognized processes in terms of equipment, techniques and procedures do not guarantee that the 'wanted' signal has not been damaged in some way. Intelligibility is assessed on a subjective basis, so without knowing what the original signal should be the results will depend on who is to say one enhancement is better than another.

Less is often more in terms of successful enhancements. If we have a recording that has poor bandwidth and a very low signal-to-noise ratio (SNR), then a speech signal could be so fragile that any form of processing is likely to degrade it further and may be best left in its original form.

1.2.3.11 Video Submissions

Viewing of video material serves as a preliminary assessment of the video exhibits, which may lead to a submission for more complex work such as editing, image comparison, enhancement or data recovery. Common outcomes of viewings include capturing and printing still images, creation of edit lists for court copies and creation of working copies.

Details present in footage can sometimes be enhanced to reveal more information. The capabilities for enhancing video are often very limited, particularly if the original footage is of low quality. Enhancement can range from simply cropping and zooming in on an image, adjusting brightness and contrast or more complex processes such as frame averaging in static components of an image.

If video footage cannot be accessed by conventional means, or has potentially been removed from a device, the officer can submit the exhibit for data recovery. Data recovery is diverse and can include recovering data from damaged media, extracting deleted files, rebuilding corrupt files, reverse engineering video formats and

CCTV systems for playback of footage and extracting time and date information from proprietary file systems. Processes used in data recovery are often developed on an ad hoc, case-by-case basis, as new technology emerges.

1.2.3.12 Statements

Each statement must include the following phrase[4]:

Witness statement

(CJ Act 1967, s.9; MC Act 1980, ss.5A(3) (a) and 5B; MC Rules 1981, r.70)

The aforementioned title is followed by the following:

- The analysts name along with qualifications, occupation and experience
- Company name
- Evidential services lab reference and company lab reference (optional)
- Details of the exhibit including exhibit number and any bag seal numbers
- Brief description of work requested and work carried out
- Details of the copies produced
- Brief description of any enhancement process
- Details of resealing and storage
- Disclosure details, that is 'A full record of work carried out in this case is available at [company name]. This record will include: documents created at the time of submission; trail documents and original notes of examination'.

1.2.3.13 Basic Quality Assurance Checks

Basic quality assurance checks are required to be performed prior to the case being returned and these include the following:

1. Check that the recorded/noted reference and exhibit numbers matches the submission documents and those referred to in the statement.
2. A peer review of statements must take place.
3. A cursory listen of the final copies to establish correct copied content.
4. A final review of the documentation to pick up any typos, etc.

1.2.4 Selected Forensic Techniques Used by AV Lab

In this section, we give brief descriptions of some selected forensic techniques currently used by the AV Lab staff. This is not a complete list of techniques they use, but a personal selection by the members of staff we got information from.

---

[4] The full titles of the laws on the second line of the statement are the following: Criminal Justice Act 1967 (UK Parliament 1967), Magistrates' Courts Act 1980 (UK Parliament 1980), The Magistrates' Courts (Forms) Rules 1981 (UK Parliament 1981).

1.2.4.1 FOCUS Court Presentation System

FOCUS (Thomas 2010) is a bespoke software-based digital evidence viewer and organizer that stores all digital evidence relating to a case in a single place so it can be accessed via an easy-to-navigate front screen. This assists the investigating officer by helping speed up and streamline the presentation of evidence in court. It can be particularly useful in cases with a large quantity, range or complexity of evidence. By improving the quality of court replay systems, the presentation lab provides highly beneficial systems that assist a jury in understanding complex evidence. The system supports the presentation of all electronically held digital evidence, including video, audio, flash, documents, photographs, DVD, 3D immersive imagery, CAD, etc.

A range of presentation facilities are available, including the following:

- Annotated media, for example arrows on videos highlighting the areas of focus
- Interactive presentations, for example cell site analysis on an Adobe Flash platform
- Map animations – providing street data, geographic and bird's eye views
- Court replay, management, storyboarding, presentation system facilities and more

The final presentation can be supplied for use in court on CD or DVD or on a hard drive-based system (FOCUS) which contains all evidence types and formats. Presentations of all electronic evidence can then be shown in a professional and easily accessible manner.

1.2.4.2 Electric Network Frequency Analysis

When audio or video is recorded, the mains electrical network frequency (ENF) is often recorded in addition to the wanted audio/video signal. This happens with mains-powered recording equipment but may also occur with battery-powered equipment if it is used in proximity to mains-powered equipment or transmission cables. ENF analysis can thus be used to both determine the date and time of audio/video recordings and to help establish their integrity.

The ENF in the United Kingdom is 50 Hz, but actually varies by small amounts from moment to moment due to variations in the demands placed on the network generators. The generators are synchronized across the network so these small frequency variations are exactly the same across the United Kingdom. Over time, these variations in ENF provide a unique pattern.

The AV Lab has been recording the ENF since 2002 and is able to compare ENF traces found on evidential audio recordings with the ENF database to accurately determine the time and date of recordings.

Comparison of an evidential ENF with the database may also be used in authenticity analysis to establish whether any discontinuities occur with a recording, or whether a recording is a copy.

This technique does have some limitations which include the following:

- ENF analysis only works on digital recordings and certain VHS recordings. Variations in tape speed on analogue recordings mean that the ENF signal is distorted and cannot be used.
- It is much simpler to compare the ENF with an approximate time and date at which the recording is believed to have been made, than to determine when the recording was made without giving any indication of the alleged time and date.
- Generally a recording should be at least two and preferably five minutes long to obtain a reliable ENF result.
- It is not always possible to extract an ENF signal from recordings either because it is not present or because there is too much noise.

The AV Lab co-authored a document for the Forensic Speech and Audio Analysis Working Group (FSAAWG) of the European Network of Forensic Science Institutes (ENFSI) for ENF analysis which was released in 2009.

The AV Lab have developed a set of practitioner competency tests for ENF analysis which should meet the requirements as laid down by the Forensic Regulator (ISO/IEC 17025), and participated in an ENFSI competency test for ENF analysis in 2009.

1.2.4.3   Image Source Identification

Recent research has shown that it is possible to identify the source camera of a digital image based on intrinsic noise patterns called Photo Response Non-Uniformity (PRNU) (Kot and Cao 2013). The AV Lab is actively researching on PRNU-based image source identification (Cooper 2013) and has developed its own source camera identification system. This system allows specific images to be tied to its source cameras (phones/compacts) and casework requiring this form of analysis is on the increase. The AV Lab passed two PRNU source camera identification competency tests run by the Netherlands Forensic Institute (NFI) for the ENFSI Digital Imaging Working Group (European Network of Forensic Science Institutes 2011) in 2010 and 2013.[5]

1.2.4.4   Challenges

There are a number of challenges identified by the AV Lab staff in the written document the authors received. They are not necessarily a complete list, but can show key areas the research community needs to pay more attention to.

The Met's DEFS is still dealing with the legacy of the high cost per bit/low storage capacity of earlier generation of digital media where content was 'thrown away' in an

---

[5] The PRNU source camera identification competency test was devised by Zeno Geradts of the NFI in 2009.

attempt to save space. The compression algorithms used are largely perception-based (i.e. lossy) and may not optimize the reproduction of forensic features.

Image enhancement potential for video material is still limited because of both the recording system and its configuration, for example time lapse capture. Quality of digital AV material is not optimized for forensic purposes such as vehicle identification number (VIN) identification, image comparison and analysis. For video and image compression schemes, lack of resolution, compression artefacts, fields captured instead of full frames and other issues all reduce the effective number of lines for forensic examiners to work with. Audio compression schemes are often optimized for music with a quality level of 16-bit 44 kHz but the actual sound precision can be much less for forensic purposes. On the other hand, there are exaggerated expectations from the general public regarding the precision and capability of digitally stored imagery and sound, so how to manage this is a (not only technical) challenge as well.

Another concern is about anti-forensics which is about manipulations of multimedia evidence by criminals to mislead investigations. There is a greater opportunity for criminals to do so at low cost on readily available equipment (i.e. Adobe Photoshop on digital still images). Access to knowledge to do it is more easily available across the internet. The DEFS and the whole digital forensics community need to continually update their techniques to meet this anti-forensics challenge.

1.2.5   Acknowledgements

The authors would like to thank the following informants from the Met's DEFS: Paul Groninger (Manager of the AV Lab) and Robin P. How (Senior Manager of the DEFS). The authors would also like to thank Mark A. Stokes, the Head of DEFS, for his help in coordinating the communications between the authors and the DEFS staff for this section.

1.3   Digital Forensics Team (Including Affiliated AV Team), Surrey Police, UK

The information about the Digital Forensics Team (including an affiliated AV Team) of Surrey Police was gathered through a structured interview which took place on 5 July 2013. Some further email communications took place after the interview to get more information and to clarify some unclear issues raised during the interview. The published content of this section was reviewed by the informants.

1.3.1   Background: Surrey Police

Surrey Police is the territorial police force responsible for policing Surrey, one of the so-called 'home counties' surrounding London with a population of around 1.1 million (Surrey County Council 2012). The force's headquarters is in Guildford, the same town where the University of Surrey is located. Surrey Police originally started with the much smaller Surrey Constabulary (along side with a number of separate borough police forces) since 1851, and gradually evolved into the current county wide force in the 1940s (Surrey Police 2014b). Surrey Police had around 2 000 police officers, over 1 600 police staff, around 200 PCSOs and over 200 special constables as of September 2013 (UK Home Office 2014). Although Surrey Police is much smaller compared with the Metropolitan Police Service, it is a typical medium-sized territorial police force in the United Kingdom in terms of both its size and the population it covers.[6] In 2013–2014 fiscal year, there were in total 48 486 notifiable offences (Surrey Police 2014a).

1.3.2   Structure of Surrey Police's Digital Forensics Team and AV Team

As a smaller police force, Surrey Police does not have a large body for digital forensics like the Met's DEFS. Instead, it has a medium-sized Digital Forensics Team ('DF Team' hereinafter) composed of mainly the following units (as of July 2013):

- A team leader supported by an administrative support officer
- Seven computer forensics analysts
- Five mobile phone forensics examiners

The mobile forensics examiners are assisted by 38 volunteers (detective constables and investigating officers from Guildford, Reigate, Staines police stations–the three largest police stations in Surrey Police) who were trained to do basic forensic analysis on mobile phone cases for triage purpose.

Surrey Police also has a separate AV Team[7] which operates independently but is affiliated with the DF Team and under the same management line 'Forensic Technology Team.' The AV Team has two technicians focusing fully on multimedia (audio/video) evidence. Occasionally the AV Team passes a submission to the DF Team if the submission is outside its area of work, but normally the two teams do not work with each other directly on submissions.

Although the DF Team does not particularly focus on multimedia data and devices like the AV Team, it routinely handles cases involving multimedia data for cases such as those related to child pornography. Therefore, in this section we cover both the DF Team and the AV Team to give a bigger picture of digital forensics practices at Surrey Police.

---

[6] Surrey is the 12th most populated ceremonial county in England out of 48 ones as of mid 2011, and if metropolitan counties and unitary authorities are counted separately Surrey is the 11th most populated county (UK Office for National Statistics 2012).

[7] This AV Team is also called the Video Audio Laboratory, but to align with the name of the Digital Forensics Team, we use the former name throughout the whole chapter.

1.3.3   Training and Certification

When recruiting new staff to the DF Team, it is expected that the applicants have gone through the 'Core Skills in Data Recovery and Analysis' training course provided by the College of Policing (CoP) (UK College of Policing 2014), but other equivalent training programs and experiences are also counted. However, if a newly recruited staff have not gone through the Core Skills training course, he/she will be asked to attend it, so the CoP training course can be considered as a compulsory qualification. In addition to the CoP training courses, technical staff is normally expected to attend product training programs among which the EnCE (EnCase Certified Examiner) (Guidance Software, Inc. 2014) is of particular importance because Guidance Software's EnCase Forensics (Guidance Software, Inc. 2014a) is the mostly used digital forensics software toolkit at Surrey Police. Some other training courses include AccessData BootCamp (AccessData Group, LLC 2013) for another important digital forensics toolkit FTK (AccessData Group, Inc. 2014a) used at Surrey Police and courses for general techniques such as internet artefacts, networking technologies, Apple Mac computer forensics, Linux computer forensics, PDA forensics, satellite navigation system forensics, etc.

The two technicians of the AV Team are both Certified Forensic Video Technicians (CFVT) through the Law Enforcement and Emergency Services Video Association International, Inc. (LEVA International, Inc. 2014). This is however not a formal requirement to be certified as a CFVT although it has proved very beneficial in practice.

1.3.4   Standard Procedure

Regarding the standard procedure, both the DF and the AV Teams are following mainly the ACPO Good Practice Guide for Digital Evidence (Association of Chief Police Officer of England, Wales & Northern Ireland 2011). In the reports of both the DF and AV teams, it is stated that ACPO guidelines are followed, which can normally make sure judges will accept the results. The AV Team also follows the guidelines defined by ACPO and/or the former Home Office Scientific Development Branch (HOSDB)[8] on multimedia evidence especially on CCTV evidence and digital imaging (Home Office Scienfitic Development Branch 2010).

Similar to the case of the Met's DEFS, the DF Team and the AV Team will need to pass the ISO/IEC 17025 (ISO/IEC 2005) lab accreditation required by the UK Forensic Science Regulator (2011). As of June 2014, Surrey Police is still in the planning phase for this task.

---

[8] The HOSDB was renamed to the Centre for Applied Science and Technology (CAST) in April 2011 (Professional Security Magazine 2011).

1.3.5   Routine Tasks Involving Multimedia Data and Devices

The AV Team is mainly handling video and audio footage on different kinds of storage media such as DVDs, CDs, SD cards, USB sticks, DVR CCTV recorders, VHS tapes, dictaphones and compact cassettes. All submissions come via Surrey Police's Submissions Department which ensures that each submitted exhibit is security sealed, that the submission form has been correctly completed, and that continuity signatures on the exhibit are up to date. The average work load for the whole team is around 50 cases per week, and the most common type of multimedia data is video (around 80%), followed by still images and audio (around 10% each). The routine tasks conducted mainly include

- copying and enhancing video sequences onto DVD (Video), CD or VHS videotape,
- copying and enhancing of audio onto CD or cassette,
- exporting/downloading of video data from CCTV digital video recorders,
- enhancement and printing of still images from video,
- editing video and/or audio,
- highlighting and/or masking of persons or items in video sequences,
- decoding of multiplex-encoded videotapes,
- recovery of data from corrupt or damaged discs, and
- repairing damaged media.

The DF Team does not have a focus on multimedia data and devices, but it is common to encounter multimedia data and devices. For computer forensics cases (over 100 TB in 2013), it is estimated that around 80% of them involve some kind of multimedia data. For mobile phone forensics cases (around 5000 and 1500 before and after triage) the percentage is around 40%. The multimedia data and devices include audio files, digital still images, digital video clips, CDs, DVDs, digital tapes, digital cameras, digital audio recorders, video camcorders, GPS devices, PDAs, etc. Among all the multimedia data, digital still images are more common than video and audio files, which is partly because video and audio files are often processed by the AV Team and cases with the DF Teams are often about still images (e.g. child pornography cases).

1.3.6   Submission Formats

For this section we discuss the AV Team only because the cases of the DF Team is much broader.

For video submissions, the AV Team needs to provide the results in DVD video format for court. Referring to instructions on the submission form, the technicians normally use screen grabber software to digitize the relevant analogue footage which is then authored to DVD Video and sometimes they have to play the footage from the PC direct to a DVD recorder via composite/S-Video cables. Sometimes the AV Team is asked to grab still images from CCTV footage and print them or burn them to CD.

For audio submissions, the AV Team provides them in Audio CD format for court, for which the technicians usually create WAV files first and then convert them to Audio CD.

After the CD or DVD is made, the technician then duplicates it as required by the standard procedure and uses a disc printer to print a label onto the disc. If the submission is for court, then one of the copies would be sealed and the technician handling the case will include a statement. The job is then returned to the Submissions Department.

1.3.7   Triage

For the DF Team, mobile forensics cases go through a triage process where 38 trained volunteering officers first conduct some basic-level forensic analysis to identify cases requiring more advanced analysis by the forensic examiners of the AV Team. All triage officers focus on file systems only, but not on file content because the latter would require the triage officers to have more advanced skills. The whole process is coordinated by the administrative support officer of the DF Team. Computer forensics cases do not have a triage process, but are prioritized by the computer forensic analysts of the team.

The AV Team does not have a triage process for cases it receives, either. Normally cases are processed in the order they are received, unless a submission is authorized as urgent by a senior police officer.

1.3.8   Software and Hardware Tools Used for Handling Multimedia Data

The DF Team is mainly using EnCase scripts to extract multimedia files from storage media which mostly depend on signatures in file headers. Other software tools that are sometime used include C4ALL (also known as C4P – Categorizer for Pictures) developed by the Ontario Provincial Police Technological Crime Unit[9] in Canada (Ontario Provincial Police Technological Crime Unit 2010), ZiuZ VizX2 (ZiuZ B.V. 2014), Digital Detective's Blade (Digital Detective Group Ltd 2014) and Microsoft PhotoDNA Microsoft (2013). The DF Team also uses the open-source software tool VLC player to extract frames from video sequences and crop regions of interests if needed.

The AV Team uses various PCs and audio/video players to play audio and video footage. For CCTV there are hundreds of different formats so it is often the case that proprietary player software has to be installed to play a piece of CCTV footage. When such software is not included on the disc, the AV Team technicians will try to download player software from the manufacturer's website. The AV Team technicians also often edit CCTV to show just the relevant footage, for example they often create a sequence where a suspect is 'followed' from camera to camera rather than showing each camera separately in its entirety and sometimes they are asked to highlight or blur out areas of

---

[9] http://www.e-crime.on.ca./.

the picture. The AV Team uses mainly Avid Media Composer (Avid Technology, Inc. 2014a) and Sony Vegas (Sony Creative Software, Inc. 2014) to edit the footage.

1.3.9   Cases Involving Encryption and Child Pornography

Such cases are mainly handled by the DF Team. Cases involving encrypted files are not common, and probably less than 10% but occasionally cases involving a lot of encrypted files are encountered. In those cases, some software tools for password cracking are normally used such as AccessData PRTK (AccessData Group, Inc. 2014b) which uses distributed computing and GPU. There are also special software tools developed for cracking passwords on Apple iOS and Blackberry devices.

The DF Team handles a lot of child pornography cases. There are two sub-teams for handling such cases: online abuse materials and computer-based analysis. The process can be partly automated using the software tool C4ALL (Ontario Provincial Police Technological Crime Unit 2010) because it has a hash (traditional hashing) database of child pornography pictures. The DF Team realized that perceptual hashing (Monga 2005) (hashing methods robust to benign signal processing operations that do not change the perceptual content of the processed image) will be helpful for handling those cases but it has not been used yet.

1.3.10   Cases Involving Source Device Identification

The DF Team had cases where the sources of digital images need identifying which were handled by looking into the EXIF metadata which is unfortunately prone to manipulation. Advanced techniques like PRNU-based camera source identification have not been used by Surrey Police yet but will be welcomed by the DF Team. There were some cases about source printer identification as well which were handled by looking into the computer spool files. There were two cases about counterfeiting printed documents, one about a banking statement and the other on an ID card, which were analyzed by looking at the material and typos.

1.3.11   Challenges

In our interview with staff members of Surrey Police, only a few challenges were highlighted. We understand this was due to their concerns on more practical issues.

For the DF Team, perceptual hashing seems to be the major technique still missing from existing digital forensics tools, and they feel other multimedia forensic techniques will be very helpful as well. The DF Team members also feel that HD videos cause trouble because of the lack of tools supporting the new format. A similar issue exists for mobile phone forensics because tools are normally updated only once a quarter, but new models of mobile phones are released much more frequently and more new brands/models are emerging in the market.

For the AV Team, one major challenge is the lack of a software tool that can play all types of CCTV footage and output the results to DVD. Currently there seems to be only one product in this space, Siraview (Sira Defence & Security Ltd 2014), but it is not a complete solution yet although the AV Team is indeed using it.

1.3.12   Acknowledgements

The authors would like to thank the following interviewees and informants from Surrey Police: Jill Wheeler (Forensic Technology Team Leader), Garry England (member of the DF Team) and David Lawrence (Principal Technician of the AV Team). The authors would also like to thank Jill Wheeler for her help in coordinating the whole information gathering process.

1.4   Shanghai Stars Digital Forensic Centre, Third Research Institute of China's Ministry of Public Security

The information about the Shanghai Stars Digital Forensic Centre ('SSDFC' hereinafter) in China was gathered through its official websites,[10] a written document provided by the Director of SSDFC, and a structured interview which took place on 5 July 2013. Some further email communications took place after the interview to get more information and to clarify some unclear issues raised during the interview. The published content of this section was reviewed by the informants.

1.4.1   Background: Third Research Institute of China's Ministry of Public Security

The institute was founded in 1978 and is one of 15 research organizations directly managed by the Ministry of Public Security of China.[11] Its main research areas include information and network security, Internet of Things (IoT), special operations communications, illegal drugs control, anti-terrorism and explosion protection, image processing and transmission, and public security protection techniques. It has around 1700 members of staff including over 1300 researchers among which more than 500 have a doctoral or a master's degree. In 2005 the institute passed the ISO 9001 quality management test (ISO 2008) (Third Research Institute of Ministry of Public Security 2014).

---

[10] The Shanghai Stars Digital Forensic Centre's official website is http://www.stars.org.cn/, but there is another website http://202.127.0.199/ which provides some supplementary information to the first one.

[11] The other nine research institutes are First Research Institute, Forensic Analysis Center (formerly known as Second Research Institute), Fourth Research Institute, Traffic Management Research Institute, Southern Research Institute, Disc Production Source Identification Center, four fire research institutes and four police dog research organizations (one research institute and three dog bases conducting research activities). There are several universities and colleges conducting research activities as well, but they are not counted in the 15 research organizations.

1.4.2 Background: Related Legislations and Regulations

In China, only certified persons and bodies can provide forensic services as laid out in China's NPC (National People's Congress) Standing Committee's Decision regarding Management of Judicial Appraisals (National People's Congress Standing Committee, People Republic of China 2005).[12] According to how the certification is done for digital forensic services, there are two types of service providers: those serving police forces and certified by the Ministry of Public Security,[13] and those open to the public and certified by the Ministry of Justice. Although following different regulations and certification processes, both types of digital forensic service providers abide by the same basic laws, technical standards and procedures, digital forensic hardware and software equipment, so their expert reports are considered technically equivalent.

1.4.3 Overview of SSDFC

The SSDFC was created by the Third Research Institute of China's Ministry of Public Security and certified by the Shanghai Judicial Bureau (the provincial branch of the Ministry of Justice in Shanghai) in 2007. It provides forensic services on digital data and AV material to law enforcement agencies, law offices, other bodies in public and private sectors, and natural persons as well (as permitted by the certification it received from the Ministry of Justice). The SSDFC is one of the leading service providers nationwide on forensic analysis of digital data and AV material, and the first one passing both lab accreditation and metrology accreditation by national accreditation authorities.

The SSDFC currently has 32 certified digital forensics experts including 15 experts with a senior professional title (associated researcher or researcher)[14] of the Third Research Institute (Shanghai Stars Digital Forensic Center 2014a). Some digital forensics experts of the SSDFC are members of different technical committees in the digital forensics field and have received awards at both provincial and national levels. Digital forensics, experts of SSDFC also conduct research in digital forensics, and they publish research papers at national and international venues. The SSDFC also has several forensic analysis assistants. All digital forensics experts are required to go through annual training organized by the Shanghai Judicial Bureau and the SSDFC.

---

[12] There is no official translation of the document's tile and the word 'Judicial Appraisals' is not an accurate translation of the original Chinese word, which actually covers judicial appraisals, authentication, forensic examination, etc.

[13] The Ministry of Public Security is the central administration of police forces in China.

[14] These two titles are direct translations from Chinese. Their seniority is actually close to associate professor and full professor, respectively, as in universities.

1.4.4 Services Provided

The SSDFC provides a variety of digital forensic services mainly in two categories: forensic analysis of digital data, and forensic analysis of audio/video data and devices. There is no internal allocation of digital forensics experts to the two categories, but there are two appointed experts who are in charge of the two categories, respectively. In addition to the two normal categories of services, digital forensics experts of the SSDFC also provide other services such as field inspection, court expert witnesses, printed documents inspection and consultancy.

The SSDFC's forensic services on audio/video data and devices including mainly the following (Shanghai Stars Digital Forensic Center 2014c):

- Forensic analysis of audio and video material such as denoising, restoration, enhancement and feature extraction
- Identification of languages, persons or objects in audio and video material
- Content analysis of languages and images in audio and video material
- Authentication of audio and video material including event (e.g. start and end) detection and forgery detection
- Speaker authentication from audio material
- Identification of audio and video recording devices such as brands, models, vendors, manufacturing sites, and also identification of the source device of audio and video material
- Forensic analysis and examination of audio and video material on the Web
- Forensic comparison of audio and video material to determine if the sample under analysis is a pirate copy
- Preservation of audio and video material on read-only storage media.

1.4.5 Procedure

Different from in-house digital forensics laboratories run by law enforcement agencies, the SSDFC is open to any legal bodies and natural persons, so it gets cases directly from their customers who provide digital evidence and requirements of the wanted forensic analysis.[15] The customer receives two copies of the expert report and the SSDFC keeps a third copy for archival purpose (Shanghai Stars Digital Forensic Center 2014b). Each case is normally handled by two digital forensics experts who either work independently or jointly. One of the two digital forensics experts has more responsibilities and will be the person who appears in court as the expert witness (if required).

For the digital forensic process itself, the SSDHC follows international and national standards, and also its own in-house best practices for each type of digital evidence.

---

[15] In-house digital forensics laboratories of law enforcement agencies in China mainly focus on crime scene investigations and simple digital forensics cases. Complicated digital forensics cases are normally outsourced to independent digital forensics laboratories like the SSDFC.

In 2009 the SSDHC passed the ISO/IEC 17025 lab accreditation test (ISO/IEC 2005). In 2014 it passed the ISO/IEC 17043 test (ISO/IEC 2010),[16] becoming the first proficiency testing provider (on digital forensics) in China's public security (police) system. The national standards followed by the SSDFC include national standards published by China's Standardization Administration and General Administration of Quality Supervision, Inspection and Quarantine (GB/T series), industrial standards published by Ministry of Public Security (GA/T series), forensics-related specifications published by Ministry of Justice (SF/Z series), and lab accreditation standards published by the China National Accreditation Service for Conformity Assessment (CNAS-CL series). All the first three standard series cover digital forensics in general and also forensic analysis of AV data and systems. There are also a large number of standards about forensic imaging and photography. For more details about different standards, readers are referred to Chapter 2 of the book.

1.4.6   Workload and Typical Cases

In 2013 the average workload of digital forensics experts of the SSDFC was around 20 cases per person. Most cases can be completed within 30 working days, but some special cases may require up to 60 working days. Since the workload is not very heavy, the SSDFC does not have a triage process, but will prioritize cases based on customers' needs.

Around 10% of all cases involved multimedia data and devices such as recorded speech, music, digital images in computers and mobile phones, digital video sequences (including movie and TV programs), computer-generated pictures, multimedia discs, digital cameras and camcorders, sound recording devices, GPS devices, and so on. The SSDFC also perform speaker comparison and identification based on recorded speech.

The SSDFC had cases where the authenticity of multimedia data was validated. There are national standards for digital forensics experts to follow for such tasks (GA/T and SF/Z standards).

All digital forensics experts of SSDFC are technically qualified to do forensic photographs, but the SSDFC has a dedicated forensic photographer for such tasks.

1.4.7   Software and Hardware Tools Used

Software tools used by digital forensics experts of the SSDFC include EnCase (Guidance Software, Inc. 2014a), Recover My File (GetData Pty Ltd 2014), X-Ways Forensics (X-Ways Software Technology AG 2014), Adobe Audition (Adobe Systems Inc. 2014) (for audio forensics), Adroit Photo Forensics (Digital Assembly, LLC 2014) (for photo evidence handling), the Chinese edition of IMIX IMPRESS (IMIX Vision

---

[16] What the SSDFC passed is actually a national standard CNAS-CL03 (China National Accreditation Service for Conformity Assessment 2010), which is effectively equivalent to the ISO/IEC 17043:2010 standard (ISO/IEC 2010).

Support Systems 2014)[17] (for forensic image/video restoration and enhancement), VS-99 Computer Speech Processing Workstation (also known as voice print meter) (Beijing Yangchen Electronic Technology Company 2014) (for speaker recognition), etc. The SSDFC also uses password cracking software to break encrypted files and is currently planning to use Microsoft PhotoDNA software (Microsoft 2013) to detect child pornography images automatically. Most software tools are running under Microsoft Windows platform.

The SSDFC is also supported by the rest of the Third Research Institute of China's Ministry of Public Security to develop bespoke software tools for handling special file formats, extracting data from network streams and uncommon digital devices such as GPS.

1.4.8 Challenges

A number of challenges were identified by staff of SSDFC in the written document provided.

1.4.8.1 Standardization Lagging behind IT Technology Development

The rapid development of IT technologies has led to the emergence of new digital devices, storage media, software and hardware tools, and networking protocols. By contrast, the development of digital forensics standards would normally take a long time and actions cannot be taken before any new technology and the corresponding forensic solution becomes available. How to improve the efficiency of digital forensics standardization activities is a key challenge in this field.

1.4.8.2 Preservation of Digital Evidence Compatible with Electronic Signature Law

Digital evidence in its original format is normally prone to manipulation, so it must be preserved following a forensically sound process to make it admissible to the court according to China's Electronic Signature Law (China National People's Congress Standing Committee 2004). Therefore, digital forensics standards must consider the problem of how to preserve digital evidence so that the expert reports can be accepted in court.

1.4.8.3 Diverse Storage Media and Device Interfaces

In a digital forensic analysis process, various types of storage media (e.g. floppy disks, CDs, hard disks, USB sticks, MMC/SD/CF storage cards and SIM cards) and device interfaces (e.g. IDE, SATA, SAS, SCSI, fibre channel, USB, Firewire and ISO/IEC

---

[17] The Chinese edition has a different name 'Imaging Doctor'.

7816) need to be handled effectively by using proper hardware devices and to follow forensically sound procedures and methods.

1.4.8.4 Data Loss in Live Forensics

Forensics of live systems (live forensics) can change the status of the live system such as volatile memory, processes, network communications, which can in turn influence extraction of digital evidence and even the results of the digital forensic analysis. This influence normally differs from system to system. Therefore, different live forensic methods and standard procedures are needed for various operating systems and devices.

1.4.8.5 Reverse Engineering and Binary Code Comparison

In cases involving malicious code, destructive computer program, commercial secrets, digital rights violation, reserve engineering and/or binary code comparison are often required. This can be very complicated to do due to different computer architectures, CPU instruction sets, results produced by different advanced programming languages compilers, operating systems, and run-time environments. To help handle such cases, standard methods and procedures are needed for reverse engineering and binary code comparison.

1.4.8.6 Cloud Forensics

In digital forensics cases involving data stored in the cloud, it becomes very challenging to extract and preserve data due to the difficulties of localizing data. In addition, data in cloud is not necessarily stored physically on a single physical server, but could distribute on many servers which are physically located in different countries and regions. This means that it can be extremely difficult to collect complete data for digital forensic analysis. Even if complete data collection is possible, the huge amount of data stored in cloud requires a lot of more computations. The complexity of data in the cloud can also raise concerns on the originality, integrity and reproducibility of digital evidence extracted, thus making it less admissible in court. More research is needed on this topic to provide information for standard methods and procedures on cloud forensics.

1.4.9 Acknowledgements

The authors would like to thank the following interviewees and informants from the Shanghai Stars Digital Forensic Center: Dr. Bo Jin and Dr. Hong Guo who are the Director and Director Assistant of the Center, respectively.

1.5 Discussions

Looking at the descriptions of the three digital forensics laboratories covered earlier, we can see a common feature shared among them: they all separate cases involving forensic analysis of audio and video material from 'normal cases' involving more 'traditional' digital forensics such as computer forensics and mobile phone forensics. Although not all laboratories have a separate internal group focusing on audio and video forensic analysis (e.g. the SSDFC in China does not have a separate group), they all offer separate services on audio and video forensic analysis. This fact has its root in separate standards and best practice guides on audio and video forensic analysis in some countries such as in China (specific standards for audio, image and video processing) and in the United States (a separate working group on digital imaging, SWGIT), and can be understood based on the fact that very different techniques (e.g. audio and image processing/analysis) are needed for audio and video forensic analysis, while traditional digital forensics are more focused on computer architectures, file systems and specific software/hardware tools.

It is however common for a computer and mobile forensic examiner to use advanced multimedia file carving tools such as Digital Assembly's Adroit Photo Forensics (Digital Assembly, LLC 2014) and X-Ways Forensics (X-Ways Software Technology AG 2014) to recover digital images and video sequences out of a disk image, although those tools are based on advanced image analysis techniques (which however are not necessarily what users of the tools need to know). The boundary between the two classes of cases is not always a clear cut especially when digital still images are involved. It is common that child pornography cases are normally handled by computer and mobile forensic examiners rather than image/video forensic analysis experts in the AV team of a digital forensics laboratory because it involves more file carving than signal processing tasks. Multimedia devices are also lying between the boundary: when file extraction or recovery is the actual task, they are normally handled by computer and mobile forensic examiners, but if some level of content analysis is required they are often handled by the AV team.

There are three interesting observations in our interviews and communications with the three digital forensics laboratories covered. The first observation is about forensic photography. Although the SSDFC has a dedicated forensic photographer and all their experts are trained for this task, this topic is not covered by the two UK laboratories but more with other departments/teams of the police forces. Although we did not attempt to cover any forensic photography team, we are aware that forensic photography share many techniques with image and video forensic analysis especially image enhancement. The second observation is about relationship between biometrics and digital forensics. Biometric techniques such as fingerprint and face recognition (and even DNA matching) are almost only considered 'non-digital' forensic services although the whole process is nearly fully digitized except for the first step (data capturing). This is however not surprising because fingerprint and face recognition have been handled routinely by traditional forensic laboratories even before the systems become fully digitized and manual inspection/matching is still sometimes required even in the digital era for difficult cases or false positive/negative cases identified through other evidence. The third observation is about the use of photogrammetry in (digital) forensic analysis. We noticed only the AV Lab of the Met's DEFS is conducting some photogrammetric analysis (height and speed estimation). Surrey Police's DF and AV Teams did not have cases requiring such analysis, and when asked about how they would handle such cases they said they would outsource the task. The SSDFC in China did not handle such cases and they did not do research on this topic either. This is also not surprising because of the same reason as in the biometrics case: photogrammetry is independent of the digitization process, and it has existed for long time before digital cameras and camcorders were invented, so it may be seen as a more traditional 'non-digital' forensic technique. For the last two observations, we refer the readers to another two chapters of the book which cover the use of biometrics and photogrammetry in digital forensics.

Comparing the three digital forensics laboratories covered in this chapter, we can see clear differences between the two in-house laboratories of local police forces and the SSDFC which is a public-facing laboratory (even though it is managed by a law enforcement research institute). The in-house laboratories are more driven by the needs of the law enforcement bodies they serve and focus on what are the most important tasks (also simpler or more routine tasks), but public-facing laboratories have to provide a broader range of services and be able to handle more complicated cases. This is reflected from the fact that all digital forensics experts of the SSDFC can do both computer/mobile and AV forensics (and even forensic photography). During the interview with the SSDFC, the two interviewees mentioned that local police forces in China also have their in-house digital forensics teams, but they normally handle crime scene investigations and simple forensic analysis but outsource complicated cases to external laboratories like the SSDFC. Surrey Police has the same strategy since it has a smaller digital forensics team, but the Met's DEFS seems to be less concerned about this probably because it is large enough to cover nearly everything on its own.

When comparing the two digital forensics laboratories in the United Kingdom and the one in China, we also noticed a clear difference in the standard procedures they follow. Neither of the the UK laboratories have passed the required ISO/IEC 17025 lab accreditation test yet, although they both are preparing for it, but the SSDFC in China passed the ISO/IEC 17025 lab accreditation test in 2009 (just 2 years after its establishment in 2007) and the more advanced 17043 test as a proficiency test provider in 2014. In the United Kingdom the ACPO Good Practice Guide for Digital Evidence (Association of Chief Police Officer of England, Wales & Northern Ireland 2011) seems to be the only one followed by most in-house laboratories of police forces, although some digital forensic service providers in private sector have passed the ISO/IEC 17025 test (CCL-Forensics Ltd. 2014; Forensic Telecommunications Services Ltd. 2015).[18] While the United Kingdom does not have any national standards

---

[18] Some private-sector digital forensic service providers in the United Kingdom also passed some other relevant ISO tests such as ISO 9001 (ISO 2008) for general quality management and ISO/IEC

on digital forensics yet, a large number of national standards have been published by different governmental bodies on digital evidence and forensics including audio and video forensic analysis since many years ago. The ISO standards-based lab accreditation test and the proficiency test required are also more established in China. For a more detailed description on standards and best practice guides in digital and multimedia forensics, readers are referred to Chapter 2 of this book.

Although this chapter only covers three digital forensics laboratories in two nations (the United Kingdom and China), many observations can be generalized to other nations such as the US. Note that the United Kingdom and China are representatives of two most widely adopted law systems in the world (the common law system and the civilian law system), so we expect they reflect more about practices in their respective law systems. For instance the civilian law systems tend to depend more on established laws and regulations rather than important cases, which may explain why China has more standards, while the United Kingdom has more or less only one best practice guide. Despite the possibility to generalize the observations in this chapter to other nations following the two common law systems, we call for caution when readers want to apply such generalizations to their own nations. We plan to cover more digital forensics laboratories from more other nations on the website of this book. The main targets include the United States, more European and Asian nations, Russia, Canada, Australia, New Zealand, major Latin American nations like Brazil and Mexico, South Africa and some major African nations, and also Islamic nations some of which follow a complete differently law system. It will be of interests to look at some very special regions such as Hong Kong, Macau and Taiwan.[19]

One encouraging fact we noticed is that many new multimedia forensic techniques developed in research community have been used and some are even standardized in China. It is also interesting to see the Met's DEFS and the SSDFC are both actively conducting research themselves by publishing papers at research venues. The SSDFC is also involved in organization of research events such as scientific conferences (Shanghai Stars Digital Forensic Center 2013). While the use of multimedia forensic techniques are more known to people in digital forensics field or the wider information security field (which are both more in computer science and engineering), researchers working in multimedia forensics are less informed because they are often from a signal processing background (which belongs to electronic engineering more than computer science). One of the motivations of the book is to fill the gap between digital forensics and multimedia forensics fields so that people in the two fields can work together more often and more effectively to meet the real-world needs of digital forensics laboratories.

27001 (ISO/IEC 2013) for information security management (CCL-Forensics Ltd. 2014; Forensic Telecommunications Services Ltd. 2015).

[19] We use 'Taiwan' as this is the mostly used term in academia without any particular political position. Depending on the political positions of readers, they should interpret this word as the 'Republic of China', 'Taiwan, People Republic of China' or 'Taiwan (Republic of China)' or any other terms more acceptable to them.

Regarding the challenges identified by the three digital forensics laboratories, we can see the most common challenge is the diversity of media storage, file formats and device models, which require digital forensics software and hardware tools to be updated more frequently. This is a challenge more for software developers and less so for researchers, although research on reconfigurable software frameworks such as the MPEG Reconfigurable Video Coding (RVC) standard (ISO/IEC 2011) may find opportunities. The three digital forensics laboratories are less converged on other challenges, but some important areas for further research are highlighted which include live forensics, cloud forensics, binary code analysis, anti-forensics, perceptual hashing, advanced image enhancement techniques. More standardization activities are also mentioned by staff of the SSDFC, which is also a concern of the UK Forensic Science Regulator as laid out in its Codes of Practice and Conduct for forensic science providers and practitioners in the Criminal Justice System (UK Forensic Science Regulator 2011).

1.6 Summary

This chapter samples three digital forensics laboratories in two nations (the United Kingdom and China) to reflect how multimedia data and devices are handled in real-world digital forensics laboratories in the context of more general digital forensics. The focus is more on the operational side, but some key techniques used and main technical challenges are also covered so that readers can have a better understanding on the positions of those multimedia forensic techniques covered in real world. A comparison of the three digital forensics laboratories has allowed us to discuss some key observations and led to some future research directions. The content of this chapter will be further enriched by supplemental material on the book's website where we plan to cover more digital forensics laboratories from other nations.

1.A Appendix: Questionnaires for Interviewing Surrey Police and Shanghai Stars Digital Forensic Centre

In this appendix, we give the questionnaires we used to interview staff of Surrey Police and the Shanghai Stars Digital Forensic Centre. Some questions were changed slightly to reflect different contexts of the two digital forensics laboratories.

- What is the structure of your digital forensic team? Do you have a single unit or is it composed of a number of independent groups who are working together?
- What is the current size of your digital forensic team? How many of them are technical examiners, and how many have been certified as qualified forensic examiners? What training courses do you require technical staff to attend?
- What is the average workload of your forensic technical staff? What is the percentage of cases involving multimedia data and devices?

- What types of multimedia data and devices you have encountered in your forensic practice?
- What are the percentage of digital audio, images and videos you normally handle?
- Do you have a triage process for forensic cases? If so, how do you handle multimedia data and devices in such a process?
- Do you have a separate forensic photography team and how do you work with them? Are you using digital cameras for photographing all crime scenes or do you still use optical ones? What are the requirements of cameras used for forensic photography? Could you mention a few models of cameras you are currently using?
- How often do you outsource forensic analysis of multimedia data and devices to other digital forensic labs? If so how many cases go to other police forces and how many go to independent forensic experts in the private sector?
- How do you work with other digital forensic laboratories run by law enforcement?
- What is the standard procedure you are following for general forensic analysis? Are there any guidelines tailored for multimedia data and devices (like those defined by US Scientific Working Group on Imaging Technology (SWGIT))?
- What are the software and hardware tools you use to extract multimedia files from storage media and carve deleted files? What are the challenges of multimedia file carving from your point of view?
- How often do you encounter encrypted multimedia files? Are they always related to child pornography? Do you have a way to automate detection of (child) pornography materials?
- What software and hardware tools are you routinely using to analyze multimedia data and devices? If you have to process raw data to enhance the chance of extracting evidence, what do you know to make the results reproducible and admissible to the court?
- Do you have to often edit multimedia data to make it more understandable to the court? If so, what principles do you follow and what tools do you normally use?
- Do you often need to measure physical properties of persons and objects in digital images and videos? If so what software tools do you normally use and what are the challenges of making the result forensically sound (i.e. admissible to the court)?
- Do you work on cases where biometric data (e.g. face images, fingerprints, recorded speech for speaker recognition) are processed? What software (and hardware) tools do you use to handle biometric data?
- Have you encountered cases where you needed to identify (camera or scanner) sources of digital images and videos? If so, how did you handle those cases?
- Have you encountered cases where you needed to identify sources of recorded digital audio files? If so, how did you handle those cases?
- Have you encountered cases where you needed to identify sources of printed documents (possibly scanned)? If so, how did you handle those cases?

- Have you encountered cases where you needed to authenticate printed documents (i.e. if one piece of paper or printed document is indeed the one claimed)? If so, how did you handle those cases?
- Have you encountered cases where you needed to identify possibly forged digital multimedia files (speech , images, videos)? If so, how did you handle those cases?
- Have you encountered cases where you needed to differentiate computer-generated multimedia data from recorded ones from real world? If so, how did you handle those cases?
- Have you encountered cases where you needed to know what had happened during the whole processing chain of multimedia data (e.g. an seemingly uncompressed image was actually JPEG compressed)?
- Have you encountered cases where suspects or criminals seemed to have employed some anti-forensic techniques to make your work harder? If so are these cases related to multimedia data?
- Do you feel that existing software and hardware tools are sufficient for you to perform your work on multimedia data and devices? If not, what functionalities are missing from existing tools?
- Do you have any other information you feel useful for our book and our research on multimedia forensics in general?

References

AccessData Group, Inc. 2014a Forensic Toolkit® (FTK®): Recognized around the world as the standard in computer forensics software, http://accessdata.com/solutions/digital-forensics/forensics-toolkit-ftk (Accessed 17 February 2015).

AccessData Group, Inc. 2014b Password recovery Toolkit® (PRTK®), http://www.accessdata.com/solutions/digital-forensics/decryption (Accessed 17 February 2015).

AccessData Group, LLC 2013 AccessData BootCamp, http://marketing.accessdata.com/acton/attachment/4390/f-044b/1/-/-/-/-/ADBootCamp07-08-2013.pdf (Accessed 17 February 2015).

Adobe Systems Inc. 2014 Adobe creative cloud: Audition CC, http://www.x-ways.net/forensics/index-m.html (Accessed 17 February 2015).

Apple Inc. 2014 Final cut Pro X, http://www.apple.com/final-cut-pro/ (Accessed 17 February 2015).

Association of Chief Police Officer of England, Wales & Northern Ireland 2011 ACPO good practice guide for digital evidence, version 5.0, http://www.acpo.police.uk/documents/crime/2011/201110-cba-digital-evidence-v5.pdf (Accessed 17 February 2015).

Avid Technology, Inc. 2014a Media composer, http://www.avid.com/US/products/media-composer (Accessed 17 February 2015).

Avid Technology, Inc. 2014b Pro tools 11, http://www.avid.com/US/products/pro-tools-software/ (Accessed 17 February 2015).

Beijing Yangchen Electronic Technology Company 2014 VS-99 computer speech processing workstation, http://www.caigou.com.cn/c55382/trade_detail_406586.shtml, (in Chinese) (Accessed 17 February 2015).

CCL-Forensics Ltd. 2014 Quality standards, http://www.cclgroupltd.com/about/quality-standards/ (Accessed 17 February 2015).

Centre for Law-Enforcement Audio Research 2012 About CLEAR, http://www.clear-labs.com/about.php (Accessed 17 February 2015).

China National Accreditation Service for Conformity Assessment 2010 Accreditation criteria for proficiency testing providers (ISO/IEC 17043:2010) CNAS-CL03, http://www.cnas.org.cn/extra/col23/1295319425.pdf (in Chinese) (Accessed 17 February 2015).

China National People's Congress Standing Committee 2004 Electronic signature law of the People's Republic of China, http://www.miit.gov.cn/n11293472/n11294912/n11296092/11904895.html, (in Chinese) (Accessed 17 February 2015).

Cooper AJ 2013 Improved photo response non-uniformity (PRNU) based source camera identification. *Forensic Science International* **226**(1–3), 132–141.

Digital Assembly, LLC 2014 Adroit photo forensics 2013, http://digital-assembly.com/products/adroit-photo-forensics/ (Accessed 17 February 2015).

Digital Detective Group Ltd 2014 Blade, http://www.digital-detective.net/digital-forensic-software/blade/ (Accessed 17 February 2015).

DVD FLLC 2014 DVD format book, http://www.dvdfllc.co.jp/format/f_nosbsc.html (Accessed 17 February 2015).

Edexcel 2014 Our history, http://www.edexcel.com/Aboutus/who-we-are/our-history/Pages/Ourhistory.aspx (Accessed 17 February 2015).

European Network of Forensic Science Institutes 2011 About enfsi, structure, working groups, digital imaging, http://www.enfsi.eu/about-enfsi/structure/working-groups/digital-imaging (Accessed 17 February 2015).

European Network of Forensic Science Institutes Forensic Speech and Audio Analysis Working Group 2009 Best practice guidelines for enf analysis in forensic authentication of digital evidence, http://www.enfsi.eu/sites/default/files/documents/forensic_speech_and_audio_analysis_wg_-_best_practice_guidelines_for_enf_analysis_in_forensic_authentication_of_digital_evidence_0.pdf (Accessed 17 February 2015).

Forensic Telecommunications Services Ltd. 2015 FTS ISO standards, http://www.forensicts.co.uk/fts-isostandars.html (Accessed 17 February 2015).

GetData Pty Ltd 2014 Recover My Files data recovery software, http://www.recovermyfiles.com/ (Accessed 17 February 2015).

Guidance Software, Inc. 2014a Computer forensic software – Encase forensic https://www.guidance-software.com/products/Pages/encase-forensic/overview.aspx (Accessed 17 February 2015).

Guidance Software, Inc. 2014b EnCE® certification program, https://www.guidancesoftware.com/training/Pages/ence-certification-program.aspx (Accessed 17 February 2015).

Home Office Scienfitic Development Branch 2010 CCTV and imaging publications, http://webarchive.nationalarchives.gov.uk/20100418065544/http://scienceandresearch.homeoffice.gov.uk/hosdb/cctv-imaging-technology/CCTV-and-imaging-publications.html (Accessed 17 February 2015).

IMIX Vision Support Systems 2014 IMPRESS http://www.imix.nl/impress/ (Accessed 17 February 2015).

International Electrotechnical Commission 1999 Audio recording – Compact disc digital audio system IEC 60908, Edition 2.0, available online at http://webstore.iec.ch/webstore/webstore.nsf/artnum/023623 (Accessed 17 February 2015).

ISO 2008 Quality management systems – Requirements ISO 9001, http://www.iso.org/iso/catalogue_detail?csnumber=46486 (Accessed 17 February 2015).

ISO/IEC 2005 General requirements for the competence of testing and calibration laboratories, ISO/IEC 17025, http://www.iso.org/iso/catalogue_detail.htm?csnumber=39883 (Accessed 17 February 2015).

ISO/IEC 2010 Conformity assessment – General requirements for proficiency testing, ISO/IEC 17043, http://www.iso.org/iso/home/store/catalogue_tc/catalogue_detail.htm?csnumber=29366 (Accessed 17 February 2015).

ISO/IEC 2011 Information technology – MPEG systems technologies – Part 4: Codec configuration representation, ISO/IEC 23001-4, http://www.iso.org/iso/home/store/catalogue_ics/catalogue_detail_ics.htm?csnumber=59979 (Accessed 17 February 2015).

ISO/IEC 2013 Information technology – Security techniques – Information security management systems – Requirements ISO/IEC 27001, http://www.iso.org/iso/home/store/catalogue_ics/catalogue_detail_ics.htm?csnumber=54534 (Accessed 17 February 2015).

Kot AC and Cao H 2013 Image and video source class identification. In *Digital Image Forensics* (ed. Sencar HT and Memon N), Springer, New York, pp. 157–178.

LEVA International, Inc. 2014 LEVA certification program, https://leva.org/index.php/certification (Accessed 17 February 2015).

Metropolitan Police Service 2014a About the Met, http://content.met.police.uk/Site/About (Accessed 17 February 2015).

Metropolitan Police Service 2014b Crime mapping: Data tables, http://maps.met.police.uk/tables.htm (Accessed 17 February 2015).

Microsoft 2013 PhotoDNA newsroom, https://www.microsoft.com/en-us/news/presskits/photodna/ (Accessed 17 February 2015).

Monga V 2005 Perceptually based methods for robust image hashing, PhD thesis. The University of Texas at Austin, Austin, TX.

National People's Congress Standing Committee, People Republic of China 2005 National People's Congress Standing Committee's decision regarding management of judicial appraisals, http://www.moj.gov.cn/zgsfjd/content/2005-10/26/content_798901.htm?node=5152 (in Chinese) (Accessed 17 February 2015).

Ontario Provincial Police Technological Crime Unit 2010 C4all support forum, http://www.c4all.ca/ (Accessed 17 February 2015).

Professional Security Magazine 2011 HOSDB becomes CAST, http://www.professionalsecurity.co.uk/news/news-archive/hosdb-becomes-cast/ (Accessed 17 February 2015).

Shanghai Stars Digital Forensic Center 2013 Release of program of the 4th China Information Security Congress, http://www.stars.org.cn/node/86 (Accessed 17 February 2015). (in Chinese).

Shanghai Stars Digital Forensic Center 2014a Introduction to digital forensics experts, http://www.stars.org.cn/node/21 (in Chinese) (Accessed 17 February 2015).

Shanghai Stars Digital Forensic Center 2014b Procedure for digital forensic applications, http://www.stars.org.cn/node/6 (in Chinese) (Accessed 17 February 2015).

Shanghai Stars Digital Forensic Center 2014c Services provided, http://www.stars.org.cn/node/9 (in Chinese) (Accessed 17 February 2015).

Sira Defence & Security Ltd 2014 Siraview CCTV viewer I DVR software I DVR viewer I CCTV software SiraView, http://www.siraview.com/ (Accessed 17 February 2015).

Sony Creative Software, Inc. 2014 Vegas product family overview, http://www.sonycreativesoftware.com/vegassoftware (Accessed 17 February 2015).

Surrey County Council 2012 Mid-year estimates of population since 2001, http://www.surreycc.gov.uk/environment-housing-and-planning/surrey-data-online/surrey-data-population/mid-year-estimates-of-population-since-2001 (Accessed 17 February 2015).

Surrey Police 2014a May 2014 performance update, http://www.surrey.police.uk/Portals/0/pdf/about%20us/May-2014-Performance-Update-2.pdf (Accessed 17 February 2015).

Surrey Police 2014b Our history, http://www.surrey.police.uk/about-us/our-history (Accessed 17 February 2015).

Third Research Institute of Ministry of Public Security 2014 Brief introduction to the Third Research Institute of Ministry of Public Security, http://www.trimps.ac.cn/page/ssintroduce.html. (in Chinese) (Accessed 17 February 2015).

Thomas K 2010 Met Police bags innovation prize for its Focus media player, http://www.computing.co.uk/ctg/analysis/1863413/met-police-bages-innovation-prize-focus-media-player (Accessed 17 February 2015).

UK College of Policing 2014 Core skills in data recovery and analysis, http://www.college.police.uk/en/1262.htm (Accessed 17 February 2015).

UK Commission for Employment and Skills 2014 National Occupational Standards http://nos.ukces.org. uk/ (Accessed 17 February 2015).

UK Forensic Science Regulator 2011 Codes of practice and conduct for forensic science providers and practitioners in the criminal justice system, Version 1.0, https://www.gov.uk/government/publications/ forensic-science-providers-codes-of-practice-and-conduct (Accessed 17 February 2015).

UK Home Office 2014 Tables for 'police workforce, England and Wales, 30 September 2013', https:// www.gov.uk/government/publications/tables-for-police-workforce-england-and-wales-30-september-2013 (Accessed 17 February 2015).

UK Ministry of Justice 2011 Achieving best evidence in criminal proceedings: Guidance on interviewing victims and witnesses, and guidance on using special measures, http://www.justice.gov.uk/downloads/ victims-and-witnesses/vulnerable-witnesses/achieving-best-evidence-criminal-proceedings.pdf (Accessed 17 February 2015).

UK Office for National Statistics 2012 Population estimates for England and Wales, mid-2011 (2011 census-based), http://www.ons.gov.uk/ons/publications/re-reference-tables.html?edition=tcm: 77-262039 (Accessed 17 February 2015).

UK Parliament 1829 Metropolitan Police Act 1829, http://www.legislation.gov.uk/ukpga/Geo4/10/44 (Accessed 17 February 2015).

UK Parliament 1967 Criminal Justice Act 1967, http://www.legislation.gov.uk/ukpga/1967/80 (Accessed 17 February 2015).

UK Parliament 1980 Magistrates' Courts Act 1980, http://www.legislation.gov.uk/ukpga/1980/43 (Accessed 17 February 2015).

UK Parliament 1981 The Magistrates' Courts (Forms) Rules 1981, http://www.legislation.gov.uk/uksi/ 1981/553/contents/made (Accessed 17 February 2015).

X-Ways Software Technology AG 2014 X-Ways forensics: Integrated computer forensics software, http://www.x-ways.net/forensics/index-m.html (Accessed 17 February 2015).

ZiuZ B.V. 2014 VizX2, http://forensic.ziuz.com/en (Accessed 17 February 2015).

2

Standards and Best Practices in Digital and Multimedia Forensics

Shujun Li[1], Mandeep K. Dhami[2] and Anthony T.S. Ho[1]
[1]*Department of Computing and Surrey Centre for Cyber Security (SCCS), University of Surrey, Guildford, UK*
[2]*Department of Psychology, Middlesex University, London, UK*

2.1 Introduction

One of the main goals of digital forensics is to produce digital evidence admissible to the court, which requires that the digital forensic process or techniques used are not flawed in such a way that the evidence or intelligence generated can be questioned. This requirement is normally described as 'forensic soundness' (Casey 2007; McKemmish 2008). While the exact meaning of forensic soundness depends on the underlying jurisdiction and forensic techniques involved, there are established standard procedures and best practices around how digital forensic examinations should be conducted and managed to ensure forensic soundness for every step of the chain of custody.

This chapter provides a comprehensive review of important international, regional and national standards relevant to digital forensics and electronic evidence in general, as well as many best practice guides produced by different bodies. Some standards and best practice guides are not directly related to digital forensics; however, they are still important to digital forensics laboratories and law enforcement agencies because they define formal management procedures that help guarantee soundness of forensic examinations conducted. In addition, this chapter also covers standards and best practice guides on training and education in the digital forensics sector, with some training and certification programs which are well recognized among forensic practitioners.

Handbook of Digital Forensics of Multimedia Data and Devices, First Edition.
Edited by Anthony T.S. Ho and Shujun Li.
© 2015 John Wiley & Sons, Ltd. Published 2015 by John Wiley & Sons, Ltd.
Companion Website: www.wiley.com/go/digitalforensics

Most standards and best practice guides covered in this chapter are about digital forensics in general, but they can be applied to multimedia forensics as well since they often define steps of the general procedure rather than how a specific technique should be used in practice. There are also standards and best practice guides dedicated to digital forensics of multimedia data and devices, many of which are focused on a specific type of multimedia data or devices. We however do not cover standards and best practice guides falling more into traditional forensic sciences, such as those on fingerprint and facial image recognition systems and processes.[1]

It deserves mentioning that this chapter should not be considered as a complete list of all standards and best practice guides in digital and multimedia forensics fields, due to the fact that a large number of nations and regional/international bodies have their own standards and best practice guides. Therefore, the main areas that this chapter focuses on are a number of important regional/international bodies and representative nations such as the United States, the United Kingdom and the European Union. We plan to cover more regional/international bodies and nations on the website and future editions of this book.

The rest of this chapter is organized as follows. In the next section we will give an overview of most important standards and best practice guides covered in this chapter, in order to show a big picture of what has been happening in this space since the early 1990s when electronic evidence started becoming an important area for law enforcement and forensic practitioners to look at seriously. This section will give a complete list of all standards and best practice guides covered in this chapter. After the overview a number of sections are dedicated to different groups of standards and best practice guides according to their contents: Section 2.3 covers electronic evidence and digital forensics in general, Section 2.4 focuses on multimedia evidence and multimedia forensics, Section 2.5 looks at digital forensics laboratory accreditation, Section 2.6 focuses on general quality assurance (management) procedures important for digital forensics laboratories and finally Section 2.7 covers training, education and certification. The last section concludes this chapter with a summary of existing standards and best practices and also future trends.

2.2 Overview

Figure 2.1 provides a diagrammatic representation of the historical development of selected standards and best practice guides for digital forensics. It also illustrates how those standards and best practice guides are related to each other.[2] Largely speaking,

---

[1] Note that those systems are highly digitized as well, but we consider them less relevant for the context of digital forensics and electronic evidence due to their closer link to physical means of conducting forensic analysis and preserving the evidence.

[2] Only major dependencies among standards and best practice guides are shown to enhance readability of the diagram. It is not uncommon for one standard or best practice guide to refer to many other ones.

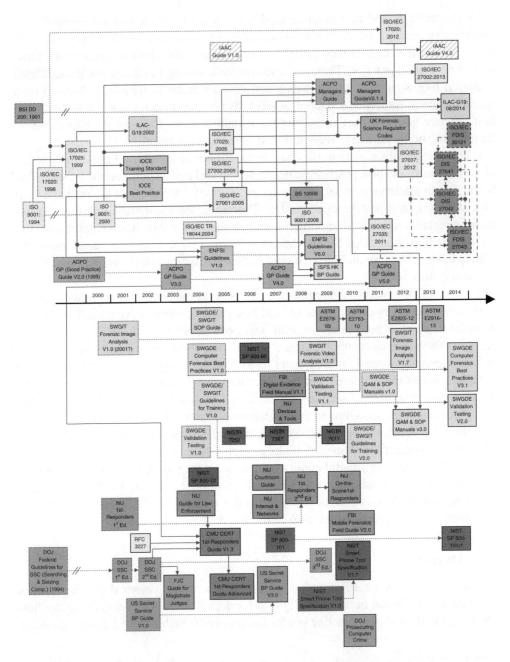

Figure 2.1 Time line and relationships of selected standards and best practice documents on digital forensics. Dotted boxes denote superseded early editions and dotted lines link these with their latest editions. The dashed boxes denote four ISO standards to be published. The Information Assurance Advisory Council (IAAC) forensic-readiness guide refers to many standards and best practice guides, so the links are omitted.

there are two subsets of standards and best practice guides: those with a closer link with ISO standards (above the time axis), and those without a link or with a very loose link with ISO standards (below the time axis). The first subset is more about quality assurance and the second is more about technical/legal/judicial processes. Most standards and best practice guides in the second subset (as covered in this chapter) are made by US bodies, which is mainly due to the leading roles of three key US bodies, National Institute of Standards and Technology (NIST) and two SWGs (Scientific Working Groups), in the digital forensics field. This partitioning has its root in the fact that ISO standards are more about quality assurance procedures, so standards and best practice guides more related to technical/legal/judicial procedures are less dependent on ISO standards.

While ISO standards are the most important ones among all digital forensics standards, the IAAC forensic-readiness guide is the most comprehensive non-standard guide and also the most recent as its latest edition was published in November 2013. The UK ACPO (Association of Chief Police Officers) 'Good Practice Guide' is probably the most cited non-standard guide, which can be explained by its long history since the 1990s.[3]

In the remaining part of this section, we list all standards and best practice guides covered in this chapter according to the following grouping:

- ISO standards
- Other international/regional standards and best practice guides
- US standards and best practice guides
- UK standards and best practice guides
- Other standards and best practice guides

The aforementioned grouping is more based on the bodies making/publishing the standards and best practice guides. In Sections 2.3–2.7 we will discuss all the standards and best practice guides in detail according to the following content-based grouping:

- Electronic evidence and digital forensics
- Multimedia evidence and multimedia forensics
- Digital forensics laboratory accreditation
- General quality assurance (management)
- Training, education and certification

It will be a very long list if we try to cover all relevant standards and best practice guides in all countries and regions. The language barrier and difficulties in accessing the fulltexts of standards from non-English-speaking regions have limited our ability to

---

[3] The authors were unable to obtain the first edition of the *ACPO Good Practice Guide*, but it must have appeared before 1999 when the second edition was published.

review other potentially relevant standards. Therefore, this chapter covers only some selected standards and best practice guides which we had access and considered more important for the digital and multimedia forensics fields. In future we plan to include a page on the book's website to (i) provide updates on new changes to standards and best practice guides covered in this chapter and (ii) cover more standards and best practice guides which are nor covered in the printed edition of this chapter.

2.2.1   ISO Standards

A number of ISO standards are important in the field of digital forensics:

* ISO/IEC 27037:2012 'Information technology – Security techniques – Guidelines for identification, collection, acquisition and preservation of digital evidence' (2012)
* ISO/IEC 27035:2011 'Information technology – Security techniques – Information security incident management' (2011)
* ISO/IEC 17025:2005 'General requirements for the competence of testing and calibration laboratories' (2005)
* ISO/IEC 17020:2002 'General criteria for the operation of various types of bodies performing inspection' (2002)
* ISO/IEC 27001:2013 'Information technology – Security techniques – Information security management systems – Requirements' (2013b)
* ISO/IEC 27002:2013 'Information technology – Security techniques – Code of practice for information security management' (2013a)
* ISO 9001:2008 'Quality management systems – Requirements' (2008)

There are also several other new standards that have not been officially published but are in the final stage of being finalized:

* ISO/IEC 27041 'Information technology – Security techniques – Guidelines on assuring suitability and adequacy of incident investigative methods' (2014a): DIS (draft international standard) as of April 2014
* ISO/IEC 27042 'Information technology – Security techniques – Guidelines for the analysis and interpretation of digital evidence' (2014b): DIS (draft international standard) as of April 2014
* ISO/IEC 27043 'Information technology – Security techniques – Incident investigation principles and processes' (2014c): FDIS (final draft international standard) as of September 2014
* ISO/IEC 30121 'System and software engineering – Information technology – Governance of digital forensic risk framework' (2014d): FDIS (final draft international standard) as of September 2014

These to-be-published standards will also be covered in this chapter because they are important new progresses and no major changes are expected in their contents (DIS and FDIS are both in voting stages).

2.2.2 Other International/Regional Standards and Guides

There are some other international/regional standards and best practice guides, although some of them (i.e. those made by ASTM International) appear to be more geared to the US digital forensics community. For regional standards and best practice guides we focused mainly on European ones.

- ASTM International Standards[4]:
  - ASTM E2678-09 'Guide for Education and Training in Computer Forensics' (2009)
  - ASTM E2763-10 'Standard Practice for Computer Forensics Guide' (2010)
  - ASTM E2825-12 'Standard Guide for Forensic Digital Image Processing' (2012)
- A best practice guide from the IETF (Internet Engineering Task Force): RFC 3227 'Guidelines for Evidence Collection and Archiving' (2002)
- International best practice guides:
  - ILAC-G19:2002 'Guidelines for Forensic Science Laboratories' (2002)
  - ILAC-G19:08/2014 'Guidelines for Forensic Science Laboratories' (2014)
  - IOCE (International Organization on Computer Evidence) 'Guidelines for Best Practice in the Forensic Examination of Digital Technology' (2002a)
  - IOCE 'Training Standards and Knowledge Skills and Abilities' (2002b)
- European best practice guides:
  - ENFSI (European Network of Forensic Science Institutions) 'Guidelines for Best Practice in the Forensic Examination of Digital Technology' Version 6.0 (2009)
  - ENFSI Forensic Speech and Audio Analysis Working Group (FSAAWG) 'Best Practice Guidelines for ENF Analysis in Forensic Authentication of Digital Evidence' (2009)

2.2.3 US Standards and Best Practice Guides

There are a large number of US standards and best practice guides. Some of these were produced by the NIST, the measurement standards laboratory of the US Department of Commerce. NIST produces Federal Information Processing Standard publications (FIPS PUB), NIST special publications (SPs), technical reports and specifications in

---

[4] ASTM International is a US-based standardization body making international voluntary consensus standards. While this body calls itself 'ASTM International', the standards it makes are more like regional standards for digital forensics practice in North America (especially in the United States).

different technical fields. Some of these can be used as guidelines for digital forensics. Many other best practice guides have been produced by the Scientific Working Group on Digital Evidence (SWGDE), which was formed by the Federal Crime Laboratory Directors in 1998[5] and the Scientific Working Group on Imaging Technology (SWGIT) formed by the US Federal Bureau of Investigation (FBI) in 1997.[6] Both SWGs produce documents regarding standard procedures for many aspects about handling digital and multimedia evidence. Finally, best practice guides have also been produced by US law enforcement bodies such as the Department of Justice (DOJ) and its research, development and evaluation agency, National Institute of Justice (NIJ). US standards and best practice guides covered in this chapter are listed in the following[7]:

- NIST special publications (SPs), interagency reports (IRs) and other publications:
  - NIST SP 800-101 Revision 1 'Guidelines on Mobile Device Forensics' (Ayers *et al.* 2014)
  - NIST SP 800-72 'Guidelines on PDA Forensics' (Jansen and Ayers 2004)
  - NIST SP 800-86 'Guide to Integrating Forensic Techniques into Incident Response' (Kent *et al.* 2006)
  - NISTIR 7387 'Cell Phone Forensic Tools: An Overview and Analysis Update' (Ayers *et al.* 2007)
  - NISTIR 7617 'Mobile Forensic Reference Materials: A Methodology and Reification' (Jansen and Delaitre 2009)
  - NIST 'Smart Phone Tool Specification' Version 1.1 (NIST 2010)
- SWGDE and SWGIT best practice guides:
  - SWGDE/SWGIT 'Recommended Guidelines for Developing Standard Operating Procedures' Version 1.0 (2004)
  - SWGDE/SWGIT 'Guidelines & Recommendations for Training in Digital & Multimedia Evidence' Version 2.0 (2010)
  - SWGDE/SWGIT 'Proficiency Test Program Guidelines' Version 1.1 (2006)
  - SWGDE 'Model Standard Operation Procedures for Computer Forensics' Version 3.0 (2012d)
  - SWGDE 'Model Quality Assurance Manual for Digital Evidence Laboratories' Version 3.0 (2012c)
  - SWGDE 'Minimum Requirements for Quality Assurance in the Processing of Digital and Multimedia Evidence' Version 1.0 (2010)
  - SWGDE 'Digital Evidence Findings' (2006)
  - SWGDE 'Focused Collection and Examination of Digital Evidence' Version 1.0 (2014h)

---

[5] See http://www.swgde.org/.

[6] See https://www.swgit.org/history.

[7] NIST published other publications related to digital forensics, but in this chapter we only consider those more relevant as standards and best practice guides.

- SWGDE 'Establishing Confidence in Digital Forensic Results by Error Mitigation Analysis' Version 1.5 (2015)
- SWGDE 'Best Practices for Computer Forensics' Version 3.1 (2014a)
- SWGDE 'Recommended Guidelines for Validation Testing' Version 2.0 (2014j)
- SWGDE 'Best Practices for Mobile Phone Forensics' Version 2.0 (2014e)
- SWGDE 'Core Competencies for Mobile Phone Forensics' Version 1.0 (2013b)
- SWGDE 'Best Practices for Handling Damaged Hard Drives' Version 1.0 (2014d)
- SWGDE 'Capture of Live Systems' Version 2.0 (2014f)
- SWGDE 'Best Practices for Forensic Audio' Version 2.0 (2014c)
- SWGDE 'Core Competencies for Forensic Audio' Version 1.0 (2011)
- SWGDE 'Mac OS X Tech Notes' Version 1.1 (2014i)
- SWGDE 'Best Practices for Vehicle Navigation and Infotainment System Examinations' Version 1.0 (2013a)
- SWGDE 'Best Practices for Portable GPS Device Examinations' Version 1.0 (2012b)
- SWGDE 'Peer to Peer Technologies' (2008)
- SWGDE 'Best Practices for Examining Magnetic Card Readers' Version 1.0 (2014b)
- SWGDE 'UEFI and Its Effect on Digital Forensics Imaging' Version 1.0 (2014k)
- SWGDE 'Electric Network Frequency Discussion Paper' Version 1.2 (2014g)
- SWGIT Document Section 1 'Overview of SWGIT and the Use of Imaging Technology in the Criminal Justice System' Version 3.3 (2010e)
- SWGIT Document Section 4 'Recommendations and Guidelines for Using Closed-Circuit Television Security Systems in Commercial Institutions' Version 3.0 (2012f)
- SWGIT Document Section 5 'Guidelines for Image Processing' Version 2.1 (2010d)
- SWGIT Document Section 6 'Guidelines and Recommendations for Training in Imaging Technologies in the Criminal Justice System' Version 1.3 (2010c)
- SWGIT Document Section 7 'Best Practices for Forensic Video Analysis' Version 1.0 (2009)
- SWGIT Document Section 11 'Best Practices for Documenting Image Enhancement' Version 1.3 (2010b)
- SWGIT Document Section 12 'Best Practices for Forensic Image Analysis' Version 1.7 (2012b)
- SWGIT Document Section 13 'Best Practices for Maintaining the Integrity of Digital Images and Digital Video' Version 1.1 (2012c)
- SWGIT Document Section 14 'Best Practices for Image Authentication' Version 1.1 (2013b)
- SWGIT Document Section 15 'Best Practices for Archiving Digital and Multimedia Evidence (DME) in the Criminal Justice System' Version 1.1 (2012a)

- SWGIT Document Section 16 'Best Practices for Forensic Photographic Comparison' Version 1.1 (2013a)
- SWGIT Document Section 17 'Digital Imaging Technology Issues for the Courts' Version 2.2 (2012d)
- SWGIT Document Section 18 'Best Practices for Automated Image Processing' Version 1.0 (2010a)
- SWGIT Document Section 19 'Issues Relating to Digital Image Compression and File Formats' Version 1.1 (2011)
- SWGIT Document Section 20 'Recommendations and Guidelines for Crime Scene/Critical Incident Videography' Version 1.0 (2012e)
- SWGIT Document Section 23 'Best Practices for the Analysis of Digital Video Recorders' Version 1.0 (2013c)
- SWGIT Document Section 24 'Best Practices for the Retrieval of Digital Video' Version 1.0 (2013d)
- Best practice guides edited/published by US law enforcement agencies (mainly DOJ and NIJ):
  - *Searching and Seizing Computers and Obtaining Electronic Evidence in Criminal Investigations*, 3rd Edition (US DOJ's Computer Crime and Intellectual Property Section 2009)
  - 'Investigative Uses of Technology: Devices, Tools, and Techniques' (US NIJ 2007c)
  - 'Investigations Involving the Internet and Computer Networks' (US NIJ 2007b)
  - 'Forensic Examination of Digital Evidence: A Guide for Law Enforcement' (US NIJ 2004)
  - 'Digital Evidence in the Courtroom: A Guide for Law Enforcement and Prosecutors' (US NIJ 2007a)
  - 'Electronic Crime Scene Investigation: A Guide for First Responders' Second Edition (US NIJ 2008)
  - 'Electronic Crime Scene Investigation: An On-the-Scene Reference for First Responders' (US NIJ 2009)
  - 'Digital Evidence Field Guide' Version 1.1 (US FBI 2007)
  - 'Mobile Forensics Field Guide' Version 2.0 (US FBI 2010)
  - 'Computer-Based Investigation and Discovery in Criminal Cases: A Guide for United States Magistrate Judges' (US Federal Judicial Center 2003)
  - 'Best Practices for Seizing Electronic Evidence: A Pocket Guide for First Responders' Version 3.0 (United States Secret Service, US Department of Homeland Security 2007)
- First responder training and education handbooks of Carnegie Mellon University (CMU) Computer Emergency Response Team (CERT):
  - 'First Responders Guide to Computer Forensics' (Nolan *et al.* 2005b)
  - 'First Responders Guide to Computer Forensics: Advanced Topics' (Nolan *et al.* 2005a)

2.2.4 UK Standards and Best Practice Guides

In the UK there is a national standard BS 10008:2008 'Evidential weight and legal admissibility of electronic information – Specification' (BSI 2008a) and a number of implementation guides of BS 10008:2008, all published by British Standard Institute (BSI):

- BIP 0008-1:2008 'Evidential weight and legal admissibility of information stored electronically. Code of Practice for the implementation of BS 10008' (BSI 2008c)
- BIP 0008-2:2008 'Evidential weight and legal admissibility of information transferred electronically. Code of practice for the implementation of BS 10008' (BSI 2008d)
- BIP 0008-3:2008 'Evidential weight and legal admissibility of linking electronic identity to documents. Code of practice for the implementation of BS 10008' (BSI 2008e)
- BIP 0009:2008 'Evidential Weight and Legal Admissibility of Electronic Information. Compliance Workbook for Use with BS 10008' (BSI 2008b)

There are also a number of best practice guides made by law enforcement including the ACPO (Association of Chief Police Officers of England, Wales and Northern Ireland), NPIA (National Policing Improvement Agency, dissolved in 2012), HOSDB (Home Office Scientific Development Branch, currently known as the CAST – Centre for Applied Science and Technology) and Forensic Science Regulator (FSR). Those best practice guides are listed in the following:

- Best practice guides on digital forensics or electronic evidence:
  - 'ACPO Good Practice Guide for Digital Evidence' Version 5.0 (UK ACPO 2011a)
  - 'ACPO Good Practice Guide for Managers of e-Crime investigation' Version 0.1.4 (UK ACPO 2011b)
  - Forensic Science Regulator (FSR) 'Codes of Practice and Conduct for forensic science providers and practitioners in the Criminal Justice System' Version 1.0 (UK Forensic Science Regulator 2011)
- Best practice guides on multimedia evidence handling:
  - 'Practice Advice on Police Use of Digital Images' (UK ACPO and NPIA 2007)
  - 'Storage, Replay and Disposal of Digital Evidential Images' (UK HOSDB 2007)
  - 'Digital Imaging Procedures' (Cohen and MacLennan-Brown 2007)
  - 'Retrieval of Video Evidence and Production of Working Copies from Digital CCTV Systems' (Cohen and MacLennan-Brown 2008)
  - 'Practice Advice on The Use of CCTV in Criminal Investigations' (UK ACPO and NPIA 2011)

In addition to the above, the Information Assurance Advisory Council (IAAC), a UK-based not-for-profit organization, also publishes a comprehensive guide on digital investigations and evidence since 2005 (Sommer 2005) and the latest edition is the fourth edition published in 2013. The guide's title was originally 'Directors and Corporate Advisors Guide to Digital Investigations and Evidence' but was changed to 'Digital Evidence, Digital Investigations, and E-disclosure: A Guide to Forensic Readiness for Organizations, Security Advisers and Lawyers' since its third edition published in 2012.

2.3 Electronic Evidence and Digital Forensics

In this section, we describe standards and best practice guides about electronic evidence and digital forensics in general (excluding those focusing on multimedia evidence and multimedia forensics, which will be covered in the next section). While for most standards we only give a very brief description, for some very important standards we provide more details to reflect their importance for the digital forensics community.

2.3.1 International Standards

2.3.1.1 ISO/IEC 27037:2012 'Guidelines for identification, collection, acquisition and preservation of digital evidence'

This standard provides guidelines for the identification, collection, acquisition and preservation of digital evidence. The scope of the guidance provided concerns general circumstances encountered by personnel during the digital evidence handling process. This standard is part of the ISO/IEC 27000 series of standards on information security management, and should be used as an accompaniment to ISO/IEC 27001 and ISO/IEC 27002 (the two most important standards in the ISO/IEC 27000 series) since it provides supplementary guidance for the implementation of control requirements for digital evidence acquisition.

According to the standard, digital evidence is usually regulated by three central principles:

1. Relevance: digital evidence proves or disproves an element of a case, and be relevant to the investigation.
2. Reliability: digital evidence serves its purpose, and all processes used in the handling of it should be repeatable and auditable.
3. Sufficiency: a digital evidence first responder (DEFR) should gather enough evidence for effective investigation and examination.

In addition, all the tools to be used by the DEFR should be validated prior to use and the validation evidence should be available when a challenge of the validation technique is encountered.

As part of its coverage on the whole process of digital evidence handling, this standard also covers issues related to personnel, roles and responsibilities, technical and legal competencies (core skills in Annex A), documentation (minimum requirements in Annex B), formal case briefing session, prioritization of potential digital evidence, among other important aspects.

The standard contains a detailed discussion on concrete instances of digital evidence identification, collection, acquisition and preservation. Such instances include computers, peripheral devices digital storage media, networked devices and also CCTV systems.

2.3.1.2   ISO/IEC 27035:2011 'Information security incident management'

This international standard provides techniques for the management of security incidents that are highly related to digital forensics, especially network forensics. It is also part of the ISO/IEC 27000 series and relies on the terms and definitions in ISO/IEC 27000. The standard overviews basic concepts related to security incidents, and describes the relationship between objects in an information security incident chain. It states that there should be a well-structured and planned approach to handle security incidents, and states the objectives of such an approach; which are beneficial for an organization to plan and establish its own security incident management approach. Moreover, it discusses the various benefits of having a structured approach. One of the key benefits is strengthening evidence and rendering it forensically sound and legally admissible. The standard also states that the guidance provided is extensive, and some organizations may vary in the need to deal with all of the issues mentioned depending on the size, nature of business conducted in the organization and complexity of mechanisms implemented within the organization.

Five main phases are identified that should constitute any information security incidence management. These are as follows:

1. Plan and prepare
2. Detection and reporting
3. Assessment and decision
4. Responses
5. Lessons learnt

The various procedures performed as part of the incident response to security incidents should handle and store digital evidence in a way to preserve its integrity in case it is later required for further investigation and legal prosecution.

The rest of the standard overviews key activities of each phase mentioned earlier. Moreover, Annex A provides a cross-reference table of ISO/IEC 27001 versus ISO/IEC 27035. Annex B provides examples of information security incidents and possible causes, while Annex C gives examples of sample approaches to categorization of security events and incidents. Annex D provides examples of incident and vulnerability reports and forms, while Annex E deals with legal and regulatory aspects.

This standard is to be split into three parts in future editions as currently planned by ISO/IEC JTC 1/SC 27 (the expert group editing ISO/IEC 27000 series standards): ISO/IEC 27035-1 'Principles of incident management', ISO/IEC 27035-2 'Guidelines to plan and prepare for incident response', ISO/IEC 27035-3 'Guidelines for incident response operations'. All the three parts are still in CD (committee draft) stage, so it is still too early to introduce them in this chapter.

2.3.1.3 ISO/IEC DIS 27041 'Guidance on assuring suitability and adequacy of incident investigative methods' (2014a)

This standard is also part of a set of new ISO/IEC standards to be published on investigation of information security incidents. As at the time of this writing, it is still in DIS stage, but it is expected that it will be officially published soon.

This standard is about providing assurance of the investigative process used and results required for the incident under investigation. It also describes the abstract concept of breaking complex processes into smaller atomic components so that simpler and robust investigation methods can be developed more easily. This standard is considered important for any person involved in an investigation ranging from authorizer, manager and the actual conductor. It is required that the standard is applied before an investigation starts so that all other relevant standards including ISO/IEC 27037, ISO/IEC 27042, ISO/IEC 27043 and ISO/IEC 27035 are all properly considered.

2.3.1.4 ISO/IEC DIS 27042 'Guidelines for the analysis and interpretation of digital evidence' (2014b)

This standard is also part of a set of new ISO/IEC standards to be published on investigation of information security incidents. Its current status is DIS.

This standard provides guidance on analysis and interpretation of potential digital evidence for identifying and evaluating digital evidence that may be used to investigate an information security incident. It is not a comprehensive guide, but provides some fundamental principles for ensuring that tools, techniques and methods can be selected and justified appropriately. This standard also aims to inform decision makers who need to determine the reliability of digital evidence presented to them. It is assumed to be used together with ISO/IEC 27035, ISO/IEC 27037, ISO/IEC 27041 and ISO/IEC 27043 in order to achieve compatibility.

2.3.1.5 ISO/IEC FDIS 27043 "Incident investigation principles and processes" (2014c)

This standard is also part of a set of new ISO/IEC standards to be published on investigation of information security incidents. Its current status is FDIS, the final phase of an international standard so we do not expect any major changes to its contents once published.

This standard provides guidelines for common investigation processes across different investigation scenarios, covering pre-incident preparation up to and including returning evidence for storage or dissemination. It also provides general advice and caveats on processes and appropriate identification, collection, acquisition, preservation, analysis, interpretation and presentation of digital evidence.

A basic principle of digital investigations highlighted in this standard is repeatability, which means the results obtained for the same case by suitably skilled investigators working under similar conditions should be the same. Guidelines for many investigation processes are given to ensure clarity and transparency in obtaining the produced results. The standard also provides guidelines to achieve flexibility within an investigation so that different types of digital investigation techniques and tools can be used in practice. Principles and processes are specified and indications are defined for how the investigation processes can be customized for different scenarios. Guidelines defined in this standard help justify the correctness of the investigation process followed during an investigation in case the process is challenged.

This standard covers a rather wide overview of the entire incident investigation process. It it is supposed to be used alongside some other standards including ISO/IEC 27035, ISO/IEC 27037, ISO/IEC 27041, ISO/IEC 27042 and ISO/IEC 30121.

2.3.1.6 ISO/IEC FDIS 30121 'Governance of digital forensic risk framework' (2014d)

This standard is a forensic-readiness standard for governing bodies (e.g. owners, senior managers and partners) to prepare their organizations (of all sizes) for digital investigations before they occur. It focuses on the development of strategic processes and decisions relating to many factors of digital evidence disclosure such as availability, accessibility and cost efficiency. Currently this standard is in the FDIS stage, so its content can be considered stable.

2.3.1.7 IETF RFC 3227 'Guidelines for Evidence Collection and Archiving' (2002)

This IETF RFC (Internet standard) defines best practice for system administrators in collecting electronic evidence related to 'security incidents' as defined in RFC 2828 'Internet Security Glossary' (Shirey 2000). It is a short document covering some general guidelines on 'order of volatility', 'things to avoid', legal and privacy considerations, and discusses evidence collection procedure and evidence archiving procedure separately. It also lists a number of tools system administrators need to have for evidence collection and archiving procedures.

2.3.2 National Standards

2.3.2.1 US Standards

In this section we review two standards made by ASTM International. We categorize them as US standards because they are more US-facing and were developed based on some national best practice guides.

ASTM E2763-10 'Standard practice for computer forensics' (2010)
This standard is a best practice document briefing methods and techniques for seizure, proper handling, digital imaging, analysis, examination, documentation and reporting of digital evidence in the scope of criminal investigations. The standard comprises 11 brief sections, each providing direct steps on general guidelines for a specific process. Section 1 contains the scope of the document, mentioned above, while Section 2 mentions the reference documents that are the ASTM Guide for Education and Training in Computer Forensics, and the SWGDE Recommended Guidelines for Validation Testing. Section 3 mentions the significance and the use of this document, most importantly that the examiner should be trained in accordance with the previous ASTM guide. Section 4 provides very general guidelines on evidence seizure. Section 5 concerns evidence handling, and Section 6 outlines equipment handling. Section 7 outlines steps needed for forensic imaging, while Section 8 handles guidelines on forensics analysis and examination. Section 9 goes over the documentation process, and Section 10 briefly outlines the report and its main function. Section 11 covers review policy of the forensic examiners' organization.

2.3.2.2 UK Standards

BS 10008:2008 'Evidential weight and legal admissibility of electronic information – Specification'
This British standard, which has evolved from the early 1990s,[8] covers the requirements for the implementation and operation of electronic information management systems, including the storage and transfer of information with regard to this information being used as potential digital evidence. It focuses on 'potential' evidence as opposed to other ISO/IEC standards which normally focus on digital material already labelled 'evidence'. The standard states that the requirements covered are generic and can be used by any kind of organization regardless of size or nature of business and can be applied to electronic information of all types. Thus, the information provided in this standard is to a minor degree more generic compared with other ISO/IEC standards.

---

[8] The first of such early standards we identified is BSI DD 206:1991, titled 'Recommendations for preparation of electronic images (WORM) of documents that may be required as evidence'. It defines 'procedures for the capture and storage of electronic images of hardcopy documents to ensure the preservation and integrity of information recorded on them'.

This standard covers the scope of three BIP 0008 codes of practice on the same topic that have been published by the BSI Group since 1990s and been widely adopted. These three codes of practice are BIP 0008-1, BIP 0008-2 and BIP 0008-3. They cover evidential weight and legal admissibility of information stored electronically, transferred electronically, and linking electronic identity to documents, respectively. The latest version of the BIP 0008 provides more thorough guidance that will assist in the effective application of this standard. They are reviewed in the following.

BIP 0008-1:2008, BIP 0008-2:2008, BIP 0008-3:2008 and BIP 0009:2008

The British Standards Institution (BSI) issued codes of practice BIP 0008-1:2008 'Evidential Weight and Legal Admissibility of Information Stored Electronically', BIP 0008-2:2008 'Evidential Weight and Legal Admissibility of Information Transferred Electronically', and BIP 0008-3:2008 'Evidential Weight and Legal Admissibility of Linking Electronic Identity to Documents', to be used for the implementation of the British Standard 10008, which covers the scope of all these documents. BIP 0009:2008 'Evidential Weight and Legal Admissibility of Electronic Information' is a workbook that needs to be completed, and it aids in the assessment process of compliance with the BS 10008:2008 standard.

The BIP 0008-1:2008 code of practice explains the application and actions of information management systems that store information electronically via any storage media and using any type of data files, where the legal admissibility and evidential weight requirements include authenticity, integrity and availability (referred to 'CIA' in literature). Moreover, this code deals with features of information management processes that influence the usage of information in regular business operations where legal admissibility does not constitute an issue. This widens the applicability of this code of practice. Issues such as accuracy and wholeness of stored information and how information is transferred to other systems are covered, although in more detailed in the next code of practice.

The BIP 0008-2:2008 code of practice explains methods and procedures for transferring information between computer systems where confidentiality, integrity and authentication are required by the legal admissibility and evidential weight of the sent and/or received documents. This is especially when the transfer process occurs between organizations. The code is applied to any type of computer files containing all sorts of data ranging from text and images to video and software. The transmission media can be circuit-switched networks, telephone circuits, cable, radio or satellite technologies, or any other form of transmission networks.

The BIP 0008-3:2008 code of practice explains methods and processes that are associated with four authentication principles, namely electronic identity verification, electronic signature, electronic copyright, and linking the electronic identity and/or electronic signature and/or electronic copyright to the particular electronic document. This is useful when the identity of the sender needs to be proven in identity theft cases. Moreover, the code provides guidelines on digital signatures and electronic copyright protection systems.

BIP 0009:2008 is a compliance workbook to be used with BS 10008:2008, which aids in the assessment of an information management system for compliance with the BS 10008:2008 standard. It should be completed and stored on the information management system under identical conditions by which other information on the system is stored. Moreover, it provides guidance on how to complete the workbook itself.

2.3.2.3   HB 171-2003 'Guidelines for Management of IT Evidence' (Australia)

This Australian standard provides guidance on managing electronic records that might be used as potential evidence in judicial and administrative procedures. It ensures the evidential value of records processed electronically.

It provides an overview of the management of IT evidence and the different uses of IT evidence for judicial, administrative and criminal proceedings. Moreover, it provides different principles for the management of IT evidence, and the IT evidence management lifecycle that consists of six stages: design for evidence, production of records, collection of evidence, analysis of evidence, reporting and presentation, and evaluating evidentiary weight.

2.3.3   Best Practice Guides

2.3.3.1   IOCE 'Guidelines for Best Practice in the Forensic Examination of Digital Technology' (2002a)

This document was made by the International Organization on Computer Evidence (IOCE). It presents requirements for systems, personnel, procedures and equipment in the whole forensic process of digital evidence. It provides a structure of standards, quality principles, and methods for processing digital evidence for forensic purposes in compliance with the requirements of an accreditation body or a prominent organization in the digital forensics community. It also promotes greater consistency in the methodology of processing digital evidence, which can yield equivalent results, thus enabling the interchange of data.

2.3.3.2   ENFSI 'Guidelines for Best Practice in the Forensic Examination of Digital Technology' Version 6.0 (2009)

This guide, produced by the European Network of Forensic Science Institutions (ENFSI), presents a structure for the standards and methods to be used for the detection, recovery and inspection of digital evidence for forensic objectives conforming to the standards of the ISO/IEC 17025 standard. Moreover, the guideline seeks to promote the level of efficiency and quality assurance in laboratories conducting forensic investigations using digital evidence that can produce consistent and valid results as well as increase cooperation among laboratories.

The guide covers most of the procedures and phases in the digital forensic process from evidence recovery and examination to report presentation for the court. Moreover, the document states that participating laboratories should have achieved or be in the process of achieving ISO/IEC 17025 accreditation for laboratory testing events (by following ILAC G19:2002 implementation guidance). The document defines different terms and concepts necessary for forensic laboratory operation, and outlines the different types of personnel involved in the process, along with the qualifications and requirements to perform their respective roles. Moreover, it encourages the active participation of personnel in seminars, workshops and training sessions in order to maintain their level of competence if not increase it. It also mentions the importance of proficiency testing, and outlines the procedure of administering such tests and the personnel responsible for this testing.

This guide also includes details of complaint procedures, and outlines general procedures and guidelines for complaints, stressing the importance of prompt action when dealing with complaints and anomalies.

2.3.4   US Guides

2.3.4.1   NIST

NIST SP 800-101 Revision 1 'Guidelines on Mobile Device Forensics' (2014)
The National Institute of Standards and Technology (NIST) published this guide on mobile device forensics, a relatively new and growing area in digital forensics. It covers a wide range of examination techniques, including examining operation and features of cellular networks (e.g. GSM), that helps forensics examiners better understand cellular phone operation and subsequently improve their forensic analysis skills. Its first edition was published in 2007 with a slightly difference title 'Guidelines on Cell Phone Forensics' and the revision in 2014 contains mainly updated and augmented to reflect new development in this field especially smart phones (which also caused the change of the title to 'Guidelines on Mobile Device Forensics').

This guide covers procedures for the preservation, acquisition, examination, analysis and reporting of mobile device evidence. Moreover, one of the main aims of the guide is to enable organizations and personnel to make more educated decisions about mobile forensics and provide support for personnel performing this task. It also provides information that can be used as a basis for standard operating procedures (SOPs) and policy development in the organization. In fact, the guide expands on this issue by stating that an organization must have its own forensics procedures, which can be tailored to the nature of the business of that organization. To support this, the guide provides information regarding the level of detail needed for policy in order to maintain a chain of custody. In addition, detailed information is provided for procedures and examinations of mobile phones to ensure successful policy production for an organization dealing with this task, along with how to perform proper evidence handling. Furthermore, different methods of data acquisition are discussed in detail, and the use of a cable is recommended in most situations. However, when this is not possible, the risk level of performing wireless acquisition is discussed.

This guide provides both coverage on physical and logical examinations, with detail on how to find evidence in memory. It also provides additional information on how to find and examine evidence on subscriber identity modules (SIM) cards, and discusses different types of data that can be extracted.

Legal practices are also considered, and the UK ACPO best practice guide (UK ACPO 2011a) is mentioned, along with the four principles that serve the goal of ensuring the integrity and accountability of evidence during an investigation. The Daubert standard originating in the United States (Project on Scientific Knowledge and Public Policy 2003) is also mentioned as a guide to be referred to when presenting evidence in a court of law.

Finally, this guide presents a variety of forensic tools evaluation: producing test results, reference data and proof of concept implementations and analysis. This tools evaluation has also resulted in the implementation of a test description and requirements document for a tool to be tested. The guide can provide aid to tool manufacturers so that they can improve their products. It also addresses the need for manufacturers to continuously update their products, since new mobile devices are being released very frequently.

NIST SP 800-72 'Guidelines on PDA Forensics' (2004)

This publication focuses on personal digital assistants (PDAs) and provides guidance on the preservation, examination and analysis of digital evidence on PDAs. It focuses on the properties of three families of PDAs:

1. Pocket PC
2. Palm OS
3. Linux-based PDAs

This guide also outlines actions to be taken during the course of evidence handling, device identification, content acquisition, documentation and reporting, in addition to forensic tools needed for such activities. The two main objectives of the publication are to aid organizations in gradually developing SOPs and strategies for forensic actions involving PDAs, and to prepare forensic examiners and first responders to effectively tackle any obstacles and challenges that might arise with digital evidence on PDAs. This guide is also intended to be used in addition with other guidelines to present more detailed insight into the issues associated with PDAs.

NIST SP 800-86 'Guide to Integrating Forensic Techniques into Incident Response" (2006)

This special publication is a detailed guide for organizations to help them build a digital forensic capability. It focuses on how to use digital forensic techniques to assist with computer security incident response. It covers not only techniques but also the development of policies and procedures. This NIST guide should not be used as an executive guide for digital forensic practices. Instead, it should be used together with guidance provided by legal advisors, law enforcement agencies and management.

The guide provides general recommendations for the forensic process in four phases: collection, examination, analysis and reporting. For the analysis phase, it covers four major categories of data sources: files, operating systems, network traffic and applications. For each category, it explains basic components and characteristics of data sources and also techniques for the collection, examination and analysis of the data. It also provides recommendations when multiple data sources need analyzing for a better understanding of an event.

The document also highlights four basic guidelines to organizations:

1. Organizations should establish policies with clear statements addressing all major forensic considerations and also conduct regular reviews of such policies and procedures.
2. Organizations should create and maintain digital forensic procedures and guidelines based on their own policies and also all applicable laws and regulations.
3. Organizations should ensure that their policies and procedures support use of the appropriate forensic tools.
4. Organizations should ensure that their IT professionals are prepared for conducting forensic activities.

NISTIR 7387 'Cell Phone Forensic Tools: An Overview and Analysis Update' (2007)

This interagency report (IR) of NIST is an overview of cell phone forensics tools designed for digital evidence acquisition, examination and reporting, which is an updated edition of NISTIR 7250 published in 2005. This document is not supposed to be a step-by-step guide, but serves to inform the digital forensic community about available tools and their performance.

NIST Smart Phone Tool Specification Version 1.1 (2010)

The purpose of this document is to specify requirements for mobile phone forensic tools that can be performed by obtaining internal memory from GSM smart phones and SIM cards and the internal memory of CDMA smart phones. Moreover it specifies test methods to verify if a certain tool meets the requirements.

The requirements which are specified are used to obtain test assertions. These are conditions that can be verified after tool testing, and these assertions produce at least one test case which consists of a test protocol and associated parameters. The protocol provides measures for performing the test along with the associated results to be expected from the test. The test cases and associated assertions are specified in the 'Smart Phone Acquisition Tool Test Assertions and Test Plan' (2010) document published by NIST.

NISTIR 7617 'Mobile Forensic Reference Materials: A Methodology and Reification' (Jansen and Delaitre 2009)

This report from NIST considers validation of mobile forensics tools, focusing on the use of reference test data. It describes a computer program and a data set for populating mobile devices. The data set was used to analyze some existing mobile forensics tools and a variety of inaccuracies were identified. It highlights the importance and difficulties of conducting proper tool validation and testing.

2.3.4.2 SWGs (Scientific Working Groups)

SWGDE/SWGIT 'Recommended Guidelines for Developing Standard Operating Procedures' Version 1.0 (2004)

This document considers SOPs that refer to the procedures followed regularly by law enforcement agencies in their activities. Thus, when considering digital forensics and evidence, SOPs can include all the activities related to digital evidence ranging from evidence recovery and examination to analysis and court presentation. Since SOPs can basically cover the whole forensics process, SWGDE has recommended guidelines regarding developing such SOPs that should be reviewed annually, and should contain all the information needed for a specific task in relation to a case being investigated, type of evidence collected, and the type of agency conducting the investigation. General guidelines include the purpose of the SOP, definitions, equipment to be used and its type and limitations, detailed steps of the SOP, references, authorization information and any additional material required for the SOP (e.g. safety instructions). Two example SOPs are given, one for wiping media and the other for video processing.

SWGDE 'Model Standard Operation Procedures for Computer Forensics' Version 3.0 (2012d)

This document provides a model of SOPs for digital forensics that can be used by organizations as a template. It is designed to be functional for both a single person operation, multiple person units and laboratory organizations. The model was developed by SWGDE based on a variety of SOPs from a broad selection of federal, state and local organizations. The model follows a modular approach so that a digital forensics lab can include sections they want to implement. The focus of each module is the methodology to conduct a forensic examination properly, under the assumption that the examiner is properly trained and competent in digital forensic analysis. The template SOPs are only examples and should not be used as mandatory step-by-step guides, and their contents must be revised to reflect an organization's policies and procedures.

SWGDE 'Minimum Requirements for Quality Assurance in the Processing of Digital and Multimedia Evidence' Version 1.0 (2010)

This document describes the minimum requirements for quality assurance when examining digital evidence as part of a forensic investigation. It outlines minimum requirements for the purposes of training, examiner certification, examination requirements, and laboratory requirements. Section 3 deals with educational aspects and training and discusses employment qualifications, training in areas related to duties, apprenticeship, on-going training, competency assessment and resources. Section 4 handles requirements for certification of digital evidence practitioners. Section 5 outlines laboratory standards in terms of personnel, facility design, evidence control, validation, equipment performance, examination procedures, examination review, documentation and reporting, competency testing, audits, deficiencies, health and safety. Policies for handling customer complaints, document control and disclosure of information are also mentioned in this section. Section 6 considers how to handle examination requests, examination and documentation.

SWGDE 'Digital Evidence Findings' (2006)

This is a very short document with the aim of ensuring that digital examination findings are presented to interested parties in an easily understandable format. It briefly lists what to include in a findings report and highlights the need to make such reports readable in non-technical terms and delivered on commonly accepted media supported by appropriate software. It also suggests that digital evidence laboratories educate people who review findings' reports. The document, however, does not cover laboratory specific topics and issues about providing digital evidence to defence representatives.

SWGDE 'Recommended Guidelines for Validation Testing' Version 2.0 (2014j)

This document is designed for all bodies performing digital forensic examinations, and it provides recommendations and guidelines for validation testing which is crucial for the results and conclusion of the examination process. It is stated that the testing should be applied to all new, revised and reconfigured tools and methods prior to their initial use in digital forensic processes. Such a validation testing process will guarantee the integrity of all the tools and methods used. The guideline outlines the process of validation testing, and also offers a sample test plan, a sample test scenario report and a sample summary report.

SWGDE 'Best Practices for Computer Forensics' Version 3.1 (2014a)

This document aims to outline best practices for computer forensics (forensics of normal computers like desktop PCs), starting from evidence collection and handling to report presentation and policy review. The sequence of information presented is very similar to that defined in ISO/IEC 27037, but it is more general in nature. However, an additional section of forensic imaging is included that briefly discusses how images should be taken.

SWGDE 'Focused Collection and Examination of Digital Evidence' Version 1.0 (2014h)

This document provides forensic examiners with a list of considerations when dealing with the review of large amounts of data and/or numerous devices so that they can focus their investigation on more relevant types of evidence. The main goal of this focused approach is to maximize efficiency and utilization of resources, so it can be considered as part of a generalized triage process.

SWGDE 'Establishing Confidence in Digital Forensic Results by Error Mitigation Analysis' Version 1.5 (2015)

This document provides a process for recognizing and describing errors and limitations associated with digital forensics tools. It starts with an explanation to the concepts of errors and error rates in digital forensics and highlights the differences of those concepts from other forensic disciplines. This document suggests that confidence in digital forensic results can be enhanced by recognizing potential sources of error and applying mitigating techniques, which include not only technical means but also trained and competent personnel using validated methods and following recommended best practices.

SWGDE 'Capture of Live Systems' Version 2.0 (2014f)

This document provides guidance on acquiring digital evidence from live computer systems. A primary concern is to capture and save data in a usable format. Factors that a forensic examiner should consider include the volatility or the volume of data, restrictions imposed by legal authority, and the use of encryption. It covers both volatile memory and data from mounted file systems as stored in a computer.

SWGDE 'Best Practices for Handling Damaged Hard Drives' Version 1.0 (2014d)

This document supplements the more general guide on best practices for computer forensics (SWGDE 2014a) by describing how to handle magnetic media hard drives when the data cannot be accessed using normal guidelines. It does not cover all storage media (e.g. solid-state drives, flash media and optical media). This document highlights that hard drive data recovery should be conducted by properly trained personnel only because traditional computer forensic software tools may destroy data stored on such hard disks.

SWGDE 'UEFI and Its Effect on Digital Forensics Imaging' Version 1.0 (2014k)
This document provides a general overview and guidance on Unified Extensible Firmware Interface (UEFI) used in media imaging. UEFI and its implementations are currently evolving so this document is expected to change as this technology and its standards become maturer. This document is for trained forensics professionals who may encounter UEFI for the first time.

SWGDE 'Best Practices for Examining Magnetic Card Readers' Version 1.0 (2014b)
This document describes best practices for seizing, acquiring and analyzing data contained within magnetic card readers used for illegal purposes (commonly called skimmers) to store personally identifiable information (PII). This document discusses different types of skimmers and explains the technical approaches to handling such devices for forensic investigation purposes.

SWGDE 'Best Practices for Mobile Phone Forensics' Version 2.0 (2014e)
This document provides best practice guidelines for the examination of mobile phones using hardware and software tools, including physical and logical acquisition. The target audiences include examiners in a laboratory setting and first responders encountering mobile phones in the field. It lists the most common limitations of mobile phones, ranging from dynamic and volatile data, to passwords and SIM cards. Guidelines for evidence collection are given in terms of evidence seizure and handling. The procedure for processing mobile phones in the laboratory is outlined in terms of equipment preparation, data acquisition, examination/analysis, documentation and archiving. Reporting and reviewing are also briefly mentioned at the end.

SWGDE 'Mac OS X Tech Notes' Version 1.1 (2014i)
This document describes the procedures for imaging and analyzing computers running Mac OS X (an operating system from Apple Inc.). It includes a discussion of OS X but does not cover iOS used by mobile devices and smart home appliance such as Apple TV. As a collection of technical notes, this document largely focuses on technical explanations to Mac OS and applications typically running from this platform.

SWGDE 'Best Practices for Portable GPS Device Examinations' Version 1.0 (2012b)
This document describes the best practices for handling portable GPS device examinations and provides basic information on the logical and physical acquisition of GPS devices. It also covers other steps of the whole evidence handling process including archiving and reporting.

SWGDE 'Best Practices for Vehicle Navigation and Infotainment System Examinations' Version 1.0 (2013a)
This document describes best practices for acquiring data contained within navigation and information and entertainment (Infotainment) systems installed in motor vehicles. It provides basic information on the logical and physical acquisition of such systems after physical access is obtained. It should be used in conjunction with the SWGDE document 'Best Practices for Portable GPS Devices'. It is limited to user data and does not cover information such as crash data.

SWGDE 'Peer-to-Peer Technologies' (2008)
This document is intended to provide guidelines when attempting to extract/recover evidence from peer to peer systems and associated files used in such systems for forensic purposes. The document provides results from a methodology used for testing peer to peer systems on different operating systems that was conducted by the US National White Collar Crime Center (NW3C), and the methodology is briefly outlined in the document.[9]

2.3.4.3   Department of Justice and Other Law Enforcement Agencies

DOJ 'Searching & Seizing Computers and Obtaining Electronic Evidence in Criminal Investigations' 3rd Edition (2009)
This document deals with the laws that preside over digital evidence in criminal investigations that stem from two key sources: the Fourth Amendment and the statutory privacy laws like the Stored Communications Act, the Pen/Trap statute and the Wiretap statute. Moreover, it focuses on matters that arise in drafting search warrants, forensic analysis of seized computers, and post-seizure obstacles posed to the search process. Finally, in addition to discussing the applications of the above mentioned laws for searching and seizing digital evidence, it also deals with issues of seizure and search with and without a warrant.

NIJ Special Report 'Investigative Uses of Technology: Devices, Tools, and Techniques' (2007c)
This publication is intended to serve as a resource for law enforcement personnel dealing with digital evidence, including relevant tools and techniques. As most NIJ reports, it focuses on three pillars:

1. Preserving the integrity of digital evidence during collection and seizure.
2. Adequate training of personnel examining digital evidence.
3. Full documentation and availability of procedures involving seizure, examination, storage or transfer of digital evidence.

---

[9] It states that the results of the NW3C research can be found in a report titled 'Peer to Peer: Items of Evidentiary Interest' at the SWGDE website. However, our search into both SWGDE and NW3C websites did not provide any link to this report.

In addition to the above, care must be taken when seizing electronic devices since inappropriate data access may violate federal laws including the Electronic Communications Privacy Act of 1986 and the Privacy Protection Act of 1980. The report is structured into three chapters: Chapter 1 covers techniques, Chapter 2 covers tools and devices, and Chapter 3 covers legal issues for the use of high technologies.

NIJ Special Report 'Investigations Involving the Internet and Computer Networks' (2007b)

This report deals with investigations involving the Internet and other computer networks. It focuses on tracing an Internet address to a source, investigating emails, websites, instant message systems, chat rooms, file sharing networks, network intrusions, message and bulletin boards and newsgroups, and legal issues associated with performing these activities. The guide provides technical knowledge of how to handle digital evidence found on computer networks and on the Internet, ranging from chat rooms to information stored on the Internet Service Provider's records.

NIJ Special Report 'Forensic Examination of Digital Evidence: A Guide for Law Enforcement' (2004)

This report presents a guide for law enforcement on forensic digital examination. It also repeats the three pillars observed in most NIJ reports. It overviews how digital evidence is processed covering four essential steps, that is, assessment, acquisition, examination, and documentation and reporting. It discusses whether an agency is ready to handle digital evidence based on appropriate resources needed for that task. The report presents five basic steps for conducting evidence examination as follows: Policy and procedure development, evidence assessment, evidence acquisition, evidence examination, and documenting and reporting.

NIJ Special Report 'Digital Evidence in the Courtroom: A Guide for Law Enforcement and Prosecutors' (2007a)

This report provides guidance on preparation of digital evidence for presentation in court. It details how to effectively present a case involving digital evidence, and suggests how to deal with a Daubert[10] gate-keeping challenge. In addition, it discusses challenges that face the process of maintaining the integrity of digital evidence along discovery and disclosure of the evidence. There is a chapter dedicated to child pornography cases that provides recommendations for law enforcement personnel on how to have a better understanding of the subculture of child pornographers to help with their investigation, since child pornographers are likely to be knowledgeable with the Internet, computers and technology.

---

[10] This refers to the Daubert Standard (Project on Scientific Knowledge and Public Policy 2003) following by US courts of law on scientific evidence. This is mentioned earlier in this chapter when NIST SP 800-101 (Ayers *et al.* 2014) is introduced.

NIJ Special Report 'Electronic Crime Scene Investigation: A Guide for First Responders' Second Edition (2008)

This report serves as a guide for law enforcement agencies and first responders in the tasks of recognition, collection and protection of digital evidence. In the introduction it defines digital evidence, and outlines how an agency can prepare for handling digital evidence investigations. It provides detailed information on different types of electronic devices on which potential evidence can be found, including computer networks and any investigative tools and equipment used in this process. Moreover, it outlines the process of securing and evaluating a crime scene, and instructs first responders on how to act when they arrive at the scene. There is also consideration of documentation, especially documenting the scene, and a list of what pieces of information to include. Evidence collection and transportation are also discussed towards the end, with the last chapter providing examples on categories and considerations given for specific crimes that can involve digital evidence.

NIJ Special Report 'Electronic Crime Scene Investigation: An On-the-Scene Reference for First Responders' Second Edition (2009)

This booklet style report is extracted from the above reviewed special report.

FBI 'Digital Evidence Field Guide' Version 1.1 (2007)

The guide starts out by mentioning five important facts about digital evidence that every law enforcement officer should be aware of including that most crimes involve digital evidence so that most crime scenes are digital crime scenes. Moreover, it states that digital evidence is usually volatile and can be easily modified if not handled with care. It also states that most digital evidence can be recovered even from damaged computers and other devices.

The guide focuses on explaining the role of a computer (or any other digital device in general) in a crime:

• Computers as the target of a crime, that is, related to the notion of unauthorized access to computer systems and hacking for specific aims like espionage, cyber terrorism and identity theft among others.
• Computers as the instrument of a crime, that is, using the computer to commit a specific crime, as above depending on the intention of the attacker including all of the above cases and additional cases like credit card frauds and child solicitations.
• Computer as the repository of evidence, that is, evidence on a computer can be found in various forms, like files, images, logs, etc. This is usually associated with crimes like frauds, child pornography, drug trafficking and email accomplices in traditional crimes.

Finally, the guide differentiates between the two types of digital evidence found on computers: universal (e.g. emails, logs, images) and case specific (e.g. for cyber terrorism it would include computer programs making the user anonymous, IP addresses, source code, among others).

FBI 'Mobile Forensics Field Guide' Version 2.0 (US FBI 2010)

This guide is available to law enforcement only, and we were unable to get a copy. From its title we can however guess the contents are about guidelines for mobile forensic examiners in the field. If we get a copy of this guide later, we will add a more detailed description on the book's website (permission will be sought from the FBI).

United States Secret Service 'Best Practices for Seizing Electronic Evidence V.3: A Pocket Guide for First Responders' (2007)

This field guide is intended to provide law enforcement personnel and investigators with an explanation of the methods in which computers and electronic devices may be used in committing crimes or as an instrument of a crime, or being a storage medium for evidence in a crime. It also provides guidance for securing evidence and transporting it for further analysis and examination to be performed by a digital evidence forensic examiner.

US FJC 'Computer-Based Investigation and Discovery in Criminal Cases: A Guide for United States Magistrate Judges' (2003)

This guide is actually a mixture of slides for a workshop presentation, a list of annotated case laws, excerpts from the guide 'Searching and Seizing Computers and Obtaining Electronic Evidence in Criminal Investigations' (US DOJ's Computer Crime and Intellectual Property Section 2002), a review of research on unresolved issues in computer searches and seizures (Brenner and Frederiksen 2002), and a draft report and recommendations for a working group on electronic technology in criminal justice system. The excerpts from the US DOJ guide and the review paper form the main body of the guide.

Compared with other guides and best practice documents, this one is less formal, and this may be explained by the fact that judges are much less familiar with electronic technologies than digital forensic examiners. In the draft report for the working group on electronic technology in criminal justice system, it is acknowledged that electronic data is pervasive and the lack of resources and training to handle electronic data. For the trial stage, it is also recommended that appropriate means should be taken to identify, preserve and present electronic evidence for the appellate record in an appropriate form.

2.3.4.4   CMU CERT

CMU CERT 'First Responders Guide to Computer Forensics' V1.3 (2005b)

This document highlights a serious training gap common to the fields of information security, computer forensics and incident response, which is performing basic forensic data collection. It comprises four modules:

- Module 1: Cyber laws and their influence on incident response
- Module 2: Understanding file systems and building a first responders toolkit
- Module 3: Volatile data, including tools for its collection, methodologies and best practices
- Module 4: Collecting persistent data in a forensically sound fashion

Of interest is that it quotes two related standards: ISO/IEC 17025 and RFC 3227. In addition, it also refers to the US DOJ guide 'Searching and Seizing Computers and Obtaining Electronic Evidence in Criminal Investigations' (Version 2.0 published in 2002), NIJ guide 'Forensic Examination of Digital Evidence – A Guide for Law Enforcement' (2004) and the UK ACPO good practice guide (Version 3.0 published in 2003).

CMU CERT 'First Responders Guide to Computer Forensics: Advanced Topics' (2005a)
This guide builds on the technical data presented in the previous guide, with a special focus on advanced technical operations and procedures instead of methodology. It comprises five modules:

- Module 1: Log file analysis
- Module 2: Process characterization, analysis and volatile data recovery
- Module 3: Image management, restoration and capture, and a tool called dd
- Module 4: Capturing a running process
- Module 5: Identifying spoofed email and tracing it using various techniques

2.3.5   European Guides

2.3.5.1   UK ACPO Good Practice Guide for Digital Evidence (2011a)

This guide is the latest edition of a series of such guides started in the 1990s (UK ACPO 1999, 2003, 2007, 2011a). Its title was originally 'Good Practice Guide For Computer Based Evidence' and changed to 'Good Practice Guide for Digital Evidence' in its latest edition, which reflects a major change from focusing on normal computers to diverse computing devices that can generate evidence in digital form. It is the current 'gold standard' for digital forensic examiners in UK police forces and widely accepted in UK court.

The guide presents instructions on the handling and examination of digital evidence found on computer devices. It ensures the collection and recovery of evidence in an efficient and timely fashion. The guide presents four principles for the handling of digital evidence by law enforcement personnel, as follows (note that we reproduce wording from the guide itself to be more accurate):

- **Principle 1**: No action taken by law enforcement agencies, persons employed within those agencies or their agents should change change data which may subsequently be relied upon in court.
- **Principle 2**: In circumstances where a person finds it necessary to access original data, that person must be competent to do so and be able to give evidence explaining the relevance and the implications of their actions.
- **Principle 3**: An audit trail or other record of all processes applied to digital evidence should be created and preserved. An independent third party should be able to examine those processes and achieve the same result.
- **Principle 4**: The person in charge of the investigation has overall responsibility for ensuring that the law and these principles are adhered to.

The main body of the guide covers planning, evidence capturing, analysis and presentation. Online evidence and mobile phone forensics are covered in this guide as well. For the presentation part, it covers different forms of presentation of digital evidence including verbal feedback, formal statements and reports, and as witness evidence. It also has a section on general issues such as training and education, welfare in the workplace, digital forensics contractors, disclosure and relevant legislation in the United Kingdom. The guide refers to another ACPO guide 'Good Practice Guide for Managers of e-Crime investigation' (UK ACPO 2011b) which covers more about managerial issues around e-crime investigation.

The guide also contains four appendices, one dedicated to network forensics covering wired and wireless networking devices and also live forensics, one dedicated to crime involving online evidence (websites, forums, blogs, emails, covert communications on the Internet, etc.), one dedicated to crime scene investigations and the last discussing how a law enforcement agency can develop a digital investigation strategy.

2.3.5.2   UK ACPO Good Practice and Advice Guide for Managers of e-Crime Investigation Version 0.1.4 (2011b)

This ACPO guide was produced in the context of the latest edition of the aforementioned best practice guide for digital forensic examiners. While the earlier one is for technical staff, the current one is mainly for managers of e-crime investigation laboratories/teams.

The guide consists of six sections. The first involves the initial setup, and it outlines issues such as role definitions: Roles of staff within e-crime units relating to the activities conducted that fall under two categories, that is forensic and network. This section also outlines key issues including training, budget, personnel and skill profiles, line managers within specialist investigation units, security of data and general points for consideration including accreditation. It encourages units to obtain ISO 9001 accreditation in the medium term, and ISO/IEC 17025 in the long term.

The second section of the guide involves management matters: where key issues such as business continuity are considered, including those for personnel and data, health and safety, and an example risk assessment policy developed by Sussex Police. Moreover, it also provides general advice on presentation of evidence, and on archiving and disposal.

The third section of the guide focuses on investigative matters, providing advice on balancing intrusion and privacy when conducting investigations and also outlining issues concerning intelligence acquisition and dissemination. Moreover, quality of process is covered, and the guide encourages units to implement different ISO standards including ISO 9001, ISO/IEC 17025, ISO/IEC 27001, and ISO/IEC 20000, advising managers to achieve this through the UKAS (United Kingdom Accreditation Service) while also considering the costs of such actions. In addition, it examines the different aspects related to defence 'experts', their instructions, and their use in the prosecution process, including the motivation for potential meetings to take place between defence experts and prosecution experts, with a description of the potential benefits.

The fourth section is about general issues, briefly providing recommendation for DNA profiling from keyboards, and also a review of recent changes in legislation and its impacts, including the Computer Misuse Act 1990 Amendments, Fraud Act 2006, and Part III RIPA 2000 – Investigation of Electronic Data Protected by Encryption, Powers To Require Disclosure. Moreover, it includes proposed amendments to the Obscene Publications Act 1959. It also mentions sources of advice including the NPIA, Serious Organised Crime Agency (SOCA) e-Crime, ACPO High-Tech Crime Sub-Group, Digital Evidence, The Centre for Forensic Computing, First Forensic Forum (F3), and the Internet Watch Foundation.

The fifth section is brief and focuses on forensic matters. It is linked to Appendix G in the guide, and focuses on issues such as peer review, dual tool verification, horizon scanning, and preview of machines and triage.

The last section is involved with training; including where and when to do training, what courses to be provided. It provides a training matrix for digital evidence recovery personnel, network investigators and mobile phone examiners.

2.3.5.3　UK Forensic Science Regulator 'Codes of Practice and Conduct for Forensic Science Providers and Practitioners in the Criminal Justice System' Version 1.0 (2011)

This document defines codes of practices for providers of forensic services to the criminal justice system in the United Kingdom. The Forensic Science Regulator expects that all forensic laboratories handling digital data recovery should pass ISO/IEC 17025:2005 accreditation (supplemented by ILAC-G19:2002) by October 2015.[11] The codes are basically an implementation of ISO/IEC 17025 and

---

[11] For other types of forensic activities, there are different deadlines.

ILAC-G19:2002 for forensic laboratories providing services to criminal justice system. The main difference is that ISO/IEC 17025 and ILAC-G19:2002 are voluntary, but the codes are mandatory: 'All practitioners and providers offering forensic science services to the CJS are to be bound by these Codes' (Clause 2.3). In the codes, it is also made clear that the UKAS will be the accreditation body assessing all forensic laboratories. However, note that the official role of the UKAS is not defined by the Forensic Science Regulator or the Home Office. Rather, this is done by the Accreditation Regulations 2009 (No. 3155)[12] as the UK's response to Decision No 768/2008/EC of the European Parliament and of the Council of 9 July 2008.[13]

2.3.5.4 Information Assurance Advisory Council (IAAC, Based in the United Kingdom) Digital Evidence, Digital Investigations, and E-disclosure: A Guide to Forensic Readiness for Organizations, Security Advisers and Lawyers (Fourth Edition, 2013)

This document is the fourth edition of a guide published by the IAAC whose first edition appeared in 2005. It stresses the importance of having a corporate forensic readiness program for an organization, and it is aimed at three types of audience who are involved in this process: Owners and managers of organizations, legal advisors and computer specialists. It seeks to provide information for the target audience on the various issues involved in evidence collection, analysis and presentation. This guide offers a rich amount of information about many standards and best practice documents, and also UK law enforcement resources and structures. Appendix 2 of this guide provides detailed individual procedures for the preservation of different types of evidence.

2.3.5.5 Information Security and Forensic Society (ISFS) Hong Kong 'Computer Forensics Part 2: Best Practices' (2009)

This document provides techniques and requirements related to the whole forensic process of providing digital evidence. It presents a deep level of examination of computer forensics from a technical aspect, along with an explanation underlying different procedures. It claims to be written in a neutral technological and jurisdictional manner, but with considerations in mind for readers from Hong Kong. The document is composed of the following five sections: Introduction to computer forensics, quality computer forensics, digital evidence, gathering evidence, and considerations

---

[12] The Accreditation Regulations 2009, No. 3155, Regulation 3, Appointment of UKAS as national accreditation body, 2009, http://www.legislation.gov.uk/uksi/2009/3155/regulation/3/made.

[13] Regulation (EC) No 765/2008 of the European Parliament and of the Council of 9 July 2008 setting out the requirements for accreditation and market surveillance relating to the marketing of products and repealing Regulation (EEC) No 339/93, Official Journal of the European Union, L218, 30–47, 13 August 2008, http://eur-lex.europa.eu/LexUriServ/LexUriServ.do?uri=OJ:L:2008:218:0030:0047: EN:PDF.

of law. It also has four appendices providing further information: A sample statement of findings, a list of sources of data (potential evidence), additional evidence considerations covering admissibility of digital evidence in HK, and relevant selections from HK electronic transaction ordinance.

2.4 Multimedia Evidence and Multimedia Forensics

Compared to general digital forensics, we did not find any international standards focusing on multimedia evidence or multimedia forensics only. This is not surprising because most international standards look at procedural aspects covering digital evidence and digital forensics in general which can also be applied to multimedia evidence and multimedia forensics. There are however some national standards available. Particularly, we noticed that there are a large number of national standards published by Chinese standardization bodies and authorities. These standards cover different aspects of multimedia evidence and multimedia forensics such as crime-scene photography and videography, digital forensic imaging, multimedia evidence extraction, forensic audio/image/video evidence enhancement, image authenticity detection, photogrammetry, multimedia evidence recovery from media storage, etc.[14] We were unable to obtain most of these China standards as they are not freely available, so we will not cover them in this chapter.

2.4.1 ASTM E2825-12 'Standard Guide for Forensic Digital Image Processing' (2012)

This more US-facing standard addresses image processing and related legal considerations in image enhancement, image restoration, and image compression. It provides guidelines for digital image processing to ensure quality forensic imagery is used as evidence in court. It also briefly describes advantages, disadvantages and potential limitations of each major process. This standard is partly based on the best practice guides of SWGDE and SWGIT.

2.4.2 US SWGs (Scientific Working Groups)

A large number of best practice guides on multimedia evidence and multimedia forensics are published by SWGDE and SWGIT, the two SWGs based in the United States. SWGDE focuses on mainly computer forensics, and it traditionally also covers forensics audio since SWGIT only covers forensic imaging (images

---

[14] There are also a large number of national standards on digital forensics published by Chinese standardization bodies and authorities, many of which are not based on any existing international standards. The "national vs. international" issue is a topic for future research as we will discuss in Section 2.8.

and video evidence). The two SWGs also work very closely with each other to produce joint best practice guides on common issues covering both digital and multimedia evidence/forensics particularly on training (see Section 2.7 for some joint SWGDE/SWGIT guides in this area). In the following, we review selected best practice guides from SWGDE and SWGIT on multimedia evidence and multimedia forensics.

2.4.2.1 SWGDE 'Best Practices for Forensic Audio' Version 2.0 (2014c)

This document provides recommendations for the handling and examination of forensic audio evidence for successful introducing such evidence in a court of law. It covers best practices for receiving, documenting, handling and examining audio evidence, independent of the tools and devices used to perform the examination.

2.4.2.2 SWGDE 'Electric Network Frequency Discussion Paper' Version 1.2 (2014g)

This document describes the potential use of electric network frequency (ENF) analysis for forensic examinations of audio recordings. It explains the technology behind ENF analysis, what ENF analysis can address, and how to handle evidence using ENF analysis.

2.4.2.3 SWGIT Document Section 1: 'Overview of SWGIT and the Use of Imaging Technology in the Criminal Justice System' Version 3.3 (2010e)

Since digital imaging is a widely used practice in forensic science, it is important to focus on different issues arising in this field. The main objective of this document is to make readers accustomed to significant issues in the capture, preservation, processing and handling of images in digital format, analogue format or film format. The document defines each process, and mentions issues that ought to be taken into account in order to ensure the integrity and admissibility of the image in court. It also mentions that personnel should be familiar with the SOPs mentioned in the SWGDE/SWGIT 'Recommended Guidelines for Developing Standard Operating Procedures' (SWGDE and SWGIT 2004).

2.4.2.4 SWGIT Document Section 5: 'Guidelines for Image Processing' Version 2.1 (2010d)

This document provides guidelines for the use of digital image processing in the criminal justice system. The main objective is to ensure the quality forensic imagery for use as evidence in a court of law. It states the position of SWGIT on image processing for forensic purposes: changes made to an image are accepted if (i) the original image is preserved, (ii) all processing steps are documented, (iii) the end result is presented as a processed or working copy of the original image and (iv) the recommendations laid out in this document are followed. This document describes advantages, disadvantages and potential limitations of each major process. It also provides guidelines for digital image processing SOPs with a sample SOP for latent print digital imaging.

2.4.2.5   SWGIT Document Section 7: 'Best Practices for Forensic Video Analysis' Version 1.0 (2009)

Forensic video analysis and the sub-discipline of forensic imaging were formally recognized by the International Association for Identification (IAI) in 2002 as a forensic science. The main purpose of this document is to establish suitable procedures for different processing and diagnostic tasks in video examination.

Forensic video analysis (FVA) consists of the examination, comparison and evaluation of video material and footage to be presented as evidence for legal investigations. The general tasks of FVA can be divided into three categories:

1. Technical preparation, which refers to the procedures and methods performed prior to examination, analysis or output (e.g. performing write-protection or visual footage inspections).
2. Examination, which involves the use of image science knowledge to obtain information from video materials (e.g. demultiplexing and decoding).
3. Analysis and interpretation, which are the use of specific knowledge to infer findings from video footage and their content.

The document describes best practice guidance for evidence management, quality assurance and control, security, infrastructure, work management, documentation, training, competency and SOPs. Moreover it describes the workflow for FVA that involves the sequence of all the events taking place during FVA. Finally, the document also presents a video submission form that contains sections necessary for all the information needed to be gathered from the scene and the victim for the investigation.

2.4.2.6   SWGIT Document Section 12: 'Best Practices for Forensic Image Analysis' Version 1.7 (2012b)

Image forensics is considered to be an important forensic discipline that has application in many domains other than the digital forensics field, including intelligence. This document provides personnel with direction regarding procedures performed when images represent the digital evidence under investigation.

Forensic image analysis involves analyzing the image and its content so it can be presented as evidence in court. Moreover, in law enforcement uses of forensic image analysis there are different sub-categories including photogrammetry and image authentication. As previously mentioned in the forensic video analysis guidelines, forensic image analysis can be divided into three phases:

1. Interpretation, which involves the use of image analysis expertise and specific knowledge to gain insight and identify images themselves, or objects in images in order to produce conclusions about the image(s) which can be used in examination step.
2. Examination, which refers to using image domain expertise to excerpt information from images, or to further characterize the different attributes of an image in order to facilitate interpretation of the image. This ought to produce valid information and interpreted results that should be admissible in a court of law.
3. Technical preparation, which refers to preparing evidence in general for further steps (e.g. examination, analysis or output).

In addition to providing best practice guidelines for evidence management, quality assurance, security, infrastructure, work management, documentation, training and SOPs, the document also provides the workflow for forensic image analysis. Finally, the document provides three examples of how the different phases of image analysis should take place and what outcomes to expect. One is an example of photogrammetric analysis, another is an example of photographic comparison analysis and the last one is an example of content analysis.

2.4.2.7 SWGIT Document Section 13: 'Best Practices for Maintaining the Integrity of Digital Images and Digital Video' Version 1.1 (2012c)

This document presents an overview of different issues affecting digital media files, and lists different methods of maintaining and demonstrating the integrity of such files. Moreover, it offers five workflow examples for maintaining and demonstrating the integrity of digital media files.

2.4.2.8 SWGIT Document Section 15: 'Best Practices for Archiving Digital and Multimedia Evidence in the Criminal Justice System' Version 1.1 (2012a)

This document provides best practice guidelines for archiving digital and multimedia evidence. It discusses the issues involved and provides guidelines of developing an archiving program. It starts by stressing the importance of archiving and why it is needed in organizations handling digital evidence. Furthermore, it outlines the archive creation process and lists the key elements that should be taken into consideration in this process, from security of the archived material through different types of media that can be archived, to media preservation, transmission, management and compression. It also covers archive maintenance by suggesting that new versions of software and hardware should be regularly checked to ensure they can access archived material which is usually of older versions, thus stressing the importance of reverse compatibility, interoperability and data migration. The last section of the document outlines archive retention periods that includes purging archived material.

2.4.2.9   SWGIT Document Section 16: 'Best Practices for Forensic Photographic Comparison' Version 1.1 (2013a)

This document outlines the appropriate practices to be followed when conducting photographic comparison in image analysis. It defines the purpose of photographic comparison, and emphasizes its importance as a forensic practice in various scientific fields ranging from medical applications to surveillance and intelligence. It also defines the scope of forensic photographic comparisons along with the validity of the comparison that could include a statistical model for reaching conclusions. It also discusses critical aspects of forensic photographic comparison; that include the class versus individual characteristics, the ACE-V protocol (Analysis, Comparison, Evaluation – Verification), recognition of imaging artefacts, and statistical versus cognitive evaluation. Moreover, the document provides guidelines on expertise and experience. It highlights training alone is not sufficient and translation of training into practice requires real-world expertise of qualified personnel. The document provides a rationale for best practices covering bias, selection of images for comparison, comparison processes, reconstruction, levels of findings, photogrammetry and forensic photographic comparison, and photographic documentation as a part of comparison/analysis. A brief outline of evidence management and quality assurance is provided at the end of this document.

2.4.2.10   SWGIT Document Section 17: 'Digital Imaging Technology Issues for the Courts' Version 2.2 (2012d)

This document discusses the proper use of digital imaging technology to judges and attorneys in court. It presents relevant issues in plain language to make them more understandable to the courts. It also covers case laws and research articles dealing with digital imaging technology used within the criminal justice system. In addition, it also addresses some common myths and misconceptions associated with digital imaging technologies.

2.4.2.11   SWGIT Document Section 20: 'Recommendations and Guidelines for Crime Scene/Critical Incident Videography' Version 1.0 (2012e)

This document provides recommendations and guidelines for using video camcorders to document crime scenes and critical incidents. It is suggested that videography be used a supplementary tool to still photography for investigative or demonstrative purposes. The document covers typical incidents that can be documented and equipment needed for the recording. It suggests general documentation and media-handling procedures, and also briefly covers maintenance and training.

2.4.3   ENFSI Working Groups

The only best practice guide from ENFSI on multimedia evidence and multimedia forensics we found is 'Best Practice Guidelines for ENF Analysis in Forensic Authentication of Digital Evidence' released by its Expert Working Group Forensic Speech and Audio Analysis (ENFSI-FSAAWG) in 2009. This was probably the only best practice guide on electric network frequency (ENF) analysis before SWGDE published its guide in 2014. ENF analysis is still a less mature research topic but has found its use in real-world digital forensics laboratories (see Section 1.2.4.2 of Chapter 1 of this book for its use at the Metropolitan Police Service's Digital and Electronics Forensic Service (DEFS) in the United Kingdom).

This ENFSI-FSAAWG ENF analysis guide aims to provide guidelines for FSAAWG members and other forensic laboratories for ENF analysis in the area of forensic authentication of digital audio and audio/video recordings. ENF analysis determines the authenticity of digital audio and video recordings. The document consists of four sections:

1. Quality assurance, which outlines requirements for a technical specialist dealing with ENF analysis, along with an overview of validation requirements for ENF analysis, different categories of software that can be used for forensic analysis, and the equipment to be used.
2. Case assessment, which outlines information requirements for determining authenticity of recorded evidence.
3. Laboratory examination, which outlines analysis protocols and standard procedures to be followed.
4. Evaluation and interpretation, which outlines considerations to be taken into account when handling ENF findings.

2.4.4   UK Law Enforcement

Some key UK law enforcement bodies have produced a number of best practice guides on handling multimedia evidence. They include Association of Chief Police Officers (ACPO), National Policing Improvement Agency (NPIA, ceased to exist in 2012 and most of its activities have been absorbed by the College of Policing), and the Home Office Scientific Development Branch (HOSDB, currently known as the CAST – Centre for Applied Science and Technology). Those guides are tailored to the law enforcement system in the United Kingdom, but can be reasonably generalized to other countries as well. For instance, some of the following reviewed guides are quoted by SWGIT in another guide we reviewed earlier (SWGIT 2012d).

2.4.4.1   ACPO and NPIA 'Practice Advice on Police Use of Digital Images' (2007)

This guide contains five main sections. Section 1 identifies the legal and policy framework within which digital images are managed as police information. Section 2 examines some of the police applications of digital imaging as an evidence resource (including third-party images that are given to the police for use as evidence). Section 3 defines and summarizes the functions of editing and processing images. One important principle is that any editing and processing should be done on a working copy of the original image. Both sections should be read in conjunction with another guide on digital imaging procedure (Cohen and MacLennan-Brown 2007) reviewed in the following text.[15] Section 4 describes the case preparation and disclosure of unused material relating to evidential digital images, and provides information for consideration when revealing exhibit images to the Crown Prosecution Service (CPS) and preparing for the court. Section 5 describes the decision-making process for retaining and disposing of police information, including associated images. This section should be read in conjunction with another guide 'Storage, Replay and Disposal of Digital Evidential Images' (UK HOSDB 2007) reviewed later on in this section.

2.4.4.2   HOSDB and ACPO 'Digital Imaging Procedures' Version 2.1 (2007)

This document is written for practitioners within the UK Police and Criminal Justice System (CJS) involved with the capture, retrieval, storage or use of evidential digital images. It is organized around a flowchart guiding the reader through the whole process including the following steps: (i) initial preparation and capture of images, (ii) transfer and designation of master and working copies, (iii) presentation in court and (iv) retention and disposal of exhibits. The first edition of this guide was published in 2002 and it has undergone a number of revisions. The latest edition recognizes the need to use a broader range of image capturing and storage techniques, and the allowance for the possibility that the Police can store master and working copies on a secure server rather than on physical WORM (write once, read many times) media such as CDs and DVDs.

2.4.4.3   HOSDB 'Storage, Replay and Disposal of Digital Evidential Images' (2007)

This guide focuses on the storage, replay and eventual disposal of evidential digital images generated by the police or those transferred to them from a third party. The term 'evidential' in this guide includes any image generated by, or transferred to the police, even if it is not originally generated as evidence. This document limits its coverage to police units only and leaves the wider issues of transferring images between agencies of the CJS to other guides.

   This document sets out a generic framework for thinking about how police units can store, replay and dispose of digital evidential images, and for encouraging a long-term approach to managing the technology. It also provides guidance on some technical issues and some templates for communicating requirements to the wider IT function.

---

[15] Appendix 1 of this guide also contains a diagram of the procedure (an older edition, Version 2.0).

2.4.4.4 ACPO and NPIA 'Practice Advice on the Use of CCTV in Criminal Investigations' (2011)

This document offers good practice to criminal investigators (who follow the Professionalising Investigation Programme Level 1 and 2) in the use of CCTV images as an investigative tool. Its aim is to provide a comprehensive set of fundamental processes and procedures for acquiring useful and usable CCTV material. This document does not cover roles/responsibilities, specialist techniques, real-time CCTV use or covert use of CCTV.

2.4.4.5 HOSDB and ACPO 'Retrieval of Video Evidence and Production of Working Copies from Digital CCTV Systems' Version 2.0 (2008)

This document provides guidance to technical staff in selecting CCTV methods and systems for effective retrieval and processing of video evidence. One key criterion is that the selected method should maintain evidential integrity so that maximum information is retained. The document is divided into two parts. The first part covers digital video retrieval in its native format from CCTV systems and the creation of master copies of the evidence. The second part focuses on the creation of working copies, particularly where a format conversion is required for further editing and processing purposes.

2.5 Digital Forensics Laboratory Accreditation

For most digital forensics laboratories, accreditation is an important issue because it can give the criminal justice system the needed confidence that the evidence presented in court is handled in a professional manner. According to Beckett and Slay (2011), 'laboratory accreditation is a system of peer assessment and recognition that a laboratory is competent to perform specific tests and calibrations'. The accreditation title does not automatically validate all results produced by an accredited laboratory, but it indicates that the laboratory has implemented a documented quality system so that the results of any forensic examination are repeatable. The core of the accreditation process is 'competence', and it applies to all types of forensic laboratories including the more recently developed digital forensics laboratories.

So far, there are only a few standards and best practice guides on accreditation of digital forensics laboratories. In this section we review those important ones for the digital forensics community.

2.5.1 International Standards

2.5.1.1 ISO/IEC 17025:2005 'General Requirements for the Competence of Testing and Calibration Laboratories'

This international standard is a general one for all types of laboratories, but the general principles can be applied to digital forensics laboratories. It has been followed widely in the digital forensics community and is the standard selected by the UK Forensic Science Regulator (2011) and by the US ASCLD/LAB (American Society of Crime Lab Directors/Laboratory Accreditation Board). Essentially speaking, this standard is about proving compliance and focuses on documentation of the whole of life process of any analysis performed by a laboratory (Beckett and Slay 2011).

This standard stipulates the procedures and requirements for the competence of calibration and testing laboratories, and it covers procedures that use standard methods non-standards methods, and laboratory-developed methods. It is intended to be used as a sign of competence for laboratories by accreditation bodies, and not as the basis of certification for laboratories. Laboratories or any organization conducting testing and/or calibration that fulfil the requirements stated in this standard will also be conforming to the 'principles' of ISO 9001:2008 standard. Annex A in ISO/IEC 17025:2005 provides a cross-reference between this standard and ISO 9001:2008.

The standard describes the details of management requirements for laboratories, and the associated management system that should be followed in such settings. Moreover, it reviews the process of issuing and reviewing documents for personnel working in the laboratory, and considers the issues of subcontracting services from other parties (which is becoming common as the law enforcement agencies subcontract services from external providers), and the general procedure to be followed. Corrective actions and control of records used in the laboratories are also detailed, and the emphasis of management reviews is stressed (as in the BS 10008 standard reviewed before). In addition, the standard includes recommendations on subcontracting services by laboratories.

2.5.1.2 ISO/IEC 17020:2002 'General Criteria for the Operation of Various Types of Bodies Performing Inspection' (2002)

This international standard covers inspection bodies whose activities include the examination of materials, products, installations, plants, processes, work procedures or services, and the determination of their conformity with requirements and the subsequent reporting of results of these activities to clients and, when required, to authorities. Inspection bodies are different from normal laboratories because the former can actually accredit the latter on behalf of the standardization bodies and authorities. Such bodies are important because they are involved in the laboratory accreditation process and for conducting inspection of crime scenes the laboratory accreditation is insufficient (which means that ISO/IEC 17025 cannot be used).

2.5.1.3 ILAC-G19:2002 'Guidelines for Forensic Science Laboratories' (2002)

This document provides guidance for laboratories involved in forensic analysis and examination by providing application of ISO/IEC 17025. This is useful because ISO/IEC 17025 is not particularly defined for forensic laboratories. ILAC-G19:2002 follows the clause numbers in ISO/IEC 17025 but does not re-state all clauses, so it must be read as a supplementary material to the latter. Another goal of ILAC-G19:2002 is for accreditation bodies to provide appropriate criteria for the assessment and accreditation of forensic laboratories. Since ILAC-G19:2002 covers all forensic science activities, digital forensics is just part of its coverage mainly under the heading 'Audio, Video and Computer Analysis'. Digital forensics may also be involved in some other activities such as 'Computer Simulation', 'Photography' and 'Evidence recovery' under 'Scene Investigation' heading.

2.5.1.4   ILAC-G19:08/2014 'Modules in a Forensic Science Process' (2014)

This is the latest edition of the ILAC-G19 guide, but it is not just a simple extension of the above 2002 edition. Instead, it adds coverage of a new standard ISO/IEC 17020. The title of the guide was also changed to reflect the addition of ISO/IEC 17020. The addition of ISO/IEC 17020 was due to the need for bodies performing crime scene investigation to pass ISO/IEC 17020 rather than ISO/IEC 17025 because the latter is less relevant for crime scene inspection which is better covered by ISO/IEC 17020. For digital forensics laboratories ILAC-G19:08/2014 remains largely the same but the reference to ISO/IEC 17025 was changed from its older edition in 1999 to its latest edition in 2005.

2.5.1.5   SWGDE 'Model Quality Assurance Manual for Digital Evidence Laboratories' Version 3.0 (2012c)

This document provides a model of Quality Assurance Manual (QAM) for any entity performing digital and multimedia forensic examinations. It proposes minimum requirements pertaining to all quality assurance aspects for a forensic laboratory, and it is applicable to an organization of any size including a single examiner. It follows the international standard ISO/IEC 17025:2005, ASCLD/LAB-International 2006 Supplemental Requirements for the Accreditation of Forensic Science Testing Laboratories and American Association for Laboratory Accreditation's 'Explanations for the ISO/IEC 17025 Requirements'. While this document refers to some particular accreditation bodies, it does not endorse one accreditation body over another. Not all sections of the modal QAM are required to fulfil accreditation requirements and all sections are modifiable to suit an organization's need. This document can be used in totality or partially as needed by an organization.

2.6   General Quality Assurance (Management)

In addition to laboratory accreditation, there are also some general quality assurance (management) standards widely used in the digital forensics field because they provide an additional guarantee that the digital forensic process is managed properly. They include mainly three ISO/IEC international standards which are also widely used in many other sectors.

2.6.1 ISO 9001:2008 'Quality Management Systems – Requirements'

This standard provides the requirements for a quality management system to be implemented by any type of organization, provided it is committed to showing its capability to constantly deliver services/products that conform to customer requirements and other statutory and regulatory requirements. In addition, organizations need to commit themselves to developing and advancing customer satisfaction via efficient implementation of the system. This standard mentions ISO 9000:2005 'Quality management systems – Fundamentals and Vocabulary' as an essential reference for its application.

2.6.2 ISO/IEC 27001:2005 'Information Security Management Systems – Requirements'

This standard presents a model for establishing, implementing, operating, monitoring, reviewing, maintaining and improving an information security management systems (ISMSs). The ISMSs can be tailored to the specific needs of an organization depending on the nature of business conducted, number of employees and the daily activities taking place in the organization. Moreover, the system should evolve over time to meet the changes in requirements and needs of the organization.

The standard uses the Plan–Do–Check–Act model for continual improvement and management of an ISMS, that also echoes the Organization for Economic Co-operation and Development (OECD) 'Guidelines for the Security of Information Systems and Networks' (2002). In relation to other international standards, this standard is compatible with ISO 9001:2000 reviewed earlier. ISO/IEC 27002 (formerly known as ISO/IEC 17799, see below) is crucial for the application of this standard, and Annex A in this standard derives a considerable amount of material from it.

2.6.3 ISO/IEC 27002:2013 'Code of Practice for Information Security Controls'

Historically, ISO/IEC 27002 was evolved from another standard in a different series and had a different reference number ISO/IEC 17799. After its second edition was published in 2013, the old reference number 17799 became outdated although it is still used in many other standards and documents including ISO/IEC 27001:2005 (which was published around the same time when the first edition of ISO/IEC 27002 was published based on ISO/IEC 17799).

This standard is concerned with all aspects of information security from introduction and implementation to maintenance and development of information security management in any organization. It includes control aims that are designed to be compliant with requirements resulting from a risk assessment. Moreover, it can be used for developing and implementing security management procedures for a given organization.

From a structural viewpoint, this standard is divided into 11 security control sections comprising 39 main security categories and one introductory section illustrating risk assessment and the handling of risks. The 11 security control sections refer to security policy, organizing information security, asset management, human resources security, physical and environmental security, communications and operations management, access control, information systems acquisition development and maintenance, information security incident management, business continuity management and compliance. Moreover, each of the 39 main security categories consists of a control aim outlining what the objectives are, and one or more controls that can be used to fulfil the aim.

Risk assessment plays a major part in this standard. Here, risk assessment is used for identifying and handling security risks in an organization, and should be performed in a systematic manner and regularly. Moreover, the standard contains useful sections for organizations trying to adapt their security controls, for example, Section 10.10 explains how audit logs should be used to monitor the usage of a system and also to assist in future investigations by providing reliable evidence. Finally, it is stated that such controls should be taken into account during the systems and projects requirement specification phase in order to optimize future costs and effective security solutions.

2.7 Training, Education and Certification on Digital and Multimedia Forensics

Training, education and certification of digital forensics examiners and any personnel working on cases involving digital and multimedia evidence are important for digital and multimedia forensics because of the need to ensure legal admissibility of evidence. Some standards and best practice guides particularly focus on these areas in order to provide guidance on how such programs can be managed. In the following, we first review those standards and best practice guides, and then briefly cover existing training, educational and certification programmes known in the digital and multimedia forensic community.

2.7.1 Standards and Best Practice Guides

2.7.1.1 ASTM E2678-09 'Standard Guide for Education and Training in Computer Forensics' (2009)

This standard provides guidance for individuals and students seeking academic and professional qualifications in the field of computer forensics, as well as academic institutions concerned with introducing forensics programs, and employers who are interested in the academic background of graduates from the computer forensics field. The guide outlines the knowledge, skills, and abilities (i.e. KSAs) necessary for a career in computer forensics, and how academic institutions (in the United States) should aim to provide such elements, by providing a list of model curricula for a variety of degrees including 2-year associate degrees, bachelor degrees, and master degrees in addition to professional certifications. According to SWGDE (2012a), this standard was developed from an NIJ publication.

2.7.1.2   IOCE 'Training Standards and Knowledge Skills and Abilities' (2002b)

This brief document provides principles of training in terms of minimum recommendations for training, minimum training topics, costs and cooperation. It also outlines the core training standards in terms of personnel, qualifications, competence and experience, and the recommended knowledge base. Moreover, it mentions specialized training in terms of court training/legal issues, partnerships and management awareness. The document lists general recommendations regarding training in terms of recommendations for cooperation in training and training for the G-8 24/7 points of contact.

2.7.1.3   Best Practice Guides from US SWGs

SWGDE and SWGIT also published a number of guides on training, proficiency testing and definition of core competencies.

SWGDE/SWGIT 'Guidelines & Recommendations for Training in Digital & Multimedia Evidence' Version 2.0 (2010)
This document provides guidelines for building a suitable training program in forensic digital and multimedia evidence for personnel engaged in evidence handling, collection, analysis and examination. It defines the various job categories involved in such processes, and divides the categories of training into six main ones as follows:

1. Awareness
2. Skills and techniques
3. Knowledge of processes
4. Skills development for legal proceedings
5. Continuing education
6. Specialized applications and technologies

The document provides individual areas focused on training for each specific job category. Moreover, the document outlines aspects for consideration in respect to training needs such as on the job training, continuing education, testimony training, certifications, higher education and training documentation. The last section of the document focuses on competency and proficiency testing, where it stresses that the examiner should be tested continuously whenever acquiring new skills and techniques for competency. The guidelines states that completed competency tests demonstrate proficiency in a given branch of knowledge.

SWGIT Document Section 6: 'Guidelines and Recommendations for Training in Imaging Technologies in the Criminal Justice System' Version 1.3 (2010c)

Different from SWGDE and SWGIT (2010), this document provides guidelines in proper training on imaging technologies for personnel or laboratories that are *not* performing image analysis or video analysis. This document defines various categories of training and different user categories that include ones unique to the criminal justice system. Moreover, it defines specific areas for focused training for each user category, and basically follows the same flow of information as in SWGDE and SWGIT (2010) for addressing aspects unique to training needs such as on the job training, continuing education, testimony training and certifications.

SWGDE/SWGIT 'Proficiency Test Program Guidelines' Version 1.1 (2006)

This document provides a proficiency test program for digital and multimedia evidence (DME) that would ascertain whether the technical methods used by a forensic examiner are valid and subsequently if the results produced by the examiner conform to a certain quality standard. It can be applied to the following fields: computer forensics, forensic audio, video analysis, and image analysis. This document provides advice on test preparation and design, and then briefly outlines test approval, distribution, testing process, review of test results, documentation and corrective action.

SWGDE 'Core Competencies for Mobile Phone Forensics' Version 1.0 (2013b)

This document identifies the core competencies necessary for handling and forensic processing of mobile phones. It applies to both first responders and laboratory personnel. It discusses different levels of cell phone analysis and the basic skills required at each level, but it does not address core competencies for chip-off or micro-read analysis. The elements covered in this document provide a basis for training, certification, competency and proficiency testing programs.

SWGDE 'Core Competencies for Forensic Audio' Version 1.0 (2011)

Similar to the aforementioned document, this one identifies core competencies necessary for conducting forensic audio functions. It covers the whole process of forensic audio from audio laboratory configuration, to audio evidence collection, to result reporting following needed legal standards. The elements covered in this document also provide a basis for training, certification, competency and proficiency testing programmes on forensic audio.

2.7.2   Certification, Training and Educational Programs

In addition to the above reviewed standards and best practice guides, there are many established training and certification programmes. Some such programmes are run by vendors of digital forensics software such as the EnCE Certification Program run by Guidance Software, Inc., the vendor of the widely used EnCase Forensics software tool, and AccessData BootCamp run by AccessData Group, LLC, the vendor of another widely used software tool Forensic Toolkit (FTK). Some other programs are run by law enforcement bodies such as the Core Skills in Data Recovery and Analysis program run by the UK College of Policing. Another example is the courses offered by the European Cybercrime Training and Education Group (ECTEG) which are for European law enforcement only. For multimedia evidence and forensics, one well-known program is the LEVA Certification Program (LEVA International, Inc. 2014) for forensic examiners handling video evidence.

For UK law enforcement, Appendix C of the ACPO guide (UK ACPO 2011b) contains a list of courses for assisting e-crime managers to train their staff, which include many digital forensics related training courses including higher education degree programs, general training course and product-oriented training programs. The list of courses is not exhaustive, but can reflect what UK law enforcement agencies are doing. Among all the training programs, those provided by National Policing Improvement Agency (NPIA), Centre for Forensic Computing (CFFC) of the Cranfield University, QinetiQ and 7Safe are the main highlights. NPIA was abolished in 2012 and its main activities were absorbed by the College of Policing (CoP) which is currently running former NPIA training courses. The NPIA/CoP Core Skills training courses are among the most fundamental ones in UK law enforcement, for example Surrey Police Digital Forensics Team requires all its technical staff to go through such training courses (see Section 1.3.3).

2.8   Conclusions

Digital and multimedia forensics is a fast evolving field both in terms of technologies involved and legal requirements forensic examiners need to follow. While a lot of international efforts have been made to provide international standards and best practice guides that can be applied to different jurisdictions, the majority of digital forensics practitioners and law enforcement bodies still follow a more national or local approach. For instance, most best practice guides covered in this chapter are made by UK or US law enforcement bodies and criminal justice system themselves, and to some extent the most followed international standards are those for general quality assurance (e.g. ISO/IEC 17025, ISO 9001 and ISO/IEC 27001/27002). This can be explained by the lack of international standards specially made for digital evidence and digital forensics.

The new international standards on digital forensics, that is ISO/IEC 27037, 27041, 27042, 27043, 30121 (most have been officially published as previously reviewed) have started making a difference as the first set of ISO/IEC standards dedicated to digital evidence and digital forensics. For instance, the Cloud Security Alliance recently published a report on mapping the elements defined in ISO/IEC 27037 to cloud computing (Cloud Security Alliance 2013). We expect after all the above ISO/IEC digital forensics focused standards are officially published, they will play a more active role in digital forensics practice at the national and international levels. We also predict more interactions between international standardization bodies and national bodies which will further develop the above ISO/IEC standards and may also create new standards covering other areas which are currently left out, for example network and cloud forensics.

Four of the above five new standards on digital forensics are in the ISO/IEC 27000 series which is largely about information security management. Those standards are made in the context of information security incident investigation, which however does not cover all areas of digital and multimedia forensics. In addition, as far as we are aware of, currently we are still lacking an international standard on multimedia evidence and multimedia forensics (even though some national standards do exist). We therefore call for more standardization activities in broadening the scope of current/forthcoming ISO/IEC standards, potentially creating a new series covering digital and multimedia forensics which can be based on the rich set of best practice guides as reviewed in this chapter. Furthermore, as acknowledged in the report (UK Forensic Science Regulator 2011), the lack of a more relevant forensic laboratory accreditation standard led to the debate of if ISO/IEC 17025 (plus ILAC-G19) is the 'right' standard the community should go for. This calls the community to work more closely with each other to produce something more tailored towards the needs of digital forensics practitioners and law enforcement.

Another important area to look at is how national jurisdiction and digital forensics practice interact with international ones. Since national best practice guides are currently more accepted at the national level, we do not foresee a rapid change of such practice in the near future. Harmonizing national and international systems can be hard especially for countries following the civil law system. Since most countries following the civil law systems are non-English-speaking countries, we did not focus on any of such countries in this chapter. We plan to focus on non-English-speaking countries in our future research, especially on major European, Asian, Latin American and African countries such as Germany, France, China, Russian, Indian, Japan, Brazil and South Africa. Another interesting group of countries are those in the Middle East and Islamic countries in general as they have very different legal systems but many of them are active in accepting digital evidence. Our current plan is to make use of the book's website to report more about our future research along this line. We also hope this book will have printed future editions so that we can further extend this chapter to cover more national standards and best practice guides on digital and multimedia forensics.

Acknowledgements

Part of this chapter was funded by the UK Home Office's Centre for Applied Science and Technology (CAST) through its research project 'Digital Forensics: Scenarios and Standards' between 2011 and 2012. The authors would like to thank Mr. Ahmad al-Natour who was a research assistant of the UK Home Office funded project and contributed to its final report which covers many standards and best practice guides in this chapter.

References

AccessData Group, LLC 2013 AccessData BootCamp, http://marketing.accessdata.com/acton/attachment/4390/f-044b/1/-/-/-/-/ADBootCamp07-08-2013.pdf (Accessed 17 February 2015).

ASTM International 2009 Standard guide for education and training in computer forensics, ASTM E2678-09, http://www.astm.org/Standards/E2678.htm (Accessed 17 February 2015).

ASTM International 2010 Standard practice for computer forensics, ASTM E2763-10, http://www.astm.org/Standards/E2763.htm (Accessed 17 February 2015).

ASTM International 2012 Standard guide for forensic digital image processing, ASTM E2825-12, http://www.astm.org/Standards/E2825.htm (Accessed 17 February 2015).

Ayers R, Jansen W, Cilleros N and Daniellou R 2005 Cell phone forensic tools: An overview and analysis, NIST Interagency Report 7250, http://csrc.nist.gov/publications/nistir/nistir-7250.pdf (Accessed 17 February 2015).

Ayers R, Jansen W, Moenner L and Delaitre A 2007 Cell phone forensic tools: An overview and analysis update, NIST Interagency Report 7387, http://csrc.nist.gov/publications/nistir/nistir-7387.pdf (Accessed 17 February 2015).

Ayers R, Brothers S and Jansen W 2014 Guidelines on mobile device forensics, NIST Special Publication 800-101 Revision 1, http://www.nist.gov/manuscript-publication-search.cfm?pub_id=915021 (Accessed 17 February 2015).

Beckett J and Slay J 2011 Scientific underpinnings and background to standards and accreditation in digital forensics. *Digital Investigation* **8**(2), 114–121.

Brenner SW and Frederiksen BA 2002 Computer searches and seizures: Some unresolved issues. *Michigan Telecommunications and Technology Law Review* **8**(12), 39–114.

Brezinski D and Killalea T 2002 Guidelines for evidence collection and archiving, IETF RFC 3227, http://tools.ietf.org/html/rfc3227 (Accessed 17 February 2015).

BSI 2008a Evidential weight and legal admissibility of electronic information – Specification, BS 10008:2008, http://shop.bsigroup.com/ProductDetail/?pid=000000000030172973 (Accessed 17 February 2015).

BSI 2008b Evidential weight and legal admissibility of electronic information. compliance workbook for use with BS 10008, BIP 0009:2008, http://shop.bsigroup.com/ProductDetail/?pid=000000000030186720 (Accessed 17 February 2015).

BSI 2008c Evidential weight and legal admissibility of information stored electronically. code of practice for the implementation of BS 10008, BIP 0008-1:2008, http://shop.bsigroup.com/ProductDetail/?pid=000000000030186227 (Accessed 17 February 2015).

BSI 2008d Evidential weight and legal admissibility of information transferred electronically. code of practice for the implementation of BS 10008, BIP 0008-2:2008, http://shop.bsigroup.com/ProductDetail/?pid=000000000030186228 (Accessed 17 February 2015).

BSI 2008e Evidential weight and legal admissibility of linking electronic identity to documents. code of practice for the implementation of BS 10008, BIP 0008-3:2008, http://shop.bsigroup.com/ProductDetail/?pid=000000000030186229 (Accessed 17 February 2015).

Casey E 2007 Editorial: What does "forensically sound" really mean? *Digital Investigation* **4**(2), 49–50.

Cloud Security Alliance 2013 Mapping the forensic standard ISO/IEC 27037 to cloud computing, ISO/IEC 27037:2012, https://downloads.cloudsecurityalliance.org/initiatives/imf/Mapping-the-Forensic-Standard-ISO-IEC-27037-to-Cloud-Computing.pdf (Accessed 17 February 2015).

Cohen N and MacLennan-Brown K 2007 Digital imaging procedures, UK HOSDB Publication No. 58/07, Version 2.1, in association with ACPO, http://tna.europarchive.org/20100413151426/http://scienceandresearch.homeoffice.gov.uk/hosdb/publications/cctv-publications/DIP_2.1_16-Apr-08_v2.3_%28Web%2947aa.html?view=Standard&pubID=555512 (Accessed 17 February 2015).

Cohen N and MacLennan-Brown K 2008 Retrieval of video evidence and production of working copies from digital CCTV systems, UK HOSDB Publication No. 66/08, Version 2.0, in association with ACPO, http://tna.europarchive.org/20100413151426/http:/scienceandresearch.homeoffice.gov.uk/hosdb/publications/cctv-publications/66-08_Retrieval_of_Video_Ev12835.pdf?view=Binary (Accessed 17 February 2015).

ENFSI 2009 Guidelines for best practice in the forensic examination of digital technology, Version 6.0, http://www.enfsi.eu/sites/default/files/documents/forensic_speech_and_autio_analysis_we_best_practice_guidence_for_enf_analysis_in_forensic_authentication_of_digital_evidence_0.pdf (Accessed 17 February 2015).

ENFSI-FSAAWG 2009 Best practice guidelines for ENF analysis in forensic authentication of digital evidence, http://www.enfsi.eu/sites/default/files/documents/forensic_speech_and_audio_analysis_wg_-_best_practice_guidelines_for_enf_analysis_in_forensic_authentication_of_digital_evidence_0.pdf (Accessed 17 February 2015).

European Cybercrime Training and Education Group 2014 E.C.T.E.G courses, http://www.ecteg.eu/courses.html (Accessed 17 February 2015).

Guidance Software, Inc. 2014 EnCE® certification program, https://www.guidancesoftware.com/training/Pages/ence-certification-program.aspx (Accessed 17 February 2015).

ILAC 2002 Guidelines for forensic science laboratories, ILAC-G19:2002, unavailable on ILAC website any longer, but still downloadable from http://www.nat.hu/dokumentumok/ilac-g19.pdf as of the date of this writing (28 September 2014).

ILAC 2014 Modules in a forensic science process, ILAC-G19:08/2014, http://ilac.org/?ddownload=805 (Accessed 17 February 2015).

Information Security and Forensics Society, Hong Kong, China 2009 Computer forensics Part 2: Best practices, http://www.isfs.org.hk/publications/ISFS_ComputerForensics_part2_20090806.pdf (Accessed 17 February 2015).

IOCE 2002a Guidelines for best practice in the forensic examination of digital technology, Draft V1.0, http://web.archive.org/web/20070812213853/http://www.ioce.org/fileadmin/user_upload/2002/Guidelines\%20for\%20Best\%20Practices\%20in\%20Examination\%20of\%20Digital\%20Evid.pdf (Accessed 17 February 2015).

IOCE 2002b Training standards and knowledge skills and abilities, Draft V1.0, http://www.ioce.org/.

ISO 2008 Quality management systems – Requirements, ISO 9001:2008, http://www.iso.org/iso/catalogue_detail?csnumber=46486 (Accessed 17 February 2015).

ISO/IEC 2002 Conformity assessment – Requirements for the operation of various types of bodies performing inspection, ISO/IEC 17020:2002, http://www.iso.org/iso/home/store/catalogue_ics/catalogue_detail_ics.htm?csnumber=52994 (Accessed 17 February 2015).

ISO/IEC 2005 General requirements for the competence of testing and calibration laboratories, ISO/IEC 17025:2005, http://www.iso.org/iso/catalogue_detail.htm?csnumber=39883 (Accessed 17 February 2015).

ISO/IEC 2011 Information technology – Security techniques – Information security incident management, ISO/IEC 27035:2011, http://www.iso.org/iso/catalogue_detail?csnumber=44379 (Accessed 17 February 2015).

ISO/IEC 2012 Information technology – Security techniques – Guidelines for identification, collection, acquisition and preservation of digital evidence, ISO/IEC 27037:2012, http://www.iso.org/iso/catalogue_detail?csnumber=44381 (Accessed 17 February 2015).

ISO/IEC 2013a Information technology – Security techniques – Code of practice for information security controls, ISO/IEC 27002:2013, http://www.iso.org/iso/catalogue_detail?csnumber=54534 (Accessed 17 February 2015).

ISO/IEC 2013b Information technology – Security techniques – Information security management systems – Requirements, ISO/IEC 27001:2013, http://www.iso.org/iso/catalogue_detail?csnumber=54534 (Accessed 17 February 2015).

ISO/IEC 2014a Information technology – Security techniques – Guidance on assuring suitability and adequacy of incident investigative methods, ISO/IEC DIS 27041, http://www.iso.org/iso/catalogue_detail.htm?csnumber=44405 (Accessed 17 February 2015).

ISO/IEC 2014b Information technology – Security techniques – Guidelines for the analysis and interpretation of digital evidence, ISO/IEC DIS 27042, http://www.iso.org/iso/catalogue_detail.htm?csnumber=44406 (Accessed 17 February 2015).

ISO/IEC 2014c Information technology – Security techniques – Incident investigation principles and processes, ISO/IEC FDIS 27043, http://www.iso.org/iso/catalogue_detail.htm?csnumber=44407 (Accessed 17 February 2015).

ISO/IEC 2014d System and software engineering – Information technology – Governance of digital forensic risk framework, ISO/IEC FDIS 30121, http://www.iso.org/iso/home/store/catalogue_tc/catalogue_detail.htm?csnumber=53241 (Accessed 17 February 2015).

Jansen W and Ayers R 2004 Guidelines on PDA forensics, recommendations of the National Institute of Standards and Technology, NIST Special Publication 800-72, http://www.nist.gov/manuscript-publication-search.cfm?pub_id=150217 (Accessed 17 February 2015).

Jansen W and Delaitre A 2009 Mobile forensic reference materials: A methodology and reification, NIST Interagency Report 7617, http://csrc.nist.gov/publications/nistir/ir7617/nistir-7617.pdf (Accessed 17 February 2015).

Kent K, Chevalier S, Grance T and Dang H 2006 Guide to integrating forensic techniques into incident response, NIST Special Publication 800-86, http://www.nist.gov/manuscript-publication-search.cfm?pub_id=50875 (Accessed 17 February 2015).

LEVA International, Inc. 2014 LEVA certification program, https://leva.org/index.php/certification (Accessed 17 February 2015).

McKemmish R 2008 When is digital evidence forensically sound? In *Advances in Digital Forensics IV* (ed. Ray I and Shenoi S), vol. 285 of *IFIP – The International Federation for Information Processing*, pp. 3–15, Springer.

NIST 2010 Smart phone tool specification, Version 1.1, http://www.cftt.nist.gov/documents/Smart_Phone_Tool_Specification.pdf (Accessed 17 February 2015).

Nolan R, Baker M, Branson J, Hammerstein J, Rush K, Waits C and Schweinsberg E 2005a First responders guide to computer forensics: Advanced topics, Carnegie Mellon Software Engineering Institute, CERT Training and Education Handbook CMU/SEI-2005-HB-003, http://resources.sei.cmu.edu/library/asset-view.cfm?assetid=7261 (Accessed 17 February 2015).

Nolan R, O'Sullivan C, Branson J and Waits C 2005b First responders guide to computer forensics, Carnegie Mellon Software Engineering Institute CERT Training and Education Handbook, CMU/SEI-2005-HB-001, http://resources.sei.cmu.edu/library/asset-view.cfm?assetid=7251 (Accessed 17 February 2015).

OECD 2002 Guidelines for the security of information systems and networks: Towards a culture of security, http://www.oecd.org/internet/ieconomy/15582260.pdf (Accessed 17 February 2015).

Project on Scientific Knowledge and Public Policy 2003 Daubert: The most influential supreme court ruling you've never heard of, http://defendingscience.org/sites/default/files/upload/Daubert-The-

Most-Influential-Supreme-Court-Decision-You-ve-Never-Heard-Of-2003.pdf (Accessed 17 February 2015).

Shirey R 2000 Internet security glossary, IETF RFC 2828, http://tools.ietf.org/html/rfc2828 (Accessed 17 February 2015).

Sommer P 2005 Directors and corporate advisors' guide to digital investigations and evidence, 1st edition, published by Information Assurance Advisory Council (IAAC), http://www.iaac.org.uk/media/1146/evidence-of-cyber-crime-v12-rev.pdf (Accessed 17 February 2015).

Sommer P 2012 Digital evidence, digital investigations, and e-disclosure: A guide to forensic readiness for organisations, security advisers and lawyers, 3rd edition, published by Information Assurance Advisory Council (IAAC), http://cryptome.org/2014/03/digital-investigations.pdf (Accessed 17 February 2015).

Sommer P 2013 Digital evidence, digital investigations, and e-disclosure: A guide to forensic readiness for organisations, security advisers and lawyers, 4th edition, published by Information Assurance Advisory Council (IAAC), http://www.iaac.org.uk/media/1347/iaac-forensic-4th-edition.pdf (Accessed 17 February 2015).

SWGDE 2006 SWGDE digital evidence findings, Version 1.0, https://www.swgde.org/documents/Current%20Documents/2006-04-12%20SWGDE%20Digital%20Evidence%20Findings (Accessed 17 February 2015).

SWGDE 2008 Peer to peer technologies, Version 1.0, https://www.swgde.org/documents/Current%20Documents/2008-01-30%20SWGDE%20Peer%20to%20Peer%20Technologies%20v1.0 (Accessed 17 February 2015).

SWGDE 2010 SWGDE minimum requirements for quality assurance in the processing of digital and multimedia evidence, Version 1.0, https://www.swgde.org/documents/Current%20Documents/2010-05-15%20SWGDE%20Min%20Req%20for%20QA%20in%20Proc%20Digital%20Multimedia%20Evidence_v1 (Accessed 17 February 2015).

SWGDE 2011 SWGDE core competencies for forensic audio, Version 1.0, https://www.swgde.org/documents/Current%20Documents/2011-09-15%20SWGDE%20Core%20Competencies%20for%20Forensic%20Audio%20v1.pdf (Accessed 17 February 2015).

SWGDE 2012a Foundational forensic science annotated bibliographies requested by RDT-E IWG SWGDE's reply to RDT&E IWG letter, https://www.swgde.org/documents/Current%20Documents/Foundational%20Forensic%20Science%20Annotated%20Bibliographies%20Requested%20by%20RDT-E%20IWG (Accessed 17 February 2015).

SWGDE 2012b SWGDE best practices for portable gps device examinations, Version 1.1, https://www.swgde.org/documents/Current%20Documents/2012-09-12%20SWGDE%20Best%20Practices%20for%20Portable%20GPS%20GPS%20Devices%20V1-1 (Accessed 17 February 2015).

SWGDE 2012c SWGDE model quality assurance manual for digital evidence laboratories, Version 3.0, https://www.swgde.org/documents/Current%20Documents/SWGDE%20QAM%20and%20SOP%20Manuals/2012-09-13%20SWGDE%20Model%20QAM%20for%20Digital%20Evidence%20Laboratories-v3.0 (Accessed 17 February 2015).

SWGDE 2012d SWGDE model standard operation procedures for computer forensics Version 3.0, https://www.swgde.org/documents/Current%20Documents/SWGDE%20QAM%20and%20SOP%20Manuals/2012-09-13%20SWGDE%20Model%20SOP%20for%20Computer%20Forensics%20v3 (Accessed 17 February 2015).

SWGDE 2013a SWGDE best practices for vehicle navigation and infotainment system examinations, Version 1.0, https://www.swgde.org/documents/Current%20Documents/2013-02-11%20SWGDE%20Best%20Practices%20for%20Vehicle%20Navigation%20and%20Infotainment%20System%20Examinations%20V1-0 (Accessed 17 February 2015).

SWGDE 2013b SWGDE core competencies for mobile phone forensics, Version 1.0, https://www.swgde.org/documents/Current%20Documents/2013-02-11%20SWGDE%20Core%20Competencies%20for%20Mobile%20Phone%20Forensics%20V1-0 (Accessed 17 February 2015).

SWGDE 2014a SWGDE best practices for computer forensics Version 3.1, https://www.swgde.
    org/documents/Current%20Documents/2014-09-05%20SWGDE%20Best%20Practices%20for%
    20Computer%20Forensics%20V3-1 (Accessed 17 February 2015).
SWGDE 2014b SWGDE best practices for examining magnetic card readers, Version 1.0,
    https://www.swgde.org/documents/Current%20Documents/2014-06-11%20SWGDE%20Best%
    20Practices%20for%20Credit%20Card%20Skimmers (Accessed 17 February 2015).
SWGDE 2014c SWGDE best practices for forensic audio, Version 2.0, https://www.swgde.org/
    documents/Current%20Documents/2014-09-08%20SWGDE%20Best%20Practices%20for%
    20Forensic%20Audio%20V2 (Accessed 17 February 2015).
SWGDE 2014d SWGDE best practices for handling damaged hard drives, Version 1.0,
    https://www.swgde.org/documents/Current%20Documents/2014-09-05%20SWGDE%20Best%
    20Practices%20for%20Handling%20Damaged%20Hard%20Drives (Accessed 17 February 2015).
SWGDE 2014e SWGDE best practices for mobile phone forensics, Version 2.0, https://www.swgde.
    org/documents/Current%20Documents/2013-02-11%20SWGDE%20Best%20Practices%20for%
    20Mobile%20Phone%20Forensics%20V2-0 (Accessed 17 February 2015).
SWGDE 2014f SWGDE capture of live systems Version 2.0, https://www.swgde.org/documents/
    Current%20Documents/2014-09-05%20SWGDE%20Capture%20of%20Live%20Systems%20V2-0
    (Accessed 17 February 2015).
SWGDE 2014g SWGDE electric network frequency discussion paper, Version 1.2, https:
    //www.swgde.org/documents/Current%20Documents/2014-02-06%20SWGDE%20Electric%
    20Network%20Frequency%20Discussion%20Paper%20v1-2 (Accessed 17 February 2015).
SWGDE 2014h SWGDE focused collection and examination of digital evidence, Version 1.0,
    https://www.swgde.org/documents/Current%20Documents/2014-09-05%20SWGDE%20Focused%
    20Collection%20and%20Examination%20of%20Digital%20Evidence (Accessed 17 February 2015).
SWGDE 2014i SWGDE Mac OS X tech notes, Version 1.1, https://www.swgde.org/documents/Current%
    20Documents/2014-09-05%20SWGDE%20Mac%20OS%20X%20Tech%20Notes%20V1-1
    (Accessed 17 February 2015).
SWGDE 2014j SWGDE recommended guidelines for validation testing, Version 2.0, https://www.
    swgde.org/documents/Current%20Documents/2014-09-05%20SWGDE%20Recommended%
    20Guidelines%20for%20Validation%20Testing%20V2-0 (Accessed 17 February 2015).
SWGDE 2014k SWGDE UEFI and its effect on digital forensics imaging, Version 1.0,
    https://www.swgde.org/documents/Current%20Documents/2014-02-06%20SWGDE%20UEFI%
    20Effect%20on%20Digital%20Imaging%20V1 (Accessed 17 February 2015).
SWGDE and SWGIT 2004 SWGDE/SWGIT recommended guidelines for developing standard
    operating procedures, Version 1.0, https://www.swgit.org/pdf/Recommended%20Guidelines%20for%
    20Developing%20Standard%20Operating%20Procedures?docID=59 (Accessed 17 February 2015).
SWGDE and SWGIT 2006 SWGDE/SWGIT proficiency test program guidelines, Version 1.1,
    https://www.swgit.org/pdf/Proficiency%20Test%20Program%20Guidelines?docID=58 (Accessed 17
    February 2015).
SWGDE and SWGIT 2010 SWGDE/SWGIT guidelines & recommendations for training in
    digital & multimedia evidence, Version 2.0, https://www.swgit.org/pdf/Guidelines%20and%
    20Recommendations%20for%20Training%20in%20Digital%20and%20Multimedia%20Evidence?
    docID=57 (Accessed 17 February 2015).
SWGIT 2009 Best practices for forensic video analysis, SWGIT Document Section 7, Ver-
    sion 1.0, https://www.swgit.org/pdf/Section%205%20Guidelines%20for%20Image%20Processing?
    docID=49 (Accessed 17 February 2015).
SWGIT 2010a Best practices for automated image processing, SWGIT Document Section 18, Ver-
    sion 1.0, https://www.swgit.org/pdf/Section%2018%20Best%20Practices%20for%20Automated%
    20Image%20Processing?docID=41 (Accessed 17 February 2015).

SWGIT 2010b Best practices for documenting image enhancement, SWGIT Document Section 11, Version 1.3, https://www.swgit.org/pdf/Section%2011%20Best%20Practices%20for%20Documenting%20Image%20Enhancement?docID=37 (Accessed 17 February 2015).

SWGIT 2010c Guidelines and recommendations for training in imaging technologies in the criminal justice system, SWGIT Document Section 6, Version 1.3, https://www.swgit.org/pdf/Section%206%20Guidelines%20and%20Recommendations%20for%20Training%20in%20Imaging%20Technologies%20in%20the%20Criminal%20Justice%20System?docID=50 (Accessed 17 February 2015).

SWGIT 2010d Guidelines for image processing, SWGIT Document Section 5, Version 2.1, https://www.swgit.org/pdf/Section%205%20Guidelines%20for%20Image%20Processing?docID=49 (Accessed 17 February 2015).

SWGIT 2010e Overview of swgit and the use of imaging technology in the criminal justice system SWGIT Document Section 1, Version 3.3, https://www.swgit.org/pdf/Section%201%20Overview%20of%20SWGIT%20and%20the%20Use%20of%20Imaging%20Technology%20in%20the%20Criminal%20Justice%20System?docID=35 (Accessed 17 February 2015).

SWGIT 2011 Issues relating to digital image compression and file formats, SWGIT Document Section 19, Version 1.1, https://www.swgit.org/pdf/Section%2019%20Issues%20Relating%20to%20Digital%20Image%20Compression%20and%20File%20Formats?docID=42 (Accessed 17 February 2015).

SWGIT 2012a Best practices for archiving digital and multimedia evidence (DME) in the criminal justice system, SWGIT Document Section 15, Version 1.1, https://www.swgit.org/pdf/Section%2015%20Best%20Practices%20for%20Archiving%20Digital%20and%20Multimedia%20Evidence%20%28DME%29%20in%20the%20Criminal%20Justice%20System?docID=55 (Accessed 17 February 2015).

SWGIT 2012b Best practices for forensic image analysis, SWGIT Document Section 12, Version 1.7, https://www.swgit.org/pdf/Section%2012%20Best%20Practices%20for%20Forensic%20Image%20Analysis?docID=38 (Accessed 17 February 2015).

SWGIT 2012c Best practices for maintaining the integrity of digital images and digital video, SWGIT Document Section 13, Version 1.1, https://www.swgit.org/pdf/Section%2013%20Best%20Practices%20for%20Maintaining%20the%20Integrity%20of%20Digital%20Images%20and%20Digital%20Video?docID=54 (Accessed 17 February 2015).

SWGIT 2012d Digital imaging technology issues for the courts, SWGIT Document Section 17, Version 2.2, https://www.swgit.org/pdf/Section%2017%20Digital%20Imaging%20Technology%20Issues%20for%20the%20Courts?docID=56 (Accessed 17 February 2015).

SWGIT 2012e Recommendations and guidelines for crime scene/critical incident videography, SWGIT Document Section 20, Version 1.0, https://www.swgit.org/pdf/Section%2020%20Recommendations%20and%20Guidelines%20for%20Crime%20Scene%20and%20Critical%20Incident%20Videography?docID=44 (Accessed 17 February 2015).

SWGIT 2012f Recommendations and guidelines for using closed-circuit television security systems in commercial institutions, SWGIT Document Section 4, Version 3.0, https://www.swgit.org/pdf/Section%204%20Recommendations%20and%20Guidelines%20for%20Using%20Closed-Circuit%20Television%20Security%20Systems%20in%20Commercial%20Institutions?docID=48 (Accessed 17 February 2015).

SWGIT 2013a Best practices for forensic photographic comparison, SWGIT Document Section 16, Version 1.1, https://www.swgit.org/pdf/Section%2016%20Best%20Practices%20for%20Forensic%20Photographic%20Comparison?docID=40 (Accessed 17 February 2015).

SWGIT 2013b Best practices for image authentication, SWGIT Document Section 14, Version 1.1, https://www.swgit.org/pdf/Section%2014%20Best%20Practices%20for%20Image%20Authentication?docID=39 (Accessed 17 February 2015).

SWGIT 2013c Best practices for the analysis of digital video recorders, SWGIT Document Section 23 Version 1.0, https://www.swgit.org/pdf/Section%2023%20Best%20Practices%20for%

20the%20Analysis%20of%20Digital%20Video%20Recorders?docID=117. The publication date of the document shown in the document itself (11 June 2012) does not match the one shown on the SWGIT general document page, https://www.swgit.org/documents/Current%20Documents (11 January 2013). We use the latter one because the PDF file was generated in June 2013.

SWGIT 2013d Best practices for the retrieval of digital video, SWGIT Document Section 24, Version 1.0, https://www.swgit.org/pdf/Section%2024%20Best%20Practices%20for%20the% 20Retrieval%20of%20Digital%20Video?docID=141 (Accessed 17 February 2015).

UK ACPO 1999 ACPO good practice guide for computer based evidence, Version 2.0, http://www.swgit. org.history.

UK ACPO 2003 ACPO good practice guide for computer-based electronic evidence, Version 3.0, http://web.archive.org/web/20050525143019/http://www.acpo.police.uk/asp/policies/Data/gpg_ computer_based_evidence_v3.pdf (Accessed 17 February 2015).

UK ACPO 2007 ACPO good practice guide for computer-based electronic evidence, Version 4.0, published by 7Safe, http://www.7safe.com/electronic_evidence/ (Accessed 17 February 2015).

UK ACPO 2011a ACPO good practice guide for digital evidence, Version 5.0, http://www.acpo.police. uk/documents/crime/2011/201110-cba-digital-evidence-v5.pdf (Accessed 17 February 2015).

UK ACPO 2011b ACPO good practice guide for managers of e-crime investigation, Version 0.1.4, http: //www.acpo.police.uk/documents/crime/2011/201103CRIECI14.pdf (Accessed 17 February 2015).

UK ACPO and NPIA 2007 Practice advice on police use of digital images, http://www.acpo.police. uk/documents/crime/2011/20111014%20CBA%20practice_advice_police_use_digital_images_ 18x01x071.pdf (Accessed 17 February 2015).

UK ACPO and NPIA 2011 Practice advice on the use of CCTV in criminal investigtions, http://www. acpo.police.uk/documents/crime/2011/20110818%20CBA%20CCTV_Final_Locked.pdf (Accessed 17 February 2015).

UK College of Policing 2014 Core skills in data recovery and analysis, http://www.college.police.uk/en/ 1262.htm (Accessed 17 February 2015).

UK Forensic Science Regulator 2011 Codes of practice and conduct for forensic science providers and practitioners in the criminal justice system, Version 1.0, https://www.gov.uk/government/publications/ forensic-science-providers-codes-of-practice-and-conduct (Accessed 17 February 2015).

UK HOSDB 2007 Storage, replay and disposal of digital evidential images, Publication No. 53/07, Version 1.0, in association with ACPO and NPIA, http://library.college.police.uk/docs/APPref/ storage-replay-and-disposal.pdf (Accessed 17 February 2015).

United States Secret Service, US Department of Homeland Security 2007 Best practices for seizing electronic evidence: A pocket guide for first responders, Version 3.0, http://www.forwardedge2.com/ pdf/bestPractices.pdf (Accessed 17 February 2015).

US DOJ's Computer Crime and Intellectual Property Section 2002 Searching and seizing computers and obtaining electronic evidence in criminal investigations, 2nd edition, http://cdn.ca9.uscourts.gov/ datastore/library/2013/02/26/CDT_cyber.pdf (Accessed 17 February 2015).

US DOJ's Computer Crime and Intellectual Property Section 2009 Searching and Seizing Computers and Obtaining Electronic Evidence in Criminal Investigations OLE Litigation Series 3rd edn.. Office of Legal Education (OLE) & Executive Office for United States Attorneys (EOUSA). http://www.justice. gov/criminal/cybercrime/docs/ssmanual2009.pdf (Accessed 17 February 2015).

US FBI 2007 Digital evidence field guide, Version 1.1, http://www.rcfl.gov/downloads/documents/ digital-evidence-field-guide/ (Accessed 17 February 2015).

US FBI 2010 Mobile forensics field guide, Version 2.0, http://www.rcfl.gov/continuing-education-series/ products/field-guides (Accessed 17 February 2015).

US Federal Judicial Center 2003 Computer-based investigation and discovery in criminal cases: A guide for united states magistrate judges, National Workshop for Magistrate Judges II, http://www.fjc.gov/ public/pdf.nsf/lookup/CompInve.pdf/$file/CompInve.pdf (Accessed 17 February 2015).

US NIJ 2004 Forensic examination of digital evidence: A guide for law enforcement, Special Report NCJ 199408, https://www.ncjrs.gov/pdffiles1/nij/199408.pdf (Accessed 17 February 2015).

US NIJ 2007a Digital evidence in the courtroom: A guide for law enforcement and prosecutors, Special Report NCJ 211314, https://www.ncjrs.gov/pdffiles1/nij/211314.pdf (Accessed 17 February 2015).

US NIJ 2007b Investigations involving the internet and computer networks, Special Report NCJ 210798, https://www.ncjrs.gov/pdffiles1/nij/210798.pdf (Accessed 17 February 2015).

US NIJ 2007c Investigative uses of technology: Devices, tools, and techniques, Special Report NCJ 213030, https://www.ncjrs.gov/pdffiles1/nij/213030.pdf (Accessed 17 February 2015).

US NIJ 2008 Electronic crime scene investigation: A guide for first responders, 2nd edition, Special Report NCJ 219941, http://www.nij.gov/publications/ecrime-guide-219941/ (Accessed 17 February 2015).

US NIJ 2009 Electronic crime scene investigation: An on-the-scene reference for first responders, Special Report NCJ 227050, https://www.ncjrs.gov/pdffiles1/nij/227050.pdf (Accessed 17 February 2015).

3

A Machine Learning-Based Approach to Digital Triage

Fabio Marturana[1], Simone Tacconi[2] and Giuseppe F. Italiano[1]

[1]Department of Civil Engineering and Computer Science Engineering,
University of Tor Vergata, Rome, Italy
[2]Computer Forensic Unit, Postal and Communication Police, Rome, Italy

3.1 Introduction

Nowadays the pervasiveness of high-tech multimedia devices and fast mobile networking is unprecedented, and this trend is expected to grow in the future. Terabytes of image files, audio and video footages are downloaded or exchanged every day over the Internet, contributing to the global diffusion of cyber threats and cybercrimes.

As a consequence, the impact of multimedia forensics, the sub-field of digital forensics whose task is to extract evidence from multimedia data, is dramatically gaining momentum due to the central role played by multimedia evidence in modern investigations.

Unfortunately, the road to build a general understanding of a digital investigation is often paved with challenges such as time restrictions, huge volumes of data requiring in depth analysis, legal constraints to excessive search and seizure and suspect's privacy protection. The situation is even worsened by the short innovation cycle of digital devices, which makes it difficult to keep up with innovation and technology (Poisel and Tjoa 2011).

Further, dealing with most digital forensics frameworks in use today (Palmer 2001; Kent *et al.* 2006), the seizure of an entire hard disk drive or a mobile phone's memory, and consequent creation of a forensically sound image is the best practice followed

Handbook of Digital Forensics of Multimedia Data and Devices, First Edition.
Edited by Anthony T.S. Ho and Shujun Li.
© 2015 John Wiley & Sons, Ltd. Published 2015 by John Wiley & Sons, Ltd.
Companion Website: www.wiley.com/go/digitalforensics

by law enforcement responders to preserve evidence integrity. Later on, sized devices and forensic images are sent to the Digital Forensic Laboratory (DFL) for in-depth examination with time-consuming and computationally intensive procedures which allow to reconstruct event timeline and extract evidence of suspect's guilt. As a consequence of the average workload of a typical DFL, spanning from 9 to 12 months, such a framework may cause unacceptable delays in prosecution of criminals (Gomez 2012). Further, due to the inherent nature of the incriminatory information which normally refers to a minimal part of a digital device's memory, most forensic frameworks currently in use may even violate suspect's privacy (Hong *et al.* 2013).

Nevertheless, dealing with crimes such as terrorist attacks, murders, kidnapping, human trafficking, just to name a few, the need for a timely identification, analysis and interpretation of the evidentiary material found on the crime scene is crucial (Rogers *et al.* 2006; Pearson and Watson 2010). In such cases, on-site selective pre-examination of available data sources with digital triage tools, for instance, could provide investigators with actionable intelligence to guide the search and seizure process. Further, adopting a digital triage framework to classify digital sources as relevant or not to the case at hand could be a viable solution to lighten DFL's workload (Gomez 2012). Finally, digital triage could address the issue of selectively identifying and collecting digital evidence on the crime scene in those cases where excessive search and seizure is not allowed by the legal system, as it happens in Republic of Korea, for instance, where some courts have even ruled out inadmissible evidence gathered out of the scope of the search warrant (Hong *et al.* 2013).

Taking into account the preceding considerations, the need arises to look for new forensic frameworks based on digital triage by which identifying relevant information and isolating probative content (i.e. evidence) in a timely manner among the huge amount of data pertaining to a single crime investigation.

Therefore, digital triage is emerging as a practical solution to cope with time-critical investigations, huge examination workloads and legal restrictions to excessive search and seizure. In this regards, methods for rapid data extraction have been developed so far for specific purposes, such as the identification of relevant hard drives in fraud cases on the basis of common credit card numbers and email addresses retrieved (Garfinkel 2009). However, despite some data retrieval and matching functions, the classification of a digital evidence as relevant or not to a criminal investigation still remains a mostly manual process in charge of experienced individuals. More recently, drawing inspiration from earliest digital triage models (Rogers *et al.* 2006; Pearson and Watson 2010), research has focused on the need for automating such process by means of crime templates (Cantrell *et al.* 2012), similarity digest (Roussev and Quates 2012), automated tools for bulk data extraction and analysis (Garfinkel 2013), machine learning-based supervised (Grillo *et al.* 2009; Garfinkel *et al.* 2010; Gomez 2012), and unsupervised approaches (Drezewski *et al.* 2012).

Dealing with machine learning-based digital triage, this chapter presents a framework for selective pre-examination and statistical classification of digital data sources that may be deployed both on the crime scene and at DFLs. It is aimed at (i) providing investigators with quick actionable intelligence on the crime scene when time is a critical factor, (ii) lightening the workload at DFLs and (iii) protecting suspect's privacy where excessive search and seizure of data is not allowed by the legal system. Based on the opinion of surveyed forensic experts, and the exploitation of machine learning classification algorithms, such framework presents two main advantages with respect to most examination techniques in use today as it (i) requires limited manual intervention and (ii) produces measurable and reproducible error rates. The latter important aspect makes it compliant, for instance, with the U.S. Supreme Court's Daubert ruling (Heilbronner 2011), which states that a scientific method, in order to be admissible in a trial, must have a 'known or potential' error rate and must be subject to empirical testing.

3.1.1   Chapter Outline

The remainder of the chapter is organized as follows: related work on digital triage is summarized in Section 3.2, whereas a machine learning-based digital triage framework for use in data source supervised classification is illustrated in Section 3.3. Subsequently, a case study on the crime of exchange of child pornography is presented in Section 3.4 as an example of the described framework. Concluding remarks are drawn in Section 3.5 and challenges and future directions for the forensic community finally explored in Section 3.6.

3.2   Related Work on Digital Triage

This section provides an overview about history and state-of-the-art research on digital triage.

3.2.1   Triage in the Medical Field

Derived from the French verb trier, meaning to separate, sift or select, the term *triage* may have originated during the Napoleonic Wars from the work of Dominique Jean Larrey.[1] The term was used further during World War I by French doctors treating the battlefield wounded at the aid stations behind the front.

Those responsible for the removal of the wounded from a battlefield or their care afterwards would divide the victims into three categories:

1. Those who are likely to live, regardless of what care they receive;
2. Those who are likely to die, regardless of what care they receive;
3. Those for whom immediate care might make a positive difference in outcome.

In the medical field, triage is referred to as the process of determining the priority of patients' treatments based on the severity of their condition. This allocates medical

---

[1] French surgeon in Napoleon's army and an important innovator in battlefield medicine.

resources to patient treatment efficiently when those are insufficient for all to be treated immediately. Triage may result in determining the order and priority of emergency treatment, the order and priority of emergency transport, or the transport destination for the patient. Triage may also be used for patients arriving at the emergency department, or telephoning medical advice systems, among others.

As medical technology has advanced, so have modern approaches to triage which are increasingly based on scientific models. The victim's categorization process is frequently the result of algorithm-based triage scores based on specific physiological assessment findings. As triage concepts become more sophisticated, triage guidance has evolved into both software and hardware decision support systems for use by caregivers in both hospitals and the battlefield.

3.2.2   Early Digital Triage Models

The term *triage* is being adopted in the context of digital devices and technology as well to describe the process of prioritizing digital data sources according to their importance in relation to an investigation (Chipman *et al.* 1980).

Digital triage may be referred to as the prioritization process that attempts to both guide the search and seizure process on the crime scene and reduce the volume of data that need to be exhaustively examined at DFLs (Pollitt 2013).

Digital triage has experienced an outstanding development in recent years, and it is still evolving at a rapid pace. Research on the subject matter, initially at least, attempted to provide a solution to the delay introduced by most forensic frameworks currently in use which resulted to be inappropriate to time-sensitive criminal cases such as kidnapping or danger to a human being. In this regards, Rogers *et al.* (2006) proposed a pioneering field triage model, called computer forensics field triage process model (CFFTPM), aimed at gathering actionable intelligence on the crime scene in investigations where time was a critical factor. The authors, motivated by the need to provide investigative leads in time-critical situations, defined a workflow for identification, analysis and interpretation of digital evidence on the crime scene, without the need for acquiring a forensic copy of the device or taking it back to the lab for in-depth examination. Entailing a real risk of exhibit tampering, the CFFTPM assumes that the crime scene has been properly secured and controlled and includes integrity protection tasks (e.g. hardware write blocker and forensic software usage) to assure that any potential evidence found could undertake traditional examination and analysis back at the lab in a controlled environment.

After proper prior planning by which quantifying the various possibilities of the crime scene, and qualifying the expertise of the various investigators on the investigation team, the CFFTPM enables the triage phase, a process in which things are ranked in terms of importance or priority. Essentially, those items, pieces of evidence or potential containers of evidence that are the most important or the most volatile are dealt with first. Once a system or storage media has been identified and prioritized during the triage phase, the actual examination and analysis are conducted.

Links between the evidence found and a specific, identifiable suspect are shown to ascertain which individual or individuals are responsible for, or even had knowledge of, incriminating data found on the storage media. Thorough knowledge of user profiles and artefacts relating to usage, are essential to accomplishing this goal. User profiles, home directories, file permissions, MAC times (i.e. Modification, Access, Creation), Internet activity, emails and web browsing are analyzed to place artefacts in context with verifiable real world events. Doing that as quick as possible may result in a suspect presented with clear evidence indicating that he or she, and no other person is responsible for evidence recovered who may feel compelled to admit their guilt.

Drawing its inspiration from the CFFTPM and considering the challenges posed by the battlefield crime scene, Pearson and Watson (2010) defined the so-called digital triage forensics (DTF) process, a methodology tailored for terroristic attacks, or post-blast investigations which allows the collection of actionable intelligence and potential evidence in the battlefield. The DTF model is in line with the CFFTPM with the exception that triage processing is done at the forward operating base (FOB), before in-depth analysis, rather than on the crime scene. The reason is obvious: the forensic team, who is tasked of collecting evidence in the battlefield, must always operate in a timely manner and under safe conditions. It is noteworthy to mention that both Rogers *et al.* and Pearson and Watson focused their attention on the crime scene rather than the DFLs since, at the time, the issue of heavy workload sat DFLs was not as prevalent in research.

As the issue of DFLs being unable to cope with heavy workloads became more widespread, the need for digital triage methodologies to reduce backlogs arose (Gomez 2012).

In conclusion, digital triage methodologies could be used to address the following problems:

- Timely identification of case-relevant items on the crime scene in time-critical situations;
- Examination of large volumes of data and lighten workloads at DFLs;
- Suspect's privacy protection by reducing the scope of search from the entire device to only some area of interest more likely to turn into incriminatory data and evidence.

3.2.3   Machine Learning-Based Digital Triage

Dealing with machine learning-based digital triage it is noteworthy to mention the pioneering *Five Minutes Forensics* (5MF) technique described by Grillo *et al.* (2009). In this regards, the cited authors defined a framework for fast computer user profiling based on five predefined user categories (i.e. *occasional, chat-internet, office worker, experienced* and *hacker*), and a set of features related to each of them. Dealing with the feature set or feature vector, the authors assumed that, for instance, an experienced computer user owns operating system configuration skills, while a hacker could own particular software packages, cracking tools and so on. Based on previous observations the statistical analysis of such features extracted from a seized hard disk allows to label each computer user as belonging to one of the five predefined category, and thus define examination priorities at DFLs.

In a child pornography exchange investigation where many computers are found on the crime scene, for example, a computer user classified as occasional will be treated with less priority than a chat-Internet user due to the limited amount of evidentiary material that is likely to be found by investigators.

Dealing with machine learning-based digital triage, the 5MF technique (Grillo *et al.* 2009) has influenced several authors. In this regards, Marturana *et al.* (2011a, 2011b) have built their model for mobile handsets classification on the basis of the 5MF technique. In particular, Marturana *et al.* (2011a) have analyzed a corpus of data extracted from handsets and smartphones in court cases of exchange of child pornography, corporate espionage, murder and human trafficking, and used multiclass categorization for classifying objects on the basis of owner's usage profile (i.e. base, medium and expert). Further, Marturana *et al.* (2011b) have processed the same data corpus with binary categorization to identify test objects allegedly used for exchanging child pornography material. Finally, Marturana and Tacconi (2013) have discussed findings of a case study on court cases of copyright infringement with a dataset extracted from desktop and laptop computers as an application of the aforementioned methodology.

Grillo *et al.* (2009) have influenced Cantrell *et al.*'s (2012) work as well in their definition of a partially-automated and crime-specific digital triage process model under development at Mississippi State University. In such work, the authors dealt with crime templates to identify relevant items in a given crime context. In child pornography, for instance, a template included the search of file names within a dictionary of commonly used words by pornographers or by mathematical hash values compared with a repository of well-known pornographic material. In the latter case, even if the name was forged, its hash value was likely to remain the same. Such templates are made available to the investigators during on site search and seizure and to the examiners at DFLs. Dealing with digital triage, it is not the first time that the concept of crime-related template emerges since some authors (Marturana *et al.* 2011a, 2011b; Gomez 2012) has already identified the need to train machine learning supervised classifiers on the basis of crime-related features which is very similar to the idea of searching a digital source with a template. The two aforementioned approaches are different in that, with templates, the pre-examination phase should focus *a priori* on the search for information pertaining to the crime under investigation (e.g. specific files, file types, strings, logs and executables) whereas, with a machine learning-based approach, it is necessary to define a reasonable (i.e. in terms of classification accuracy) number of likely crime-related features to be extracted during the pre-examination stage and let the algorithm do the rest of the work. In the latter case it is noteworthy to mention that classification accuracy is influenced by factors such as the total number of features, their correlation or linear dependence, the training set dimensionality etc.

With the aim of solving the multiuser carved data ascription problem, Garfinkel *et al.* (2010) have followed a similar approach to the one presented in this chapter. The authors used a machine learning-based technique to ascribe files extracted from multiuser personal computers. The approach taken to extract file system, file placement and embedded file metadata relies upon specific metadata extraction tools (i.e. fiwalk's metadata extraction system) according to the following workflow:

- Ascribable (i.e. with known owner) exemplar file are recovered from the disk image.
- For each exemplar, metadata are extracted.
- The data is pre-processed.
- The exemplars are used to train the classifier.
- Classifier accuracy is calculated.
- The carved files (i.e. with unknown owner) are processed and metadata extracted.
- The carved data is processed and classified.

3.2.4   *Other Multimedia Source Classification Techniques*

Dealing with multimedia forensics in child pornography investigations, a technique similar to digital triage allows for the extraction of crime-relevant information (e.g. nudity) from specific files (e.g. photos and video footages) with custom algorithms (e.g. skin detection) and consequent classification of the digital data source. In this particular context, two areas of interest concerning, namely, image and video media analysis, arise. Regarding image content classification, on the one hand, works regarding the detection of pornographic material refer, for instance, to the detection of skin regions in an image and matching with human bodies by applying geometric grouping rules (Fleck *et al.* 1996) or the estimation of the 'skin probability' of a pixel based on its colour (Jones and Rehg 2002). Dealing with video footages, on the other hand, Janshon *et al.* (2009) proposed a new approach on automatic detection of pornographic content in video databases which adds motion analysis and periodicity detections to key frames analysis (i.e. extraction of representative key frames and application of image classification techniques on them).

3.3   A Machine Learning-Based Digital Triage Framework

This section describes a machine learning-based digital triage framework for use in crime scene search and seizure (*on-scene digital triage*) and DFL examination (*off-site digital triage*). Such framework is laid out to integrate with other forensic frameworks currently in use (e.g. based on *acquisition, retrieval* and *analysis*), preserve digital evidence integrity guarantee analysis repeatability,[2] and selectively identify relevant (e.g. requiring in-depth analysis) digital sources.

---

[2] ACPO. Good Practice Guide for Computer-Based Electronic Evidence. Available at http://www.7safe. com/electronic_evidence/ACPO_guidelines_computer_evidence_v4_web.pdf.

3.3.1 Machine Learning Terminology

Being the described framework based on machine learning, a description of related terms that will be frequently used along the chapter is mandatory.

Machine learning is related to the construction and study of systems that can learn from data. It adopts the same principles that humans use to learn, that is to say repetition and experience. Among the various types of machine learning schemes reported in literature we need to mention supervised and unsupervised learning.

Supervised learning is the process of learning from a training set of correctly identified exemplars or *training instances*. This is achieved by observing present behaviour and comparing it with past observations to discover structural patterns which can be examined, reasoned and used to inform decisions (Witten *et al.* 2011). Supervised machine learning is the classification technique adopted in this work. Typically, the process requires many exemplars to be able to find such structural patterns. This is known as the learning data or training set. Defining the training set is crucial to the machine learning process as it provides the known properties upon which future predictions of unknown instances (i.e. test instances) can be made. In binary supervised machine learning, for instance, during the training phase each instance is manually given a binary classification known as class (e.g. yes or no, relevant or non-relevant). In this regards, the training is defined as *controlled*, or, *supervised*. The corresponding unsupervised procedure, out of the scope of the present work, is known as clustering or cluster analysis, and involves grouping data into categories based on some measure of inherent similarity (e.g. the distance between instances, considered as vectors in a multi-dimensional vector space).

3.3.1.1 State Vector

Supervised learning requires data to be structured into a *state vector*. Such vector encompasses a set of variables related to the classification problem. Once the state vector has been defined, a set of values corresponding to such variables is populated each time a new *instance* of data is created. In the context of this study, the state vector encompasses a number of *features* whereas the instances are populated with the corresponding values.

3.3.1.2 Features

Features (or *attributes*) are key elements of the framework which will be used interchangeably throughout the chapter. A feature or attribute, in this context, is defined as *a quantitative measure of a potentially crime-related item*. In other words, a feature x is something potentially related to the investigation that one can extract from a digital source and measure. An example could be the number of .jpg files recovered from a digital source (e.g. number of .jpg files $= 24$). It is noteworthy to mention that, being the described framework based on a statistical model, there is no causal relationship between a single feature or set of features and the investigated crime. In other words, whenever a feature x takes a value y doesn't necessarily mean that the digital source is 100% related to the investigated crime. We can only predict the likelihood that each digital source is related to the crime at hand and provide the correspondent error rate based on the occurrence of a specific value for that feature or group of features.

3.3.1.3   Instances and Exemplars

The set of values corresponding to the feature vector extracted from each digital data source is called *instance,* which may be represented as a vector of data. Instances represent the content mapping of each digital data sources into model data according to the *state vector.* For example, if a concept can be modelled using two integer variables, x and y, the corresponding state vector will be (x, y) and $(3, 7)$ an instance. Instances derived from training set devices, whose relation with the investigation is known, are called *exemplars.* The *training set* is a collection of exemplars given as input to the classification algorithm. This provides the basis upon which a machine learning algorithm may be trained to determine a device's relevance or non-relevance to an investigation. Features and exemplars are important to the overall model as they are given as input to the processing algorithms.

3.3.2   *The framework in Detail*

Once the machine learning terminology has been defined, we can provide the framework details and describe the process that we used to validate the methodology.

Exploiting some concepts derived by other digital triage models (Garfinkel *et al.* 2010; Marturana and Tacconi 2013), such framework can be generalized as encompassing four phases, namely *data extraction, attributes* (or *features*) *extraction, processing* and *classification,* corresponding to the tasks of data *collection, processing* and *presentation,* as illustrated in Figure 3.1.

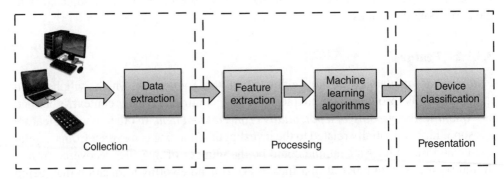

Figure 3.1   Digital triage workflow.

An overview of the phases included in the workflow is as follows:

- Collection[3]
  - A digital data source (*test object*) is collected/identified on the crime scene for forensic analysis.
  - A forensic image of the test object is optionally created.[4]
  - Data is extracted, alternatively, from the test object or its forensic copy either on-site or at DFLs using appropriate tools and possibly adopting forensically sound procedures.
- Processing
  - A supervised machine learning algorithm is trained on the basis of a knowledge base of training exemplars pertaining to the crime and already classified (training set).
  - The data extracted from the test object turns into features according to the state vector and an observation (test instance), which is the machine-learning equivalent of the test object, is populated accordingly.
  - The test instance is provided as input to the algorithm and processed.
- Presentation
  - As a result of the training phase, the test instance is given a binary classification and the corresponding test object is categorized as relevant or negligible to the investigation (i.e. data source's relevance to the case).
  - The data source's relevance to the case is presented to the investigator on the crime scene, who can proceed with the search and seizure, or to the examiner, who may decide to perform further analysis at DFLs.

3.3.3 Collection–Data Extraction

Data extraction is tasked to collect relevant information from the digital data sources in a timely manner. Data extraction can be carried out using both open source and commercial tools. Once extracted from the media, each piece of information (e.g. file information or metadata) is ready to be presented as a feature. Concerning guidelines for extraction of crime-relevant data from computer systems, Kent *et al.* (2006) suggested to:

- look at the last change date of critical files,
- examine configuration and start-up files,
- search for hacking tools (e.g. password crackers, key loggers etc.),

---

[3] Prior to extracting data from multimedia devices, it is mandatory to define a state vector with relevant crime-related attributes.

[4] In time-critical situations, when the preservation of the discovered device is less important than finding quick intelligence to guide the search and seizure process, investigators can opt to extract data directly from the target system without imaging it first.

- examine the password file for unauthorized accounts,
- search for keywords appropriate to the incident,
- search for hidden areas, slack space and cache,
- look for changes to files, critical file deletions and unknown new files,
- collect a list of all e-mail addresses as well as visited and bookmarked URLs,

whereas, regarding mobile phones and smartphones, Jansen and Ayers (2007) identified the following areas of interest:

- Subscriber and equipment identifiers,
- Date/time, language, and other settings,
- Phonebook information,
- Calendar information,
- SMS and MMS history,
- Dialed, incoming, and missed call logs,
- E-mail history,
- Photos, audio recordings and video footages,
- Instant messaging App and Web browsing activities,
- E-documents,
- GPS information.

3.3.4 Processing–Feature Extraction, Dataset Creation and Processing Algorithms

As a result of the data extraction phase, all the features related to a searched digital device are collected within an instance. The collection of all such instances forms the dataset used for classification tasks that may be represented as the two-dimensional matrix called input matrix as illustrated in Table 3.1.

Looking at the input matrix by rows we find the value taken by each instance in the dataset corresponding to a given feature whereas instances (i.e. one for each digital device) are represented by the columns. Features are discrete variables with values from countable sets that may be both finite, in case a test or a multiple selection is made (true/false, yes/no/maybe, etc.) or infinite (i.e. integer), in case we count the number of variable occurrences. A test on a particular attribute (i.e. are there installed P2P clients on the digital device? Yes/No) is an example of a multiple selection whereas the number of image files and video footages extracted from a device is an example where the number of feature occurrences is counted. In our model, *nominal* features assume values from finite countable sets, whereas *numeric* features assume values from infinite countable sets.

Following the dataset creation, the processing phase relies on machine learning algorithms tasked of classifying each instance in the dataset. The following is a list of processing tasks, illustrated in the flow chart in Figure 3.2, pertaining to the described framework:

Table 3.1 Input matrix.

	Phone #1	Phone #2	Phone #3	Phone #M-1	Phone #M
Feature #1–Number of phonebook contacts	5	25	12	42	23
Feature #2–Number received calls	148	12	57	25	128
Feature #3–Number dialled calls	57	85	45	12	78
Feature #4–Number missed calls	12	47	120	75	49
Feature #5–Number received SMS	12	78	23	0	0
Feature #6–Number sent SMS	42	15	46	8	7
Feature # N-1–Mean duration received calls	120	40	60	15	12
Feature #N–Mean duration dialled calls	125	12	42	12	5
Class (predicted)	relevant	relevant	non-relevant	non-relevant	non-relevant

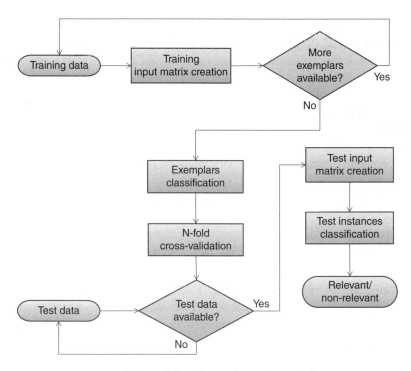

Figure 3.2 Processing tasks workflow.

- *Training set creation.* The collection of a consistent set of exemplars or past observations whose class is known, is the preliminary step of supervised learning. Exemplars are collected on the basis of their relevance to the investigated crime. The *training input matrix* is populated by merging all the available exemplars and then given as input to selected classifiers.
- *Training set classification.* Training set exemplars are processed with supervised learning classification algorithms. Two classification methods are available for the purpose: *multiclass categorization*, where each instance is given the class related to the most likely crime in the target set (e.g. exchange of child pornography, copyright infringement, hacking, murder and terrorism, with regards to our data corpus) and *binary categorization*, where a classifier per each target crime is trained. In multi-class categorization, the instance is associated with the most likely crime among the aforementioned targets and the class variable is given the correspondent value. In binary categorization the class is given a binary value (i.e. relevant or non-relevant etc.), depending on the likelihood that a connection between the digital media and the target crime exists. In the case study on exchange of child pornography with mobile phones described later on, binary categorization will be tested. It is noteworthy to mention that the training step is crucial for the whole classification process as training exemplars must be chosen accurately to avoid both *overfitting*, when training data is limited, and the *imbalanced dataset* problem, when the number of training samples pertaining to one (or more) of the classes is small with regards to the others (Weiss 2004).
- *Learning accuracy evaluation.* Once the training phase is over, learning accuracy is evaluated with a validation technique on the basis of a set of performance indicators, such as *Confusion Matrix, Precision, Recall* and *F-measure*. In this specific context learning accuracy evaluation is crucial to build an effective framework for class prediction that can be used in a court room.
- *Test set creation.* Upon creation, classification and validation of the training set, the model is ready to be used in *test mode*. Features are extracted from those digital data sources requiring crime-related classification; Related test set instances and *test input matrix* are populated accordingly.
- *Test set classification and performance evaluation.* The *test input matrix* is given as input to selected supervised learning algorithms and test set instances are classified accordingly. Testing accuracy is evaluated according to the aforementioned technique and performance indicators

In Section 3.4 an example of training set creation, classification and validation related to child pornography exchange is provided; conversely it was not possible to provide an example of test set creation and classification due to the limited number of available instances. The reason is simple, being already used for training purpose, it was not possible to use the same instances for testing.

3.3.5   Presentation

Once algorithmic processing has identified structural patterns in training data, a dependent variable, called *class*, is predicted for each instance in the dataset. Each digital source in the dataset is therefore classified as relevant or not to the crime at hand and its relevance to the investigation estimated. In other words, the classification output is represented by a binary nominal variable assuming the following two values: *relevant* or *non-relevant*. The classification output is then presented to law enforcement responders and forensic examiners for further decisions.

3.3.6   Model validation

Evaluation of any scientific method is essential to validate its effectiveness and performance. For the proposed methodology to be able to accurately classify test objects, evaluation and validation must be carried out. This will also assist in building legal confidence in the methodology through placing a value on its accuracy.

Ideally, the best way to evaluate a classifier would be to have huge amounts of training and test data for evaluation purpose. However, this is not a realistic scenario where obtaining data can be a major issue, as it happens for instance with satellite remote sensing where a mind-numbing and time-consuming manual process is mandatory to collect any usable data. Digital forensics faces an even more difficult problem with acquiring data (e.g. legal constraints and privacy preservation). Nevertheless, evaluating and validating performance with limited data is still necessary and the most popular method used in such situations is cross-validation (Witten *et al.* 2011).

Cross-validation is a statistical method for evaluating and comparing algorithms by dividing available data into two sets, namely *training* and *test sets*, respectively used to train and validate the model. In typical cross-validation, training and test sets must cross-over in successive rounds such that each data point has a chance of being validated against. Estimating learning performance using one algorithm, and comparing performance of different algorithms to find out the best one for the available data are scientific goals of cross-validation (Refaeilzadeh *et al.* 2009). A commonly used method for evaluating and comparing learning algorithms is *N-fold cross-validation* where N represents a fixed number of folds, or data partitions. At each step of the validation process, it is important that classes in the original dataset are proportionally represented in both training and test sets sin order to avoid data overfitting. It is difficult, for instance, for an algorithm to predict future observations when the training set consists of true instances in a binary classification problem. Therefore, the data should be rearranged (e.g. randomized) so each fold contains a good representation of the binary classes. This procedure is called stratification and is important prior to N-fold cross-validation. In the case of three fold, for example, two-thirds of the data is reserved for training and one-third is held out for testing. The procedure is

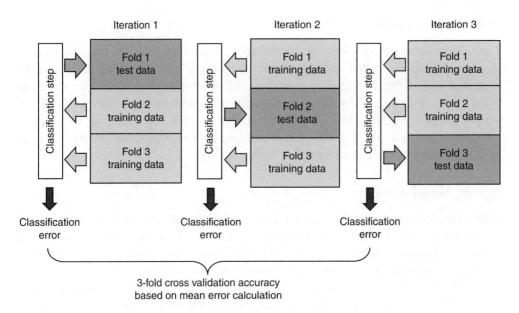

Figure 3.3   Procedure of three-fold cross-validation.

repeated three times until every instance has been used for testing (Witten *et al.* 2011), as illustrated in Figure 3.3.

The standard way to evaluate a learning algorithm given a single, fixed and limited sample of data is to use 10-fold cross-validation. Ten fold validation is the best practice as extensive tests have been done, and it was evaluated that ten is about the correct number of folds to get the best estimate (Witten *et al.* 2011). Grillo *et al.* (2009), Marturana *et al.* (2011a, 2011b), Gomez (2012) and Marturana and Tacconi (2013), all used 10-fold cross-validation.

The results of 10-fold cross-validation can be presented in several formats. The aim of the validation technique is to estimate the classification accuracy of the method on the available dataset, therefore analysis of correct and incorrect classifications is required.

In classification problems, results may be represented in the form of a two-dimensional *confusion matrix* (Witten *et al.* 2011). Consisting of positive and negative classifications represented by *true positive* (TP), *true negative* (TN), *false positive* (FP) and *false negative* (FN). TP and TN represent correctly labelled items as belonging to the positive and negative class respectively. For example, in a binary classification of yes or no, TP = real "yes", whereas TN = real "no". Conversely, FP and FN are incorrect classifications. A FP is an item classified as yes, but it should have been listed as no whereas a FN is an item classified as no which should have been listed as yes.

The *precision* rate is calculated as follows:

$$\text{Precision} = \frac{\text{Number of correctly labelled positives}}{\text{Total number of labelled positives}} = \frac{\text{TP}}{\text{TP} + \text{FP}} \qquad (3.1)$$

The *recall* rate is calculated as:

$$\text{Recall} = \frac{\text{Number of correctly labelled positives}}{\text{Total number of positives}} = \frac{TP}{TP + FN} \quad (3.2)$$

The *F-measure*, representing a weighted harmonic mean of *precision* and *recall*, is calculated as follows:

$$\text{F-measure} = \frac{2*\text{Recall}*\text{Precision}}{\text{Recall} + \text{Precision}} \quad (3.3)$$

The resultant two-dimensional *confusion matrix* is displayed as such:

$$\begin{bmatrix} TP & FN \\ FP & TN \end{bmatrix}$$

Note that, in binary classification problems, the diagonal values represent the correct and incorrect classifications. For clarity, this is outlined in the following text:

$$\begin{bmatrix} \mathbf{TP} & FN \\ FP & \mathbf{TN} \end{bmatrix}$$

This represents positive results, e.g. correct yes and no results.

$$\begin{bmatrix} TP & \mathbf{FN} \\ \mathbf{FP} & TN \end{bmatrix}$$

This represents negative results, e.g. incorrect yes and no results.

Often a row and column for each class is added to the Confusion Matrix. In the below example, 'a' and 'b' represent the original classes (e.g. 'yes' and 'no' respectively), whereas a_p and b_p are the predicted ones.

	a_p	b_p
a	TP	FN
b	FP	TN

Below is reported another example of how a Confusion Matrix may be represented.

Confusion Matrix		
a_p	b_p	<—classified as
4	3	a = Relevant
9	7	b = Non relevant

A two-dimensional confusion matrix encompassing two-classes is appropriate for this approach as the output is a binary classification (e.g. relevant, non-relevant). As shown, classification accuracy can be determined using confusion matrix precision, recall and

F-measure. Therefore, average values of the aforementioned performance indicators will be presented when demonstrating results.

3.4   A Child Pornography Exchange Case Study

This paragraph describes a case study on mobile phones classification in court cases of child pornography exchange as an application of the methodology described in Section 3.3. We provide a definition of the crime in Section 3.4.1, define the state vector and related features list in Section 3.4.2, introduce the data corpus in Section 3.4.3, describe the strategy adopted for selecting the proper machine learning algorithms (i.e. classifiers and data pre-processing filters) in Section 3.4.4 and finally illustrate experiment details and results in Section 3.4.5.

3.4.1   Definition of Child Pornography Exchange

In general, child pornography can be defined as depictions of children in a sexual act or in a sexual way. In this regards, child pornography contains the same kind of sexual activities depicted in adult pornography, except that media contain images or descriptions involving children, children and adults, or children and animals or objects (Osanka and Johann 1989).

Likewise, the crime of child pornography exchange may be defined as the diffusion, under any form (e.g. download, upload, peer-to-peer and file sharing), of images and video footages depicting children in a sexual act or in a sexual way for personal or commercial purposes through networks of sexual exploitation of children.

Several national and international organizations exist that are dedicated to paedophiles and child sexual exploitation. Links between organizations members serve as a communication network for the exchange of child pornography. This underground system was maintained in the past through group associations, by mail and telephone. Among the many solutions deployed so far to address such problem, the vigilance of postal service inspectors allowed to almost entirely wipe out the exchange of child pornography via traditional mail in the past decade. Unfortunately Internet e mail and the World Wide Web have given trafficking in child pornography new life by providing almost instantaneous access to child pornography and a nearly infinite number of Web-based repositories. As a consequence, paedophiles who used to be limited to the geographic area in which they lived or travelled have changed their habits by exploiting clandestine exchange computer networks on the Internet to share child pornography among them (Durkin 1997). With Internet chat rooms and e mail, such problem has become so widespread to include an almost unlimited population and geographic scope.

Even though unlike other forms of obscenity or pornography, child pornography laws are actively prosecuted, a variety of factors complicate the enforcement of child pornography laws, including detection, and the distribution networks. Detecting child pornography is difficult because those who actively possess and/or distribute the material take great care to hide their crimes. Many times the pornography is not discovered until after the individual has been arrested for a far more serious sexual offence involving a child, at which point the child pornography is used as evidence but is not likely to be prosecuted separately. Further, the perceived anonymity provided by the Internet has made detection and prosecution of child pornography exchange even more problematic than in the past. Finally, law enforcement efforts in this area are problematic and characterized by legal definitional issues, overlapping jurisdictions and efforts, and entrapment problems.

3.4.2   Child Pornography Exchange–State Vector

Searching cell phones for child pornography (e.g. possession or exchange) involves the forensic analysis of phones' memory for child nudity in files downloaded from the Internet, as well as photos and video footages taken with the embedded camera. Most forensic frameworks currently in use imply the manual intervention of a forensic examiner who is tasked of analyzing extraction tool's output. Regarding content classification, as already mentioned in Section 3.2.4, research has focused on automatic detection of pornographic material in image files and digital video footages (Fleck *et al.* 1996; Jones and Rehg 2002; Janshon *et al.* 2009). The methodology described in this chapter, which may be considered a valid alternative to automated content classification, allows to identify crime-relevant phones without the need for manual in-depth files content analysis.

As outlined in Section 3.3.3, the following is a list of data pertaining to phonebook contacts, call logs, Internet browsing and multimedia logs, that may be extracted from mobile phones with most forensic tools currently in use and converted into corresponding features according to the state vector:

- *Phone model* (Smartphone, GSM);
- *Number of phonebook contacts* (stored both on SIM and phone);
- *Number of dialed/received/missed calls;*
- *Percentage of dialed/received/missed calls* (with regards to the specific time frame: Morning, Afternoon, Evening and if generated or received from phonebook contacts or not);
- *Average duration of dialled/received calls* (with regards to the specific time frame: Morning, Afternoon, Evening and if generated or received from phonebook contacts or not);
- *Number of received/sent SMS/MMS;*
- *Percentage of received/sent SMS/MMS* (with regards to the specific time frame: Morning, Afternoon, Evening and if they are sent or received from phonebook contacts or not);
- *Number and percentage of visited URLs* (with regards to the specific time frame: Morning, Afternoon, Evening and if they are bookmarked or not;

Table 3.2 Survey child pornography feature category results.

Feature category	Average weighting*
Number of phonebook contacts	1,1
Call log, e.g. call time (morning, afternoon, evening) and total number of calls (received, dialed)	1,4
SMS/MMS statistics, e.g. time (morning, afternoon, evening) and total number (sent, received)	2,1
Number of picture files on device	3,7
Number of video files on device	4,0
Number of music files on device	1,1
Frequency of visited URLs (morning, afternoon, evening)	3,0
Number of visited URLs	2,7
Email statistics (sent, received)	2,6
Frequency of call time (morning, afternoon, evening)	1,7

* A discreet value is assigned to features according to their importance (e.g. $0 =$ low importance/exclude, $5 =$ high importance). The reported value is calculated on average from received answers.

- *Number and percentage of downloaded images and videos files* (downloaded or created by the embedded camera);
- *Number of sent/received E mail;*
- *Number of stored notes.*

Taking into account the guidelines outlined in Section 3.3.3, in order to make data collection more focused, we gathered forensic experts' opinion about which feature were more important in the criminal case of child pornography exchange. In this regards, we carried out interviews and surveys with forensic professionals and law enforcement responders. This resulted in identifying the data of particular interest that they perceived important in this specific context.

In particular, answers received to Question 1 of the survey 'From your experience and opinion, please rank the following attribute CATEGORIES according to importance, i.e. those you feel would pertain to child pornography', have been summarized in the following table with average values (Table 3.2).

Answers to Question 2 'More specifically, please rank the following attributes according to importance, in your opinion', have been summarized in the following table again with average values (Table 3.3).

As expected, according to field experts'opinion, the presence of multimedia files, time-related Internet browsing and email statistics is perceived as of above-average importance in relation to child pornography exchange with mobile phones. Specifically, the number of downloaded and produced picture and video files were rated as having the most importance.

Table 3.3   Survey child pornography feature results.

Feature	Average weighting
Number received calls	1,3
Number dialed calls	1,6
Number missed calls	0,7
Mean duration received calls	1,1
Mean duration dialed calls	1,1
Time of call (morning, afternoon, evening)	1,9
Higher % of calls to non-contacts vs. contacts in phonebook	2,1
Number read SMS	1,6
Number sent SMS	1,4
Number read MMS	2,3
Number sent MMS	2,3
Time of SMS/MMS (morning, afternoon, evening)	2,1
Number downloaded picture files	4,4
Number produced picture files (camera or other)	4,3
Number downloaded video files	4,3
Number produced video files (camera or other)	4,1
Number downloaded audio files	1,3
Number produced audio files	1,6
High % of visited URLs in morning	2,6
High % of visited URLs in afternoon	2,3
High % of visited URLs in evening	3,1

As a result of the survey, we identified the complete state vector listed in the table below as the set of offence-related features pertaining to a child pornography exchange case.

Mobile phones dataset
State vector = [1, 2 ... 114]

Phone model,	Number_read_sms,	Number_dialed_calls_contacts_morning,
Number_phonebook_contacts,	Number_sent_sms,	Number_dialed_calls_contacts_afternoon,
Number_received_calls,	Percentage_read_sms,	Number_dialed_calls_contacts_evening,
Number_dialed_calls,	Percentage_sent_sms,	Percentage_dialed_calls_contacts_morning,
Number_missed_calls,	Number_read_sms_ morning,	Percentage_dialed_calls_contacts_afternoon,

Continued

Mobile phones dataset
State vector = [1, 2 ... 114]

Percentage_received_calls,	Number_read_sms_ afternoon,	Percentage_dialed_calls_ contacts_evening,
Percentage_dialed_calls,	Number_read_sms_ evening,	Number_missed_calls_ contacts_morning,
Percentage_missed_calls,	Percentage_read_sms_ morning,	Number_missed_calls_ contacts_afternoon,
Number_received_calls_ morning,	Percentage_read_sms_ afternoon,	Number_missed_calls_ contacts_evening,
Number_received_calls_ afternoon,	Percentage_read_sms_ evening,	Percentage_missed_calls_ contacts_morning,
Number_received_calls_ evening,	Number_sent_sms_ morning,	Percentage_missed_calls_ contacts_afternoon,
Percentage_received_calls_ morning,	Number_sent_sms_ afternoon,	Percentage_missed_calls_ contacts_evening,
Percentage_received_calls_ afternoon,	Number_sent_sms_ evening,	Mean_duration_received_ calls,
Percentage_received_calls_ evening,	Percentage_sent_sms_ morning,	Mean_duration_dialed_ calls,
Number_dialed_calls_ morning,	Percentage_sent_sms_ afternoon,	Mean_duration_received_ calls_morning,
Number_dialed_calls_ afternoon,	Percentage_sent_sms_ evening,	Mean_duration_dialed_ calls_morning,
Number_dialed_calls_ evening,	Number_read_mms,	Mean_duration_received_ calls_afternoon,
Percentage_dialed_calls_ morning,	Number_sent_mms,	Mean_duration_dialed_ calls_afternoon,
Percentage_dialed_calls_ afternoon,	Percentage_read_mms,	Mean_duration_received_ calls_evening,
Percentage_dialed_calls_ evening,	Percentage_sent_mms,	Mean_duration_dialed_ calls_evening,
Number_missed_calls_ morning,	Number_read_mms_ morning,	Number_produced_picture_ files,
Number_missed_calls_ afternoon,	Number_read_mms_ afternoon,	Number_produced_video_ files,
Number_missed_calls_ evening,	Number_read_mms_ evening,	Number_produced_audio_ files,
Percentage_missed_calls_ morning,	Percentage_read_mms_ morning,	Percentage_produced_ picture_files,
Percentage_missed_calls_ afternoon,	Percentage_read_mms_ afternoon,	Percentage_produced_video_ files,
Percentage_missed_calls_ evening,	Percentage_read_mms_ evening,	Percentage_produced_audio_ files,

Mobile phones dataset
State vector = [1, 2 ... 114]

Number_received_calls_ contacts,	Number_sent_mms_ morning,	Number_URL_visited,
Number_dialed_calls_ contacts,	Number_sent_mms_ afternoon,	Number_URL_visited_ morning,
Number_missed_calls_ contacts,	Number_sent_mms_ evening,	Number_URL_visited_ afternoon,
Percentage_received_calls_ contacts,	Percentage_sent_mms_ morning,	Number_URL_visited_ evening,
Percentage_dialed_calls_ contacts,	Percentage_sent_mms_ afternoon,	Percentage_URL_visited_ morning,
Percentage_missed_calls_ contacts,	Percentage_sent_mms_ evening,	Percentage_URL_visited_ afternoon,
Number_received_calls_ contacts_ morning,	Number_downloaded_ picture_files,	Percentage_URL_visited_ evening,
Number_received_calls_ contacts_afternoon,	Number_downloaded_ video_files,	Number_URL_bookmarks,
Number_received_calls_ contacts_evening,	Number_downloaded_ audio_files,	Number_sent_email,
Percentage_received_calls_ contacts_morning,	Percentage_downloaded_ picture_files,	Number_received_email,
Percentage_received_calls_ contacts_afternoon,	Percentage_downloaded_ video_files,	Number_notes _memo,
Percentage_received_calls_ contacts_evening,	Percentage_downloaded_ audio_files,	Class

A sample *training input matrix* derived from such state vector is illustrated in Table 3.4.

3.4.3   Data Corpus

The dataset used in this case study is based on real data extracted from digital devices classified by forensic examiners of the Italian Postal and Communication Police (Marturana *et al.* 2011a, 2011b; Marturana and Tacconi 2013). Such dataset encompasses 23 mobile phones and smartphones pertaining to different crimes (e.g. exchange of child pornography, violation of trade secret, human trafficking and extortion). In particular the dataset encompasses 7 child porn exemplars and 16 pertaining to other crimes. We labelled the first set as *pedo* and the latter as *non-pedo* exemplars. Moreover, some realistic data have been added to the dataset with WEKA's pre-processing algorithms as well to mitigate two inherent issues, namely *imbalanced dataset* and *overfitting*, described in detail in the following section.

Table 3.4   Example of training input matrix.

Attribute name	Attribute type	Nokia 5200	Nokia Lumia
Phone model	{GSM, Smartphone}	GSM	GSM
Number_phonebook_contacts	numeric	14	35
Number_received_calls	numeric	45	0
Number_dialed_calls	numeric	54	0
Number_missed_calls	numeric	36	0
Mean_duration_received_calls	numeric	?	0
Mean_duration_dialed_calls	numeric	?	0
Number_read_sms	numeric	57	37
Number_sent_sms	numeric	0	36
Percentage_read_sms	{Low, Medium, High}	High	High
Percentage_sent_sms	{Low, Medium, High}	Low	High
Number_read_mms	numeric	0	0
Number_sent_mms	numeric	0	0
Percentage_read_mms	{Low, Medium, High}	Low	Low
Percentage_sent_mms	{Low, Medium, High}	Low	Low
Number_downloaded_picture_files	numeric	36	56
Number_downloaded_video_files	numeric	0	0
Number_downloaded_audio_files	numeric	2	1
Number_URL_visited	numeric	0	0
Number_URL_bookmarks	numeric	0	0
user_class	**{Pedo, Non-pedo}**	**Pedo**	**Non-pedo**

3.4.4   Learning from Available Data

In this case study we have tested a very popular machine learning workbench called Waikato Environment for Knowledge Analysis (WEKA)[5] (Witten *et al.* 2011), encompassing a collection of machine learning algorithms for classification tasks. WEKA is a collection of state-of-the-art machine learning algorithms and pre-processing filters available as a standalone application with the intuitive graphical user interface illustrated in Figure 3.4, or as a Java-based class libraries for use in customized environments.

The most appropriate classification algorithms among those available in WEKA have been selected to build an accurate and forensically sound framework that may be used to motivate digital triage findings in a courtroom.

In this regards, we have selected both K-nearest neighbours (KNNs), one of the most influential data mining algorithms (Wu *et al.* 2008), which is considered a legally superior data mining classifier (Garfinkel *et al.* 2010), whose predictions are easy to argue in a trial, and neural network (NN) a powerful, flexible, general-purpose

---

[5] WEKA from the Machine Learning Group at the University of Waikato, New Zealand, is downloadable at http://www.cs.waikato.ac.nz/ml/weka.

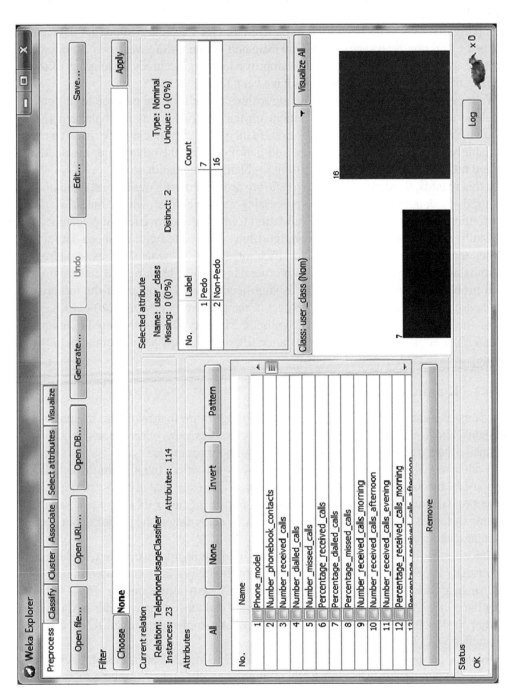

Figure 3.4 WEKA Explorer's GUI.

technique readily applied to prediction, estimation and classification problems (Witten *et al.* 2011).

In particular, KNN has been used to benchmark NN's accuracy to predict device's relevance (or non-relevance) to the investigated crime. Moreover, two automatic attribute selection techniques aimed to improve learning accuracy, belonging to the families of *wrapper* and *filter* methods have been tested.

We have tested some pre-processing algorithms as well for the purpose to mitigate two inherent issues of the dataset, which, on the one hand, was not balanced (i.e. *imbalanced dataset* problem) as fewer exemplars of the relevant class (i.e. *pedo*) were available with regards to the other (i.e. *non-pedo*), and, on the other hand, included a limited number of samples (i.e. *overfitting* problem during validation).

In this regards, to address the *data overfitting* problem during training set validation we used *randomize*, a WEKA pre-processing instance filter which modifies the distribution of exemplars within the validation folds, and improve learning accuracy.

Finally, since most machine learning algorithms work better with balanced data sets as they tend to optimize the overall classification accuracy, the need arises to address the imbalanced dataset problem, which is crucial in machine learning and data mining. Such problem occurred in our dataset as significant fewer exemplars of the minority (i.e. *pedo*) class were available with respect to the majority (i.e. *non-pedo*) class.

For the purpose, we have selected Synthetic Minority Over sampling TEchnique (*SMOTE*), a WEKA pre-processing instance filter to add new *pedo* exemplars to the training set.

3.4.4.1   K-nearest Neighbours

K-Nearest Neighbours (KNN) is a data mining algorithm that maps the training set elements into a multidimensional space and classifies new observations (e.g. test instance or test object) according to nearest neighbour exemplars.

In other words, whenever a new test instance is classified, the algorithm finds the group of k objects (i.e. the neighbourhood)in the training set that are closest to it with respect to a distance function. As illustrated in Figure 3.5, after locating the k nearest training instances, KNN classifies the test instance t according to the predominant class in the neighbourhood.

Choosing the best k, which is usually an odd value to ensure that a class is numerically superior to the other in the neighbourhood, is a key issue affecting KNN performance. If k is too small the result can be sensitive to single noise points. On the other hand, if k is too large the neighbourhood may include too many points pertaining to other classes. In the case study we will test WEKA's implementation of KNN, called *IBk*, with $k = 1, 3, 5$.

Defining the right distance function, a relatively simple task in case of numeric attributes, is another key aspect of KNN (Witten *et al.* 2011). Let's assume to have (i) the state vector S made up of k features ($f_j, j = 1, 2, \ldots, k$), (ii) n training exemplars ($x_i, i = 1, 2, \ldots, n$) which form the training set D and (iii) the test instance t whose

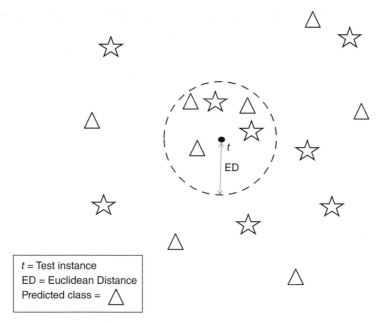

Figure 3.5   KNN neighbourhood with K = 5.

class is to be predicted; we can thus recall the *Minkowski distance* (MD) to calculate the distance between t and x_i as follows:

$$MD_p(t, x_i) = \left(\sum_{j=1}^{k} |t_j - x_{ij}|^p \right)^{\frac{1}{p}} \tag{3.4}$$

The formula is a generalization of the more common Manhattan distance (obtained with p = 1) where the difference between attribute values is added up (after taking the absolute value), and the standard Euclidean distance (ED), which is the obtained result when p = 2. Although other metrics (e.g. Levenshtein, Chebyshev, etc.) may be obtained by taking higher powers (p > 2) which increase the influence of large differences between attributes at the expense of small ones, generally, the ED represents a good compromise. The following is the formula for the ED:

$$ED(t, x_i) = \sqrt[2]{\sum_{j=1}^{k} (t_j - x_{ij})^2} \tag{3.5}$$

Different attributes are often measured on different scales, so if the previous formula were used directly, the effect of some attributes might be negligible with regards to others that had larger scales of measurement. Consequently, it is a good practice to normalize all attribute values to lie between 0 and 1 by calculating the following formula:

$$t_j = \frac{t_j - \min(x_{ij})}{\max(x_{ij}) - \min(x_{ij})} \tag{3.6}$$

where $\max(x_{ij})$ and $\min(x_{ij})$ are, respectively, the maximum and minimum value of x_{ij} over all instances in the training set. Even though it becomes less discriminating as the number of attributes increases, we have selected the Euclidean distance to calculate distance between instances in the dataset. The reason is that calculating Euclidean distance in WEKA is computationally simple and fit to our low-dimensional dataset.

3.4.4.2 Neural network

The idea behind a Neural Network (NN) is that very complex behaviour can arise from relatively simple units, the *artificial neurons*, acting in concert. Used in domains such as money laundering and credit card fraud detection, a network of artificial neurons is designed to mimic the human ability to learn from experience (Linoff and Berry 2011).

Based on NN history that merges biology and machine learning concepts, we can introduce the terms *single-layer perceptron* (SLP) which comes from machine learning terminology, and *artificial neuron*, derived by biology glossary, to describe the basic component of a NN which represents at the same time the simplest existing NN model.

As illustrated in Figure 3.6, a SLP or simply *perceptron*, takes a number of inputs and produces the corresponding output on the basis of the *activation function* (AF).

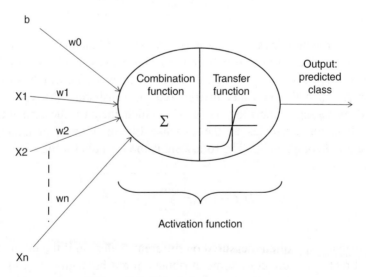

Figure 3.6   Single layer perceptron.

This function is split into two parts: *combination function* (CF) and *transfer function* (TF).

The CF is generally represented by a weighted sum, which is the default in data mining problems, where each input is multiplied by its weight and these products are added together, whereas the TF serves to model perceptron's behaviour. The *linear* function, for instance, is suitable for representing *linear regression models*, the *step* function which has the value 1 when the weighted sum of the inputs is above some threshold, and 0 otherwise mimics the human brain's 'all-or-nothing' response, the *sigmoid* functions (i.e. *logistic* and *hyperbolic tangent*) produce output values from limited intervals (i.e. $[0, 1]$ or $[-1, 1]$) and are appropriate to estimating variables with non-linear behaviour.

Since the SLP is able to solve simple classification problems with the limitation that classes must be linearly separable (e.g. a typical example of SLP limitations is represented by the XOR problem), a more complex structure called *multi-layer perceptron* (MLP) has been defined in literature as the typical NN model.

As illustrated in Figure 3.7, a MLP may be represented by a network of interconnected SLPs, with an *input layer*, where data enters the network, one or more *hidden layers*, comprised of SLPs, each of which receives multiple inputs from the

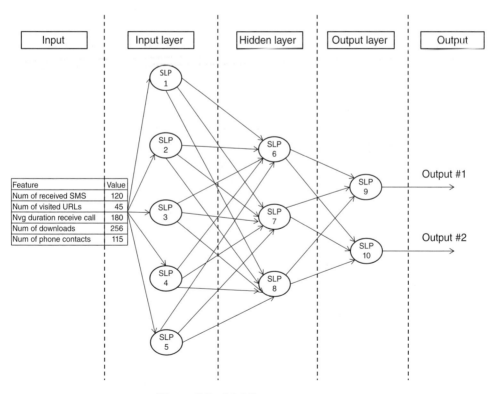

Figure 3.7   Multilayer perceptron.

input layer, and a *output layer*, which includes a variable number of SLPs, based on the number of predictable classes. In binary prediction, for instance, the output layer comprises a single SLP.

In this work we will test WEKA's implementation of the NN algorithm, called *multilayer perceptron*, to classify the dataset.

3.4.4.3   Learning from Imbalanced Dataset

This section deals with the so-called *imbalanced dataset problem*, which occurs during the training stage when significant fewer exemplars of a class are available with regards to others. The lack of data related to a class may result in inaccurate construction of decision boundaries between classes and consequent minority class (i.e. *pedo*) misclassifications. The problem can occur as well within a majority class formed of more clusters of data as one of them may contain significantly fewer exemplars than others, which may lead to misclassifications. Standard machine learning algorithms, indeed, tend to treat exemplars from a minority class or cluster of data within a class as noise with consequent performance degradation (Nguyen *et al.* 2009).

Imbalanced data set is an intrinsic property in investigations related to credit card frauds, money laundering and financial crimes, for instance, where very few illegal transactions are carried out compared to the huge number of normal operations. The problem, however, is not limited to the aforementioned cases as imbalanced data sets may be present in areas that don't have an inherent imbalance problem as well. Legal, economic, privacy limitations in collecting data, for instance, and the large effort required to obtain a representative data set may result in imbalanced data sets, as it occurred to the framework described in this chapter where collecting real data related to the class *pedo* from seized devices was the main difficulty that we encountered.

Data sampling (e.g. under-sampling, over-sampling and advanced sampling) with exemplars pre-processing to minimize discrepancy among the classes is a common approach to solve the class imbalance problem. Under-sampling, for instance, removes exemplars from majority class whereas over-sampling adds exemplars to minority class before processing the training set. In this regards, we have tested SMOTE, an advanced sampling algorithm (Chawla *et al.* 2002) which introduce new (e.g. synthetic), non-replicated *pedo* exemplars in the training set.

3.4.4.4   Data Overfitting

Dealing with machine learning, *overfitting* occurs when a statistical model describes random error or noise instead of the underlying relationship. Overfitting generally occurs when a model is excessively complex, such as having too many parameters relative to the number of training data points and begins to memorize training data rather than learning to generalize from them. A model which has been overfit will generally have poor performance on new data, as it can exaggerate minor fluctuations in the training set. For example, if the number of parameters is the same as or greater than the number of available training instances, a learning process can perfectly predict the training data simply by entirely memorizing them, but it will typically fail drastically when making predictions about unseen instances, since it has not learned to generalize at all.

One way to avoid overfitting is to use a lot of data thus forcing the algorithm to generalize and come up with a good model suiting all the points. Using a small training set, we have an intrinsic limitation in generalizing the model.

Therefore we run 10-fold cross-validation with an unsupervised pre-processing filter implemented by WEKA called *randomize*, which randomly shuffles the order of training set exemplars passed through it.

3.4.4.5   Automatic Feature Selection Techniques

Besides selected algorithms, number of training set exemplars, and related order in validation folds, even the number of features in the state vector, affects classification accuracy as well. If this number is too large, indeed, classification accuracy may be negatively affected as the inclusion of irrelevant, redundant and noisy attributes in the model can result in poor predictive performance and increased computation.

In this regards, among the available techniques (Hall and Holmes 2003), we have tested two algorithms for selecting the most effective features from our state vector, namely belonging to the family of *wrapper* e *filter* methods.

The *wrapper* method creates all possible subset of features in the state vector and then uses an evaluator to choose the best performing subset according to the selected classifier. It uses a search technique, such as best first, random search, exhaustive search, etc., to find a subset of features and then selects the one with which the classification algorithm performs the best according to the performance indicators outlined in Section 3.3.6.

Conversely, the *filter* method uses a combination of attribute evaluation and ranker algorithms to assign a rank to all the features in the state vector regardless of the selected classifier. Once attribute ranking is over, lower ranking features are removed one at the time from the state vector and related classification accuracy evaluated.

In comparison, due to the utilization of a classification algorithm to evaluate the accuracy resulting from the selected subset of features, rather than an attribute ranker usually makes the wrapper method produce better result than the filter method.

3.4.5   *Experiment Setup, Results and Discussion*

The case study described hereafter is based on the data corpus outlined in Section 3.4.3, encompassing $M = 23$ training exemplars extracted from just as many cell phones and smartphones from the case repository of the Italian Postal and Communications Police and concerning different crimes such as exchange of child pornography, violation of trade secret, human trafficking and extortion. As outlined in Section 3.4.3, such training set encompasses seven child porn exemplars (i.e. *pedo*) and 16 exemplars concerning other crimes (i.e. *non-pedo*). The *state vector*, defined in Section 3.4.2, encompasses $N = 114$ attributes. As a consequence, we have a rectangular $N \times M$ *training input matrix*.

The data was extracted using *CellDEK 1. 11, Paraben's Device Seizure 4.3.4092.33445, XRY 4.5* and *5.10, UFED 1.1.38, FTK 3.1* and stored in a database. The database was processed using functions and procedures to count feature occurrences and, finally, normalization was carried out to ensure data integrity so that classification could take place.

In this case study we will use results from Marturana *et al.* (2011a, 2011b); Marturana and Tacconi (2013) as a reference baseline for benchmark purposes. Class predictions obtained with KNN and NN, respectively described in Sections 3.4.4.1 and 3.4.4.2, will be compared with such baseline.

Pre-processing data techniques, such as synthetic over-sampling (i.e. SMOTE), and training exemplars randomization, described in Sections 3.4.4.3 and 3.4.4.4, will be explored to address the imbalanced dataset and overfitting issues, and related results discussed.

The feature selection techniques described in Section 3.4.4.5 will be tested as well and results discussed.

As outlined in Section 3.3.6, classifiers' learning accuracy will be calculated on the basis of *confusion matrix, precision, recall* and *F-measure*.

3.4.5.1　Baseline results

This section summarizes the reference baseline from Marturana *et al.* (2011a, 2011b); Marturana and Tacconi (2013) which includes the following WEKA's classifiers: Bayesian networks (BNs), decision tree (DT) and locally weighted learning (LWL), namely *BayesNet, J48* and *LWL* in WEKA. Training accuracy has been validated by the cited authors using 10-fold cross-validation and Table 3.5 illustrates the related output which forms the reference baseline against which this case study will be compared:

To summarize, the DT algorithm resulted to have the greater accuracy and, on average, all the classifiers were able to classify correctly more than half of the 23 phones. It is noteworthy to mention that the baseline tests are not based on pre-processing and are used as-is for benchmark purposes.

Table 3.5　Reference baseline results.

Performance parameter	Machine learning schemes		
	BN	DT	LWL
Weighted_avg_Precision	0.553	**0.68**	0.644
Weighted_avg_Recall	0.579	**0.684**	0.632
Weighted_avg_F_Measure	0.56	**0.644**	0.636

3.4.5.2 KNN and NN Without Pre-Processing (Complete State Vector)

This section illustrates findings of the training experiment with KNN and NN (see Sections 3.4.4.1 and 3.4.4.2), using the training input matrix outlined in Section 3.4.5. Basic algorithms have been tested, without further pre-processing. Training accuracy has been validated using 10-fold cross-validation and classification results are summarized in Table 3.6:

KNN ($k = 1$) resulted to have the highest average precision rate whereas NN had a higher precision rate concerning the majority class (i.e. non-pedo), resulting in better recall and F-measure; As a consequence, without further processing, NN is more likely to identify non-relevant devices (i.e. true negative) than KNN. On average, all the classifiers were able to classify correctly more than half of the 23 phones, which is not good enough for the purpose of the proposed method.

3.4.5.3 KNN and NN with Training Exemplars Randomization (Complete State Vector)

This section illustrates findings of the training experiment with KNN and NN (see Sections 3.4.4.1 and 3.4.4.2), obtained with the training input matrix outlined in Section 3.4.5, after randomizing the 23 training exemplars to address the overfitting problem. Training accuracy has been validated using 10-fold cross-validation and classification results are summarized in Table 3.7.

The proposed methodology showed better results in terms of classification accuracy and less data overfitting as a result of randomizing exemplars within the training set.

Table 3.6 KNN and NN without pre-processing (i.e. 23 non-randomized exemplars and 114 features).

Performance parameter	Machine learning schemes			
	KNN ($K = 1$)	KNN ($K = 3$)	KNN ($K = 5$)	NN
Weighted_avg_Precision	**0.581**	0.452	0.455	0.525
Weighted_avg_Recall	0.478	0.348	0.435	**0.565**
Weighted_avg_F_Measure	0.496	0.365	0.445	**0.542**

Table 3.7 KNN and NN with Randomize only (i.e. 23 randomized exemplars and 114 features).

Performance parameter	Machine learning schemes			
	KNN ($K = 1$)	KNN ($K = 3$)	KNN ($K = 5$)	NN
Weighted_avg_Precision	0.632	0.43	0.515	**0.665**
Weighted_avg_Recall	0.565	0.348	0.478	**0.696**
Weighted_avg_F_Measure	0.582	0.373	0.493	**0.667**

Table 3.8 KNN and NN with SMOTE and Randomize (i.e. 30 over-sampled and randomized exemplars and 114 features).

Performance parameter	Machine learning schemes			
	KNN ($K = 1$)	KNN ($K = 3$)	KNN ($K = 5$)	NN
Weighted_avg_Precision	**0.806**	0.775	0.767	0.767
Weighted_avg_Recall	0.667	0.567	0.533	**0.767**
Weighted_avg_F_Measure	0.635	0.487	0.43	**0.766**

On average, all the tested algorithm performed better than without randomization and NN outperformed KNN with every value of k. Classification accuracy is comparable with the baseline results described in Section 3.4.5.1 and it is still not good enough for the purpose of the proposed method.

3.4.5.4 KNN and NN with Training Exemplars Randomization and Over-Sampling (Complete State Vector)

This section illustrates findings of the training experiment with KNN and NN (see Sections 3.4.4.1 and 3.4.4.2), obtained with the training input matrix outlined in Section 3.4.5, and:

- over-sampling the minority class (i.e. pedo) from 7 to 14 exemplars, to get a more balanced training set,
- randomizing the training exemplars to avoid data overfitting during validation,

with the aim of further improving accuracy and performance. Training accuracy has been validated using 10-fold cross-validation and classification results are summarized in Table 3.8:

The combined effect of balancing and randomizing training exemplars turns into a general improvement in classifiers accuracy. In particular KNN ($k = 1$) was able to correctly classify all the minority class instances (i.e. pedo) resulting in the highest average precision rate whereas NN had a more balanced behaviour, resulting in better recall and F-measure; As a consequence, without further processing, KNN ($k = 1$) is more likely to identify relevant devices (i.e. true positive) than other algorithms whereas NN is more likely to identify non-relevant devices (i.e. true negative).

3.4.5.5 KNN and NN with Pre-Processing and Automatic Feature Selection (Reduced Training Input Matrix)

This section illustrates findings of the training experiment with KNN and NN (see Sections 3.4.4.1 and 3.4.4.2), obtained with the training input matrix outlined in Section 3.4.5, and:

Table 3.9   KNN and NN with SMOTE and Randomize (number of attributes reduced with *wrapper* method).

Performance parameter	Machine learning schemes			
	KNN ($K = 1$)	KNN ($K = 3$)	KNN ($K = 5$)	NN
Weighted_avg_Precision	0.834	0.933	0.767	**0.941**
Weighted_avg_Recall	0.833	**0.933**	0.767	**0.933**
Weighted_avg_F_Measure	0.833	**0.933**	0.766	**0.933**

Table 3.10   KNN and NN with SMOTE and Randomize (number of attributes reduced with *filter* method).

Performance parameter	Machine learning schemes			
	KNN ($K = 1$)	KNN ($K = 3$)	KNN ($K = 5$)	NN
Weighted_avg_Precision	0.867	0.781	**0.874**	0.781
Weighted_avg_Recall	**0.867**	0.767	**0.867**	0.767
Weighted_avg_F_Measure	**0.867**	0.766	**0.867**	0.766

- over-sampling the minority class (i.e. pedo) from 7 to 14 exemplars, to get a more balanced dataset,
- randomizing the training exemplars to avoid data overfitting during validation,
- reducing the number of attributes with automatic feature selection techniques (i.e. wrapper and filter methods), to improve classification accuracy.

Training accuracy has been validated using 10-fold cross–validation. Classification results are summarized in Tables 3.9 and 3.10:

As the wrapper method tends to choose the best performing subset of attributes according to the selected classifier, a different subset of attributes is selected according to every algorithm. In particular, after testing the wrapper method with KNN ($k = 1$) a subset of 2 attributes is selected, with KNN ($k = 3$) a subset of 6 attributes is chosen, whereas with KNN ($k = 5$) a total of 5 attributes are selected. In this experiment, both KNN ($k = 3$) and NN have correctly classified 93.3% of the training exemplars, a considerable achievement with regards to the initial 47.8% of KNN ($k = 1$) without pre-processing the training set. The slightly different average precision depends on the number of minority class exemplars incorrectly identified by each of them (i.e. 1 exemplar with KNN and 0 exemplars with NN) (Table 3.10).

The same subset of attributes is selected for using with each classification algorithm as a consequence of training set pre-processing with filter method, which is a suboptimal choice. In this case a total of 9 attributes out of 114 have been selected resulting in a general improvement in classification accuracy with regards to the baseline. Further, the wrapper method outperforms filter method as far as KNN ($k = 3$) and NN are concerned whereas the filter method resulted in greater classification accuracy regarding KNN ($k = 1, 5$). For this reason, we warmly suggest to take both attribute selecting techniques into account.

3.5 Conclusion

This chapter has dealt with the application of dealt with the digital forensics triage to solve the problem of classifying digital data sources as relevant or not to an investigation. In particular, we have described a digital triage framework based on Machine Learning algorithms which allows to infer the relevance of a target object (e.g. a digital device) to the case at hand on the basis of a set of *features*. Such features, concerning suspect's activity likely related to the investigated crime, are included in a *state vector* representing a set of features that may be extracted from target objects either on the crime scene or at DFLs.

The aforementioned framework encompasses the following three phases: *data extraction, feature extraction and processing, presentation of results*. A collection of a consistent set of exemplar data sources (i.e. the *training set*) is the preliminary step of the learning process. Exemplars are collected on the basis of their relevance to the investigated crime. Such training set is given as input to selected Machine Learning algorithms for classification purpose and the resulting model, validated using 10-fold cross-validation, is thus ready for future predictions of unknown data instances. Our goal is to provide investigators and forensic examiners with a tool that can be used both during the search and seizure on the crime scene and examinations at DFLs.

The described solution has been tested in a real-life scenario concerning child pornography exchange with mobile phones in which two machine learning algorithms, namely k-nearest neighbour (KNN) and neural network (NN) have been tested and related findings compared. In particular, KNN is considered a legally superior data mining classifier whose predictions are easy to argue in a trial, whereas NN can be readily and effectively applied to digital triage classification tasks.

Further, three inherent issues of the available data corpus, namely *imbalanced dataset, overfitting,* and *linear dependence* of the available features have been addressed with just as many data pre-processing techniques, namely *exemplars oversampling, randomization* and *automatic feature selection*.

As a result of the aforementioned case study we have shown that, with the right combination of pre-processing techniques and classification algorithms, it was possible to get a classification accuracy of 93.3%, with regards to the initial 47.8%, obtained with KNN ($k = 1$) and no pre-processing.

3.6 Challenges and Future Directions for the Digital Forensics Community

Digital triage is emerging as a forensic discipline and new models and methodologies are being developed at an outstanding pace. Nevertheless, research on digital triage is still in its infancy as further experiments are underway, and several issues must be taken into great consideration such as the lack of experimental data resulting in potential limitations that must be evaluated on a case-by-case basis. Further, with regards to specific types of crime, it is hard to infer the relevance of a digital source to the case at hand without first inspecting each file's content. For example the only evidence of a fraud on a computer may be the presence of a certain document (e.g. a spreadsheet) containing a list of credit card numbers. Such evidence is highly case-specific and not very different from normal documents that the person creates during normal working hours as part of his daily activities. For this reason, as file content analysis is out of the scope of the statistical approach undertaken in this research, whose aim is to keep digital data source pre-examination as simple and less time-consuming as possible, the inability to infer crime-related features from files' content is another potential limitation.

Nevertheless, the digital triage framework presented in this chapter resulted in very good predictions related to exchange of child pornography cases. However, to enable the described digital triage framework to be generalized to other crime types it is necessary to (i) acquire domain knowledge about the investigated crime and (ii) build up a valid dataset by which correctly train selected Machine Learning algorithms. It is noteworthy to mention that the proposed methodology is thought to be integrated with other forensic frameworks currently in use rather than replacing them, speeding-up the identification of relations between target objects and the investigated crimes on a commonality basis that, if done manually, would be a mind-numbing and time-consuming activity.

In this regards, being the described framework easily extendable to other crimes, we believe that it will serve as an inspiration to forensic practitioners and researchers. In this regard, interested readers who want to try their own implementations should, first of all, identify a set of crime-related features (i.e. the model's independent variables represented by the feature vector) and then collect a consistent set of training exemplars pertaining to the investigated crime, which is the core activity of the whole process on which classification accuracy strongly depends.

The higher the number of exemplar digital data sources (e.g. hard drives, smartphones, handsets, tablets and PDAs) pertaining to the investigated crime are used to train the classification algorithms, the better new data sources (i.e. test instances) will be classified.

Further, since a classifier may outperform others with regards to a specific training set, it is noteworthy to mention the importance of making a benchmark study on different classifiers to find the one(s) that perform(s) better than others on the available dataset. In the child pornography exchange case study, for instance, the limited amount of available training data made it impossible to determine whether a classifier was better or worse than others on average, across all possible training sets that can be drawn from the domain.

Moreover more effective solutions to the problem of classifying a digital data source would be to enable attribute weighting through automated and manual methods. In this regards, several techniques to improve classification accuracy such as attribute manipulation and weighting can be explored. Having the ability to place more importance on specific attributes by means of weights could possibly improve classification performance. Automatic methods for weighting attributes such as fully automating the classification process by using measures to calculate attribute weights without human intervention may be tested. Alternatively, weighting attributes manually may allow forensic investigators to provide invaluable input which can come from their experience or the context of the investigation.

Acknowledgements

This chapter is based on a research conducted in collaboration with the Computer Forensics Unit of the Italian Postal and Communications Police.

References

Cantrell G., Dampier D., Dandass Y. S., Niu N., Bogen C. Research Toward a Partially-Automated, and Crime Specific Digital Triage Process Model. *In Computer & Information Science*, vol. 5, no. 2, pages 29–38, 2012.

Chawla N. V., Bowyer K. W., Hall L. O., Kegelmeyer W. P. SMOTE: Synthetic Minority Oversampling Technique. *In Journal of Artificial Intelligence Research*, vol. 16, pages 321–357, 2002.

Chipman M., Hackley B.E., Spencer T.S. Triage of Mass Casualties: Concepts for Coping With Mixed Battlefield Injuries. *In Military Medicine*, vol. 145, no. 2, pages 99–100, 1980.

Drezewski R., Sepielak J., Filipkowski W. System Supporting Money Laundering Detection. *In Digital Investigation*, vol. 9, pages 8–21, Elsevier, 2012.

Durkin K. Misuse of the Internet by Pedophiles: Implications for Law Enforcement and Probation Practice. *In Federal Probation*, vol. 61, pages 14–18, Sage Publications, 1997.

Fleck M., Forsyth D., Bregler C. Finding Naked People. In Proceedings of the 4th European Conference on Computer Vision, Cambridge, UK. Lecture Notes in Computer Science, vol. 2, pages 593–602, Springer-Verlag, 1996.

Garfinkel S. L. Automating Disk Forensic Processing with SleuthKit, XML and Python. In Proceedings of 4th International IEEE Workshop on Systematic Approaches to Digital Forensic Engineering, Berkeley, CA, 2009.

Garfinkel S. L., Parker-Wood A., Huynh D., Megletz J. An Automated Solution to the Multiuser Carved Data Ascription Problem. *In IEEE Transactions on Information Forensic and Security*, vol. 5, no. 4, pages 868–882, 2010.

Garfinkel S. L. Digital Media Triage with Bulk Analysis and Bulk_Extractor. *In Computers & Security*, vol. 32, pages 56–72, Elsevier, 2013.

Gomez L. Triage in-Lab: Case Backlog Reduction with Forensic Digital Profiling. In Proceedings of Simposio Argentino de Informática y Derecho, La Plata, Argentina, pages 217–225, 2012.

Grillo A., Lentini A., Me G., Ottoni M. Fast User Classifying to Establish Forensic Analysis Priorities. In 5th IEEE International Conference on IT Security Incident Management and IT Forensics, Stuttgart, Germany, 2009.

Hall M. A., Holmes G. Benchmarking Attribute Selection Techniques for Discrete Class Data Mining. *In IEEE Transactions on Knowledge and Data Engineering*, vol. 15, no. 3, 2003.

Heilbronner R. L. *Daubert vs. Merrel Dow Pharmaceuticals, Inc.* (1993). In Encyclopedia of Clinical Neuropsychology, pages 769–770, Springer, New York, 2011.

Hong I., Yu H., Lee S., Lee K. A New Triage Model Conforming to the Needs of Selective Search and Seizure of Electronic Evidence. In Digital Investigation, vol. 10, no. 2, pages 175–192, Elsevier, 2013.

Jansen W., Ayers R. Guidelines on Cell Phone Forensics. In Recommendations of the National Institute for Standard and Technology (NIST), NIST Special Publication 800–101, Gaithersburg, MD, 2007.

Janshon C., Ulges A., Breuel T. M. Detecting Pornographic Video Content by Combining Image Features with Motion Information. In Proceedings of 17th ACM International Conference on Multimedia, Beijing, China, pages 601–604, 2009.

Jones M. J., Rehg J. M. Statistical Color Models with Application to Skin Detection. *In International Journal of Computer Vision*, vol. 46, no. 1, pages 81–96, 2002.

Kent K., Chevalier S., Grance T., Dang H. Guide to Integrating Forensic Techniques into Incident Response. In Recommendations of the National Institute for Standard and Technology (NIST), NIST Special Publication 800–86, Gaithersburg, MD, 2006.

Linoff G. S., Berry M. J. *Data Mining Techniques: For Marketing, Sales, and Customer Relationship Management*, 3rd Edition, Wiley, Indianapolis, IN, 2011.

Marturana F., Bertè R., Me G., Tacconi S. Mobile Forensics 'triaging': new directions for methodology. In Proceedings of VIII Conference of the Italian Chapter of the Association for Information Systems (ITAIS), Springer, Rome, Italy, 2011a.

Marturana F., Bertè R., Me G., Tacconi S. A Quantitative Approach to Triaging in Mobile Forensics. In Proceedings of International Joint Conference of IEEE TrustCom-11/IEEE ICESS-11/FCST-11 (TRUSTCOM 2011), Changsha, China, pages 582–588, 2011b.

Marturana F., Tacconi S. A Machine Learning-Based Triage Methodology for Automated Categorization of Digital Media. *In Digital Investigation*, vol. 10, no. 2, pages 193–204, Elsevier, 2013.

Nguyen G. H., Bouzerdoum A., Phung S. L. Learning Pattern Classification Tasks with Imbalanced Data Sets. In P.-Y. Yin (Ed.), *Pattern Recognition*, pages 193–208, InTech, Rijeka, 2009.

Osanka F. M., Johann S. L. *Sourcebook on Pornography*. Lexington Books, Lexington, MA/Toronto, 1989.

Palmer G. A road map for digital forensic research. In Report from the first Digital Forensic Research Workshop (DFRWS), 2001. http://www.dfrws.org/2001/dfrws-rm-final.pdflast (Accessed 18 February 2015).

Pearson S., Watson R. *Digital Triage Forensics-Processing the Digital Crime Scene*. Syngress, Boston, 2010.

Poisel R., Tjoa S. Forensics Investigations of Multimedia Data: A Review of the State-of-the-Art. In Proceedings of 6th IEEE International Conference on IT Security Incident Management and IT Forensics, Stuttgart, Germany, 2011.

Pollitt M. M. Triage: A practical solution or admission of failure. *In Digital Investigation*, vol. 10, no. 2, pages 87–88, Elsevier, 2013.

Refaeilzadeh P., Tang L., Liu H. Cross-Validation. In L. Liu and M. Tamer Özsu (Eds), *Encyclopedia of Database Systems*, pages 532–538, Springer, New York, 2009.

Rogers M. K., Goldman J., Mislan R., Wedge T., Debrota S. Computer Forensics Field Triage Process Model. In Journal of Digital Forensics, Security and Law, vol. 1, no. 2, pages 19–38, Association of Digital Forensics, Security and Law, Maidens, VA, 2006.

Roussev V., Quates C. Content Triage with Similarity Digests: The M57 Case Study. *In Digital Investigation*, vol. 9, pages S60–S68, Elsevier, 2012.

Weiss G. M. Mining with Rarity: A Unifying Framework. *In SIGKDD Explorations and Newsletters*, vol. 6, pages 7–19, 2004.

Witten I. H., Frank E., Hall M. A. *Data Mining Practical Machine Learning Tools and Techniques*. 3rd Edition, Elsevier, Burlington, MA, 2011.

Wu X., Kumar V., Quinlan J. R., Ghosh J., Yang Q., Motoda H., McLachlan G. J., Ng A., Liu B., Yu P. S., Zhou Z.-H., Steinbach M., Hand D. J., Steinberg D. Top 10 Algorithms in Data Mining. *In Knowledge and Information Systems*, vol. 14, pages 1–37, 2008.

4

Forensic Authentication of Digital Audio and Video Files

Bruce E. Koenig and Douglas S. Lacey
BEK TEK LLC, Stafford, VA, USA

4.1 Introduction

Throughout the world, claims of altered, edited or contrived digital audio and video recordings have become commonplace, with, for example, politicians asserting they were misquoted, even when confronted with conflicting high-definition, network television audio/video recordings; criminal drug suspects testifying that the federal police have altered on-the-body audio recordings of undercover narcotics buys; and parties in civil suits using local recording studio employees as witnesses to discredit recordings proffered by opposing parties. In the United States, the scientific examination of the 18.5-min effaced portion of the recording between President Richard M. Nixon and his former Chief of Staff caught the public eye, especially after President Nixon's secretary, Rose Mary Woods, claimed (inaccurately) that she had accidently erased the portion (Bolt *et al.* 1974). In Europe, the claims and counterclaims regarding the recordings from Ukrainian President Leonid Kuchma's office implicating him in the death of journalist Georgiy Gongadze became front-page news (Encyclopædia Britannica 2014). In Asia, video footage of the alleged extrajudicial executions of Tamil civilians in Sri Lanka has produced a number of forensic authenticity analyses (Wikipedia 2014). In Canada, the disputed, recorded statement made by Prime Minister Stephen Harper in a defamation civil suit against the Liberal Party of Canada was a major piece of evidence in the case (The Star 2008). In these listed cases, and in numerous others, forensic authenticity examinations of

Handbook of Digital Forensics of Multimedia Data and Devices, First Edition.
Edited by Anthony T.S. Ho and Shujun Li.
© 2015 John Wiley & Sons, Ltd. Published 2015 by John Wiley & Sons, Ltd.
Companion Website: www.wiley.com/go/digitalforensics

recordings have produced conclusive scientific results useable by the courts, law enforcement agencies, private investigators, attorneys, and other individuals involved in criminal, civil, security, terrorism and administrative matters.

Larger governmental and private forensic facilities have been conducting authenticity examinations of analogue media since the 1960s for audio (Koenig 1990) and the early 1980s for video recordings. Starting in the mid-1990s, the submissions began shifting to digital formats (Koenig and Lacey 2009), and today, most authenticity examinations are conducted of digital audio and video files. The usual purpose of these complex analyses is fourfold: to determine if a file (i) is an original/clone (i.e. a bit-for-bit copy) or a re-encoded or transcoded copy, (ii) contains any alterations, (iii) has any discontinuities due to stop or start events and (iv) matches the characteristics of a specified recording system, if known. These authenticity examinations, probably more than any other forensic laboratory analysis, require a more conceptual rather than a purely 'cookbook' protocol, since every case is at least somewhat different from any one preceding it, due to differences in the audio/video material, metadata, compression effects and various forms of possible duplication and alteration (Koenig and Lacey 2009; Grigoras *et al.* 2012; Koenig and Lacey 2012a). The major forensic laboratories who regularly conduct these authenticity analyses have developed laboratory protocols for their training and procedures based on peer-reviewed articles, appropriate formal education, in-house apprenticeships, attendance at applicable short courses and the examiners' collective experience. This chapter provides an overview of those protocols gleaned from the authors' experience and training and regular contact with scientists and engineers from other laboratories in the field.

This chapter will begin by listing the reasons that these examinations are conducted, types of common recording devices, digital file formats and evidence requirements in Section 4.2. Next, the physical laboratory facility (Section 4.3), requisite software and equipment (Section 4.4) and the protocols for conducting scientific authenticity examinations of digital audio and video files, including critical listening and visual reviews, data analysis and temporal/frequency analyses (Section 4.5), are discussed. At the end of the chapter are sections on preparing work notes and laboratory reports (Section 4.6), expert testimony recommendations (Section 4.7) and case examples (Section 4.8).

4.2 Examination Requests and Submitted Evidence

4.2.1 Examination Requests

Requests of a forensic laboratory to conduct authenticity examinations of digital audio and video files are usually based on one or more of the following reasons: legal, investigative or administrative.

Legal issues include chain-of-custody inconsistencies; the death, disability or flight to a foreign country of the witness who produced the recording; credibility problems with the recorder's operator; the recorder's operator being a non-testifying criminal defendant; differences between witness statements and what is seen or heard in a digital recording; and unexplainable differences between known crime scene events and the recorded information. However, the most common legal reason is initiated by an audio or video consultant or expert alleging, for example, that a file has been edited to add or remove information, is a duplicate, has been 'deliberately' stopped or started by an operator or has been produced in an illegal manner. An example of illegal activity would be listening to a conversation on a federal wiretap in the United States, without actually recording the information. The backgrounds of these consultants or experts can vary from preeminent practitioners in the field to individuals with no relevant education, experience or training; therefore, their examination findings can vary from highly accurate to technically impossible (Koenig and Lacey 2009).

Requests for authenticity examinations involving investigative matters can originate from all levels of law enforcement and by private investigative firms and include problems when an activated recorder fails to produce a recording; a recorded conversation has unexplained visual or aural dropouts; there is a lack of synchronization between the audio and video information; there are discrepancies between the on-screen date/time information and the law enforcement officer's notes or report; and visual or aural inconsistencies exist in an internal affairs interview.

Administrative matters often come from the business sector and can include suspicious or anonymously submitted recordings regarding alleged employee misconduct; recordings of important board meetings with visual or aural losses; internal recordings of client telephone calls that are being challenged; video surveillance of suspected employee theft; and questioned video recordings of allegedly napping 'workers'.

4.2.2 Submitted Evidence

Since an authenticity examination is often based on allegations of alteration, equipment misuse, chain-of-custody deficiencies and illegal monitoring techniques, it is often necessary for the contributor to supply specific information, media, software and equipment to properly conduct the examination, including the following, as appropriate:

1. Sworn testimony, witness statements, laboratory reports and judicial documents reflecting information regarding the evidence alterations; missing recordings; unexplained sounds and visual content; contradictory statements; and prior expert analyses. The descriptions of the alleged problems should be as complete as possible, including the exact locations within the recording, types of suspected editing, scientific tests performed and so on.
2. The original/native recording, a bit-for-bit copy or an appropriate digital image (see Section 4.5.2), along with an explanation of the process by which the file/clone/image was produced. A digital data image of the entire contents of the storage medium is also acceptable. Meaningful scientific authenticity results are normally not possible when examining non-bit-for-bit representations of the original

recording, including audio and video files which have either been transcoded from their native format to another format or re-encoded to the same format.

3. Any separate, simultaneous recordings, which contain pertinent audio or video information that is also present on the original recording.

4. If needed, proprietary software to access and play back the native file format, or sufficient information to contact the manufacturer or other source of the software.

5. The digital recorder and related components used to produce the recording, including proprietary hardware and cables for data access. If the specific equipment is no longer available or cannot be provided, then the unit's make and model plus the software version should be supplied, if available.

6. Any written records of damage to or maintenance of the recorder, accessories and other submitted equipment.

7. A detailed statement from the person or persons who made the recording, describing exactly how it was produced and the conditions that existed at that time, including:
   a. Power source, such as alternating current (AC), rechargeable batteries, a vehicle electrical system or a portable generator
   b. Input, such as a cellular telephone, radio frequency (RF) transmitter/receiver system, internal microphone(s) or miniature external microphone(s)
   c. Environment(s), such as a landline telephone transmission, restaurant or small office
   d. Known background noises, such as television, radio, unrelated conversations and computer games
   e. Foreground information, such as the number and genders of individuals involved in the conversation, general topics of discussion and proximity to microphones
   f. Original storage media format, if a clone file is submitted
   g. Recorder operations, when known, such as the number of times turned on and off in the record mode (i.e. record events); type of recorder functions or remote operations employed for all known record events; use of voice-activated features; and video settings, such as the time-lapse image rates, pixel dimensions, motion detection parameters and compression characteristics

8. A typed transcription of the entire recording, if it contains recorded speech, or, if that is not available, transcriptions of the portions in question.

The above items should not be considered a complete list, but are examples of information, software, and equipment often needed to conduct digital authenticity examinations (Koenig 1990; Koenig and Lacey 2009).

4.2.3 Digital Recording Devices

Digital audio and video files submitted for examination are produced on a wide variety of recorders, with most devices grouped into three general forensic categories: (i) law enforcement/investigative/judicial, (ii) consumer and (iii) business. The first grouping of units includes recordings from proprietary on-the-body devices, telephone wiretap systems, surveillance recorders, police vehicle dashboard cameras, emergency hot line systems, police radio traffic devices, semi-professional camcorders and interview room recording systems. The consumer devices include small 'hand-held' audio recorders, consumer camcorders, cellular telephones, tablets, webcams, video-sharing websites, telephone answering machines and digital cameras. Business recording units consist of, for example, exterior and interior building surveillance video recorders, aircraft cockpit voice recorders, video/audio recorders on buses and trains, broadcast television recordings, office telephone recording units and incoming telephone recording systems. The quality of the recordings in each of these three groups can vary widely from excellent to very poor in both aural intelligibility and video resolution, based on a number of factors including the degree and type of compression; microphone and camera locations; the physical environment; and time of day/night.

4.2.4   Digital File Formats

The range of digital audio and video files received for examination is quite large, and expanding, as the last of the analogue devices are being discontinued by manufacturers. Digital media files are generally characterized by two features: their container format and their content streams. A container format (also referred to as a 'wrapper') defines the overall structure of the file, including how the file's metadata, video/audio information and index information are multiplexed together, but does not explicitly define how the video/audio information is encoded. Common container formats include Audio Video Interleave (AVI), Waveform Audio File Format (WAV), QuickTime File Format and 3rd Generation Partnership Project (3GPP). Content streams refer to the actual recorded video, audio and other data (e.g. subtitles) stored in a container format. The video and audio content streams contain information which has been encoded by one of many codecs (compress/decompress algorithms) and which is specified in the metadata of the container format such that the playback device can properly decode the content. For example, an AVI file may contain a video content stream encoded by a Moving Picture Experts Group (MPEG)-based codec, a Motion Joint Photographic Experts Group (M-JPEG) codec or one of many other video codecs presently available. The overall structures of the AVI files containing differently encoded video content streams will remain essentially the same, however, with possible differences in what information needs to be written into the files' headers to accommodate the codecs. Similarly, a WAV file may contain uncompressed or compressed audio data, but the overall file structures would be alike.

An overall characterization of the container file types normally received in forensic authenticity laboratories can be grouped into the following areas: (i) proprietary law enforcement, (ii) proprietary multiplexed telephone audio and surveillance video, (iii) proprietary consumer audio, (iv) standard professional audio and video and (v) standard consumer audio and video.

Proprietary container formats are only produced on specific recording hardware, usually with no available written standards or other detailed structure descriptions; the exceptions are law enforcement container formats, which have written file protocols, but they are often unavailable to non-law enforcement laboratories. Though proprietary container formats may require the use of the manufacturer's software for complete access and playback, they may actually contain standardized video/audio content streams that can be identified and extracted through detailed analyses (Lacey and Koenig 2010). However, the associated metadata, such as recording date/time, camera/channel number, etc., may not be as readily accessible without the use of the manufacturer's software or extensive byte-level analyses.

The standard consumer and professional container formats have written digital protocols, which are usually available through standards organizations; most of these standards' documents are free of charge, or at a minimal cost, but a few can be expensive to purchase.

4.3 Laboratory Space

Authenticity laboratories are configured or constructed in an attempt to optimize the complex examinations of submitted audio and video recordings. These analysis facilities normally use appropriate building materials and designs to attenuate outside noise and vibration; reduce bright sunlight from entering the space; minimize operational noise from laboratory computers and other equipment; lessen interference from communication transmissions, such as cellular telephone use outside the laboratory; and minimize stronger electric and magnetic fields, both within and outside the facility. Whenever possible, these forensic spaces are located well away from noisy indoor and outdoor environments and do not have large window exposure; but if that is not possible, the laboratory walls can be isolated and soundproofed, and the windows appropriately covered. Equipment and ventilation sounds within the laboratory are reduced with the use of low-noise fans, enclosed equipment racks and properly designed and maintained heating and air-conditioning systems. Whenever possible, the facility should be located far from radio, television and other high-wattage communication transmitters. Cellular telephones, two-way pagers, tablets and other electronic devices which transmit and receive information wirelessly are normally turned off in the laboratory, since they can add noise and produce dropouts in recordings. Non-essential devices that produce electric and magnetic fields are placed in storage outside the laboratory, and whenever possible, incandescent instead of fluorescent lighting is utilized (Koenig *et al.* 2003; SWGDE 2008; Koenig and Lacey 2009).

4.4 Laboratory Software and Equipment

The types of laboratory equipment and software used to authenticate digital audio and video files are diverse but can generally be categorized into the following areas (Koenig *et al.* 2003; Koenig and Lacey 2009):

1. High-speed computers, external sound cards, high-resolution monitors and printers
2. Media readers/writers devices
3. Hardware and software write blockers
4. Professional headphones, amplifiers, cables and connectors
5. Proprietary audio and video playback software
6. Digital data imaging and analysis software
7. High-resolution audio waveform analysis software
8. Fast Fourier transform (FFT) analyzers and software
9. Spectrographic analysis software
10. Scientific computing software
11. Professional audio editing and playback software
12. Professional non-linear video editing and processing software
13. Media conversion/transcoding software
14. Professional image measurement and processing software

Following is a description of the above software and equipment.

4.4.1   High-Speed Computers, Computer Peripherals, Media Readers/Writers, Hardware/Software Write Blockers, Professional Headphones, Amplifiers, Cables and Connectors

Since the processing of large audio and video files can be time consuming, especially when rendering a video file or executing a complex procedure in scientific computing software, the computer systems normally contain fast CPUs, the maximum useable RAM, hard drive storage in the terabytes, high-speed digital interfaces, large high-resolution monitors (i.e. displaying pixel dimensions of 2560 by 1600 or larger) and laser or inkjet printers (with resolutions of at least 600 by 600 dots per inch and legal size or larger paper capability). High-resolution monitors are often needed to simultaneously show multiple windows for two or more video images, audio waveforms, work notes and scientific analysis graphs. Media readers/writers for compact discs (CDs), DVDs, Blu-ray discs and memory cards are often included as internal devices in the laboratory computer systems, but separate external devices can also be utilized through the USB computer connections. External sound cards are often used for critical listening to avoid the internal noise generated by the computer motherboard, hard drives and other internal components. A hardware write blocker is a device that, when installed between a computer and USB- or Fire Wire-enabled hardware, allows the computer to 'read' the information on a recorder or storage device, but does not permit the computer to change or 'write' information to the media/device; there are also software programs available to replicate the effects of the hardware write blocker. Professional quality, 'around-the-ear', stereo headphones (without noise-reduction circuitry), amplifiers, cables, and connectors are part of the standard accessories used in forensic audio authenticity examinations. As noted in the

Figure 4.1 Example of manufacturer-specific software for playback of proprietary video surveillance recordings.

'Critical Listening' portion in Section 4.5.5.1, loudspeakers are not used in forensic audio authenticity examinations.

4.4.2 Proprietary Audio and Video Playback Software

Proprietary formats are uniquely configured for specific manufacturer's audio and video recording systems and do not typically have standardized file structures. These formats need specialized software, available from the manufacturer, to play back the files and to export still images and standardized audio and video files. Figure 4.1 illustrates an example of manufacturer-specific software for the playback of proprietary video surveillance recordings.

Since manufacturers seldom update this software for newer computer operating systems, the available playback software for a particular recording may only be installable or useable on older operating systems. To overcome this limitation, 'virtual' computing software, in conjunction with an installed prior-generation operating system, can be utilized on a newer computer, so that the older playback software can be installed and executed. Since some virtual algorithms can process information quite sluggishly, especially on computers without high-speed CPUs, care is taken to determine the best software for a particular laboratory setup. It is noted that capturing the original file in one format and then transcoding to another format are not acceptable, since the digital data and audio/video information will always be changed in a way to make the authenticity meaningless, as is discussed in later sections.

Unique cables or other peripheral equipment may sometimes be necessary for interfacing between a recorder's flash memory or internal memory card and a laboratory computer. With regard to removable media, a memory card reader or other appropriate interface mechanism is always accessed with applicable write protection implemented in hardware or software. In the case of proprietary consumer recorders, the software and cables are commonly supplied with the recorder or are available for download or purchase from the manufacturer's website. For certain specialized law enforcement recorders, the proprietary playback and conversion software is often transferred onto the same medium (such as a data CD or DVD) that holds the evidentiary recording(s). If the software is not supplied on the digital medium itself, contact should be made with either the contributor of the evidence or the manufacturer of the recording device for the software to be provided, either on a separate disc or through an Internet download.

4.4.3   Digital Data Imaging and Analysis Software

Digital data imaging and analysis software can perform a number of functions, including the creation of bit-for-bit copies of submitted media (digital data images); calculation of mathematical representations (hashes) of digital data images and individual files; extraction of individual files from digital data images; comparisons of the matching and non-matching bytes of two files/digital data images; viewing and searching the contents of data images, files, memory cards and solid-state devices; editing of file copies (if needed); writing and application of automated scripts for analyzing data (Lacey and Koenig 2008); and saving of edited recordings as separate files. Modifications or edits of the digital data may be required, for example, in cases where non-audio information in a file has become corrupted and prevents proper playback of the audio recording.

This software displays various parameters characterizing digital data images and files, such as the type of file system, file name, file size and disc cluster and sector information. The software normally portrays the digital information in both hexadecimal and American Standard Code for Information Interchange (ASCII, which is supported by the American National Standards Institute) or Unicode (supported by The Unicode Consortium) representations and is able to read and interpret a variety of file systems. As background, ASCII is a set of 7-bit values that designates 128 characters – the numbers 0–9, the letters a–z and A–Z, some basic punctuation symbols, some control codes that originated with Teletype machines and a blank space. Unicode is an extension of ASCII and consists of more than 110 000 characters covering 100 scripts; the ASCII characters are the first 128 symbols in the Unicode

Offset	0	1	2	3	4	5	6	7	8	9	10	11	12	13	14	15		
00000000	49	44	33	03	00	00	00	00	07	76	58	4F	4C	59	00	00	ID3	vXOLY
00000016	03	EC	00	00	02	64	73	73	01	00	01	00	02	00	00	00	ì	dss
00000032	56	4E	38	31	30	30	50	43	20	20	20	20	20	20	20	20	VN8100PC	
00000048	66	00	00	00	FE	FF	FF	FF	FF	FF	31	34	31	30	31	36	f	þÿÿÿÿÿ141016
00000064	32	31	31	32	30	37	31	34	31	30	31	36	32	31	31	32	2112071410162112	
00000080	32	37	30	30	30	30	31	39	FF	07	FF	FF	FF	FF	FF	FF	27000019ÿ ÿÿÿÿÿÿ	
00000096	FF	FF	FF	FF	FF	FF	FF	FF	FF	FF	FF	FF	FF	FF	FF	FF	ÿÿÿÿÿÿÿÿÿÿÿÿÿÿÿÿ	
00000112	FF	FF	FF	FF	FF	FF	FF	FF	FF	FF	FF	FF	FF	FF	FF	FF	ÿÿÿÿÿÿÿÿÿÿÿÿÿÿÿÿ	
⋮																	⋮	
00000896	00	00	00	00	00	00	00	00	00	00	00	00	00	00	00	00		
00000912	00	00	00	00	00	00	FF	FF	FF	FF	FF	FF	FF	FF	FF	FF		ÿÿÿÿÿÿÿÿÿÿ
00000928	FF	FF	FF	FF	FF	FF	FF	FF	FF	FF	FF	FF	FF	FF	FF	FF	ÿÿÿÿÿÿÿÿÿÿÿÿÿÿÿÿ	
00000944	FF	FF	FF	FF	FF	FF	FF	FF	FF	FF	FF	FF	FF	FF	FF	FF	ÿÿÿÿÿÿÿÿÿÿÿÿÿÿÿÿ	
00000960	FF	FF	FF	FF	FF	FF	FF	FF	FF	FF	FF	FF	FF	FF	FF	FF	ÿÿÿÿÿÿÿÿÿÿÿÿÿÿÿÿ	
00000976	FF	FF	FF	FF	FF	FF	FF	FF	FF	FF	FF	FF	FF	FF	FF	FF	ÿÿÿÿÿÿÿÿÿÿÿÿÿÿÿÿ	
00000992	FF	FF	FF	FF	FF	FF	FF	FF	FF	FF	FF	FF	FF	FF	FF	FF	ÿÿÿÿÿÿÿÿÿÿÿÿÿÿÿÿ	
00001008	FF	FF	FF	FF	FF	FF	FF	FF	FF	FF	FF	FF	FF	FF	FF	FF	ÿÿÿÿÿÿÿÿÿÿÿÿÿÿÿÿ	

Figure 4.2 A hexadecimal/ASCII display of portions of the ID3 header information from an MP3 audio file.

character set (7-bit ASCII, 2007; Unicode Standard 2012). Regarding file systems, for example, submitted memory cards may use a FAT32 file system (Microsoft 2000), which must be read properly for accurate file extraction and analysis. A typical hexadecimal/ASCII display of commonly encountered ID3 information in an MP3 (MPEG-2 Audio Layer III) compressed audio file is represented in Figure 4.2 (Koenig *et al.* 2014).

4.4.4   *High-Resolution Audio Waveform Analysis Software*

High-resolution waveform analysis software creates an on-screen display of the graphic relationship between time (horizontal axis) and amplitude (vertical axis) of recorded audio (Jayant and Noll 1984). This software allows the identification of the beginning and end of a displayed time range (in seconds, samples or other user-selected measurements); the separate maximum and minimum amplitudes [in quantization levels, decibels (dB) or percent of maximum]; the non-compressed graphic format of a saved waveform display file; and administrative data such as laboratory number, specimen number and event description. The print output of some of these specialized software programs utilizes the full resolution of the attached printer and is not a 'screen dump', which is the print method, unfortunately, utilized by most professional audio editing and some scientific graphic programs. The resolution of a screen dump is limited to the maximum resolution of the computer monitor. Figure 4.3 reflects the high-resolution waveform printout of a 50-ms segment of an audio file using this

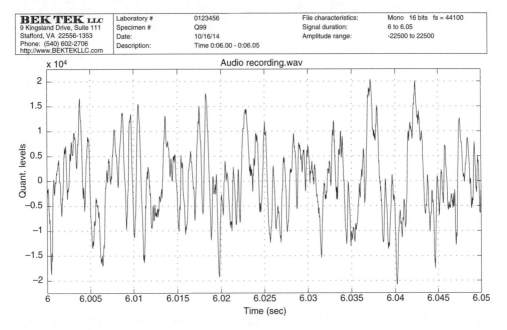

BEK TEK LLC	Laboratory #	0123456	File characteristics:	Mono 16 bits fs = 44100
9 Kingsland Drive, Suite 111	Specimen #	Q99	Signal duration:	6 to 6.05
Stafford, VA 22556-1353	Date:	10/16/14	Amplitude range:	-22500 to 22500
Phone: (540) 602-2706	Description:	Time 0:06.00 - 0:06.05		
http://www.BEKTEKLLC.com				

Figure 4.3   High-resolution waveform print for a 50-ms portion of a 16-bit PCM WAV file, with the vertical axis set to $\pm 22\,500$ quantization levels.

specialized software, which displays all the data points of the file up to the printer's maximum resolution (for illustrative purposes, the waveform line has been thickened).

4.4.5   FFT Analyzers and Software

A real-time FFT analyzer or appropriate software produces either a continually updating or a time-defined, visual representation of audio information. The standard FFT display reflects frequency on the horizontal axis and amplitude on the vertical, with the waterfall format having multiple, closely spaced, horizontally stacked FFT displays to reflect changes in time. As examples, Figure 4.4 reflects a standard FFT display of a 50-hertz (Hz) sine wave plus its odd harmonics, and Figure 4.5 shows a waterfall display of a dropout occurring during an otherwise constant 400-Hz sine wave signal. See the enhanced e-book for the audio files accompanying these two figures.

As background, the Fourier transform converts a signal from the time to the frequency domain through a mathematical relationship based on the periodicity inherent in sine and cosine functions, which allows any continuous, periodic signal to be represented as a summation of these trigonometric functions and an imaginary component, which is usually ignored in forensic authenticity applications. This conversion is adapted to computer operations using the discrete Fourier transform (DFT), which is a version suitable for converting a discrete time signal to a frequency

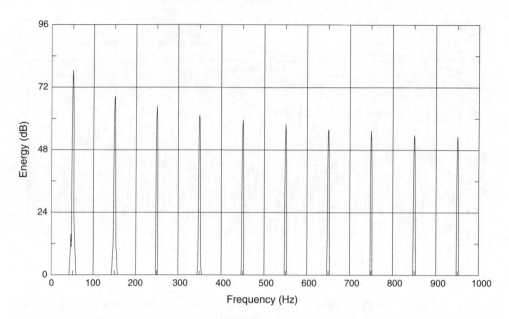

Figure 4.4 FFT display of a 50-Hz sine wave plus its odd harmonics, over a frequency range of 0–1000 Hz on the horizontal (with gridlines every 100 Hz) and an amplitude range of 0–96 dB on the vertical (with gridlines every 24 dB).

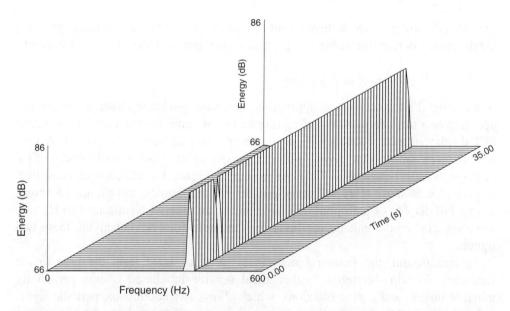

Figure 4.5 Waterfall FFT display of a 400-Hz sine wave with a 0.5-s audio dropout occurring near the beginning of the 35-s sequence, over a frequency range of 0–600 Hz on the horizontal and an amplitude range of 66–86 dB on the vertical. The FFT displays are stacked chronologically from the front (0 s) to the rear (35 s) on the diagonal axis.

representation. The DFT is, in turn, improved into a faster implementation called the FFT, which requires fewer computations, thus providing an appreciable increase in processing speed. A detailed description of FFT theory is beyond the scope of this chapter, but many excellent texts are available on the subject, some of which are included in the listed references (Wallace and Koenig 1989; Bracewell 2000; Kammler 2000; Owen 2007; James 2011).

As a group, FFT analyzers and software programs used in the forensic laboratory have parameters that normally include selectable transform sizes of 32 to over 8192 (equating to a range of resolution lines, at 80% of the Nyquist frequency, of approximately 12 to over 3200); two or more separate channels; 100-kilohertz (kHz) or greater real-time processing rates for the analyzers; linear, peak and exponential averaging; 90-dB or greater signal-to-noise ratios; adjustable frequency ranges of 100 kHz or higher; zoom functions allowing processing of any user-selected frequency range; interactive cursor controls; correlation and comparison analyses; and laser or inkjet printer and digital plotter outputs. As background, the Nyquist theorem states that in order to capture the desired bandwidth (referred to as the Nyquist frequency) of a signal, it must be sampled at a rate which is at least twice the desired bandwidth; for example, the Nyquist frequency of a digital audio signal sampled at 44.10 kHz would be 22.05 kHz (Pohlmann 2011).

4.4.6   Spectrographic Analysis Software

A sound spectrogram reflects a graphical display of time (on the horizontal axis) versus frequency (on the vertical axis) versus amplitude (in either grey or colour scaling); Figure 4.6 shows a sound spectrogram of a male talker saying 'speech

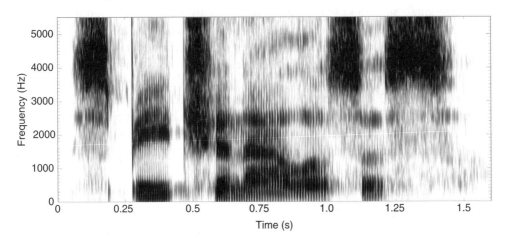

Figure 4.6   Sound spectrogram of the phrase 'speech analysis' spoken by a male talker, over a range of 0–5.5 kHz on the vertical axis, 1.6 s on the horizontal axis and a 30-dB amplitude range represented as greyscale values.

analysis', with the horizontal axis representing 1.6 s, the vertical axis displaying a frequency range of 0–5.5 kHz and the grey scaling an amplitude range of 30-dB, where the highest-energy areas are represented by the blackest shades, the lowest by white and the middle levels by varying shades of grey. See the enhanced e-book for the audio file accompanying this figure. The darker, generally horizontal bands represent vocal tract resonances of vowel and vowel-like sounds, while the broader, higher-level energy in the upper frequencies corresponds to vocal fricatives, such as 's' and 'ch' sounds. These programs allow real-time review and usually have selectable transform sizes of 32 to over 8192; 90-dB or greater signal-to-noise ratios; adjustable frequency ranges of 50 kHz or higher; zoom functions allowing processing of user-selected frequency ranges; adjustable grey- and colour-scale palettes; interactive cursor controls; modifiable pre-emphasis; and high-quality printing capabilities. A detailed description of sound spectrogram theory is beyond the scope of this chapter, but many excellent texts are available on the subject, some of which are included in the listed references (Potter *et al.* 1966; Cooke *et al.* 1993; Tanner and Tanner 2004; Koenig and Lacey 2009).

4.4.7   *Scientific Computing Software*

Scientific computing software programs, typified by MATLAB® (MathWorks, Inc., Natick, Massachusetts, United States), use advanced programming languages and pre-written scripts, or toolboxes, to produce numerous mathematical functions useful for the analyses of audio and video signals. These algorithms are flexible, with multiple parameters and outputs, such that temporal, frequency and image data can be transformed, plotted, viewed, listed statistically, filtered and so on. Since every authenticity examination has its own unique characteristics, these software programs allow audio and video data to be adapted for the analysis requirements of a specific digital file. Common examples of the use of this software is the production of high-resolution waveform charts, specialized frequency domain transforms, correlation of audio signals, analysis of direct current (DC) offsets (see Section 4.5.5.2) and comparison of video images. This software is fully capable of producing spectrograms and FFT representations, such as those available in specialized software; however, most non-academic, forensic laboratories limit its usage to analysis needs unavailable in the specialized software, due to the time-consuming process of fully learning these general scientific programs and implementing specific applications.

4.4.8   *Professional Audio and Non-linear Video Editing and Playback Software*

These software programs, which are also used in nearly all production sound and video studios, allow for playback, processing and analysis of most standard audio and video files. These basic programs are normally supplemented with 'plugins' or other algorithms that can perform specific functions useful in forensic analyses. Specifically, these programs can display the time waveform and the individual video images in great detail; play back selected portions; transcode to other audio and video formats; resample and re-quantize (i.e. change the bit depth) audio; correct audio DC offsets; increase or decrease the overall or selected audio amplitude; enhance the sound and visual quality; stabilize camera movement; and provide measurement and statistical information.

4.4.9   Media Conversion/Transcoding Software

These programs allow video and audio files to be converted or transcoded to standardized formats that are useable by the audio and video editing software. Specifically, this software is frequently used with files that cannot be directly imported into audio/video editing programs.

4.4.10   Professional Image Measurement and Processing Software

Professional image measurement and processing programs, typified by Photoshop CC (Adobe Systems, Inc., San Jose, California, United States), allow detailed analyses of still and video images to compare characteristics, identify similarities and differences (Lacey and Koenig 2012), filter, measure visual objects, and list certain metadata.

4.5   Audio/Video Authentication Examinations

The following sections provide the protocol details of the scientific analyses conducted during forensic audio and video authenticity examinations. Starting with an overview, the playback, digital data, critical listening, temporal, frequency, visual and other analyses are discussed.

4.5.1   Overview of Examinations

The protocol used when conducting forensic authenticity examinations of digital audio and video files is based on a number of different analysis steps, as applicable, which are generally categorized as follows: (i) hashing and cloning; (ii) playback and conversion optimization; (iii) digital data analysis; (iv) audio analyses (critical listening, temporal, frequency, temporal vs. frequency and miscellaneous); (v) video analyses (visual review, simultaneous visual and aural and miscellaneous); and (vi) preparation of work notes. The overall goals of the authenticity protocol are to determine whether a digital file (i) is either an original/bit-for-bit clone or a converted copy, where an original/clone file is defined as the recording that was produced simultaneously with the occurrence of the acoustic, visual and system information received by the microphone(s) and camera sensors, and not a subsequent copy; (ii) contains discontinuities from the original recording process, including stops, starts, stops/starts and amplitude-activated pauses; (iii) has been altered after the original recording process was completed through digital, analogue, physical or data manipulation editing; and (iv) is consistent with a particular recording system, if that information is available. During these procedures, detailed work notes are prepared and, if requested, a formal laboratory report is provided to the contributor at the end of the examination (Koenig and Lacey 2009; Koenig and Lacey 2012a).

Submitted files that have been re-encoded into the same format or transcoded into a different format usually cannot be definitively authenticated, since many types of alterations or changes could have occurred that are not scientifically detectable. Considerations regarding originality include consistency between the submitted container file format and content streams and those produced by the claimed original recording system; the ability to convert to and from the native and standard file formats; metadata present in the submitted recording and in test recordings produced on the purported original recorder (or a similar model); and so on. In addition, the lack of multiple environments, dropouts consistent with editing, unnatural speech or visual interruptions, phase discontinuities and discrete electric network frequencies (ENF; see Sections 4.5.5.5 and 4.5.6.6) produced by a duplication process may be further indications of originality.

If the file name and a preliminary review of the metadata of an audio and/or video recording reveal no obvious indications of being a converted copy, then the examination continues with additional data, listening, visual, signal analysis and other scientific examinations. The digital data examination consists of displaying the raw digital data to identify the file structure and to determine whether the non-audio/video data contains embedded information pertinent to the production of the submitted recording (such as date, start and stop times of the recording, model and serial number of the recording device, and storage location of the submitted recording on the device). Critical listening encompasses four major steps to ensure that all questionable events in the recording are catalogued and to allow for a smooth transition to the other tests. This is followed by signal analysis examinations, as detailed in later sections of this chapter, including temporal, frequency, temporal versus frequency and spatial analyses. High-resolution temporal representations of sounds in a time-versus-amplitude display can reveal stop and start events, partial dropouts, DC offsets, zero (DC) amplitude areas, questioned transient-like events, repeated sounds, phase discontinuities and variances with exemplar recordings. Frequency displays are useful for analyzing background sounds, aliasing effects, bandwidth, general spectral characteristics and discrete tones. Temporal versus frequency versus amplitude analyses can identify certain voice-activated stops/starts, telephone signalling, speech abnormalities, ENF data, aliasing, editing and discontinuities. The visual review can identify the image rate(s), visual discontinuities, the visual environment, unnatural scenes, questionable transitions and obvious editing. Simultaneous visual and aural analysis can accurately detect synchronization issues between the audio and video information, word changes when facial close-ups are present and the identification of complex environmental events. The miscellaneous techniques include phase continuity, statistical analysis, identification of duplicate and near-duplicate video images, and automated/semi-automated software applications. After all the necessary tests have been performed, conclusions are normally reached regarding the authenticity of the recording based upon the examination of the submitted file, comparison with test recordings, review of appropriate manufacturers' literature and prior experience gained from examination of similar recording systems (Koenig and Lacey 2009; Koenig and Lacey 2012a).

Forensic authenticity examinations are aided by the complexity of editing 'real-world' recordings, compared to professional studio sessions, since the submitted files often contain characteristics such as background voices, music, television broadcasts, traffic noise and other environmental sounds; constant movements of individuals, vehicles or other objects; limited signal-to-noise ratios; limited resolution; obvious non-linear distortion; video and audio compression artefacts; effects of concealed or poorly placed microphones and cameras; sounds of clothing rubbing across the microphone faces; and long-time-delay reverberation. These factors, and others, greatly complicate an editing process, making it difficult or even impossible to add or delete words or visual content in many portions of forensic recordings without producing unnatural effects, audible and/or visible to even a naive listener or viewer. For example, a young woman states 'I buried my husband under the backyard barbeque' in a packed local bar on a Friday night, with a jazz trio playing in the corner, which is recorded on a proprietary, stereo, on-the-body audio/video recorder. If someone then wants to edit in 'Jan' for 'I', consideration would have to be given to not only imitating the particular acoustic environment, background talkers, ventilation sounds, dual microphone characteristics and compression artefacts, at that particular time, but also the exact music and lyrics of the live musical performance and ensure there is proper alignment with the video information; additionally, the editing process must not change the structure and order of the data fields within the header metadata in a manner detectable in a laboratory analysis.

Authenticity examinations cannot always be quantified into a standard set of analysis procedures for a particular case. The analysis requires conceptualizing possible hypotheses for each questioned event or claim of alteration and then through a detailed laboratory process determines which, if any, of the explanations are correct. This process is repeated until the conclusions can be directly supported by scientific tests and digital recorder theory. Areas of interest should always be specifically described in the work notes and, when appropriate, in a laboratory report by their aural, temporal, frequency, visual or other specific characteristics, always avoiding the use of non-specific terms such as 'anomalies' or 'alterations'. Descriptions such as 'suspicious', 'purposeful', 'accidental', 'untrustworthy', or other non-scientific words which reflect the state of mind of the recorder operator or other involved individuals should normally be avoided (Koenig and Lacey 2009; Koenig and Lacey 2012a).

Following are descriptions of many of the specific techniques and procedures used to authenticate digital audio and video files, but the interrelationships between them are too numerous and complex to be completely set forth.

4.5.2   Hashing and Imaging

Upon receipt, all digital media containing audio and video files requiring examination are opened, if possible, in digital data imaging and analysis software, with write blockers if appropriate. Unique numerical representations of the media and/or individual files are computed through the use of a specialized algorithm, which is frequently referred to as a 'hash'. Two of the most common hashes used in the audio/video forensic field are message-digest algorithm 5 (MD5) (Rivest 1992) and Secure Hash Algorithm 1 (SHA1) (Jones, 2001); however, both of these algorithms have been broken using collision search attack methods which found completely different files producing the same hash value (Marcella and Menendez 2008; Sencar and Memon 2013). It is recommended that examiners use SHA-256 (Secure Hash Standard, 2012) or comparable algorithms in authenticity examinations.

Typically, when an audio or video file is submitted on a physical medium (i.e. data CD, DVD, and Blu-ray disc; flash memory card; internal and external computer hard drive), a hash and a bit-for-bit copy (digital data image/clone) are prepared of the entire medium; then the same hash algorithm is applied to the image file, and that result is compared to the medium's hash and verified as being identical. As an example of the robustness of a hashing process, the SHA-256 (256-bit or 32-byte) hash value for a Windows Media Video (WMV) file having a size of 3 395 572 872 bytes, or 27 164 582 976 bits, was computed as the following: 4902475E84DDA6C30F3EC5086E6063A100FBA83406577F84B66EA2855D92EB 07. After changing the value of one byte by one bit (the smallest possible change), the new SHA-256 hash value was computed as the following: 4859699C2A4031B5E 77CBE6E578C201009059244522D243D756F4ED592E1C42B.

Further analyses of the evidentiary recording, as described in the following, are then conducted using the digital data image of the evidence; the submitted digital media is accessed again only in the event that another digital data image or a final hash value is needed at the completion of the examination (Caloyannides 2001; SWGDE best practices for computer forensics 2006; SWGDE data integrity within computer forensics 2006; Marcella and Menendez 2008; Koenig and Lacey 2009; SWGDE and SWGIT 2013).

4.5.3   Playback and Conversion Optimization

Optimal playback and conversion of digital audio and video recordings begins with determining their exact container file format and content stream codecs and identifying the proper software for playback and conversion. Proper playback, by using the appropriate software with the correct settings, ensures that critical listening, visual review and other examinations are conducted with the best audio and video quality with no added playback artefacts. For many of the examinations to follow, conversion of the submitted recording to a standard, uncompressed audio format such as a pulse code modulation (PCM) WAV file, and/or an uncompressed video file format or high-quality, lossy compressed formats such as Digital Video (DV) or M-JPEG, is required for the production of hard copy waveforms, spectrograms, uncompressed/high-quality images, and other scientific charts and graphs. If lossy video compression is used, visual reviews of the resulting file should be conducted to verify that no additional compression artefacts, degradation or other playback issues have been introduced.

Digital audio and video files submitted for authenticity analysis are generally categorized as either a standard or proprietary container format and may contain information encoded through the use of commonly available or manufacturer-specific codecs. Standard audio container file formats include WAV files and common compressed formats/encoding such as MP3 and Windows Media Audio (WMA), and standard video container files include WMV and MOV (QuickTime), which normally do not require the use of specialized playback software (but may require the download of freely available codecs depending on the format of their content streams). Proprietary files include specialized law enforcement, video surveillance and consumer recording formats for which a codec is not always freely available, or which require the use of unique software or hardware for playback, conversion or access. These files may be submitted on a wide variety of media, including data CDs, data DVDs, memory cards, hard drives and non-removable flash memory; in the last case, specialized cables and software may be required to access the recordings contained in the fixed memory of a device. Special considerations often have to be made for digital logging recorders, such as video surveillance and law enforcement emergency hot line systems, which often contain multiple noncontiguous files for a single telephone conversation or a specific video camera's capture.

Fixed memory devices with no digital interface capability are sometimes encountered on older consumer digital audio recorders used for dictation or personal recording purposes and older video surveillance systems. While they may contain recordings in either a standard or a proprietary format, they lack a digital interface which prevents the recordings from being conventionally transferred to external media in their native format. For such devices, the only outputs typically available are analogue line level outputs, headphone connections and/or small built-in loudspeakers for audio and composite or S-video outputs for video. If a direct analogue output from the headphone jack serves as the most optimal output for playback, the recorder's volume controls should be set appropriately such that the output distortion is minimized and the signal-to-noise characteristics are maximized (Koenig *et al.* 2007). If no line output or headphone jacks exist on the device, a hardwired connection to the terminals of the built-in loudspeaker is the only useable choice, but often this method requires that the device be dismantled to some extent, which may not be allowed by the contributor of the evidence. As with the headphone output, the recorder's volume controls may affect the distortion and signal-to-noise characteristics and therefore should be set accordingly.

An 'acoustic' copying process, which generally consists of playing the submitted recording through the built-in loudspeaker of the device and recording this through-the-air signal via a microphone and a separate recording device, is not an acceptable method of playback and recording for authenticity examinations. The acoustic copying process itself will introduce deleterious effects to the content of the produced copy, which can mask indications of editing or conversely add signals that may be false indicators of editing or reverberation. Similarly, the analogue video outputs may also need to be utilized as the most favourable method for video playback output for devices with no digital output capability. Optically rerecording the video displayed on a monitor with another device such as a camcorder is not a suitable method for preserving the evidence for authenticity examinations.

4.5.4 Digital Data Analysis

Digital data analysis is conducted using data imaging and analysis software that allows for the visual review of the actual bits and bytes that comprise both the audio/video and administrative (metadata) information of the submitted recording. This analysis is only conducted using the digital data image, files exported from the digital data image or an intermediate copy that completely captures the native file format. Examinations of re-encoded or transcoded files usually provide limited authenticity-specific information regarding the digital data content of the original recording, because these conversions modify the format of the audio data and change or exclude most of the original metadata (Koenig and Lacey 2012b; Koenig et al. 2014; Koenig and Lacey 2014b). Initially, software applications that list some of the metadata information for specific file formats can be used to provide general information regarding the container file format and audio/video characteristics of a media file, if the format is standard and/or the file consists of recorded video/audio information with standard codecs. However, manual digital data examinations of the file's structure and contents may be required for non-standard files or files containing manufacturer-specific information not otherwise recognized.

The submitted audio/video file is normally analyzed by simultaneously viewing all of the digital data in both hexadecimal and Unicode/ASCII text formats. The first step in this examination is a general review of the file to determine its overall structure (the separate locations of the audio/video data and metadata) and if it is consistent with a standard or known proprietary container format. Since the structure of standard audio and video container formats, for a particular recording system, can be determined from test recordings, changes made by editing programs due to re-encoding are usually readily detectable (Koenig and Lacey 2012b; Koenig et al. 2014). If the file is in an unknown proprietary container format, a detailed examination of its structure is undertaken to identify the subdivisions of audio/video and metadata. Though metadata is usually present at the beginning of a file (the 'header'), it can also occur at the end (the 'footer') and within audio/video blocks. In some digital recordings, the main

Offset	0	1	2	3	4	5	6	7	8	9	10	11	12	13	14	15	
00000000	41	55	44	49	4F	5F	42	45	47	49	4E	53	19	50	5D	AF	AUDIO_BEGINS P]¯
00000016	35	C9	0E	8B	01	61	E0	F8	1C	CA	FA	DD	A9	5C	20	50	5É ▌ aàø ÉúÝ©\ P
00000032	E8	47	C5	76	1C	44	33	2F	45	5E	97	D9	7D	ED	C5	DB	èGÅv D3/E^▌Û}iÅÛ
00000048	FB	EE	E1	BE	18	3A	6F	FF	D8	5C	54	F4	BC	F5	D6	54	ûîá¾ :oÿØ\Tô¼õÖT
00000064	FE	F8	AC	53	6A	31	9C	57	76	DD	49	B7	53	3E	22	10	þø¬Sj1▌WvÝI·S>"
					●							●					
					●							●					
					●							●					
00000496	34	1C	0E	C3	CD	E3	0E	31	D6	4E	68	60	63	8D	A5	95	4 Ãíã 1ÖNh`c ¥▌
00000512	41	55	44	49	4F	5F	42	45	47	49	4E	53	EF	A4	0E	53	AUDIO_BEGINSï¤ S
00000528	CE	93	56	68	FF	9F	E1	CD	BB	02	08	E9	16	4A	2A	4C	Î▌Vhÿ▌áÍ» é J*L
00000544	39	B0	54	B5	E4	E0	19	D9	CC	E7	2F	4D	92	4D	DA	2A	9°Tµää ÙÌç/M'MÚ*
00000560	6E	8D	35	4B	D0	6E	E1	3D	85	F3	A5	41	32	E6	11	FB	n 5KÐná=▌ó¥A2æ û
00000576	28	2F	28	A6	DF	80	A8	F1	B2	07	6C	D7	8C	26	05	19	(/(¦ß▌¨ñ² 1×▌&
					●							●					
					●							●					
					●							●					
00001008	1A	0F	7B	25	62	00	02	D4	B1	E1	1B	F0	DE	2B	38	CA	{%b Ò±á ðÞ+8Ê
00001024	41	55	44	49	4F	5F	42	45	47	49	4E	53	43	A9	7B	E8	AUDIO_BEGINSC©{è
00001040	59	50	95	5C	B0	9E	E5	F1	1F	59	F5	8A	AC	46	BC	34	YP▌\'▌åñ Yõ▌¬F¼4
00001056	66	A7	E8	46	8B	18	4E	4D	84	AB	31	12	7C	18	43	62	f§èF▌ NM▌«1 ▏ Cb
00001072	AA	57	99	68	93	1B	EE	C1	D1	E2	ED	9C	2A	9D	6C	86	ªW▌h▌ îÁÑâí▌* 1▌
00001088	05	2C	C1	C4	8B	24	9A	4D	F9	A5	11	C1	F2	31	D5	03	,ÄÄ▌$▌Mù¥ Áò1Õ

Figure 4.7 Hexadecimal/ASCII display of portions of an audio file, showing a 512-byte pattern of blocks within the file, as represented by the presence of the text 'AUDIO_BEGINS'.

data file contains only, or mainly, raw audio/video information and is accompanied by separate metadata files; telephone logging recorders, proprietary law enforcement and video surveillance systems often utilize this method of storing recordings, giving them an additional layer of security, since modification of a recording requires a corresponding change to the separately stored administrative data.

Identification of the structured, internal arrangement of an audio/video file is typically made by noting the repetition of byte sequences that can be identical, increasing sequentially (such as a time stamp) or in a semi-random pattern. For example, if a 12-byte sequence having a hexadecimal value of 0x415544494F5F424547494E53 ('AUDIO_BEGINS' in ASCII) is present every 512 bytes in a file, it is indicative of an internal arrangement of the data in 512 byte 'blocks', as illustrated in Figure 4.7 (Koenig and Lacey 2009).

If a structured arrangement of the audio/video and metadata is observed, a determination is then made of its consistency throughout the file. Beyond the purely authenticity aspects, understanding the structure of a file may be crucial for properly opening and playing back a recording in standard or proprietary software, since deviations in the required structure may render a file inaccessible in part or in whole and may be indicative of data corruption or alterations.

A confirmation (if already known from other material) or estimation of the data rate of the recording, in bits or bytes per second, can be determined by dividing the total number of bits or bytes attributed only to the audio/video information by the total length of the recording in seconds. For an estimation of the total data rate of the file, the metadata would also be included in this calculation.

Once the overall structure of the data is more clearly understood or confirmed as being in a standard container format, the metadata is further analyzed for the presence of information regarding the production of the recording, such as the (i) start and end date/time; (ii) date/time of each segment of audio/video data; (iii) duration of the recording; (iv) channel configuration (i.e. mono or stereo); (v) audio sampling rate and/or video frame rate; (vi) quantization/encoding scheme and rate; (vii) video pixel dimensions; (viii) group of pictures (GOP) structure; (ix) make, model and serial number of the recording device; (x) case-related information; (xi) caller ID information or telephone number dialled; (xii) total number of recordings on the device at the time of download; (xiii) position of the present recording within the total number of recordings; (xiv) GPS or other location data; and (xv) cyclic redundancy check code (CRCC) or other internal data verification values (Koenig and Lacey 2012b; Koenig and Lacey 2014b; Koenig *et al.* 2014). The identification of these segments of administrative information may not be clear-cut upon initial review of the metadata and may require that a series of test recordings be made with systematically different audio/video record settings, times and dates, using the submitted recording device or one of the same make and model. Analyses of these test recordings may then establish which bytes of information within the metadata relate to the various characteristics of the recording and permit documentation of the hexadecimal values of these bytes as they correspond to changes in a particular recording parameter.

When identified in the metadata, the preceding information can be used to confirm observations made through other laboratory analyses, provide additional information that was not obvious through examination of the audio/video data itself and corroborate or refute information regarding the production of the recording (i.e. dates, times and number of recordings made on the recorder) (Cooper 2005; Lacey and Koenig 2008; Perraud 2008). Analysis of the presence, location and structure of the metadata in test recordings may also provide insight into the consistency between the submitted file and the identified brand and model of the recording device. Additionally, comparisons between the metadata information directly related to the recording device (such as the serial number and header format) may be used to determine if a submitted recording is consistent with having been produced on a particular recording device. To the extent that this metadata may be easily altered irrespective of the video/audio data, the metadata analysis should not be solely relied upon when authenticating digital video/audio files.

In some cases, proprietary audio or video files may contain information encoded with commonly encountered codecs but have file structures which are not standard. For example, a video/audio surveillance system installed in an interview room at a law enforcement agency provides exported files in a proprietary format, but the recorded audio information is encoded as conventional PCM data. The exported files cannot be played back in software such as Windows Media Player because the overall structure of the files cannot be properly interpreted outside of the software supplied by the manufacturer. Through digital data analyses, the segments of recorded audio can be located, extracted into a new file and made playable in standard media playback/editing software (Lacey and Koenig 2010).

A detailed analysis of the audio/video data itself is then performed to reveal indications of alterations, deletions, or additions to the information. Corrupt or unwritten segments of data may also be identified, often by the presence of DC audio samples or broadband noise portions in the audio stream without relation to captured acoustic information, or in blank frames of video with no relation to captured visual information. These corrupt or unwritten segments may not be the result of alteration, but rather a fault of the recording system during the original recording process or subsequent data transfer. When possible, the exact locations within the data corresponding to possible stops, starts or other events, found through other analyses, are determined using the time/sample location of the events and the known or estimated data rate. Based on this information, the relative offsets of the data bits or bytes can be calculated.

Recording devices developed specifically for law enforcement often employ 'integrity' measures within the digital data, which verify the transfer of the recording from the device to a removable medium, validate play back of the removable medium and alert the user when any data has been changed since the transfer. These measures may include non-standard structuring of the data within the file, hash values on a block-by-block level and/or for the entire recording, date/time stamping, the presence of separate metadata files, file name hash verification, watermarking and proprietary encryption/decryption by the manufacturer. The alteration of even one bit in these types of recordings may result in an access or playback error, signifying that the recording has been modified from its original state. Watermarking refers to a process where the original digital audio and/or video information is altered slightly with embedded data, which can subsequently be extracted to aid in the authentication of the recording. One of the common techniques is called a fragile watermark, where the embedded data is unrecoverable if the file has been changed in any manner (Cox *et al.* 2002). Proprietary encryption of a file requires a decryption key for playback or accessing the file, thus limiting the individuals who can listen to/view the file contents. Most of these law enforcement-specific recorders, and some consumer units, possess a further level of integrity, given that there is no existing capability for the direct digital conversion of standard files (such as PCM audio WAV files) into their proprietary file formats. This means that the proprietary recordings cannot be converted to a standard file format, altered through the use of conventional audio/video editing software, converted back into the proprietary format and then represented as being original, unaltered recordings. Other than proprietary file formats, consumer recording devices seldom employ 'integrity' measures, making them more susceptible to partial data deletion, movement of data within a file and other changes.

In an effort to discern whether or not 'integrity' measures are incorporated into a proprietary digital audio/video file and to observe the effects of breaching those measures, test recordings or separate bit-for-bit copies of the submitted file can be subjected to modifications, ranging from changing the value of one bit within the file to deleting or moving large numbers of bytes within the file. Attempts to access and play back the modified versions of the file are then made and the changes documented. The effects of these changes may include complete or partial inaccessibility to the file; sudden stops in the linear play back of the file; the addition of transient sounds, dropouts, distortion, or other audible indicators at the points of modification; the addition of visual artefacts or video playback issues at or following the points of modification; and error notifications in the proprietary playback software (Koenig and Lacey 2009).

4.5.5   Audio Analyses

The following sections include details of the four steps used in critical listening; production of waveform, FFT and spectrogram displays; use of ENF discrete frequency tones for comparison to known databases; and miscellaneous audio analyses.

4.5.5.1   Critical Listening

After creating a digital data image, optimizing playback/conversion and conducting at least a preliminary digital data analysis, the next examinations are often listening tests when audio is present in the submitted file, using the bit-for-bit copy or an optimally-made copy/conversion prepared in the forensic laboratory. These aural evaluations, though not conclusive, can provide considerable direction in locating areas requiring specific signal analysis or additional digital data review. These tests are first performed without digital processing, using professional headphones and high-quality playback systems, external computer sound cards or external devices connected to the computer via high-speed cabling; because of internal computer noise (discussed in Section 4.4.1), a direct review via internal sound cards is not usually recommended. If needed, additional reviews are conducted with appropriate enhancement procedures, which are set forth in the references (Koenig et al. 2007; Koenig and Lacey 2010). Loudspeakers are not utilized since even the best playback environments will convolute the recorded audio characteristics. Other advantages to the use of headphones are better isolation from laboratory background noise, more accurate amplitude and frequency characteristics that are not dependent on the examiner's head position, stable stereo imaging and identical inputs to both ears with monaural recordings (Berriman 2000; Koenig and Lacey 2009).

    The critical listening analysis has four separate but overlapping areas of interest: (i) preliminary overview, (ii) possible record and edit events, (iii) background sounds and (iv) foreground information. In the preliminary overview, the entire file is reviewed, or in the case of formats that contain an extensive amount of recorded material such as digital audio or video loggers with multiple channels of information, each lasting for weeks or even months, one or more designated recordings or portions identified by the contributor are reviewed. In this review, a chronology, or time log, is begun to generally describe the recording conditions and to list pertinent events that will be of assistance in other examinations; normally, the log will list the elapsed times referenced to the beginning of the recording or other convenient orientation point, usually in an hour:minute:second.millisecond format (e.g. 1 h, 14 m and 6.235 s would be 1:14:06.235) or sample location (e.g. 1 h, 14 m and 6.235 s would have a sample location of 35,569,880 at a sampling rate of 8000 Hz). Thereafter, in the authenticity analysis, all events in question are normally referred to by their time or sample location, and not by real-time 'counters', or other playback recorder or software timing. Events often noted include obvious digital record stop and start occurrences, analogue-like record events (from an earlier or intermediate copy of the original recording), environmental changes (Howard and Angus 2009; Kuttruff 2009), unrecorded portions, playback speed fluctuations, louder transients with fast rise times, obvious background noise differences and other audible variations in the character of the recording that will require further study. This preliminary aural overview is an important first step, since it often limits the portions of the file that require the most detailed examinations; allows the identification of obvious record stops and starts; identifies acoustic environments contrary to the information provided by contributor; and guides additional authenticity analyses.

The remaining three aural tests are conducted, and repeated, at various times and in no particular order, both before and after other appropriate analyses, depending on the information developed in the overview, the type of alterations alleged, the signal analysis findings and the preference of the examiner. The record and edit events examination requires prior listening to known stops and starts on the evidence media, test recordings prepared on submitted recording equipment and/or a general knowledge of editing functions and their results using consumer and professional editing software. Substantial experience in examining record events from a wide variety of other systems and identifying known edits on previously submitted evidence media is also essential. This critical listening test and the signal analyses permit close scrutiny of the signatures left by known digital recorders and editing processes during stopping, starting, sound activation, adding, deleting, file combining and other functions. These signatures can vary considerably in amplitude, shape and duration, not only between different formats and recorder brands but sometimes between identical units. Any complete dropout or non-environmental transient-like signal, whether consistent with a record or editing event or not, is listed on the chronology for further study, unless it is known to be innocuous (Koenig and Lacey 2009).

Analyses of background sounds include documenting unnatural sounding changes in the background; repetitive noise sources; music, recorder, and system noise; subtle or obvious changes in the digital noise floor; interruptions in background conversations (Li and Loizou 2007); the presence of low-level transient-like signals of unknown origin; multiple environmental characteristics; low-level dropouts; incongruous microphone-handling sounds; 50/60-Hz hum or buzz on battery-operated recorders; and the absence of expected environmental or transmission channel effects, reverberation and other acoustic sounds. Examples of possible concerns that would be noted are a ballad that only lasts 30 s on a radio playing in the background; a background voice cut off in midsentence; an abrupt but subtle change in the level of the background noise or the recorder noise floor; a series of unexplained clicks in a quiet room environment between spoken speech samples; no ringing, hang-up or other telephone signals or noises on a supposedly complete telephone conversation recording; and conversely telephone switching sounds within a supposed room recording made using a concealed miniature recorder (Koenig 1990).

The last general area of aural review is a detailed examination of the foreground information, including pertinent voices, the effects of digital compression, obvious bandwidth limitations, pre- and post-conversational narratives by the operator, contextual information and overriding RF transmissions. Examples of aural events requiring further signal analysis examinations include sudden, unnatural or linguistically peculiar changes in an individual's voice or cadence; inconsistency between the beginning and end times given by the recorder operator and the actual length of the recording; an abrupt and unexplained change in the topic of conversation; and digital compression artefacts and bandwidth characteristics that are inconsistent with the recorder alleged to have been used to make the recording.

The critical listening examination provides unique information in an authenticity determination due to the capability of identifying pertinent sounds in the midst of extraneous signals and noise. The examiner's experience of having reviewed numerous investigative, business, administrative and test recordings, in general and for the particular case, provides insight and direction for the various signal analysis examinations. In addition, during many of the temporal and frequency examinations, the examiner can both listen and view the real-time displays for direct aural-to-visual correlations.

4.5.5.2 Temporal Analysis

Temporal, or time-based, analyses are usually conducted with high-resolution waveform displays graphically reflecting the relationship between the time and amplitude of recorded sounds. These analyses allow for the exact timing of events identified through critical listening or other examinations, the identification and comparison of both digital and analogue record events on evidence and test recordings, the minute scrutiny of dropouts and transient-like sounds, the detection of DC offsets (Koenig *et al.* 2012a, 2013; Koenig and Lacey 2014a) and overdriven portions (such as clipped peaks), the determination of electronic signal return times (SRT), the identification of areas containing zero digital amplitude, the detection of certain editing within words or the noise floor and so on. One of the strengths of this analysis is its capability to accurately determine the relative temporal and amplitude characteristics of a digital file, ranging from a single sample point to the entire digital recording. During and after the critical listening tests, the on-screen waveform displays and printed graphs are used to identify the exact timing of recorded starts and stops, over-recordings, pertinent dropouts and transient-like events, and other sounds of interest.

While most digital systems do not produce waveform signatures when they stop and start in the record mode, a few do, though the sounds are normally of shorter duration compared to analogue events. Test recordings prepared on submitted recorders can be compared to the evidential recording for similarities and differences in timing, shape, spacing and rise times for the signatures. Record events produced on consumer-quality recorders can vary considerably among different manufacturers, models and formats and sometimes between units with closely spaced serial numbers. Even identical record-mode operations will, at times, result in slightly different waveform shaping for the same recorder. Dropouts, transient-like events, clipped peaks and zero-digital-amplitude areas are clearly detailed by the waveform analysis. Figures 4.8, 4.9, 4.10 and 4.11 illustrate waveforms of clipped peaks, a zero-amplitude (DC) area, DC offset and variable DC offsets in a test recording from a small solid-state recorder (Koenig and Lacey 2009).

The waveform measurement of SRT may reveal how long certain recorders have been turned off, prior to the start of a recording. The time difference between the start of the file and the beginning of the recorded information from a microphone or other sources is called the SRT (Koenig 1990). This is based on the circuit characteristics in many recorders, which produce a short time delay between activating the record mode and actually recording the input signal, producing a zero, or nearly zero, amplitude at

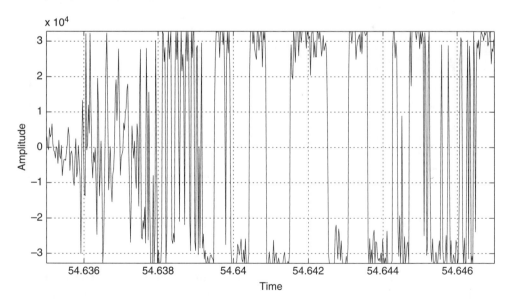

Figure 4.8 High-resolution waveform of 12-ms segment of a digital audio recording including clipped peaks.

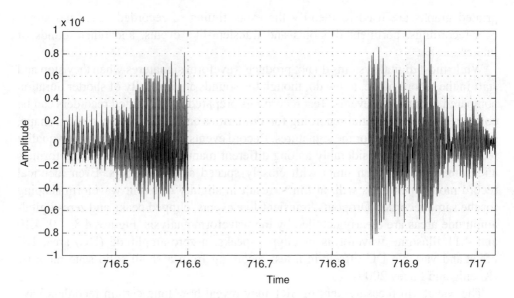

Figure 4.9 High-resolution waveform of 0.6-s segment of a digital audio recording including a 0.25-s portion of zero-amplitude (DC) samples.

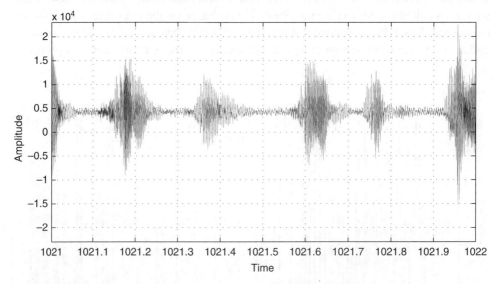

Figure 4.10 High-resolution waveform of 1-s segment of a digital audio recording reflecting a DC offset.

the beginning of the file, prior to the start of the recorded information. This delay can be directly affected by the length of time the recorder was turned off, with the longest delays for extended deactivation.

If a digital or other type of edit is made within a spoken word, the waveform display can often reflect the abrupt effect of the alteration, especially during vocal pitching.

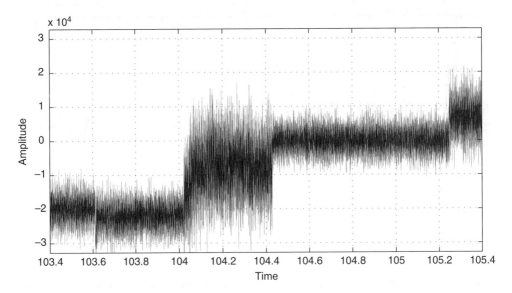

Figure 4.11 High-resolution waveform of 1-s segment of a digital audio recording from a small solid-state recorder, reflecting a variable DC offset.

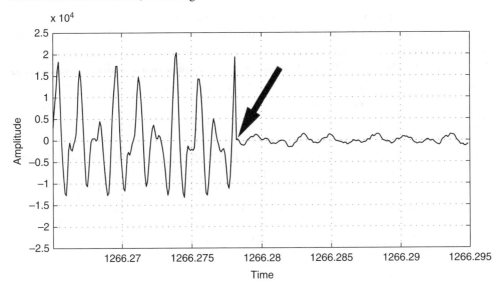

Figure 4.12 High-resolution waveform of 30-ms segment of a digital audio recording reflecting a digital edit during a spoken word, indicated by the arrow.

Figure 4.12 reflects a digital edit from a professional editing program during a spoken word. Another use of waveform analysis, during a suspected edit, is the identification of a slight change in the noise floor between words. As an additional example, Figure 4.13 shows a subtle change in the noise floor at an edit point approximately centered in the waveform display (Koenig and Lacey 2009).

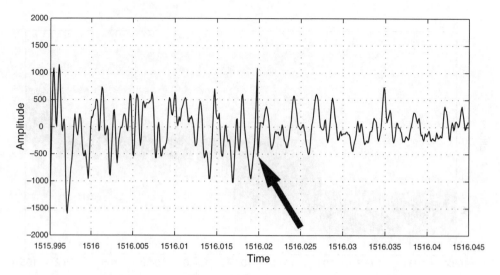

Figure 4.13   A 50-ms, high-resolution waveform of a subtle change in the noise floor at the location of a digital edit, indicated by the arrow.

DC offsets occur in audio recordings when one or more components, such as microphones, microphone preamplifiers and record circuitry, add DC voltages to the audio signal, resulting in a recorded waveform whose overall average amplitude is not the x-axis or 0. Figures 4.10 and 4.11 illustrate examples of DC offsets. Research to date in the use of average DC offsets, to identify the originating digital recorder, has found that only exemplar recordings (i) produced with the same recorder, (ii) lasting approximately 15 min or longer, (iii) having consistent audio information, (iv) using identical source and microphone locations and (v) created in the same acoustic environment allowed reliable comparisons with the audio information in the submitted file. However, except for certain older audio formats and recorders, the average DC offsets are usually quite small and have similar values between different recorders and audio formats. Therefore, the use of averaged DC offsets should only be used to exclude a digital recorder as the source of a submitted file in an authenticity examination (Fuller 2012; Koenig *et al.* 2012a, 2013; Koenig and Lacey 2014b).

4.5.5.3   Frequency Analysis

Most audio frequency analyses in authenticity examinations are conducted using narrow band spectrum transforms, principally FFT, which graphically show the overall frequency characteristics and discrete tonal signals recorded on evidential and test recordings. With this analysis technique, increasing the frequency resolution always decreases the time resolution and vice versa; therefore, with an understanding of this relationship, the display can be optimized for the specific questioned sound or event (Wallace and Koenig 1989; Bracewell 2000; Owen 2007; Koenig and Lacey 2009). Examinations include analyses of digital aliasing and imaging artefacts, sample frequency bandwidth, questioned signals, background sounds, convolutional and transmission characteristics and power-line frequency components. Digital aliasing/imaging occurs when acoustic information exceeding the Nyquist frequency is mixed in with the recorded audio due to the absence of or an improperly configured antialiasing/imaging filter during the original recording or a subsequent duplication process. This aliasing deteriorates the quality of the audio information because of the folding back of signals beyond the Nyquist frequency of the recorder into the passband, effectively adding them to the original signal (Pohlmann 2011). Additional corruption of a digital audio recording may occur when an anti-imaging filter, designed to prohibit information exceeding the original bandwidth of a recording from passing, is not used during a digital-to-analogue conversion or resampling process. Figure 4.14 shows the effect of improperly resampling an 8-kHz digital audio recording to 44.1 kHz without employing an anti-imaging filter, which clearly reflects the effects as additional mirrored frequency information beyond the original 4-kHz bandwidth. From an authenticity perspective, aliasing present on a submitted file could have been produced in the original recording process, during a duplication step or by the improper playback of a converted or processed digital file. Likewise, a frequency

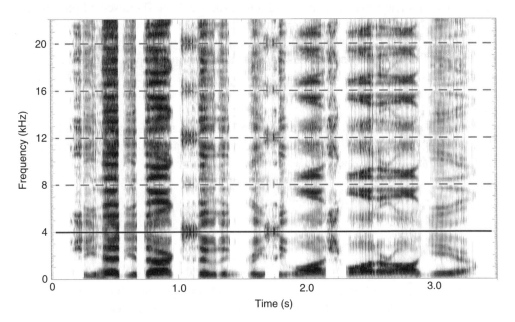

Figure 4.14 Sound spectrogram of an improperly resampled recording, having a time frame of 3.5 s, bandwidth of 22.05 kHz, analysis size of 200 points (323.00-Hz resolution), 0.8 pre-emphasis and a dynamic range of 20 dB. The black horizontal line represents the approximate Nyquist frequency of original 8-kHz recording, and the dashed horizontal lines represent the axes on which the false information is mirrored (at 8, 12, 16, and 20 kHz).

bandwidth that is inconsistent with the original recorder's sampling frequency could have been produced by a recorder defect, a duplication process or the improper playback of a converted or processed digital file (Koenig and Lacey 2009).

Analyses of questioned signals, background sounds and convolutional and transmission characteristics are conducted by comparing the frequency spectra of the events occurring on the submitted file to known sounds and system features. A waterfall display can be useful if the frequency characteristics change over time, in either a periodic or random fashion. If a particular signal or background sound in a recording is a point of contention, test recordings of the equipment at the site can be made and compared. Convolutional and transmission characteristics are normally revealed by long-term spectral averaging, which minimizes the effects of voice and other dynamic signals. They are then compared, in a general way, to the known effects of such systems. For example, a recording allegedly made on an inexpensive solid-state recorder at its lowest-quality setting [e.g. at a 5.0-kilobit-per-second (kbps) data rate], which has a bandwidth out to 8 kHz with no obvious aliasing artefacts, would be considered questionable because such units typically have sample rates at or below 8 kHz (producing maximum bandwidths at or below 4 kHz).

If a battery-operated on-the-body recording contains a power-line frequency tone (nominally 50 or 60 Hz, depending on the country), or one or more of its harmonics, this could be indicative of environmental noise, such as from fluorescent lights or of a duplication process. If the recording process takes place in more than one environment, a determination is made as to whether the power signal is present outdoors, is present in different locations and has consistent or varying amplitude in the different environments. If multiple power-line signals are present on a digital recording, then environmental sources, system noises or duplication processes could be responsible (Koenig 1992).

Frequency domain transforms other than FFT analyses, such as wavelet, Laplace and Z transforms, are sometimes utilized for authenticity examinations when specialized displays or analyses are needed. These other transforms are beyond the scope of this chapter, but there are many excellent references on the subject (Graf 2004; Mallat 2008; Widder 2010; Najmi 2012).

4.5.5.4 Spectrographic Analysis

The most common technique used in this genre is the rather distinctive time versus frequency versus amplitude displays on a sound spectrogram, based on the FFT transform, which can reveal the effects of aliasing (see Figure 4.14), certain record events, editing and varying discrete frequency components. Since increasing frequency resolution decreases time resolution (like with narrow band spectrum analysis), and with the numerous user-controlled parameters available in the software, this analysis can be optimized for specific audio events. Figure 4.15 is an example of a spectrogram showing a digital edit within voice information, identified by the arrow; this abrupt

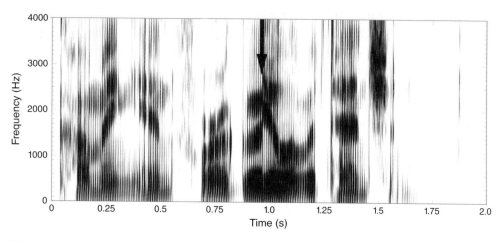

Figure 4.15 Sound spectrogram of digital edit made during male voice information, as indicated by black arrow. Duration 2 s; bandwidth 4 kHz; analysis size 75 points (312.50-Hz resolution); 0.8 pre-emphasis; and dynamic range of 20 dB.

change in the spoken vocal resonances, shown as black bands, is anatomically impossible since the human vocal cords cannot change their vibration frequency that quickly (Koenig and Lacey 2009).

4.5.5.5   ENF Analysis: Audio

When audio or video recordings are produced using equipment powered by AC or with battery-powered devices in close proximity to AC-powered equipment, a corresponding ENF may be induced into an audio recording or onto the audio track of a video file, even in cases where there is no audio input to the recording system. Nominally, these ENF values are either exactly 50 or 60 Hz, depending on the operating frequency of the power grid where the recording is made, but in reality, small frequency variations, produced by the electric utility service, occur randomly over time due to changes in the amount of power being supplied and used. These variations can be characterized through detailed examinations using one of three generally accepted methodologies, which often use signal analysis techniques to isolate the ENF frequency: (i) initial time domain measurements of the audio and then a conversion to the frequency domain, (ii) all frequency domain measurements of the audio and (iii) a time-frequency analysis of the audio where the frequencies are measured using short time periods/windows. The best protocol for a particular recording is dependent upon its quality, the amount of variance, the particular examiner's preference and other factors. The results of these examinations are then compared to those of similar analyses conducted on a verified ENF database for the same network, date and time of the submitted recording. The examiner aligns the ENF signal of the submitted recording with the corresponding ENF database signal, if possible, and then compares the two signals over time, documenting frequency and temporal differences between them, which may indicate the presence of discontinuities or alterations in the submitted recording. In addition, it may be possible to identify the date and time of a recording, to characterize the power source (electric network, uninterrupted power supply and so on) and to detect simultaneous ENF sources, indicating duplication. The applicability and veracity of ENF analysis is highly dependent on the determination of and access to the proper network database, the identification of the approximate date and time of the recording independent of any embedded date/time information (which may not be accurate), the signal-to-noise ratio of the ENF signal in the submitted recording, and, obviously, the presence or absence of a recorded ENF signal (Grigoras 2005, 2007, 2009; Brixen 2008; Cooper 2008; Koenig and Lacey 2009; Nicolalde and Apolinario 2009; Rodríguez *et al.* 2010; Archer 2012; Coetzee 2012; Grigoras and Smith 2012; Jenkins *et al.* 2012; Liu *et al.* 2012; Nash *et al.* 2012; Ng 2012; Ojowu *et al.* 2012; Yuan *et al.* 2012a,b; Sanaei *et al.* 2014).

4.5.5.6 Miscellaneous Audio Analyses

Critical listening, temporal, frequency and temporal versus frequency are the most common signal analysis laboratory procedures involving digital audio authenticity determinations (and for ENF examinations in video authenticity analyses – see Section 4.5.6.6). This section will include some examples of other techniques that do not fall neatly into the previous areas, but are useful in some cases. A complete listing is not possible since the number of potential signal analysis procedures is large; examination methods are constantly being improved and expanded by private and governmental forensic laboratories and the appropriate scientific and engineering community; emerging digital audio formats require new procedures; and unique tests will constantly need to be developed for particular forensic problems. Included in this section are descriptions of phase continuity, statistical analysis, and automated or semi-automated software applications.

The examination of phase continuity involves the direct comparison of higher-amplitude discrete tones in a submitted file, with a corresponding reference signal produced in software or hardware. A stand-alone phase metre, waveform analyzer or equivalent software application is utilized to display the value of the angular differences between the two signals over time or to produce direct statistical comparisons between them. The most commonly used discrete signal for phase continuity analysis is the ENF (50 or 60 Hz), but any discrete signal, whose frequency characteristics are known, can be used for this analysis. This technique can often identify edits, record events and other changes in continuity when there are abrupt changes in the phasing between the evidence recording and the reference. The usefulness of phase continuity analysis can be limited by poor signal-to-noise ratios and the instability of the recorded discrete signal (Koenig and Lacey 2009).

Statistical analyses can provide information regarding patterns or numerical relationships observed within the raw audio data or metadata contained in a digital file. The identification and consistency of repeated bytes of data and their relative offsets may be useful in determining whether individual time stamps, for example, are included in the data stream. These analyses not only provide information about data within a recording but can be used as points of comparison with other evidential and test recordings produced on submitted devices (Aitken and Taroni 2004; Lucy 2005; Mendenhall and Sincich 2006) and provide possible microphone identification (Malik and Miller 2012).

Automated or semi-automated software analysis tools, which make overall authenticity decisions regarding a file, may be utilized to supplement the standard analyses performed during an audio authenticity examination. Prior to their use in forensic examinations, these software tools should be formally validated regarding their accuracy through in-house testing, acceptance by appropriate standards organizations and/or peer-reviewed scientific research. These tools typically provide for faster-than-real-time analysis of the phase continuity of a recording, identification of changes in the background noise characteristics, tracking of discrete signals throughout a recording, detection of prior digitization or aliasing artefacts, and so on. While these analyses are typically performed during the conventional steps listed previously, these automated or semi-automated tools may provide for a more expedient method of analysis. However, the overall authenticity results of these automated tools should not be accepted without direct verification through other accepted analyses, since these automated algorithms can miss obvious alterations in the file, but include numerous false positives, depending on the settings of the various sensitivity options used with each analysis. For example, a recording containing severe aliasing artefacts may be identified incorrectly as having been previously digitized at a sampling rate higher than its original rate. If the results of these automated or semi-automated systems are not scrutinized very carefully and methodically, plus verified with other analysis methodology, they should not be accepted at face value (Marr and Snyder 2008; Koenig and Lacey 2009).

4.5.6 Video Analyses

The following sections include details of the visual review; simultaneous visual and aural review; multiple compression/quantization detection; duplicate frame/region detection; photo response non-uniformity (PRNU); and ENF video analyses.

4.5.6.1 Visual Review

Preliminary visual reviews are initially conducted of video files to determine or confirm various characteristics of the recorded information. These features include image capture rate, colour/greyscale, pixel dimensions and recording length, which may have been provided in previous digital data analyses or through specialized media information software. Overlaid text information, embedded into the video data by the original recording device, may be present, for example, in surveillance systems and dashboard cameras in law enforcement vehicles. This information often consists of date, time, camera name/number, location and vehicle-specific information such as car number, speed, emergency lights indicator, brake lights indicator, siren indicator, GPS readings and active microphone(s) indicator(s). Such information is noted through the visual reviews and may be used to corroborate other information. For example, if a video file contains embedded time information, the differential between the embedded end time and start time of the recording can be compared with the inherent length of the video file. If there is a discrepancy between them, then the recording may be discontinuous or have been altered. However, if the recording is time-lapse (i.e. having a lower frame rate than the 'real-time' standard definition video rates of 29.97/30.00 frames per second in the NTSC system and 25 frames per second in the PAL system), has a variable image capture rate and/or was recorded in a motion-activated mode, then the time differential between the start and end times and the length of the recording will not always be consistent.

Non-linear video editing software is usually employed to conduct these preliminary visual reviews, though standard media playback software may also be used. The timeline of the non-linear video editing software should be set to the appropriate number of frames per second of the recording to be analyzed, if possible. In cases where the video file is proprietary, use of the software supplied by the manufacturer may be required. For detailed image-by-image reviews, which often reveal subtler details regarding a recording's continuity, the non-linear video editing software should be used, when possible, as they feature greater flexibility when it comes to identifying slight visual changes between consecutive video frames.

Obvious visual discontinuities in the recording are identified through both the preliminary visual reviews and detailed image-by-image reviews and are documented in the work notes typically by their time location in hours, minutes, seconds and frames, relative to the start of the recording. Other identifying information, such as the embedded date/time information, may also be included for ease of location without using a reference timeline. Instantaneous changes in the scene (e.g. an interior environment shifting immediately to an outdoor environment); the sudden appearances/disappearances of individuals or objects; or abrupt changes in the movement or positions of individuals or objects are noted, as they may be indicative of discontinuities or alterations to the recording. Colour shifts, changes in video resolution, frame rate changes, identical or nearly identical frames, localized variations in the flow of the embedded date/time information and inconsistencies in the content of consecutive frames (e.g. light reflections, shadows, improbable body movements) are noted for further review and analysis.

It should be noted that video compression, employed by virtually all digital recording devices in the consumer and law enforcement communities, may produce visual and temporal artefacts which are not, by definition, alterations made to the recorded content. They are characteristics of the original recordings and do not represent modifications made after the fact. While such artefacts may be noted by the examiner during the visual reviews and other analyses of the recorded video, their presence alone should not be taken as indications of alterations. Test recordings made using the alleged original recording device, or one of the same make and model, may assist the examiner in discriminating between the inherent compression artefacts and record events indicating discontinuities and alteration.

4.5.6.2   Simultaneous Visual and Aural Analyses

Simultaneous visual and aural reviews are conducted in addition to independent analyses of the audio and video information, as detailed elsewhere in this chapter. Synchronization, or a lack thereof, between recorded audio and video can be affected by a number of factors within the capture, encoding, transmission, and transcoding phases, and may also be an artefact of editing one or both of the data streams. Offsets in synchronization may occur as a fixed length throughout a recording, linearly change as the playback progresses or vary from moment to moment as a result of variable video and/or audio encoding. During the analysis, the examiner notes any offsets present, describes their characteristics and indicates which data stream leads or lags the other along with the approximate offset in seconds/milliseconds. Since the temporal resolution of video content is far coarser than digital audio sampling rates (e.g. 30 frames per second compared to 44 100 samples per second), the precision of the approximate offset will be limited to the frame rate of the captured video. If the alleged recording device is also submitted for analysis, similar simultaneous reviews of test recordings made on the device can be compared to the evidentiary recording.

4.5.6.3   Multiple Compression/Quantization Detection

When a compressed video recording is re-compressed, or when two video recordings are composited together, statistical artefacts may be introduced into the video frames. These artefacts can be quantified and utilized as a measurement for the detection of video transcoding/re-encoding or alteration. When frames are removed or added, for example, from an MPEG-compressed video file, the structure of the GOP will become disrupted and may result in greater motion estimation errors. Additionally, the original intra-frames or I-frames (i.e. those frames encoded without regard to other frames within the video sequence) may undergo multiple compression steps which would manifest themselves as periodic artefacts in the images' histograms (Wang and Farid 2006, 2009).

4.5.6.4   Duplicate Frame/Region Detection

A video recording can be analyzed visually and through automated means to detect the presence of duplicate frames or regions within frames. For example, a deceitful person may wish to replace video frames in which an individual is seen walking through the scene with frames within the same recording in which the individual is not present. Similarly, an image region where a handgun is seen sitting on a tabletop may be modified through regional duplication processes, to remove the handgun from the scene. These types of analyses may be visually obvious through the frame-by-frame analyses but may be subtle enough to require mathematical correlations to be performed on entire frames or in subsets of the frames (e.g. blocks of 256 pixels by 256 pixels) (Wang and Farid 2007). Alternatively, frames can be compared pixel for pixel in image editing software such as Adobe Photoshop CC or other image processing software. It should be noted that in some cases, identical or nearly identical frames may be produced as part of video compression algorithms or other recording system functions and may not necessarily be indications that the recording has been altered or transcoded. Figure 4.16 illustrates one instance of duplicated frames introduced by a digital video/audio recorder in a cyclical pattern, which is inherent to the original recordings (Koenig *et al.* 2012b; Lacey and Koenig 2012; Maloney 2013).

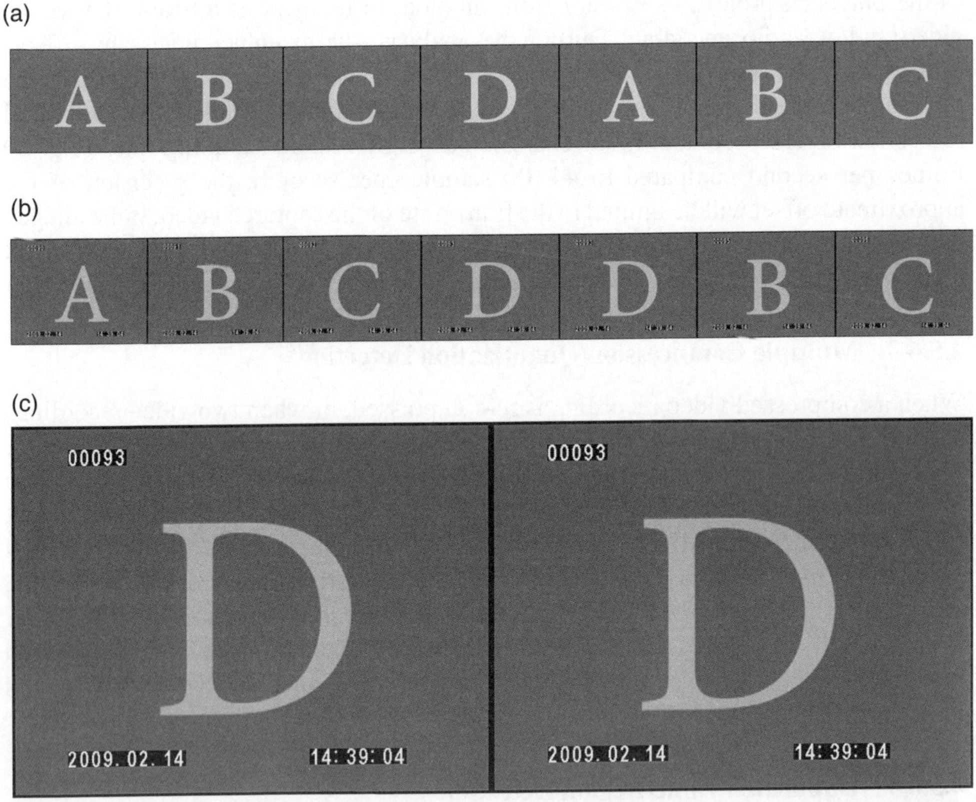

(a)

(b)

(c)

Figure 4.16 Seven-frame segments showing (a) the input video signal and (b) a recording of same on a Lawmate PV-500 digital video/audio recorder. Note that the second 'A' frame of the input sequence is replaced by an identical 'D' frame in the recording. The enlarged 'D' frames (c) show the same frame number (upper left corner) and date and time stamp in each frame, as overlaid by the recorder.

4.5.6.5   PRNU Analysis

Digital imaging sensors contain inherent variability in the light sensitivity of individual pixels, due to impurities in sensor components and imperfections introduced by the manufacturing process. This variability exhibits itself as a noise pattern within the images that is both time invariant and unique for each imaging sensor. By quantifying and comparing these patterns from multiple video frames extracted from both the evidentiary recording and test recordings of the alleged recording device, an examiner may be able to determine if the patterns are consistent and therefore whether they match the alleged recording device. Additionally, a recording may contain segments of video frames for which the PRNU analysis reveals inconsistent patterns, possibly indicating the use of more than one recording device or alteration of the recording produced by the insertion of video segments captured on a different device (Chen *et al.* 2007; Fridrich 2009; Van Houten and Geradts 2009).

4.5.6.6   ENF Analysis: Video

Preliminary research suggests that the fluctuations of the ENF signal in recorded audio discussed previously may also be detectable as corresponding intensity variations in fluorescent and incandescent light bulbs. When measured by optical sensors or recorded to video, these variations can be mapped to a frequency versus time representation, which can then be compared with a known ENF database, or with the ENF signature from the corresponding recorded audio from the file under examination. If both the audio and video content were recorded simultaneously, then their ENF signatures should exhibit a high level of consistency; inconsistent signatures may indicate that the audio and video were recorded at separate times and then later multiplexed together in an effort to make them appear as contemporaneous recordings. Lossy video compression (e.g. MPEG family) can negatively affect this visual ENF extraction, especially at lower bit rates (Garg *et al.* 2013).

4.6   Preparation of Work Notes and Laboratory Reports

Detailed typed or handwritten work notes are prepared of the major facets of the authenticity process, normally with the pages consecutively numbered. The notes usually include a description of the submitted evidence; chain-of-custody changes; the playback software or devices utilized and their settings; hash values; the length(s) of the submitted recording(s); a chronology; the lists of time, frequency and other charts/graphs, including their parameters; aural, visual and signal analysis observations; the overall examination results; and any instructions that should be provided to the contributor.

When requested, a formal laboratory report is prepared and forwarded to the contributor, listing, as a minimum, the report date, contributor's name and address, laboratory number, specimen number(s), evidence description, evidence receipt date(s), types of examinations conducted, results of the examinations, examiner's name and disposition of the originally submitted evidence. If record events are identified, they should be listed by time (usually from the beginning of the file) and type (e.g. start, stop or pause stop/start). If the authenticity examination found no evidence of alteration, but the format is such that certain specific types of changes could be made to the evidence without being scientifically detected, then the report should list those results, along with a detailed set of statements reflecting the conditions, equipment, expertise and accessibility to the evidence and equipment needed for an undetectable alteration to exist. If events are identified that are partially or completely unexplainable, then they should be listed as such. Identification of events, indications of duplication, discontinuities not matching the submitted recorders, over-recordings and other scientific findings should not be described as 'suspicious', 'likely alterations', 'anomalies', 'casting serious doubt on the authenticity or trustworthiness of the recording', 'not reliable forensic evidence', or other non-specific, negative representations. In addition, the report should never include opinions regarding the admissibility of the evidence or any other legal matters, unless requested by the client to comment on a very specific technical threshold or standard set by a legal precedent, statute or opinion. Comments regarding the state of mind of the recorder operator or the producer of an alteration or other changes to a recording should normally not be included in a report. Exceptions to the aforementioned limitations can occur when hypotheticals are required which are partially based on depositions or other legal written statements. A well-prepared, professional laboratory report should also not include pending billing information; advertising; a statement regarding the acceptance of credit cards; company brochures; religious material; and newspaper or magazine articles unrelated to the examination. Whenever possible, laboratory reports should be reviewed by a second qualified examiner to check for technical and grammatical shortcomings (Koenig and Lacey 2009).

4.7 Expert Testimony

After completion of the authenticity examination, the examiner will, at times, be asked or required to provide expert testimony regarding the scientific analysis in an appropriate legal proceeding, such as a trial, hearing or deposition. To allow the most accurate and meaningful testimony, the examiner should normally prepare a qualification list or a curriculum vitae (CV), have a pre-testimony conference with the attorney, present a proper appearance and demeanour at the proceeding and verbalize the important aspects of the examination and chain of custody in an understandable way to the judge, jury or deposing attorney. A qualification list in a trial or hearing allows the attorney to ask all the appropriate questions during direct examination to reflect the examiner's training, experience, education, peer-reviewed scientific articles, membership in professional societies and so on. The list often includes the following, as appropriate: (i) present title, organization, responsibilities and length of service; (ii) pertinent prior employment information; (iii) details of apprenticeship programs completed in the forensic authenticity field, including their length(s) and the exact nature of the training; (iv) formal college and university degrees and additional college courses; (v) pertinent technical schools, seminars and specialized classes; (vi) membership in pertinent professional societies and attendance at their meetings; (vii) peer-reviewed scientific publications directly related to or supportive of authenticity analyses; (viii) number of times previously qualified as an expert in legal proceedings; (ix) approximate number of different matters in which forensic audio/video examinations have been conducted; and (x) approximate number of different evidential recordings analyzed. A CV, instead of a qualification list, is often required in depositions and certain other legal and administrative proceedings; it is usually a narrative document containing the preceding information but in a more detailed format.

A discussion of the authenticity examination with the presenting attorney prior to testimony is useful for explaining laboratory methodology, setting forth the scientific results, providing the qualification list or CV and determining the specific questions that will likely be asked of the examiner. The attorney may provide guidance on local court procedures, the questions expected from opposing counsel and exactly when the testimony will be needed.

An examiner should always dress in proper business attire and direct explanations to the jury, when present, to allow feedback on their understanding of the answers. The examiner should maintain a proper demeanour under the stress and distractions of, for example, highly agitated opposing counsels, interruptions by court reporters, lengthy depositions and inattentive jury panel members (Koenig and Lacey 2009; Matson 2012).

4.8   Case Examples

To provide insight into how actual authentication examinations are conducted of digital video/audio files, two very short, fictional examples are detailed in the following. The first is a preliminary authenticity review of a Windows Media Audio (WMA) file to determine its originality, its consistency with its alleged recording device, and if it contains obvious alterations; the second sets forth an authenticity analysis of a proprietary video file and a transcoded copy of the file.

4.8.1   Case Example Number 1

The laboratory is telephonically contacted by Eric K., a senior partner in a boutique law firm in the state of Florida, United States. He advises that his firm is representing a department store that has been sued by a former female employee (the plaintiff) who was fired based upon poor work performance and admitting to smoking marijuana while on duty. The ex-employee, through her attorney, has provided a file made by her of the exit interview conducted by the store's Human Resources Department (HRD) using a small, hidden digital audio recorder; the provided file copy is in a WMA container format and is alleged to be an exact copy from an Olympus XYZ recorder. The original recorder and its memory card have purportedly been discarded by the plaintiff. Eric K. further states that the personnel from the HRD have listened to the file and believe that portions of the interview have been deleted and that the plaintiff has also added statements by her that were not said during the original exit interview. Eric K. is advised that only the original file or a clone can be scientifically authenticated and that a preliminary analysis could be conducted to determine the originality of the WMA file, if it is consistent with the Olympus recorder and if there are any obvious alterations.

A week later, a box is received at the laboratory via an overnight delivery service, with a recipient signature required. Opening the box revealed a cover letter and a CD-R disc with labelling reflecting, in part, that it was produced by the plaintiff attorney's office with the handwritten notation: 'Original file from SD card in Olympus XYZ recorder'. The examiner marks the submitted disc with the laboratory number, the specimen designator 'Q1', and the examiner's initials with a permanent-ink, felt-tip pen. Typed computer work notes are started by listing the date and time the evidence was received, the delivery service's tracking number and a description of specimen Q1 including its brand name, format, manufacturer's designator, the printed label information and the handwritten and other markings. A visual review of the specimen Q1 disc revealed no obvious physical alterations or damage.

Specimen Q1 is placed in a laboratory computer CD drive which revealed, based on Windows Explorer, a data CD with a volume name of 'Original File' and containing only one file: 'XYZ001.WMA'. A SHA-256 (256-bit) hash is calculated and documented for specimen Q1 using digital data imaging and analysis software, and an image is then produced of the specimen. The digital data image file is similarly hashed, and the hash values are determined to be identical. Specimen Q1 is removed from the drive and placed in a locked evidence safe. The WMA file is then extracted from the image file and used for the remainder of the authenticity examination.

A preliminary digital data analysis comparison between the WMA file and test recordings prepared on a laboratory Olympus XYZ recorder revealed a considerable number of differences in the header metadata information and structure. The specimen Q1 WMA file, compared to the test recordings, had no Olympus file identification, a different version of the WMA algorithm, 13 header objects instead of only five and a number of other changes. A review of the scientific literature revealed that the changes to the specimen Q1 file were the result of a re-encoding process in a specific audio editing program, which was readily available as a free download on the Internet. A very preliminary aural review of the specimen Q1 recording revealed a number of possible discontinuities between some of the speech segments and obvious indications of two different acoustic environments. The findings for these forensic analyses, including screen captures of the appropriate hexadecimal data matrixes for the files' header information are added to the work notes.

Attorney Eric K. is advised of the preliminary digital data and aural analyses, and he requested that no further forensic examinations be conducted. A week later, the attorney called and advised that the plaintiff had dropped the lawsuit and that the specimen Q1 disc should be returned to his office. A cover letter was prepared for specimen Q1 and it was returned to the attorney's office via an overnight delivery company with signature service.

4.8.2   Case Example Number 2

A government forensic laboratory receives an email request from Allison B., an Assistant District Attorney (ADA), to analyze a surveillance video recording from a prison facility, being offered as evidence in a criminal matter. Additionally, the laboratory is requested to review a laboratory report tendered by an expert retained by the defence counsel. Attached to the email message are an Audio Video Interleave (AVI) video file and a Portable Document Format (PDF) copy of the defence expert's report. ADA Allison B. writes in her email that an altercation occurred between the defendant and prison personnel, in a location where only one video camera captured the incident. The ADA further states that the AVI file represents only a 10-min segment of time from that camera view which was exported from the BetaZen model 1972 digital video recorder (DVR), but that if additional files or time ranges are needed, access to the DVR, which was pulled from use after the incident, can be arranged by her office.

The case is opened, given a laboratory number and assigned to a forensic video examiner, who prints the email message from ADA Allison B. and starts her typed work notes with descriptions of the AVI file (designated as specimen 'Q1') and of the PDF file (designated as item 'NE1'). She hashes the AVI file, prepares a copy of it and then hashes the copy to assure that it is identical to the submitted AVI file. She opens the prepared copy in her video workstation's software, which displays all of the pertinent information regarding the contained media streams, and prints a listing of the information. She learns that the AVI file consists of a Microsoft Video 1 video stream (with pixel dimensions of 640 × 480 and a frame rate of 29.97 frames per second). Upon preliminarily viewing the AVI file in Windows Media Player, the government's forensic examiner notices that a running date and time stamp is burned into the video data, that the images and the date/time stamps are progressing much faster than 'real time', and that there appear to be inconsistent jumps in the date/time stamps. She notes these characteristics in her work notes.

She conducts online research into the BetaZen model 1972 DVR and locates a PDF file of the user's manual. After downloading the PDF file and designating it as item NE2, she includes a description of the document in her work notes and then reviews the document. Within the section of the manual describing the process of exporting segments of video from the DVR, she discovers that there exists an option to export segments into the native format as originally recorded, in addition to the option to export to the AVI format. Additionally, the native format export can be 'validated' by the manufacturer's software, to ascertain whether or not alterations have been made to the file since the time it was exported; however, AVI file exports cannot be 'validated' by this process.

The forensic examiner reads through item NE1 (the defence expert's report) and excerpts the conclusions drawn into her work notes. The defence expert claims that the AVI file (specimen Q1) has been altered, as evidenced by the inconsistent jumps in the date/time stamp which indicate that images have been removed. The defence expert does not comment on having analyzed any video file other than the AVI file, nor does he indicate having requested access to the original DVR.

The government's forensic examiner contacts ADA Allison B. and explains the need to access the original DVR for purposes of exporting the designated portion in its native format. ADA Allison B. makes the proper arrangements, and the forensic examiner conducts the native file export of the designated portion to a forensically wiped thumb drive. In addition to the export of the video itself, the examiner notices that a separate spreadsheet file and executable file (the manufacturer's video playback software) were also saved to the thumb drive. She designates the native video file as specimen Q2, the spreadsheet file as specimen Q3 and the executable file as item K1, indicating a 'known' evidentiary item. She documents the export process in her work notes. The specimen Q3 spreadsheet is opened and printed, and upon review, the government's examiner determines that the spreadsheet includes information regarding the activity of the DVR for the exported camera view and date/time range. The particular camera view was triggered by motion within the scene, and the spreadsheet lists all portions of time during which motion was detected and the camera's output was recorded to the DVR.

Detailed frame-by-frame visual review of specimen Q2 (the native video file) using the specimen K1 software reveals numerous recording discontinuities, during which different lengths of time are not present, but no obvious indications of alteration to the content of the images. The noted recording discontinuities are compared with the specimen Q3 spreadsheet regions, and consistency was found between the 'missing' recorded images and the lack of activity of the camera view. Further, the 'jumps' in the date/time stamp observed in the specimen Q1 AVI file are also compared to specimen Q2 and the specimen Q3 spreadsheet regions, and all are found to be consistent. Lastly, the 'validate' function of the specimen K1 software is run on the specimen Q2 video file, which reveals that no alterations have been made to the file since its export. The government's forensic examiner documents these observations in her work notes, concluding that specimen Q2 (the native file export from the DVR) is not altered, but is discontinuous based on the motion-activated nature of the camera and recording system. Further, specimen Q1 is not a native file export from the DVR but is instead a transcoded copy, meaning that it cannot be relied upon solely for scientific authentication. However, through corroboration with specimens Q2 and Q3, no images have been removed from specimen Q1, as indicated by the defence expert in his report. Contact with ADA Allison B. reflects that only the allegations made by the defence expert have to be addressed at the present time. The government's examiner finalizes her work notes, prepares a laboratory report of her findings and provides ADA Allison B. with the signed original laboratory report via overnight, signature delivery service. ADA Allison B. telephonically confirms receipt of the laboratory report and advises the examiner to temporarily retain the submitted evidence and when the trial is scheduled to begin, so that she may reserve those days in the event that she is called to provide expert testimony.

4.9 Discussion

This chapter has set forth the general procedures for forensically authenticating digital audio and video files, but with each submitted recording requiring a unique protocol, their application is problematic to all but the most experienced and sophisticated scientists and engineers. These examiners must be knowledgeable regarding digital recorder theory, metadata structuring of audio and video files, FFT theory and its applications, general speech characteristics, non-linear audio/video editing, computer and image forensics, video surveillance systems, specialized law enforcement recording systems and numerous other related topics.

Many examiners are fortunate to have obtained the proper formal education and lengthy apprenticeships at major governmental or private forensic laboratories; however, other examiners practising in the field do not have these backgrounds and, as expected, often produce results that are incomplete and/or inaccurate. To date, professional societies, standards committees and the major laboratories, at least in the United States, have not found an acceptable solution to this problem, except to refute inaccurate results on a case-by-case basis by conducting the analyses properly and providing opposing testimony in depositions, hearings and trials.

A second problem area lies in there being limited scientific research in many areas of this forensic field, as seen in the shortage of peer-reviewed papers listed for this chapter. Overall, a limited number of individuals from academia plus government and private laboratories conduct and publish their research; many of the largest forensic laboratories rarely, if ever, publish papers. It is hoped that in the future more individuals and laboratories will begin to provide research support for the field, so that a better understanding can emerge for all aspects of the examination.

The future of this field will be challenging, as it adapts to the newest file formats and recorder technology, while simultaneously training the future examiners in proper forensic audio/video authentication. Development of semi- and fully automated systems will assist with specific analysis areas, but these systems and their underlying algorithms will not be able to provide all of the needed forensic tools in the near future, nor will they serve as 'push-button' solutions and supplant forensic examiners. However, even small positive steps will continue to move the forensic authentication field forward to a better understanding of the characteristics of original and altered digital recordings.

References

Aitken CGG and Taroni F 2004 Statistics and the Evaluation of Evidence for Forensic Scientists, 2nd edn. West Sussex: Wiley.

Archer H 2012 Quantifying effects of lossy compression on electric network frequency signals. Paper presentation at the Audio Engineering Society 46th International Conference, Denver, CO, USA.

Berriman D 2000 Headphones. In Audio & Hi-Fi Handbook, rev. edn., Sinclair IR editor. Boston: Newnes, 310–318.

Bolt RH, Cooper FS, Flanagan JL, McKnight JG, Stockham TG Jr. and Weiss MR 1974 Report on a technical investigation conducted for the U.S. District Court for the District of Columbia for the Advisory Panel on White House Tapes. Washington, DC: U.S. Government Printing Office.

Bracewell RN 2000 The Fourier Transform and Its Applications, 3rd edn. Boston: McGraw-Hill.

Brixen EB 2008 ENF; Quantification of the magnetic field. Paper presentation at the Audio Engineering Society 33rd International Conference, Denver, CO, USA.

Caloyannides MA 2001 Computer Forensics and Privacy. Boston: Artech House, 172–174.

Chen M, Fridrich J, Goljan M and Lukáš G 2007 Source digital camcorder identification using sensor photo response non-uniformity. Proceedings of SPIE Electronic Imaging 6505(1).

Coetzee S 2012 Phase & amplitude analysis of the ENF for digital audio authentication. Paper presentation at the Audio Engineering Society 46th International Conference, Denver, CO, USA.

Cooke M, Beet S and Crawford M (editors) 1993 Visual Representations of Speech Signals. New York: John Wiley & Sons.

Cooper AJ 2005 The significance of the serial copy management system (SCMS) in the forensic analysis of digital audio recordings. International Journal of Speech, Language and the Law 12(1), 49–62.

Cooper AJ 2008 The electric network frequency (ENF) as an aid to authenticating forensic digital audio recordings – an automated approach. Paper presentation at the Audio Engineering Society 33rd International Conference, Denver, CO, USA.

Cox IJ, Miller ML and Bloom JA 2002 Digital Watermarking. San Diego: Morgan Kaufmann Publishers, 319–358.

Encyclopædia Britannica 2014 Leonid Kuchma. Available from: http://www.britannica.com/EBchecked/topic/324312/Leonid-Kuchma [1 April, 2014].

Fridrich J 2009 Digital image forensics. IEEE Signal Processing Magazine 26(2), 26–37.

Fuller DB 2012 How audio compression algorithms affect DC offset in audio recordings. Poster presentation at the Audio Engineering Society 46th International Conference, Denver, CO, USA.

Garg R, Varna AL, Hajj-Ahmad A and Wu M 2013 "Seeing" ENF: power-signature-based timestamp for digital multimedia via optical sensing and signal processing. IEEE Transactions on Information Forensics and Security 8(9), 1417–1432.

Graf U 2004 Applied Laplace Transforms and z-Transforms for Scientists and Engineers: A Computational Approach Using a Mathematica Package. Basel/Boston: Birkhäuser.

Grigoras, C 2005 Digital audio recording analysis: the electric network frequency criterion. International Journal of Speech, Language and the Law 12(1), 63–76.

Grigoras C 2007 Applications of ENF criterion in forensic audio, video, computer and telecommunication analysis. Forensic Science International 167(2–3), 136–145.

Grigoras C 2009 Applications of ENF analysis in forensic authentication of digital audio and video recordings. The Journal of the Audio Engineering Society 57(9), 643–661.

Grigoras C and Smith JM 2012 Advances in ENF analysis for digital media authentication. Paper presentation at the Audio Engineering Society 46th International Conference, Denver, CO, USA.

Grigoras C, Rappaport D and Smith JM 2012 Analytical framework for digital audio authentication. Paper presentation at the Audio Engineering Society 46th International Conference, Denver, CO, USA.

Howard DM and Angus J 2009 Acoustics and Psychoacoustics, 4th edn. Amsterdam/London: Focal Press, 277–363.

Information systems – coded character sets – 7-bit American National Standards Code for Information Interchange (7-bit ASCII), version ANSI INCITS 4-1986 (R2007) American National Standards Institute.

James JF 2011 A Student's Guide to Fourier Transforms: With Applications in Physics and Engineering, 3rd edn. Cambridge: Cambridge University.

Jayant NS and Noll P 1984 Digital Coding of Waveforms: Principles and Applications to Speech and Video. Englewood Cliffs: Prentice-Hall, 1–251.

Jenkins C, Leroi J and Steinhour J 2012 Advances in electric network frequency acquisition systems and stand alone probe applications for the authentication of digital media. Paper presentation at the Audio Engineering Society 46th International Conference, Denver, CO, USA.

Jones PE 2001 US Secure Hash Algorithm (SHA1). Available from: http://tools.ietf.org/html/rfc3174 [31 January 2015].

Kammler DW 2000 A First Course in Fourier Analysis. Upper Saddle River: Prentice Hall.

Koenig BE 1990 Authentication of forensic audio recordings. The Journal of the Audio Engineering Society **38(1/2)**, 3–33.

Koenig BE 1992 Frequency measurement of alternating current. Journal of Forensic Identification **42(5)**, 408–411.

Koenig BE and Lacey DS 2007 Audio record and playback characteristics of small solid-state recorders. Journal of Forensic Identification **57(4)**, 582–598.

Koenig BE and Lacey DS 2009 Forensic authentication of digital audio recordings. The Journal of the Audio Engineering Society **57(9)**, 662–695.

Koenig BE and Lacey DS 2010 Evaluation of clipped sample restoration software. Forensic Science Communications **12(2)**. Available from: http://www.fbi.gov/about-us/lab/forensic-science-communications/koenig/koenig.html [24 February 2015].

Koenig BE and Lacey DS 2012a An inconclusive digital audio authenticity examination: a unique case. Jounal of Forensic Sciences **57(1)**, 239–245.

Koenig BE and Lacey DS 2012b Forensic authenticity analyses of the header data in re-encoded WMA files from small Olympus audio recorders. The Journal of the Audio Engineering Society **60(4)**, 255–265.

Koenig BE and Lacey DS 2014a The average DC offset values for small digital audio recorders in an acoustically-consistent environment. Journal of Forensic Sciences **59(4)**, 960–966.

Koenig BE and Lacey DS 2014b Forensic authenticity analyses of the metadata in re-encoded WAV files. Proceedings of the 54th Audio Engineering Society International Conference, London, UK.

Koenig BE, Lacey DS and Herold N 2003 Equipping the modern audio-video forensic laboratory. Forensic Science Communications **5(2)**. Available from: http://www.fbi.gov/about-us/lab/forensic-science-communications/fsc/april2003/lacey.htm [24 February 2015].

Koenig BE, Lacey DS and Killion SA 2007 Forensic enhancement of digital audio recordings. The Journal of the Audio Engineering Society **55(5)**, 352–371.

Koenig BE, Lacey DS, Grigoras C, Price SG and Smith JM 2012a Evaluation of the average DC offset values for nine small digital audio recorders. Paper presentation at the Audio Engineering Society 46th International Conference, Denver, CO, USA.

Koenig BE, Lacey DS, and Richards GB 2012b Video frame comparisons in digital video authenticity analyses. Journal of Forensic Identification **62(2)**, 165–182.

Koenig BE, Lacey DS, Grigoras C, Price SG and Smith JM 2013 Evaluation of the average DC offset values for nine small digital audio recorders. The Journal of the Audio Engineering Society **61(6)**, 439–448.

Koenig BE, Lacey DS and Reimond CE 2014 Selected characteristics of MP3 files re-encoded with audio editing software. Jounal of Forensic Identification **64(3)**, 304–321.

Kuttruff H 2009 Room Acoustics, 5th edn. Boca Raton: CRC Press.

Lacey DS and Koenig BE 2008 Identification of an eccentricity in the date/time metadata of a PAL MiniDV recording. Journal of Forensic Sciences **53(6)**, 1417–1423.

Lacey DS and Koenig BE 2010 Audio extraction from Silicor Technologies' digital video recorder file format. Jounal of Forensic Identification **60(5)**, 573–588.

Lacey DS and Koenig BE 2012 Identification of identical and nearly identical frames from a Lawmate PV-500 digital video-audio recorder. Jounal of Forensic Identification **62(1)**, 36–46.

Li N and Loizou PC 2007 Factors influencing glimpsing of speech in noise. The Journal of the Acoustical Society of America **122(2)**, 1165–1172.

Liu Y, Chai J, Greene B, Conners R and Liu Y 2012 A study of the accuracy and precision of quadratic frequency interpolation for ENF estimation. Paper presentation at the Audio Engineering Society 46th International Conference, Denver, CO, USA.

Lucy D 2005 Introduction to Statistics for Forensic Scientists. Hoboken: Wiley.

Malik H and Miller JW 2012 Microphone identification using higher-order statistics. Paper presentation at the Audio Engineering Society 46th International Conference, Denver, CO, USA.

Mallat S 2008 A Wavelet Tour of Signal Processing: The Sparse Way, 3rd edn. Boston: Academic Press.

Maloney A 2013 Comparing nearly identical images using "beyond compare". Jounal of Forensic Identification **63(2)**, 153–164.

Marcella AJ Jr. and Menendez D 2008 Cyber Forensics: A Field Manual for Collecting, Examining, and Preserving Evidence of Computer Crimes, 2nd edn. Boca Raton: Auerback Publ., 56–57.

Marr K and Snyder, DJ III 2008 Using automated digital tools for forensic audio examinations. Presentation at the American Academy of Forensic Sciences in Washington, DC, USA.

Matson JV 2012 Effective Expert Witnessing: Practices for the 21st Century, 5th edn. Boca Raton: CRC Press.

Mendenhall W and Sincich T 2006 Statistics for Engineering and the Sciences, 5th edn. Upper Saddle River: Pearson.

Microsoft extensible firmware initiative FAT32 file system specification, version 1.03. 2000 Hardware White Paper, Microsoft Corp.

Najmi AH 2012 Wavelets: A Concise Guide. Baltimore: John Hopkins University Press.

Nash A, Begault DR and Peltier CA 2012 Uncertainty in the measurement of the electrical network frequency. Poster presentation at the Audio Engineering Society 46th International Conference, Denver, CO, USA.

Ng N 2012 ENF database assessment. Poster presentation at the Audio Engineering Society 46th International Conference, Denver, CO, USA.

Nicolalde DP and Apolinário JA Jr. 2009 Evaluating digital audio authenticity with spectral distances and ENF phase change. IEEE International Conference on Acoustics, Speech and Signal Processing (ICASSP), 1417–1419.

Ojowu O Jr., Karlsson, J, Li J, and Liu Y 2012 ENF extraction from digital recordings using adaptive techniques and frequency tracking. IEEE Transactions on Information Forensics and Security **7(4)**, 1330–1338.

Owen M 2007 Practical Signal Processing. Cambridge: University Press, 53–57, 159–163.

Perraud D 2008 Forensic authentication of digital video tapes. Journal of Forensic Identification **58(6)**, 632–645.

Pohlmann KC 2011 Principles of Digital Audio, 6th edn. New York: McGraw-Hill, 21–28.

Potter RK, Kopp GA and Kopp HG 1966 Visible Speech. New York: Dover Publications, Inc.

Rivest RL 1992 The MD5 Message-Digest Algorithm. Available from: http://www.ietf.org/rfc/rfc1321.txt [31 January 2015].

Rodríguez DPN, Apolinário JA Jr. and Biscainho LWP 2010 Audio authenticity: detecting ENF discontinuity with high precision phase analysis. IEEE Transactions on Information Forensics and Security **5(3)**, 534–543.

Sanaei A, Toulson ER and Cole M 2014 Tuning and optimization of an electric network frequency extraction algorithm. The Journal of the Audio Engineering Society **62(1/2)**, 25–36.

Secure Hash Standard 2012 Federal Information Processing Standards Publication 180-4, Washington, DC: U.S. Department of Commerce.

Sencar HT and Memon N (editors) 2013 Digital Image Forensics: There is more to a Picture than Meets the Eye. New York: Springer, 320.

SWGDE and SWGIT digital & multimedia evidence glossary, version 2.7 2013 Scientific Working Groups on Digital Evidence and Imaging Technology.

SWGDE best practices for computer forensics, version 2.1 2006 Sci. Working Group on Digital Evidence.

SWGDE best practices for Forensic Audio, version 1.0 2008 Sci. Working Group on Digital Evidence.

SWGDE data integrity within computer forensics 2006 Sci. Working Group on Digital Evidence.

Tanner DC and Tanner ME 2004 Forensic Aspects of Speech Patterns: Voice Prints, Speaker Profiling, Lie and Intoxication Detection. Tucson: Lawyers & Judges Publishing Co., Inc., 15–45.

The Star 2008 Tape Not Altered as Harper Claimed, Expert Finds. Available from: http://www.thestar.com/news/canada/2008/10/10/tape_not_altered_as_harper_claimed_expert_finds.html [31 January, 2015].

Unicode Standard, version 6.2.0 2012 Unicode Technical Committee.

Van Houten W and Geradts Z 2009 Source video camera identification for multiply compressed videos originating from YouTube. Digital Investigation **6(1–2)**, 48–60.

Wallace A Jr. and Koenig BE 1989 An introduction to single channel FFT analysis. Crime Laboratory Digest **16(2)**, 33–39.

Wang W and Farid H 2006 Exposing digital forgeries in video by detecting double MPEG compression. Paper presentation at the Multimedia and Security Workshop 2006, Geneva, Switzerland.

Wang W and Farid H 2007 Exposing digital forgeries in video by detecting duplication. Paper presentation at the Multimedia and Security Workshop 2007, Dallas, TX, USA.

Wang W and Farid H 2009 Exposing digital forgeries in video by detecting double quantization. Paper presentation at the Multimedia and Security Workshop 2009, Princeton, NJ, USA.

Widder DV 2010 The Laplace Transform. Mineola: Dover Publications.

Wikipedia 2014 Alleged War Crimes During the Final Stages of the Sri Lankan Civil War. Available from: http://en.wikipedia.org/wiki/Alleged_war_crimes_during_the_final_stages_of_the_Sri_Lankan_Civil_War\#Sri_Lanka.27s_Killing_Fields [1 April, 2014].

Yuan Z, Liu Y, Conners R and Liu Y 2012a Using simple Monte Carlo methods and a grid database to determine the operational parameters for the ENF matching process. Paper presentation at the Audio Engineering Society 46th International Conference, Denver, CO, USA.

Yuan Z, Liu Y, Conners R and Liu Y 2012b Effects of oscillator errors on electric network frequency analysis. Paper presentation at the Audio Engineering Society 46th International Conference, Denver, CO, USA.

Part Two

Digital Evidence Extraction

5

Photogrammetry in Digital Forensics

Bart Hoogeboom[1], Ivo Alberink[2] and Derk Vrijdag[1]
[1]*Department Digital Technology and Biometrics*
[2]*Department Forensic Science, Interdisciplinary Investigations, Statistics and Knowledge Management*

5.1 Introduction

Photo and video material are often involved in forensic investigations. In some cases it is desired to extract position and/or dimension information from objects depicted in the images. The art of determining this kind of information from images is called photogrammetry.

The American Society for Photogrammetry and Remote Sensing has defined photogrammetry as the art, science and technology of obtaining reliable information about physical objects and the environment through processes of recording, measuring and interpreting photographic images and patterns of recorded radiant electromagnetic energy and other phenomena.[1]

The principles of photogrammetry go back to Leonardo da Vinci in the fifteenth century. Some pioneers on photogrammetry are Aimé Laussedat and Albrecht Meydenbauer where Meydenbauer introduced the term *photogrammetry* in a publication in 1867 (Albertz 2007). A widely used distinction in photogrammetry is far range photogrammetry versus close range photogrammetry. In forensics usually

---

[1]From: Guidelines for Procurement of Professional Aerial Imagery, Photogrammetry, Lida rand Related Remote Sensor-based Geospatial Mapping Services, December 2009, http://www.asprs.org/a/society/committees/standards/Procurement_Guidelines_w_accompanying_material.pdf.

Handbook of Digital Forensics of Multimedia Data and Devices, First Edition.
Edited by Anthony T.S. Ho and Shujun Li.
© 2015 John Wiley & Sons, Ltd. Published 2015 by John Wiley & Sons, Ltd.
Companion Website: www.wiley.com/go/digitalforensics

close-range photogrammetry is used where two types of photogrammetry can be distinguished: photogrammetry using controlled images and photogrammetry using uncontrolled images. Photogrammetry using controlled images means measuring in images which are taken especially for the purpose of performing photogrammetry. Measuring in images that were not shot for that purpose is photogrammetry on uncontrolled images.

Photogrammetry in forensic investigations was first widely accepted on controlled images: the (tire) tracks on the road caused by vehicles involved in an accident were photographed (Hugemann and Brunn 2007). The photographer knows the purpose of the photograph, and therefore can take the photographs in such a way that they are of best quality for this application including special markers. Afterwards the length of the tire track(s) and/or position of objects like glass parts are derived from the photographs. To achieve this, the simplest form of photogrammetry can be used: orthorectification of an image. This means geometrical correction of an image in such a way that the scale is uniform on a defined plane. In the case of the photographs of the accident the defined plane would be the road surface. The advantage of this is the possibility of measuring in one image with the only need of scaling information at the defined plane. The downside of orthorectification is the restriction of measurements to the defined plane. Every object which is not located at the plane cannot be localized or measured. One step further, deriving information outside the defined plane means more information is needed. This information can be provided by the parameters of a new plane, but also by a second photograph of the same scene shot from a different angle. In theory, if two photographs are available from the same scene but shot from different angles, the location of every corresponding point can be determined relative to the locations of the cameras. This is called multi-image photogrammetry and leads to a 3D point cloud. Photogrammetry using two or more controlled photographs is a common technique, and a lot of software is available to perform this. A list of available software, which is not complete but gives an impression, can be found at the Wikipedia page on photogrammetry.[2] Developments to automatically generate point clouds using multiple photographs without special markers are on-going. Accuracy also depends on the user (Alkan and Karsidag 2012; Georgantas et al. 2012; Topolšek et al. 2014), and at this moment (2014) in forensic applications the accuracy is not expected to compete with that of 3D-laser scanners.

In the case of controlled images it is possible to validate the technique and develop procedures. There is no need for validating every case if these procedures are followed. When uncontrolled images are involved things become more complicated. Lack of controlled markers in the image and lack of knowledge about the recording system makes it difficult to use more common techniques like orthorectification or multi-image photogrammetry. When there are enough straight lines in the image projective geometry might be used. Otherwise reverse projection, using the original recording equipment or a 3D model of the scene might help out. It is not straightforward to

---

[2] http://en.wikipedia.org/wiki/Photogrammetry, consulted on 7 April 2014.

estimate the reliability of the results. Measuring known objects in the images can help to gain insight in the measurement uncertainty. A statistical approach to calculate measurement uncertainties is given by Alberink and Bolck (2008). This approach is based on performing measurements on reference images from which the ground truth is known. By analyzing the differences between the ground truth and the image measurements one can quantify the systematic bias and the measurement uncertainty involved. If the reference images are comparable to the questioned images the measurement uncertainty found in the reference images can be used as an estimation of the measurement uncertainty for the questioned image.

The chapter starts with a little word about lens distortion. Before photogrammetry is conducted one should have an idea about the imperfections caused by lens distortion to the image. This is followed by a description of different methods of photogrammetry available for forensic investigations. Independent of the method used measurement errors will be made, which is discussed in Section 5.3. This is followed by some case examples from daily practise of the authors. The chapter ends with a discussion about 3D modelling and scenario testing including case examples.

5.1.1   Lens Distortion

Lens distortion will affect the results when performing photogrammetry. The most common type of lens distortion in forensic image analysis is barrel distortion. Especially wide angle lenses can have a lot of barrel distortion. This distortion is a function of the radius from the optical axis, squared. This implies that the distortion has far more effect towards the edges of the image. A recording of a checkered pattern will reveal this distortion: the straight lines become bended in the image. There are lens correction algorithms available that will correct the image. Figure 5.1 shows images of a checkered pattern: on the left-hand side the uncorrected image and on the right-hand side the corrected image.

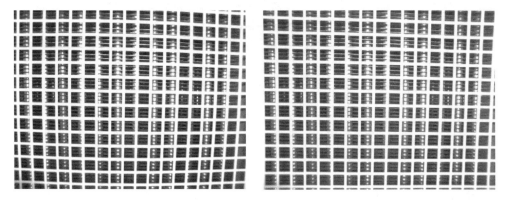

Figure 5.1   Images of a checkered pattern. Left: the uncorrected image. Right: the corrected image. Reproduced with the permission of Hoogeboom and Alberink (2010).

Despite the possibility of correcting for lens distortion the best image to use, if possible, is the image where the subject of interest is near the centre of the image. The subject will be less affected by the distortion and the approximated correction.

5.2 Different Methods

Theoretical backgrounds on close range photogrammetry can be found in Luhmann *et al.* (2006). For photogrammetry in forensic investigations several techniques based on close range photogrammetry are available:

- Projective geometry
- Space resection and multi-image photogrammetry
- Reverse projection

In Edelman *et al.* (2010) a comparison between projective geometry and reverse projection in a forensic context is described, where the 3D model needed for reverse projection is made using multi-image photogrammetry.

5.2.1 Projective Geometry or Orthorectification

In projective geometry there are two methods to be distinguished: orthorectification and vanishing points. Orthorectification, or projective rectification, is geometrical correction of an image in such a way that the scale is uniform on a defined plane. Commercial software like PC Rect are based on this principle. Figure 5.2 shows an example of the orthorectification of the road surface for the use of measuring tire tracks. The left-hand side of the image shows the original photo where the right-hand

Figure 5.2 Orthorectification of an image. Left: the original photo with the markers ('X'-s) visible. Right: rectified photo. Reproduced with the permission of Hoogeboom and Alberink (2010).

side shows the rectified photo. This photo is an example of a controlled image: the photo was intended to perform photogrammetry on and was taken after putting the markers on the road.

The geometric transformation between the image plane and the projective plane given for an object coordinate (X, Y) is given by Luhmann *et al.* (2006):

$$X = \frac{a_0 + a_1 x' + a_2 y'}{c_1 x' + c_2 y' + 1}$$

$$Y = \frac{b_0 + b_1 x' + b_2 y'}{c_1 x' + c_2 y' + 1}$$

Therefore the relation between the X and Y coordinates in the object plane with the x' and y' coordinates in the image plane of at least four points are required to determine the eight parameters. This is only possible if no three points are on one line. With the eight parameters the equations are solved and the uniform scaled plane is defined. In Figure 5.2 the four markers can be seen from which the relative position is known by five distances: four sides of the quadrangle and one diagonal. By defining one of the markers as $(0, 0)$ and one marker lying on the same line $(0, y')$ the eight parameters can be solved. If also the length of the other diagonal is known, there are more known distances than necessary to solve the equations, and therefore it is possible to estimate the accuracy of the solution. In the case of an uncontrolled image the same procedure might still be possible. For this, four natural points on the plane of interest (i.e. the road surface) which are still available at the location to be measured should be visible in the image.

In the case of uncontrolled images with no four points available it might be possible to use vanishing points and vanishing lines. Figure 5.3 shows the definition of vanishing points and lines.

When the vanishing line of the ground plane, the vertical vanishing point, and at least one reference height in the scene is known, it is possible to calculate the height of any

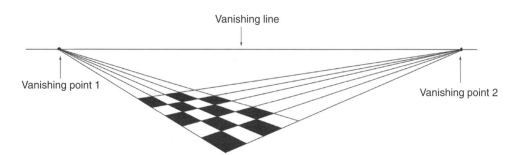

Figure 5.3   Image of a checkered pattern in perspective view with the vanishing line and vanishing points. Reproduced with the permission of Hoogeboom and Alberink (2010).

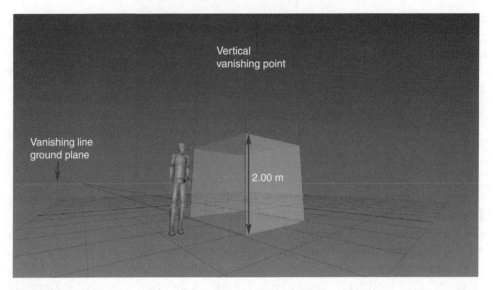

Figure 5.4 Construction of the vanishing line of the ground plane and the vertical vanishing point. The vertical vanishing point itself is not shown in this picture because this point lies too far above the drawing. The known height is the height of the box: 2.00 m.

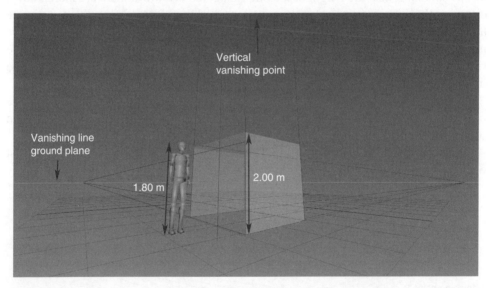

Figure 5.5 Construction of the lines that define the height of the person relative to the height of the box.

point from the ground plane. This is schematically shown in Figures 5.4 and 5.5 where the height of the person is estimated.

Through the construction of the lines the relative height of the person standing next to the box is estimated. Together with the known height of 2.00 m of the box, the

Figure 5.6   Line calculated according to the camera parameters. The line is projected as a point in the image.

height of the person is estimated at 1.80 m. In Criminisi *et al.* (1998) this technique is described in a forensic context.

5.2.2   Space Resection and Multi-image Photogrammetry

The most common technique for photogrammetry is space resection: calculation of rays from the object plane to the camera plane for a number of points. A lot of software based on these techniques are available. Well-known commercial packages are Photomodeler (Eos Systems) and Image modeler (Auto desk), but there is also open-source software available like Python Photogrammetry Toolbox.[3] If the parameters of interior orientation from the camera are unknown the equations can only be solved if the 3D-coordinates of at least five image points are known Luhmann *et al.* (2006). When the equations are solved it means the camera parameters like position, orientation and focal length are known with respect to the five given points for that image. Now, for every other point in the image the corresponding line in the real world can be calculated. Why it is a line is easy to understand if one realizes that the image point is defined by its coordinates x' and y', so two-dimensional, while a point in the real world is defined by its coordinates x, y and z, three-dimensional. There is simply not enough information in the 2D-image to make a transformation to the 3D world. However, if there is some extra information about the image point, it is possible to make the transformation to the 3D world. For instance, if a point lies on ground level, the z-component of that point is fixed. In that case the calculated camera parameters are sufficient to calculate the unknown x- and y-component of that point in the real world given the x'- and y'-components in the image. Graphically this is where the line intersects the ground plane. Figure 5.6 shows an example of this situation.

---

[3] http://www.arc-team.homelinux.com/arcteam/ppt.php.

In the enhanced e-book and the book's website a video can be found which shows the situation of Figure 5.6.

If more than one image from a different angle from the same scene is available, there is also enough information to calculate the x-, y- and z-coordinates in the real world given an image point in the images. A very powerful and accurate method for this is bundle triangulation (Luhmann *et al.* 2006). Bundle triangulation algorithms iterate to the best solution given the positions of corresponding image points in at least two images. To come to an accurate and stable solution it is important to use enough corresponding points that are well spread around the images. The minimum number of points needed depends on the angle between the images and how well the points are spread around the images. In practice a minimum of eight corresponding points are recommended. Once the bundle adjustment is calculated the location of every additional point can be calculated if the point is visible in at least two images.

It is possible to mix controlled images with uncontrolled images using bundle triangulation. This opens the possibility to make new photos at the crime scene even after a longer period and combine them with the questioned image. As long as there are enough similar points visible in both the questioned as in the new photos this is possible.

A special case of multi-image photogrammetry is photogrammetry using stereo photos: stereo-photogrammetry. Two photos shot with an angle comparable to human eyes are used to calculate the 3D-position of image points. Since there are such specific requirements according to the angle between the two photographs, this technique is not useful for uncontrolled images.

5.2.3   Reverse Projection

A technique that can be very useful in photogrammetry on uncontrolled images is reverse projection. Reverse projection means projecting images from a camera on new images from the same camera or at least on images from a camera with similar internal (i.e. focal length and distortion) and external (i.e. position and orientation) parameters. It is a technique that can be applied in two different ways:

1. Reverse projection on live video
2. Reverse projection using a 3D computer model

Reverse projection on live video means the projection of live video on the questioned image (i.e. the video still from the robber in Figure 5.6) using the video system that recorded the questioned image. On the basis of visible features from the environment the live view from the camera can be checked for changes in pan, tilt, rotation and zoom. If one or more of these parameters have changed, the camera view must be restored to meet the view from the questioned image. Once this is achieved, measurement in the images can take place by positioning people and/or objects in the scene according to the person/object in the questioned image. For instance it is possible to place a ruler at the position of the person in the questioned image to determine the body height.

Reverse projection using a 3D computer model means the projection of a 3D model on the questioned image. Software that can combine 3D computer models with video images are for instance 3ds Max (Autodesk) and Blender (open-source software). To calculate the necessary projection of the 3D model there are camera matching routines available. By pointing out similar points between the 3D model and the questioned image the camera matching algorithm calculates the camera parameters. Theoretically five corresponding points should be sufficient to solve the equations and achieve the correct projection. However, the points should then be perfectly distributed around the image and in the 3D environment. In practice a minimum of six corresponding points is needed. This means the 3D model doesn't have to consist of more than those six points plus the plane on which the object is located that needs to be measured. The advantages of reverse projection using a 3D model over live video include more flexibility, more space for control and better reproducibility. The advantages of reverse projection using live video over using a 3D model include easier to explain in court, no need to invest in expensive hardware and software, less time needed (so more cases can be done in a given time period).

For the construction of a 3D computer model there are several techniques available:

- Measurements with a ruler. This means simply to use a tapeline and measure the distances between the points of interest.
- Photogrammetry using photos of the scene. So, applying one of the techniques described before to derive 3D information of points of interest.
- Point measurements using a total station. This is a technique used in land surveying and uses a device that calculates the distance and the angle from a ground station to a certain point designated by the user. The total station uses a laser beam for the measurement.
- 3D-laser scanning. This can be seen as an automated total station. From a ground station (the 3D laser scanner) the laser beam systematically scans the environment. For every position of the laser beam the first object that is being hit the distance and the angle of that point is calculated.
- Automated point cloud extraction from several photographs. This is a technique to automatically calculate points using photogrammetry without the user pointing out the points.

Currently 3D-laser scanning is a reliable and relatively easy way to record the crime scene and deliver the 3D information needed for the 3D model. Automated point cloud extraction from several photographs is a promising technique. Especially the automation of photogrammetry using multiple photos without the need for special markers, since the use of markers at a crime scene should be minimized at all times. Developments to automatically generate point clouds using multiple photographs without special markers are on-going. Accuracy also depends on the user and how the photographs are taken. At this moment (2014) in forensic applications the accuracy is not expected to compete with that of 3D-laser scanners. However, the fact that this technique can be implemented using a consumer photo camera means it is cheap and therefore within reach of a lot more people than a 3D-laser scanner.

From all the mentioned photogrammetry methods reverse projection with the use of 3D models is the preferred method for the authors. There are two main reasons why this is the preferred method for the authors:

1. You only need 3D knowledge of a few more or less free to choose points in the questioned image to construct a measurement tool for the image.
2. The measurements are reproducible.

5.3 Measurement Uncertainty

For every measurement in real life, an error is made; that is to say, measured values are always estimations of actual values. What is then needed is a proper estimate of the accuracy and the precision of the measurements. A solid way to investigate this is by doing validation measurements given known actual values, under (more or less) identical circumstances. This will make it possible to carry out a statistical analysis on the measurement error. The practical implementation in forensic photogrammetry would result in a procedure that always consists of the following two steps:

1. carrying out a 'reconstruction' at the crime scene, and
2. performing a statistically sound analysis of the results.

The 'reconstruction' at the crime scene means making reference recordings of a known similar event under the same conditions using the same recording equipment and settings as was used for the questioned images. In this way the reference images will have the same resolution, compression and lens distortion as the questioned images and from the reference images the ground truth is known. For the similarity of the reference images it is also important that the camera is in the same position as for the questioned images. This can be checked by placing an overlay of (characteristic lines of) the scene onto the new camera image, which is the technique of reverse projection on live video. The measurements on the reference images should take place in a similar way as for the questioned image. So, the same tool and procedure should be used for both the reference and for the questioned images. The measurement results on the reference images compared with the ground truth can now be used to estimate a confidence or credible interval for the measurement on the questioned image (Alberink and Bolck 2008). The validation procedure is independent of the used technique of photogrammetry. The results of this kind of investigations are scientifically supported and likely to be accepted as evidence in court. In investigations by the police where they are still looking for a suspect and a fast result is more important than an estimation of the measurements uncertainty, the measurements and statistical analysis might be left out. Comparing the reference images with the questioned images will give an idea about the questioned images. This quantitative method is quick and might help the police to prioritize investigation leads.

How the statistical analysis takes place is described in two examples in the paragraph case studies. These are illustrations of how to quantify measurement uncertainty. The latter is required by the ISO/IEC Standard 17025,[4] which gives norms for (forensic) laboratories such as the NFI. In the situations as described it is not self-explanatory how to perform validation experiments and sample sizes are typically small, which is the reason why the process is described in detail.

5.3.1 Difficulties in Creating Reference Recordings

Measuring the speed of a vehicle by means of determination of the position of the vehicle can introduce some difficulties. The positions and the timing of the CCTV system have to be validated separately. For the measurement uncertainty in the position the above method can be used. The timing of the CCTV system has to be evaluated by means of another clock. This clock is put in front of the CCTV system and a recording is made. With this recording it is possible to determine the frame rate and the stability of the frame rate of the system.

A second difficulty may arise in casework where the questioned object is moving. In the questioned images the object can be affected by motion blur. When validation is done by positioning, the vehicle is standing still during the reference recordings. This introduces a problem with the validation. The reference images do not completely resemble the questioned image since it is not possible to recreate this motion blur. For objects that display motion blur, it is hence preferred to do the reference images with a moving vehicle and thus validate the speed of the object instead of the position.

The software to perform the measurements is a tool: the measurement error usually doesn't depend on the tool but on the questioned image and the user. The described procedure to gain insight in the measurement error is independent of the measurement technique or software used. As long as the reference images are comparable and treated the same way as the questioned image, the procedure is valid and an estimation of the measurement error can be made.

5.4 Case Studies

Three different case examples are described to show the implementation of the above-mentioned principles. The first example describes a body height measurement of a

---

[4] General Requirements for the Competence of Testing and Calibration Laboratories, ISO, Geneva, 2005.

person. This example is followed by a case example where the speed of a car was in question. The third example handles a case where the position of a motorcycle was relevant.

5.4.1 Height Measurement

Height measurements on persons in forensic investigations are measurements on uncontrolled images, usually CCTV images. Depending on the image there are several techniques available, from which the most promising are as follows:

- Projective geometry based on vanishing points
- Multi-image photogrammetry: mixing of uncontrolled images with controlled images
- Reverse projection

There is commercial forensic software available to perform height measurements which are based on one (or more) of the above mentioned techniques. The authors used 3ds Max[5] from Autodesk which is more generic for working with 3D models in combination with photo/video. This software is very common in the forensic photogrammetry community. Other capable software packages include 3D Suspect Measurement,[6] Blender[7] and Maya.[8]

An example of a height measurement is given in a case where a bank robber was captured by a CCTV system when he entered and left the bank. The images in which the robber was pointed out by the police are shown in Figure 5.7.

The geometry of the bank was captured using a 3D laser scanner. The method to measure the height of the robber was reverse projection with the use of a 3D-model. The point cloud produced by the 3D laser scanner is projected on the CCTV-image using a camera matching algorithm. Figure 5.8 shows an overview of the 3D model and the 3D model projected on a CCTV-image.

The height of the robber was measured using a 3D-model of a thin cylinder. A cylinder was placed at ground level at the position of the feet of the robber. The top of the cylinder was placed at the top of the head of the robber. Figure 5.9 shows the image of the robber entering the bank and the cylinder that was used for the height measurement. The 3D model of the bank is hidden in this picture for the purpose of a better view on the robber.

For the image of the robber entering the bank as well as for the image of the robber leaving the bank reference recordings from test persons were made. The procedure for the reference recordings consists of positioning test persons of varying known heights

---

[5] http://www.autodesk.com/products/autodesk-3ds-max/overview.

[6] http://www.delfttech.com/delfttech/markets/forensics.

[7] http://www.blender.org/.

[8] http://www.autodesk.com/products/autodesk-maya/overview.

Figure 5.7 Images of the bank robber entering (left image) and leaving (right image) the bank.

Figure 5.8 Three-dimensional model of the bank: overview of the model (left) and the model projected on a CCTV image (right).

Figure 5.9 Height measurement: projection of a cylinder at the location of the robber.

at the scene, in the same stance as the robber, using the same camera. In the lab, the height of the test persons in the images was measured using the same technique as was used for the robber, which gives an idea about the accuracy of the measurement process.

By marking the location of the feet of the robber on the overlay, the test persons are positioned so that the location and pose are more or less the same as those of the robber. The images of the test persons are captured, and their heights are measured in the same way as the robber. Figure 5.10 shows the images from the test persons for the measurement on the robber entering the bank. Figure 5.11 shows the images from the test persons for the measurement on the robber leaving the bank.

Figure 5.10 Images from the CCTV system from the test persons positioned according to the image from the robber entering the bank (bottom right).

Figure 5.11   Images from the CCTV system from the test persons positioned according to the image from the robber leaving the bank (bottom right).

The real heights of the test persons were measured on the day the reference recordings were made, using a stadiometer. Their heights in the images were also measured the same way as for the robber, using the same 3D model. Every measurement was performed independently by three operators to minimize the effect of the operator.

In Table 5.1 the results can be found of the height measurements on the person entering the bank, in Table 5.2 those on the persons leaving the bank. For both tables these are the averaged measurements over three operators.

Variation between actual and measured heights could be introduced by the following factors:

Table 5.1 Averaged height measurements over three researchers for the persons entering the bank.

Person	Measured in the image (m)	Measurement by ruler (m)	Difference (m)
Test person 1	1.814	1.871	−0.057
Test person 2	1.809	1.827	−0.018
Test person 3	1.851	1.889	−0.038
Test person 4	1.756	1.785	−0.029
Test person 5	1.983	2.013	−0.030
Test person 6	1.765	1.779	−0.014
Test person 7	1.849	1.862	−0.013
Test person 8	1.843	1.860	−0.017
Robber	1.904	−	−

Table 5.2 Averaged height measurements over three researchers for the persons leaving the bank.

Person	Measured in the image (m)	Measurement by ruler (m)	Difference (m)
Test person 1	1.872	1.871	0.001
Test person 2	1.824	1.827	−0.003
Test person 3	1.891	1.889	0.002
Test person 4	1.800	1.785	0.015
Test person 5	1.977	2.013	−0.036
Test person 6	1.779	1.779	0.000
Test person 7	1.871	1.862	0.009
Test person 8	1.865	1.860	0.005
Robber	1.938	−	−

1. Inaccuracy in creation of the 3D model;
2. Inaccurate finding of camera position, orientation, and focal length;
3. Presence of lens distortion at the location of the perpetrator in the chosen image;
4. Pose of the robber in the chosen image;
5. Presence and height of headwear and footwear; and
6. Inconsistent interpretation of head and feet in the images by the operators.

This variation may be decomposed into a systematic part and a random part. The systematic part is caused by variation in the modelling of the scene of crime (points 1, 2, 3) and difference in pose of test persons, as well as headwear and footwear (4, 5). Manual intervention by operators in the process (6) results in a random part of the variation. From a statistical point of view, it makes sense to concentrate on the 'total error' made in each particular height measurement, actual minus measured height, and assume it is normally distributed with certain mean (systematic bias) and variation (random 'error'). The mean and variation are estimated on the basis of outcomes for test persons. As the actual heights of the test persons are known, the average measurements on the test persons show whether there is a systematic bias in the results. The measured height of the perpetrator must be adjusted with this bias. The variation in the measurements is then used to determine the precision of the estimated (corrected) height, which leads to a confidence interval for the height of the perpetrator. Using 3D crime scene models, systematic errors found in casework from the authors go up to over 10 cm. However, because in these cases the perpetrator did not stand up straight, the systematic bias may, for example be explained by the loss in height caused by his pose. Hence, this does not automatically mean that the measurement is not reliable. More problematic is a high variation in the measurements, which results in a large confidence interval. In casework, for 95% confidence intervals the band widths encountered reach from 5 up to 20 cm. A more detailed description can be found in Alberink and Bolck (2008).

In the case of the example images from the robber, the average of all the differences combined indicates what the systematic bias is on a (averaged) height measurement on that position including stance. The averaged measurement on the robber should be corrected for this bias. The standard deviation over the differences signifies the natural spread of the differences and determines the width of the confidence interval per position. In Table 5.3 the means of the differences, corrected height estimations for the robber, and the corresponding standard deviations per position are given.

In the case of the image of the robber entering the bank, the standard approach is to use the fact that under appropriate statistical assumptions, the corrected height of the robber is a good (unbiased) estimator of the actual height, and the spread around this estimation is given by the appropriate quantile of the t-distribution (the degrees of freedom being the number of test persons minus 1), times the square root of $1 + 1/$(number of test persons), times the observed standard deviation over the differences. In formula form

$$X + \Delta \pm \xi_{n-1,p} \times \sqrt{(1 + 1/n)} \times s\Delta,$$

Table 5.3 Means of the differences, corrected height estimations on the robber plus the corresponding standard deviations, for both positions.

Position	Measured in the image (m)	Mean difference (m)	Corrected height (m)	Standard deviation (m)
Entering the bank	1.904	−0.027	1.931	0.015
Leaving the bank	1.938	0	1.938	0.015

where

X = estimation from the questioned image
Δ = average difference of the differences between the reference recordings and the ground truth
$\xi_{n-1,p}$ = quantile corresponding to the Student's t distribution with $n-1$ degrees of freedom and chosen level of confidence p
n = number of reference recordings (sets)
s_{Δ} = standard deviation of the differences between the reference recordings and the ground truth

The t-distribution is used instead of the usual standard normal distribution. This is done in order to compensate for the fact that the standard deviation used for the spread of the values is not the actual standard deviation but an estimation, based on a limited sample. The constant term with the square root compensates for the same phenomenon, but then for the estimation of the systematic bias.

In the example this means that based on the image of the robber entering the bank, with a confidence range of 95% the actual height of the robber is contained in the interval given by

$$1.931 \pm 2.365 \times 1.061 \times 0.015 = 1.931 \pm 0.038 \text{ m}.$$

Analogously for the image of the robber leaving the bank the interval is given by

$$1.938 \pm 2.365 \times 1.061 \times 0.015 = 1.938 \pm 0.038 \text{ m}.$$

Since they are the outcome of independent processes (different images, robber in different poses, different sets of reference recordings), the results of the two analyses may indeed be combined into one confidence interval for the actual height of the robber. The statistical details of this analysis are described in Alberink and Bolck (2008), theorem A1. The result is that with a confidence range of 95% the actual height is contained in the interval given by

$$1.935 \pm 0.027 \text{ m}.$$

This reported height is the height of the robber including shoes and headgear. In case of a comparison of the reported height with the height of a suspect, this should be taken into account.

5.4.2 Speed Measurement

To calculate the speed of a vehicle recorded by video, at least two images with the vehicle visible are needed. The position of the vehicle in the two images can be determined using photogrammetry. With knowledge about the time elapsed between the recording of the two images the average speed between the two locations of the vehicle can be derived. The most promising photogrammetry techniques for this kind of measurements are as follows:

- Multi-image photogrammetry: mixing of uncontrolled images with controlled images
- Reverse projection

Since the end result is about a velocity, only an estimation of the measurement uncertainty for the photogrammetry part is not enough. Either the uncertainty in the timing should also be determined separately, or the uncertainty in the velocity should be derived directly. These two different approaches can be solved using the earlier described method. The original recording system needs to be available though, otherwise the suggested methods cannot be used. For both approaches an example is given by a case which is described in detail by Hoogeboom and Alberink (2010). The case example is about a car that got involved in a lethal accident. The driver was suspected of exceeded speeding when he was driving along the city, which ended in a collision with another car. The car was being recorded by several cameras from a CCTV system of the city surveillance system. In the first approach in this example, travelled distance and time interval are estimated separately by using reference recordings of a positioned car and reference recordings of a calibrated stopwatch. The separate measurement uncertainties are combined to derive the measurement uncertainty for the velocity. In the second method the measurement uncertainty for the speed is derived directly using reference recordings of a car driving by at known speeds.

5.4.2.1 First Approach: Distance and Timing Separated

The car was being recorded by two cameras where the distance between the cameras is about 600 m. In Figure 5.12 the two images from the car driving by are given.

The photogrammetry technique used to extract the positions of the car was reverse projection using live video. A similar car was positioned at the location by a driver who got instructions from an operator looking at the live video of the camera while comparing the image with the questioned image. Figure 5.13 shows an image of the positioned car from both cameras.

The instructions were repeated by three different operators; each interpreted the images and gave the instructions to the driver. Once the car was positioned, the position was measured by a total station. The distance travelled by the car was derived by measuring the distance between the two positions by every operator. The measured distances by the three operators were 629.0, 629.2 and 630.0 m. The mean distance between these three measurements is 629.4 m, with a standard deviation of 0.53 m. The standard deviation can be used to calculate the borders of a 95% confidence interval.

Figure 5.12 Images from the CCTV system of the car driving by. The car of interest is indicated by the grey ellipse.

Figure 5.13 Images from the CCTV system of the positioned car.

Assuming normality of the sample, which is reasonable, a $0.95 \times 100\%$ confidence interval for the real distance is given by

$$629.4 \pm 4.30 \times 0.53/\sqrt{3} = 629.4 \pm 1.3 \text{ m}$$

In this calculation the factor 4.30 is the 95% quantile corresponding to the Student's t distribution with $3 - 1 = 2$ degrees of freedom.

For the time difference between the two images the time given by the system was used, which was 28.227 s. The accuracy of this time difference was estimated by making reference recordings of two timers. In front of both cameras timers were installed and recorded. From these recordings, differences between 21 time intervals of approximately 28.227 s given by the camera system and given by the timers in

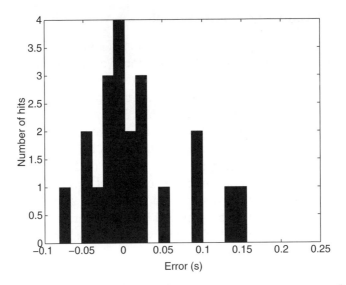

Figure 5.14   Histogram of the differences between the time interval readings from the timers and the time given by the camera system.

the images were examined. The distribution of these differences is presented as a histogram in Figure 5.14.

The mean difference of these intervals is 0.017 s, with a standard deviation of 0.061 s. A $0.95 \times 100\%$ confidence interval for the difference is given by

$$0.017 \pm 2.086 \times \sqrt{(1 + 1/21)} \times 0.061 \, \text{s} = 0.017 \pm 0.13 \, \text{s}.$$

In this calculation the factor 2.086 is the 95% quantile corresponding to the Student's t distribution with $21 - 1 = 20$ degrees of freedom. The combination of the distance and timing confidence intervals has at least 90% confidence,[9] and a straightforward, frequentistic approach leads to (an underestimation of) 90% confidence interval for the mean velocity v of the car

$$(629.4 - 1.3 \, \text{m})/(28.227 + 0.147 \, \text{s}) \le v \le (629.4 + 1.3 \, \text{m})/(28.227 - 0.113 \, \text{s}), \text{ or}$$

$$22.1 \, \text{m/s} \le v \le 22.4 \, \text{m/s}, \text{ or } 79.7 \, \text{km/h} \le v \le 80.8 \, \text{km/h}.$$

---

[9] Suppose that for certain intervals I_x and I_t, $P(\delta x \in I_x) = P(\delta t \in I_t) = 0.95$. Then the following holds:

$$P(\text{not } \{\delta x \in I_x \text{ and } \delta t \in I_t\}) = P(\{\text{not } \delta x \in I_x\} \text{ or } \{\text{not } deltat \in I_t\})$$

$$\le P(\text{not } \delta x \in I_x) + P(\text{not } \delta t \in I_t) = 0.05 + 0.05 = 0.10,$$

and hence $P(\delta x \in I_x \text{ and } \delta t \in I_t) = 1 - P(\text{not } \{\delta x \in I_x \text{ and } \delta t \in I_t\}) \ge 0.90.$

5.4.2.2 Second Approach: Velocity Directly

On another location in the city close to the location of collision, the car was recorded by one camera as it drove by. Figure 5.15 shows the two images of the car captured by that camera.

The photogrammetry technique used to extract the positions of the car was reverse projection using a 3D model. For this a (simple) 3D model of the environment was made on the basis of measurements made with a total station. A 3D (simple) model of the car was made using technical drawings of the car. A camera match was made and the position of the car was derived by positioning the 3D model of the car as an overlay on the CCTV image of the car, assuming the car was at street level. The positioning of the car was performed by three operators independently. The distance between the two derived positions was 11.66 m, averaged over the operators. Figure 5.16 shows the two locations of the 3D model of the car where the image of the second position of the car is shown in the background.

Figure 5.15   Images from one camera from the CCTV system of the car driving by.

Figure 5.16   Locations of the 3D model of the car.

Figure 5.17   Example of one set of images from a reference recording.

The time between the two images given by the CCTV system is 0.597 s. The point estimate of the velocity of the vehicle becomes

$$11.66/0.597 = 19.5\,\text{m/s} = 70.3\,\text{km/h}.$$

To estimate the measurement uncertainty, reference recordings were made from a similar car driving by the same camera under the same circumstances and measuring the speed of the car at the moment the car was driving by the camera, using a lascr gun. Figure 5.17 shows one set of images from the set of reference recordings. In total, 10 sets of different reference recordings were used.

The mean difference between the velocity derived from the images and that recorded by the laser gun is −0.18 m/s, with a standard deviation of 0.67 m/s. The mean difference −0.18 m/s is used as a correction on the derived speed of 19.5 m/s on the questioned car, whereas the standard deviation of 0.67 m/s determines a confidence band for the questioned recording. In this way a 95% confidence interval is obtained of

$$19.5 + 0.18 \pm 2.262 \times \sqrt{(1 + 1/10)} \times 0.67\,\text{m/s} = 19.7 \pm 1.6\,\text{m/s, or}$$

$$65\,\text{km/h} \le v \le 77\,\text{km/h}.$$

In this calculation the factor 2.262 is the 97.5% quantile corresponding to the Student's t distribution with $10 - 1 = 9$ degrees of freedom.[10]

The proposed methods are only valid if the reference images (sets) are comparable to the questioned images. The first method is sensitive for differences in image blur caused by the movement of the car in the questioned images and the absence of movement in the reference images. For the second method the difference in blur caused

---

[10] Note that the 97.5% quantile is needed since the interval has a minimum and a maximum value, that is to say, the procedure is two-sided. Because of this both for the minimum and for the maximum 2.5% probability is lost, and all in all a 95% confidence level is achieved.

by movement of the car will be less because the reference recordings are also from a moving car. Ideally, the test drives are operated at the same speed as the questioned drive to meet the requirement of best comparable images. In practice this means: first make an estimation of the speed of the car in the questioned images and use the estimation to set up the test drives.

In this case study the first method produced a much higher accuracy compared to the second method. The reason for this difference is not the used method, but the difference in distance and time. The mean speed determined by the first method is estimated over a long distance and therefore the relative error is small. A drawback of the long distance is that the estimated speed may be a gross underestimation of the top speed actually driven along that path.

5.4.3 Determining the Absolute Position of an Object

The absolute position of an object or person can be interesting in cases where the position of the perpetrator or the victim is in question. It can also be used when the position of a vehicle has to be determined at a certain moment. The previous example where the speed of the car was determined by positioning the car at the scene is an example of this situation. In some speed measurement cases it is not possible to get reference recordings of a moving vehicle. This can be due to the traffic situation: sometimes it is not safe to drive at the speeds that are needed for the validation experiment. In these cases the validation of the time and position measurement has to be done separately.

In general all techniques used for speed measurement or height measurement also apply to measuring the position: projective geometry (vanishing points), multi-image photogrammetry (mixing of uncontrolled images with controlled images) and reverse projection. The used method has little influence on the accuracy of the measurement. The measurement error usually depends on interpretation errors of the operator and on the image quality.

Determination of the position of an object will be illustrated with a case example. The technique used was reverse projection using a 3D model for the measurements and reverse projection with live video for the reference recordings. In this case the question was to determine the position of a motorcycle at a given moment. For the court it was important to know on which side of the road the motorcycle was driving. Figure 5.18 shows the image sequence that was used in this case.

The first decision that had to be made was: What defines the position of the vehicle? In another camera view right after this sequence, the motorcycle performs a wheelie. This sequence will not be discussed here, but it influenced the decision what part of the vehicle was used to measure the position: in this case the contact area of the rear wheel with the road was chosen. The rear wheel is the only part of the motorcycle that is in contact with the road surface at all moments. If both wheels would have stayed on the ground, any part of the motorcycle could have been used. In practice a

Figure 5.18   The image sequence. The motorcycle is indicated by a grey ellipse.

part of the vehicle must be chosen that can be measured with a total station during the recording of the reference images. Given the 3D model of the scene and the camera match, the position of the motorcycle was measured by three operators and the results were averaged. The averaged positions of the rear wheel of the motorcycle at different moments are shown in a top view in Figure 5.19.

The positions in Figure 5.19 are indications for the actual position of the motorcycle on the different images. Validation measurements were performed to determine the systematic bias and the random error. For the reference recordings in this specific case it is possible to take two different approaches: either the motorcycle is positioned as close as possible to the positions in the images in Figure 5.18, or the motorcycle is positioned on different positions on the road in a grid-like manner. In images with a high point of view, it is easier to get a rough estimate of the position and in these images one would opt for the first method. Using reverse projection on the live video, the observer behind the CCTV system can guide the object to this position and try to recreate the image. In images with a low point of view or where an object is far away, it is harder to get a rough estimate of the position. In these images one could opt for the second method. The main difference is that the first method has a higher accuracy for a single position and the second method measurements are valid over a larger area. The downside of the first method is that the validation measurements are unusable if the position of the questioned object is not near the positions of the validation object.

For the reference recordings in this case, a motorcycle was moved along five tracks that globally followed the movement of the motorcycle in the images of Figure 5.18. These five tracks were spread out over the width of the road surface. The first track on the left side of the road, the second in between the left side and the middle, the third on the middle of the road, the forth in between the middle and the right side and the

Figure 5.19 Averaged positions of the rear wheel of the motorcycle as measured.

fifth on the right side of the road. At intervals the motorcycle stood still and a reference recording was made. As a result there are five reference images per position, and these positions are spread out over the full width of the road. Figure 5.20 shows the five reference images for position 4.

The position of the motorcycle during the reference recordings, the ground truth, was measured using a total station. The location of the contact area between the road surface and the rear wheel was marked on the road surface with chalk. The position of the chalk line was measured. This can introduce an unknown error into the location measurement; however this error is assumed negligible compared to the systematic bias and the random error on the whole measurement.

At the lab, the positions of the test motorcycle in the images were measured using the same technique as was used for the motorcycle during the incident. The positions of the motorcycle in the reference recordings were measured by three operators. These measurements were used to determine the systematic bias and the random error per position. The systematic bias is used as a correction factor on the measured position; the random error is used to describe the area of possible positions. This is visualized in Figure 5.21. The grey dots are the results of the measurements on the questioned

Figure 5.20   Reference images for the fourth position.

images of Figure 5.18. The grey rectangles are the areas with possible positions after the validation experiment.

In Figure 5.21 it is visible that sometimes there is a significant difference between the position measured in a single image and the area of possible positions after the validation experiment. This is most obvious for the results of the first position at the top of Figure 5.21. The conclusion of the investigation was that after turning, the motorcycle was driving on the right half of the road.

5.4.3.1   Determining the Position of a Camera

In some cases the question may raise what the position of the camera or the position of the person holding the camera, has been. In most cases these cameras are handheld or mounted onto a vehicle. A case could be to determine the path of a car with a camera mounted behind the windscreen. Another case could be to determine the position of a ship along a coast line with someone on the ship making video recordings.

This type of investigations can be done by making use of a camera matching algorithm and a 3D model of the environment. Most camera matching algorithms work by indicating which point in the 3D models corresponds with which point on the image. The algorithm can calculate the position of the camera. If the camera match is performed by several operators, it is possible to calculate an average position of the camera. The accuracy of such a measurement is determined by the resolution and quality of the video image and by the quality of the 3D model.

As with other types of measurements, validation is needed to determine the accuracy of the measurement. In these cases it is preferred to make reference recordings. The position of the camera has to vary with each recording and this position has to be measured by other means. Measuring the position of the camera during the reference

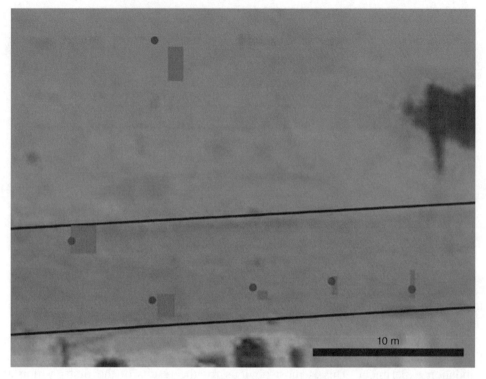

Figure 5.21 Results of the validation experiment. The grey dots are the positions measured in the sequence of Figure 5.18. The blue rectangles represent intervals for the actual positions in the sequence and are the result of the validation experiment.

recordings can be done with a total station or with a GPS-receiver, when the area with possible positions is large. The reference recordings are measured in the same way as the questioned image. From the differences between the positions measured in the reference recordings and the corresponding real positions, the systematic bias can be determined. From the spread of the positions found in the reference recordings, the random error can be determined. The end result will be a confidence area with possible positions for the camera in the questioned image given some percentage of confidence.

5.5 3D Modelling/Scenario Testing

In forensic photogrammetry problems it comes down to retrieving 3D information from images. This might involve building a (simple) 3D computer model of a crime scene which introduces the technical possibilities of 3D visualization or virtual scenario testing. In a lot of different areas in forensic investigations 3D visualizations have been presented. Some examples can be found in Schofield (2009), Thali *et al.* (2005) and Urschler *et al.* (2012). The possible level of detail of the 3D model of the crime scene depends on the available data. Are there photos? Is there 3D laser scan data? Is the crime scene still available, unchanged? The reliability of the data is important and defines the usability. Things might have changed between different methods used to record the crime scene, especially if more time has gone by. Also during recording things might change: a door can be more or less opened, a chair can be moved, etc. The resulting model therefore will consist of hard and soft evidence. Examples of hard evidence are bullet holes, bloodstain patterns, shoe marks, etc., as long as these traces can be related to the crime scene. Examples of (more) soft evidence are bullet cases (they fall on the ground and can be tossed away after the incident), location of a knife, etc. Also the absence of traces can be very interesting. In Agosto *et al.* (2008) an example is given where the absence of shoe marks was investigated given the huge amount of bloodstains. What are the possible trajectories without leaving traces? It is practically impossible to model every detail in the crime scene. Objects on a table, on a closet, on the ground, where was it at the moment of crime? When the model is made the purpose of the model must be known. What is the question that needs to be answered? This will direct the level of detail of the model and the importance of certain details. Decisions must be made about the level of detail and this implies that once the model is made, it is not straightforward to use it to answer a different question. The risk of visualization is that it can be less about truth finding and more about showing opinions. This is especially risky because people are five times as likely to remember something they see and hear rather than something they hear alone; and they are twice as likely to be persuaded if the arguments are buttressed with visual aids (Ma *et al.* 2010).

A major challenge in visualization is how to visualize uncertainties. In shooting incidents the trajectory of a bullet can be reconstructed if enough traces are left by the bullet, that is a bullet hole through a car. How to reconstruct the bullet trajectory based on the bullet hole in the car needs interpretation by a fire arms expert. The fire arms expert needs to make an estimation of the possible bullet trajectories that can result in the given bullet hole, given the bullet, the weapon, the distance and the materials of the car. The possible bullet trajectories might be visualized using a cone. The apex will be at the location of the bullet hole and the base at the location of the weapon. In Figure 5.22 an example is given from a bullet trajectory where the bullet was found in a parked car. The arrows at the end of the cone indicate that the cone stretches further to the right. The cone wasn't drawn all the way for a clearer view of the shape of the cone.

A visualization can consist of one or more still images or one or more videos. In case of one or more still images the visualization is about a situation at a certain moment in time seen from one or more viewpoints. For instance, the visualization of a shooting incident where the moment of shooting is visualized with images from one or different viewpoints. In case of visualization using video the visualization is called an animation. Two situations can be distinguished: animation from a static situation, or animation from a dynamic situation. The visualization of a shooting incident from the moment

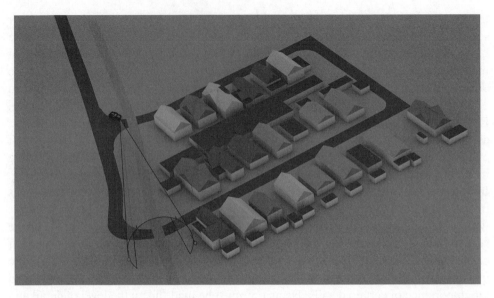

Figure 5.22   Bullet trajectory visualized by a cone to visualize the trajectory uncertainties.

of shooting where the point of view goes from the view of the shooter to the view of the victim is an example of an animation of a static situation. The virtual camera in the 3D environment is animated where the situation is static. An example of an animation of a dynamic situation is the animation of the trajectory of two cars until the moment of collision. This kind of animation is widely used in forensic accident investigations. Usually it is the result of physics using several parameters like vehicle mass, friction of the tires with the road surface, etc., in combination with the boundary conditions given by the traces on the road and the vehicles. An animation like this is based on the outcomes of a simulation; and especially in traffic accidents, research is done in error estimations (Spek 2000). The level of knowledge about the boundary conditions and parameters dictate the reliability of the simulation and thus the maximum reliability of the animation. In most traffic accident cases the animation is about the movement of the vehicles, where the persons in the vehicles are left out. If the movements of the persons in the vehicle become an issue, the animation gets to a different level. Assumptions have to be made about the behaviour of the persons and more solutions (scenarios) are possible. This kind of problems also arises in other areas of forensic investigations if animations of a dynamic situation are made. For instance the simple fact of a shooter moving from one room to another can be complicated. Did he go in straight line? Were their obstacles on his way? Does he have physical disabilities? Did he walk or run? The animation of one phase (position in Room 1) to another (position in Room 2) needs interpolation, and the assumptions for this interpolation depend on the scenario. These kinds of visualizations are therefore not unambiguous and will always visualize a certain scenario.

To show some of the difficulties that can be encountered an example of a case is given: a shooting incident has taken place where the shooter has killed the victim. The shooting was the result of a fight and the victim was killed with one bullet. The court in higher appeal asked for a 3D-visualization with the following question:

Was the victim shot when the shooter was walking towards the victim, or was the victim shot during a wrestle with the shooter?

The hard evidence in this case consisted only of a bullet-channel in the body (belly) of the victim. A reliable analysis of gun residue was not possible because the victim was first treated at the hospital where possible traces of gun residue might have been removed. The crime scene was not available anymore due to the long time between the shooting incident and the question by the court (about 2 years). Without any other information the number of possible visualizations is endless and can be misleading. If only the parts that are part of the hard evidence would be visualized, the visualization would be hard to understand. Figure 5.23 shows a possible scenario with the cone covering the possible bullet trajectories based on the wounding channel found in the victim. On the left hand side the two persons are completely visible. On the right-hand side the parts of the bodies that can be in any position, are visualized with limited opacity.

The positions of the victim and the shooter in this visualization are arbitrary. The victim is visualized in a rather resigned posture where the shooter is visualized in a rather awkward posture to fire a gun. This kind of bias has less effect in the visualization on the right-hand side, but can only be ruled out if the visibility of the parts of the body that is not known is set to zero. Although even then, the position and orientation in space will have an effect.

The basis of this visualization (the two human models and the cone) was used in a court session where the judges, the prosecutor and the lawyer came up with scenarios. These scenarios were visualized directly giving them feedback. It turned out that none of the given scenarios could be ruled out. In this case the visualization didn't help

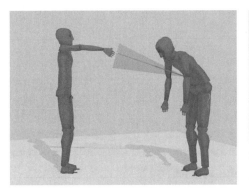

Figure 5.23 Visualization of a shooting incident where the cone depicts the possible bullet trajectories based on the wounding channel found in the victim's body.

the judges to come to a certain scenario, but it gave them arguments not to rule out scenarios based on the hard evidence.

Another application in visualization is combining 3D modelling with video. The 3D model can be used as a tool to track people and cars in CCTV images (Edelman and Bijhold 2010). The 3D model combined with camera matches for the CCTV images from different cameras can help to understand the relation between the different camera views and therefore the interpretation of the CCTV images. By extrapolating the position and movement of cars or people, it is possible to predict in which camera view they can be expected after exiting a certain camera view. Extrapolation of video information using a 3D model is also performed in the next case example. The case is about the collapse of a large number of racks in an industrial warehouse. The warehouse was filled with racks which contained among other things mainly plastic containers of about 1 m³, filled with liquids. The scenario to investigate was whether it was possible if one container was pushed between the beams of one rack, this rack would collapse and introduce a domino effect. A test was performed where one container was positioned in one of the remaining racks, and it was pushed between the beams using a digger. The test was recorded by a video camera. Afterwards a 3D model of the racks and a container was made and a camera match was made on the test video. In the 3D model an extra rack was positioned in front of the rack that contained the falling container at a distance equal to the racks in the warehouse. Figure 5.24 shows two images: on the left-hand side a still image from the video and on the right-hand side the 3D model projected on the video.

Figure 5.25 shows two different views of the 3D model after the container has fallen and the path of the container is visualized by some semi-transparent containers.

The video from the experiment with the 3D model can be found in the enhanced e-book and on the book's website. The test was performed with the container at the second storey. The path of the falling container can be extrapolated as was it falling from the third or the fourth storey. It appeared that in the test the container could just

Figure 5.24   Visualization of video combined with a 3D model. Left: still image from the video. Right: 3D model projected on the video.

Figure 5.25   Visualization of the path of the falling container. Left: view according to the video camera. Right: side view of the situation.

reach the opposite rack from the second storey. From a higher storey the container would reach the rack more firmly. The 3D visualization helped to explain what might have happened.

5.6   Summary

Photogrammetry is a widely adopted technique in forensic investigations and several methods are available to measure in images. If the result is used as evidence, independent of the method used for photogrammetry, there is the need for validation of the result. Once a proper validation is performed the result of the measurement on the questioned image is defendable in court. On uncontrolled images, validation can only be case driven, so validation experiments have to be set up for such specific cases. Paradoxically this makes it easier to use new techniques, since in these cases the whole measurement tool is validated (if the experiment is set up correctly). New techniques like automated point cloud extraction from photos are promising and might take 3D modelling closer to a wider range of forensic investigation. In some cases 3D models can be used for visualization or scenario testing. This however is an area that has a lot of pitfalls, and the admissibility in court may also depend on the legal system. However, forensic investigators should not be discouraged about using 3D visualizations; it is a promising field and the way to bring this to a higher level is by getting experience in this field.

References

Agosto, E., Ajmar, A., Boccardo, P., Tonolo, F., Lingua, A. Crime Scene Reconstruction Using a Fully Geomatic Approach, Sensors, vol 8, 2008, pp 6280–6302.

Alberink, I., Bolck, A., Obtaining Confidence Intervals and Likelihood Ratios for Body Height Estimations in Images. Forensic Science International, vol 177, 2008, pp 228–237.

Albertz, J., A Look Back, 140 Years of "Photogrammetry". Photogrammetric Engineering & Remote Sensing, vol 73, 2007, pp 504–506.

Alkan, R. M., Karsidag, G. *Analysis of the Accuracy of Terrestrial Laser Scanning Measurements*, FIG Working Week 2012, Knowing to manage the territory, protect the environment, evaluate the cultural heritage, Rome, Italy, 6–10 May 2012.

Criminisi, A., Zisserman, A., van Gool, L., Bramble, S., Compton, D., A Bew Approach to Obtain Heigt Measurements from Video. Proceedings of SPIE, vol. 3576, 1998, pp 227–238.

Edelman, G., Bijhold, J., Tracking People and Cars using 3D Modelling and CCTV, Forensic Science International, vol 202, 2010, pp 26–35.

Edelman, G., Alberink, I., Hoogeboom, B., Comparison of the Performance of Two Methods for Height Estimation, Journal of Forensic Sciences, vol. 55, no. 2, 2010. pp 358–365.

Georgantas, A., Brédif, M., Marc Pierrot-Desseilligny, *An Accuracy Assessment of Automated Photogrammetric Techniques for 3D Modeling of Complex Interiors*, 2012, http://www.int-arch-photo-gramm-remote-sens-spatial-inf-sci.net/XXXIX-B3/23/2012/isprsarchives-XXXIX-B3-23-2012.pdf (consulted 10 April 2015).

Hoogeboom, B., Alberink, I., Measurement Uncertainty When Estimating the Velocity of an Allegedly Speeding Vehicle from Images. Journal of Forensic Sciences, vol 55, no 5, 2010, pp 1347–1351.

Hugemann, W., Brunn, A., *Unfallrekonstruktion – Band 1: Grundlagen.* Chapter 1.8, 2007, Schönbach-Druck GmbH, Münster.

Luhmann, T., Robson, S., Kyle, S., Harley, I., *Close Range Photogrammetry – principles, techniques and applications.* 2006, Whittles Publishing, Dunbeath.

Ma, M., Zheng, H., Lallie, H., Virtual Reality and 3D Animation in Forensic Visualization, Journal of Forensic Sciences, vol 55, Issue 5, 2010, pp 1227–1231.

Schofield, D., Animating Evidence: Computer Game Technology in the Courtroom, Journal of Information Law & Technology, vol 1, 2009, pp 1–20.

Spek, A., *Implementation of Monte Carlo technique in a time-forward vehicle accident reconstruction program*, Proceedings of the Seventh IES Conference on Problems in Reconstructing Road Accidents, IES Publisher, Krakow, 2000, pp 233–238.

Thali, M., Braun, M., Buck, U., Aghayev, E., Jackowski, C., Vock, P., Sonnenschein, M., Dirnhofer, R., VIRTOPSY – Scientific Documentation, Reconstruction and Animation in Forensic: Individual and Real 3D Data Based Geo-Metric Approach Including Optical Body/Object Surface and Radiological CT/MRI Scanning, Journal of Forensic Sciences, vol 50, no 2, 2005, pp 1–5.

Topolšek, D., Herbaj, E. A., Sternad, M., The Accuracy Analysis of Measurement Tools for Traffic Accident Investigation, Journal of Transportation Technologies, vol 4, 2014, pp 84–92.

Urschler, M., Bornik, A., Scheurer, E., Yen, K., Bischof, H., Schmalstieg, D., Forensic-Case Analysis: From 3D Imaging to Interactive Visualization. IEEE Computer Graphics and Applications, vol. 32, no 4, 2012, pp 79–87.

6

Advanced Multimedia File Carving

Martin Steinebach, York Yannikos, Sascha Zmudzinski and Christian Winter
Media Security and IT Forensics (MSF), Fraunhofer Institute for Secure Information Technology SIT, Darmstadt, Germany

6.1 Introduction

Forensic evidence on storage media is not always directly accessible by the file system of the operating system. Whenever files have been deleted or replaced, or when the file system is corrupted, file carving is a common method to collect all remaining evidence and present the results for further investigation.

File carving works completely independently of the underlying file system so that even data[1] which was previously deleted, for example with the intention to destroy evidence data, can be recovered if it has not been overwritten physically yet. To achieve this, file carving uses a low-level perspective when analyzing stored data. It retrieves the data as stored physically on the respective media and only utilizes file system data if available to distinguish between files reachable via the file system and data not belonging to any referenced file.

Multimedia data is often the subject of file carving, and therefore a number of specific file carving strategies and algorithms solely to handle media files have been developed. On the one hand, media data like photographs or videos are valuable evidence for many types of investigations. On the other hand, the size and structure of multimedia data require and allow for methods not applied to other data types.

---

[1] Although 'data' is a plural word according to its Latin origin, we use it as singular mass noun according to common practice.

Handbook of Digital Forensics of Multimedia Data and Devices, First Edition.
Edited by Anthony T.S. Ho and Shujun Li.
© 2015 John Wiley & Sons, Ltd. Published 2015 by John Wiley & Sons, Ltd.
Companion Website: www.wiley.com/go/digitalforensics

In most countries not only the distribution but also the possession of certain image and video data like child pornography is illegal. Findings of such data during forensic investigations can be very important evidence. Therefore, the capability of recovering and analyzing image and video data is crucial for forensic investigators. Nevertheless, the identification of illegal content requires a lot of time since no reliable tools for automatic content classification are available. By now, each multimedia file has to be inspected manually by the forensic investigators.

Video files for example tend to be large. High-quality videos easily come in file sizes of 1 GB and more. Fragmentation of such files on a storage device is much more likely than of smaller files like documents. The well-defined and often independently segmented structure of media files facilitates content extraction from even small parts of a complete file: Most audio and video files, especially those derived from the the MPEG standards, are rather streams of data following a different file structure than, for example WAVE audio files, JPEG images, or non-multimedia data. Such streams mainly consist of many frames and corresponding frame headers which can be used as starting points for file carvers. Therefore, it is possible to extract even individual chunks of multimedia streams when others are already overwritten.

After the recovery of multimedia data, a time-consuming inspection and search for evidence data is currently required. To support forensic investigators during this process, several approaches exist for classifying the content by using blacklisting and whitelisting. While blacklisting and whitelisting approaches are mostly based on cryptographic hashing and therefore have a high false negative rate, robust hashing provides a promising way for detecting copies of multimedia data after content-preserving changes like lossy compression or format conversions.

In this chapter we discuss the state of the art in multimedia file carving. As we combine common file carving approaches with content identification, we call the overall result *advanced* multimedia file carving. In this context, 'advanced' means that the file carving strategy offers more assistance than the sole collection of evidence data.

The chapter is organized as follows. First we give a brief introduction to general storage strategies and file carving. Then we provide an overview of multimedia file formats. Based on this, we discuss multimedia file carving utilizing media-specific knowledge. To support the automatic analysis of evidence, we add a section on content identification based on different hashing methods. Finally, frameworks aiming to combine the previous mechanisms are examined. At the end of the chapter, we summarize it with a number of conclusions.

6.2   Digtal Data Storage

This section is a brief introduction of storage technology aspects relevant for file carving, for example storage devices and file systems. However, we expect that the reader is already familiar with these terms and technologies. The scope of this section is to specify terminology and notations of storage technology for avoiding ambiguity in this chapter and to provide some background information helping to understand the

```
00000:   FF D8 FF E0 00 10 4A 46   49 46 00 01 01 01 01 5E   |......JF IF.....^|
00010:   01 5E 00 00 FF DB 00 43   00 06 04 05 06 05 04 06   |.^.....C........|
00020:   06 05 06 07 07 06 08 0A   10 0A 0A 09 09 0A 14 0E   |................|
00030:   0F 0C 10 17 14 18 18 17   14 16 16 1A 1D 25 1F 1A   |............%..|
00040:   1B 23 1C 16 16 20 2C 20   23 26 27 29 2A 29 19 1F   |.#..., #&')*)..|
00050:   2D 30 2D 28 30 25 28 29   28 FF DB 00 43 01 07 07   |-0-(0%()(...C...|
           ⋮              ⋮                  ⋮                ⋮
09D40:   24 49 9D 3A 44 09 83 3A   74 2A 09 82 4C E9 D0 00   |$I.:D..:t*..L...|
09D50:   99 17 3A 74 0E 80 4C E9   D0 22 41 33 A7 42 20 98   |..:t..L.."A3.B.|
09D60:   24 CE 9D 2A A0 B7 10 4B   19 D3 A1 10 5C C1 2C 67   |$..*...K....\.,g|
09D70:   4E 80 3B 8C 1D C4 CE 9D   15 50 49 9C 3B CE 9D 30   |N.;......PI.;..0|
09D80:   A6 A8 96 31 89 D3 A4 53   D6 4C E9 D2 05 BC 43 CE   |...1...S.L....C.|
09D90:   9D 28 FF D9                                         |.(..|
```

Figure 6.1 Hex dump of a JPEG file (see Section 6.2.3 for an explanation of the term 'hex dump'). The sequence FF D8 denotes the beginning of the file and the sequence FF D9 denotes the end of the file. The 'JFIF' identifier at an offset of 6 bytes is a visually conspicuous indicator for JPEG files. The JPEG format is described in more detail in Section 6.4.1.

chapter. Moreover, this section provides entry points for a deeper understanding of storage media and their forensic analysis. The interested reader might delve into this topic in the book *File System Forensic Analysis* by Carrier (2005).

The basic unit for specifying storage capacities and data amounts is the *byte*, abbreviated as B. One byte is an octet of 8 *bits*. Each bit can take two values usually declared as 0 and 1. The content of a byte is commonly denoted as hexadecimal number with two digits, for example the octet 1101 1001 is denoted as D9. Hence the content of a sequence of bytes is denoted as sequence of pairs of hexadecimal digits like, for example FF D8 FF E0 00 10 4A 46 49 46 00 01, which is a typical start code of a JPEG file (see Figure 6.1 and Section 6.4.1).

For measuring amounts of data we distinguish the decimal prefixes such as 'kilo' ($k = 10^3$), 'mega' ($M = 10^6$) and 'giga' ($G = 10^9$) from binary prefixes such as 'kibi' ($Ki = 2^{10}$), 'mebi' ($Mi = 2^{20}$) and 'gibi' ($Gi = 2^{30}$). Hence 1 MB is 1 000 000 bytes while 1 MiB is 1 048 576 bytes.

6.2.1 Storage Devices

Digital data is persistently stored on various kinds of storage media such as hard disks, optical discs[2] (e.g. CD, DVD and Blu-ray) and USB flash drives. All these media present their storage area as sequence of data blocks commonly called *sectors*[3] to a

---

[2] Note that we write 'disc' with 'c' in the case of optical discs and 'disk' with 'k' in the case of hard disks and floppy disks according to common practice.

[3] Note that some technologies define a different name, for example 'block', for what we call 'sector' or use the word 'sector' with a different meaning. Even when restricting to hard disks, the word 'sector' is ambiguous as it is also used with the meaning of a geometrical (circular, three-dimensional) sector intersecting all tracks on all disk platters in the context of the legacy cylinder–head–sector (CHS)

computer. Sectors are the addressable units of the storage device. This means that data is retrieved from the device by requesting the content of a sector with a read command containing the sector address, which is usually a sequential number according to the logical block addressing (LBA) scheme. The delivered sector data is a series of bytes with a fixed total length. The size of a sector is normally a power of 2. This view of storage organization is the lowest level of data representation relevant for file carving. Lower levels of device architecture are relevant for other data recovery methods aiming at retrieving data from physically corrupted media, which can be combined with file carving.

Internally, modern hard disks and optical discs store the user data seen by the host together with error detection/correction code (EDC/ECC) data such as Reed–Solomon codes. The next deeper level applies line coding techniques such as run-length-limited (RLL) coding schemes used in hard disks, or the eight-to-fourteen modulation (EFM) used in CDs. Finally, this enhanced and transformed data is encoded physically on the storage medium together with additional physical control structures.

Mass storage devices based on flash memory also use ECC codes, but they do not apply any line coding scheme. However, a different transformation layer is present in most such devices due to the fact that flash memory cells withstand only a limited number of program-erase (P/E) cycles. Most flash memory devices internally use a flash translation layer (FTL) and over-provisioning for mitigating this problem. The FTL is a transparent, dynamic mapping of logical sector addresses to physical places. This mapping is updated by garbage collection and wear-leveling algorithms during usage of the device. In particular, unused sectors may be remapped or erased at any time.

Hard disks and floppy disks have mainly used a sector size of 512 bytes since the 1980s due to the defining influence of the IBM PC architecture. While floppy disks essentially disappeared in the mid 1990s, hard disks have remained ubiquitous devices with quickly growing capacities. Hard disk manufacturers prepared for switching to larger sectors in the 2000s. In 2010 they finalized the advanced format (AF) standard for hard disks with a sector size of 4096 bytes, that is 4 KiB. Modern hard disks use this sector size internally, but they usually emulate 512-byte sectors through their firmware for ensuring compatibility with existing computer systems. Hard disks which expose the sector size of 4 KiB to the host system currently spread into the market and will be typical investigation targets for forensic examiners in some years.

Data compact discs (CDs) , that is CD-ROMs and recordable variants[4], inherit the low-level data organization from audio CDs (Compact Disc Digital Audio, CD-DA), which provide 2352 bytes of audio data per sector. This audio data is expanded by auxiliary data and low-level coding layers to 57 624 physically encoded bits (Philips

---

addressing scheme. Hence, such a geometrical sector contains many sectors in the sense of our definition of 'sector'.

[4] The label 'ROM' is used with twofold meaning as it has been introduced for distinguishing data CDs from audio CDs although it is an acronym for 'read-only memory' defining a distinction to recordable media, too.

Table 6.1 Typical sector sizes (true size and size presented to host system).

	Magnetic disks			Optical discs			Flash devices
	Floppy disk	Old hard disk	AF hard disk	Data CD	DVD	BD	USB drive, SD card, SSD
True	512 B	512 B	4 KiB	2 KiB	2 KiB	2 KiB	Misc.
Presented	512 B	512 B	512 B, 4 KiB	2 KiB	2 KiB	2 KiB	512 B

and Sony 1980; IEC 60908 1999). Data CDs insert special data structures into the 2352 audio bytes according to one of the CD-ROM storage modes. The predominant storage mode (Sector Mode 01) uses 2048 bytes (i.e. 2 KiB) for the actual data payload and dedicates the remaining 304 bytes to another layer of header information and error correction data as a storage medium for arbitrary data must provide stronger data integrity than a medium for uncompressed audio data (Philips and Sony 1985; ECMA-130 1996). In contrast to CDs, the successor storage medium DVD[5] uses a single sector layout independently of the type of payload. It has always 2 KiB per sector for the payload, while a sector physically stores 38 688 bits (DVD Consortium 1996; ECMA-267 2001). Blu-ray Discs (BDs) also store 2 KiB of user data in each sector, but the low-level data organization differs from DVDs, and its description is only available through licensing the format specification (Blu-ray Disc Association 2002).

Flash memory devices such as USB drives, SD cards and SSDs (solid-state drives) as well as flash-based data storage modules integrated into mobile devices like smartphones present their storage space as sequence of 512-byte sectors in analogy to a hard disk. But the internal structure differs strongly from this presentation and is hidden behind the flash controller. The actual readable and programmable unit of memory inside a flash device is called a 'page', and only old 'small block' devices (many devices with a capacity smaller than 1 GB) have a page size of 512 bytes. Modern devices use a page size of 2 KiB (large blocks), 4 KiB (often called 'huge blocks'), or even more. However, unlike for hard disks, there is no indication that device manufacturers will switch to a logical sector size different from 512 bytes for new devices in the next years. Table 6.1 summarizes the typical sector sizes found for the various storage media.

6.2.2 Logical Data Organization

Users of computer systems usually do not access the data on storage devices as sequence of sectors. There are rather several layers of abstraction which allow a

---

[5] DVD has been introduced as acronym for 'digital video disc' while it has later been explained as 'digital versatile disc'.

convenient usage of storage devices. In general, memory devices are divided into *partitions*. The layout of partitions is often stored in a partition table inside a master boot record (MBR). This legacy format has been established on IBM PCs in the early 1980s, but it is still a widely used and supported format, especially for removable storage devices (e.g. USB flash drives, SD cards and external hard disks). Macintosh computers traditionally used an Apple Partition Map (APM). However, modern systems like Microsoft Windows since Windows Vista, Mac OS X since the migration to the Intel x86 architecture, and modern Linux versions support and use the organization of partitions in a GUID Partition Table (GPT) since the mid 2000s.

The system builds *logical volumes* on top of the partitions. On consumer devices a volume typically corresponds to one partition, and there is often only one partition on such a device. More complex systems like RAID[6] or logical volume management (LVM) aggregate partitions across several storage devices into one volume.

Inside a volume we usually find a *file system* which manages the storage space of the volume. A file system provides the means for creating and organizing files. Files are identified by their name and place in the directory tree. This is the view of digital memory presented to a user. Files may be created, edited, expanded, shortened, renamed and deleted. The file system must keep track of the storage space currently used, and it must know, which file resides in which part of the storage space. For this, the volume is usually divided into *clusters*[7] of fixed size, and the file system keeps a record of which clusters are occupied by which file. The cluster size is normally a power of 2 and a small multiple of the sector size.

A simple and widely supported family of file systems is File Allocation Table (FAT), which has been used for MS DOS and early Microsoft Windows versions. Especially FAT32 is still prevalent for removable storage devices. Microsoft Windows systems run nowadays on the NTFS file system. Linux systems mainly use the ext file systems, where ext4 is the current version. Mac OS systems use nowadays the HFS Plus file system. All these file systems have a default cluster size of 4 KiB on typical storage devices. Smaller clusters can be the default on volumes smaller than 1 GiB. Larger cluster sizes are often supported, too. In particular, FAT32 systems on volumes larger than 8 GiB use clusters larger than 4 KiB by default. Particular other file systems are used on optical data discs. CD-ROMs usually have an ISO 9660 file system, which is often supplemented by a Joliet file system for enabling Unicode file names. DVDs and BDs mostly employ a Universal Disk Format (UDF) file system. These file systems normally use a cluster size equal to the sector size, which is 2 KiB for optical data discs.

The clusters occupied by a file may not be in consecutive order. This phenomenon is called *fragmentation*. Each fragment of a file is a sequence of consecutive clusters.

---

[6] RAID has been introduced as acronym for 'redundant array of inexpensive disks' (Patterson *et al.* 1988), but nowadays it stands for 'redundant array of independent disks'.

[7] Note that there are various names for what we call 'cluster', for example 'block', 'allocation block' or 'allocation unit'. The term 'cluster' has been coined by the original FAT file system in the late 1970s, and it is mainly used in the context of Microsoft file systems.

Fragmentation leads to a performance loss on traditional hard disks due to seek time of the disk head assembly and rotational latency of the disk platters needed for accessing the individual file fragments. But apart from the additional access time for each fragment, fragmentation is transparent to the user of a file system. However, file carving must deal with the problem of fragmentation as file carving is applied where file system information is missing. Strategies for handling fragmentation during file carving are discussed in Sections 6.3 and 6.5.

6.2.3   Forensic Data Investigation

Forensic examiners usually create a copy of seized storage devices before they start analysis. For creating the copy, they access a device in read-only mode typically ensured by using a hardware write blocker. They perform all analysis on this copy and not on the original device. These are precautions for not altering the original evidence and to provide a complete chain of custody. The forensic copy of a device is a file called *disk image*, and the copy of a volume is a *volume image*. Any such copy is also called 'forensic image' or simply 'image'. Unfortunately, these terms can easily be confused with a visual image like a photo or graphic.

Forensic images can be analyzed with various methods. Volume images and volumes in disk images can be mounted to see what is inside the according file system. The present files can be extracted and analyzed separately. The unused space in forensic images can be inspected for further evidence, for example with a file carver. Unused space may contain remnants of previously stored content. There are many types of unused space, in particular space in a disk image not occupied by a partition, space in a volume not managed by the file system, clusters of the file system not assigned to a file, and file slack, which is space in the last cluster of a file behind the actual end of the file. If file system information or partition information is corrupted or lost, the complete image might be considered for processing by a file carver.

A very basic tool for inspecting any digital data is a *hex editor*. Such a program allows viewing the raw bytes of a file, volume, partition or device. The data is presented as hexadecimal digits as explained earlier. The hexadecimal representation of data is called *hex dump*, and there are also command line tools for creating hex dumps. Additionally, hex editors and hex dumping programs can present the data as text, that is they map the bytes to characters according to a conventional encoding like ASCII or extensions thereof such as Windows-1252. For example, the sequence 4A 46 49 46 as contained in the beginning of many JPEG files at an offset of 6 bytes (see Figure 6.1) represents the text 'JFIF'.

6.3   File Carving of Binary Data

File carving is a general method to enable accessing data lost to the file system. Before we discuss file carving designed specifically for multimedia, we briefly introduce the more general binary data file carving.

File carving describes the process of extracting file fragments from a byte stream such as a disk or volume image without file system information. This technique is mostly used to recover data from unallocated space on hard disk storage. Data in unallocated space is usually remnants of files which have been deleted from the file system. File carving can also be used where the file system is logically corrupted or where the storage area of the file system structures is physically damaged.

Basic file carvers typically identify fixed byte sequences, so-called *magic numbers* or *magic words*, that usually appear near the beginning (in the header) or the end (in the footer) of every file in a certain file format. Everything from the file header to the footer is assumed to belong to the same file, and hence the data between the identified start and end point is 'carved' out and restored into a file. This basic carving approach is called *header/footer carving*. Typical file carvers using header and footer information are *PhotoRec* (Grenier 2007), *Scalpel* (Richard III and Roussev 2005) and *Foremost* (initially developed by the Air Force Office of Special Investigations 2001).

However, if a file to be recovered is fragmented (i.e. not stored sequentially, but split up in two or more parts throughout the storage area), file carving will become difficult. A sophisticated file carver should consider each cluster boundary as potential fragment boundary. If the cluster size or the cluster alignment within a volume is unknown, it might be necessary to consider each sector boundary as potential cluster boundary, but 4 KiB clusters are often a safe assumption for file carving.

According to Garfinkel (2007), typically 10% of the files in a hard disk are fragmented. Moreover, he reported that about 17% of MPEG files and 20% of AVI files are fragmented. He introduced object validation which is a process to determine data objects using byte representations that can later be used for validating different parts of a file, hence helping in systematic reassembly of the fragments of a file. To handle file fragmentation during file carving, several approaches were proposed. Garfinkel (2007) introduced *bifragment gap carving*, a technique to validate file contents with a gap in between. However, this technique only works for files with no more than two fragments and only for small gaps. Unfortunately, files with large size tend to be fragmented more often. Multimedia video files like AVI or MPEG files as well as high-resolution image data (e.g. from digital cameras) usually have a large size.

6.4 Multimedia Data Structures

Multimedia data is commonly stored in specific file structures. Raw data files of image, audio and video in which the basic elements (e.g. audio samples or image pixels) are stored in sequence without any metadata do exist, but are rare due to their size. Today, lossy compression is applied to multimedia data before storage, requiring compression parameters and metadata for recreating the original content. This provides helpful hints for file carving, but makes it also harder to decode files which are only partially available.

As an example, we can compare raw audio samples and MP3 data. The latter features recurring header structures with standardized synchronization points. Scanning for these makes it easy to recognize an MP3 file. Conversely, raw audio data does not provide any structure. The advantage here is that if one knows that a fragment contains raw audio in a specific sampling rate, resolution and byte order, it is simple to play back the audio content as no additional information is needed. This is similar to raw image data and JPEG files. The latter can only be accessed by first learning from the JPEG header about its parameters, the former only consists of a sequence of bytes describing pixels.

Multimedia data can either come as a single media type file or in a set of multiple formats and streams. Especially video data most often is distributed in the latter form. Container data structures like those specified by the MPEG standards can contain a number of video and audio streams, all stored in one multiplexed file structure. Container file format examples are *Flash Video*[8] and *MP4*, the latter specified by the MPEG-4 standard (ISO/IEC 14496-14:2003). The individual streams can be of multiple formats if allowed by the container specifications. Audio tracks in video DVDs are often stored in multiple formats to ensure optimal support of any playback device. Table 6.2 shows a multiplexed structure of an XVID video file as an example.

While container formats often restrict the choice of stream formats, violations to this rule occur relatively often. As an example, Dolby AC-3 is not officially supported by MPEG in MP4 containers. However, MP4 video files with H.264 video stream and AC-3 audio stream can be observed from time to time. This is due to the fact that later specifications by third parties have mentioned AC-3 as a supported format and some tools followed this suggestion.

6.4.1 Digital Images

Photos, drawings, graphs, icons, scans and screenshots can be subsumed as digital images, that is digital images are digital representations of visual images[9]. While multiple types of digital images exist, one common way for distinguishing digital image types is to divide them in natural and synthetic images. Digital photographs are natural images recorded with a digital camera, and they are usually distributed in a lossy compression format. The most common one is the JPEG format, which is described in more detail in the following text. Graphs are synthetic images created on a computer and are usually stored in a lossless compression format like PNG (Portable Network Graphics) or a lossless variant of Tagged Image File Format (TIFF). For graphs with a limited amount of colours Graphics Interchange Format (GIF) is used as well.

In addition, images can come in raster or vector formats. The former are common for natural images, but are as well used for synthetic images. Vector formats are

---

[8] The Flash Video format is not related to the flash memory technology. The name 'Flash Video' refers to the brand 'Flash' owned by Adobe.

[9] Note that the terms 'digital image' and 'forensic image' (see Section 6.2.3) are used for disparate types of objects although all these objects are digital and can be relevant in forensic investigations.

Table 6.2 Excerpt from a video coded with 1324 kbit/s and featuring one XVID video and two MP3 audio streams A and B. Cluster size is 4 KiB. Multiple audio frames fit into one cluster, for example frames 2–5 all are stored in cluster 8. In this example, only P- and B-frames are included. A single I-frame form this video would need more than four clusters of storage space.

No.	Stream	Pos.	Cluster (start)		No.	Stream	Pos.	Cluster (start)	
1	Video	31 012	7/8	(28 672)	13	Audio B	40 850	9/10	
2	Audio A	33 600	8	(32 768)	14	Video	41 242	10/11	(40 960)
3	Audio A	33 992	8		15	Audio A	45 436	11	(45 056)
4	Audio B	34 336	8		16	Audio A	45 780	11	
5	Audio B	34 680	8		17	Audio B	46 124	11	
6	Video	35 024	8/9		18	Audio B	46 516	11	
7	Audio A	38 146	9	(36 864)	19	Video	47 004	11	
8	Audio A	38 490	9		20	Audio A	48 226	11	
9	Audio B	38 834	9		21	Audio A	48 618	11	
10	Audio B	39 178	9		22	Audio B	48 962	11/12	
11	Video	39 522	9		23	Audio B	49 306	12	(49 152)
12	Audio A	40 458	9		24	Video	49 650	12	

No.: Frame number in this example
Stream: Stream containing the frame
Pos.: Frame offset in bytes relative to beginning of file
Cluster: File cluster(s) storing the frame
Start: Offset of first involved cluster in bytes

solely used for synthetic images, especially when high image quality after scaling is important. JPEG, GIF, TIFF and PNG are all raster formats. Vector images are often distributed as Computer Graphics Metafile (CGM) or Scalable Vector Graphics (SVG) files. Raster formats today see an growth in stereoscopic images. Vector formats also include 3D vector images which use own file formats. There are also hybrid formats which can hold raster and vector images like Encapsulated Postscript (EPS). The number of different image formats is vast and even professional tools often fail to support all of them. Image processing software often uses own proprietary formats, and obsolete image file formats from the decades of image processing can show up during an investigation. In the mid of 2014, Wikipedia lists 55 raster image file formats.

JPEG is an acronym for 'Joint Photographic Experts Group', which is the consortium having developed the JPEG standard (ISO/IEC 10918-1:1994). JPEG files typically have the extension .jpg or .jpeg. As the JPEG standard has many options for encoding image data, ordinary encoders and decoders support only a subset of the possible options. Most image editing programs store JPEG images in the JPEG File Interchange Format (JFIF), which defines certain restrictions on the choice of encoding settings. JFIF has become Part 5 of the official JPEG standard (ISO/IEC 10918-5: 2013). A different standard for storing JPEG images is the Exchangeable image file format (Exif) widely used by digital cameras (JIETA CP-3451C 2013). Exif is mainly a standard for attaching metadata to JPEG files, but it also defines some restrictions on the choice of JPEG encoding settings. If an Exif file from a digital camera is processed with image editing software, the result is usually a JFIF file including Exif metadata.

JPEG files are internally organized as sequence of *segments* and entropy-coded data. The entropy-coded data carries the actual content of the image while the segments provide the information needed for decoding the image and auxiliary information describing the image. The entire syntactic structure is contained in the segments. Each segment starts with a so-called *marker*. A marker comprises two bytes, and the first byte is always FF. The first and last segment of a JPEG file have no payload, and they consist just of the Start-of-Image (SOI) marker FF D8 and End-of-Image (EOI) marker FF D9, respectively (see Figure 6.1). Most other segments have a payload and store the length of their payload in the two bytes immediately following the marker.

JFIF files have an application segment of type APP0 with marker FF E0 next to the SOI marker. Such a JFIF APP0 segment contains the null-terminated string 'JFIF' (4A 46 49 46 00) directly after its size information. By contrast, Exif files have an application segment of type APP1 with marker FF E1 right after the SOI marker. The payload of an Exif APP1 segment begins with the identifier 'Exif' (45 78 69 66) followed by two bytes with value 00 and one of the TIFF byte order marks ('II' encoded as 49 49 or 'MM' encoded as 4D 4D). If Exif information is present in a JFIF file, the Exif APP1 segment follows the JFIF APP0 segment.

6.4.2 Audio Data

With the evolving multimedia technologies, audio data is handled extensively in digital file formats. For example, digital audio became available for many consumers with the introduction of audio CDs in the 1980s and became viral by the popularity of the MP3 format from the middle of the 1990s. In Internet or telephone communication (e.g. VoIP, GSM and ISDN), audio data is transmitted in digital formats, nowadays. Audio data is also present as the soundtrack of a movie file or a 'video' stream, too.

Audio data can contain relevant evidence, for example on the soundtrack of a smartphone camera video. The audio file formats that are most relevant are outlined in the following.

6.4.2.1 Uncompressed Raw Audio

Unlike image and video data, audio is still extensively in use in raw, uncompressed digital representations. Well known is the audio CD, which saves up to approximately 80 min of music per disc. Here, the sound information is represented as

44 100 two-byte samples per second in each of the stereo channels. Also in professional environments like music production or digital cinema, uncompressed audio data formats are common.

Uncompressed audio files are often saved in the WAVE container format, which is also called WAV due to the file extension `.wav`. From a forensic point of view, WAV files are easy to identify by the presence of typical identifier strings of four characters length at/near the beginning of the file, for example, 'RIFF', 'data' or 'fmt␣'. The details about the WAVE syntax that can be exploited for forensic analysis can be found in the RIFF specification by Microsoft (1991).

6.4.2.2 Lossy Audio Compression

Nowadays, lossy compressed audio formats have taken the place of audio CDs for many consumers. The most popular formats are MP3 and its successor AAC. DVDs and BDs widely use miscellaneous Dolby and DTS formats.

The syntax of MP3 is described in the MPEG-1 and MPEG-2 Audio Layer III specifications (ISO/IEC 11172-3:1993c; ISO/IEC 13818-3:1998). Every MP3 file consists of hundreds of data *frames*. Each frame contains the information about the encoded sound of 1152 audio samples which represents roughly 1/38 s at a sampling rate of 44.1 kHz. Every frame begins with 12 consecutive bits (i.e. one-and-a-half bytes) set to 1 (hexadecimal FFF)[10], whereby the beginning is byte aligned in the file stream. This sequence serves as a *syncword*, which can easily be exploited for playback and for forensic inspection, too. The payload of an MP3 frame is divided into two parts called *granules*: the first and second granule each represent one half of the playing time of the frame, that is, 576 audio samples. Each granule contains a spectral representation of its audio samples using the *modified discrete cosine transform* (MDCT). Similar to other spectral representations, the MDCT coefficients represent the audio signal in terms of 'low' to 'high' frequencies in every frame.

MP3 files often contain ID3 tags with additional data fields for metadata about the MP3 file, for example, artist, song title or album title. These tags are available in two different versions at the beginning and near the end of the file and can be identified easily by their respective identifier strings 'TAG' or 'ID3'. Detail can be found in the ID3 specification hosted on http://id3.org/.

Advanced Audio Coding (AAC) data is always encapsulated in a container file format like MP4, 3GP, or ADIF/ADTS, which have their own header syntax. This applies both to soundtracks in a 'video' file and to pure music files. Unlike MP3, the syntax of raw AAC data (ISO/IEC 13818-7:2006; ISO/IEC 14496-3:2009) by itself does not allow for simple syntactic identification of AAC frames searching characteristic syncwords. Here, advanced statistical content analysis is required.

---

[10] Only 11 bits are set to one in MP3 files using the unapproved, but widely supported, MPEG-2.5 extension.

6.4.2.3   Lossless Audio Compression

Less common are lossless coded audio formats. Examples are the Free Lossless Audio Codec (FLAC) or the commercial Dolby TrueHD or DTS-HD standards. Usually, they are not relevant in forensic inspection and are not in focus of this chapter.

6.4.3   Video Data

As an example, the MPEG-1 video format is outlined. MPEG-1 serves as an explanatory example. The MPEG-1 video format is described in Part 2 of the MPEG-1 standard (ISO/IEC 11172-2:1993b). MPEG-1 video streams are mainly a sequence of still images which are called frames, similar to MP3 audio. Each MPEG-1 video stream can be divided into several *layers*. The organization of the different layers is given in Figure 6.2.

The first layer is the *sequence layer* which denotes a number of video sequences starting with a sequence header and ending with a sequence end code. The sequence header stores crucial information necessary for decoding the actual picture data, for example the resolution information. There is an arbitrary number of *groups of*

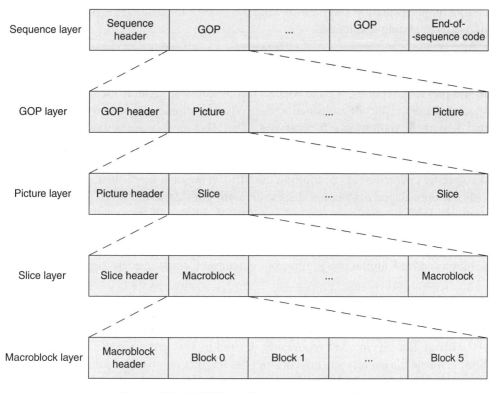

Figure 6.2   MPEG 1 video stream organization.

pictures (GOPs) between sequence header and sequence end code. Each GOP also starts with a header, after which a number of pictures (frames) follows. The first frame of each GOP is always an *I-frame*, which contains 'intraframe' coded image data that describes the image by itself completely, similar (but not identical) to a JPEG image. The others are P-frames and B-frames, which contain difference coded representations of the image data that refer to the nearest I-frame(s) in terms of 'prediction'. In the picture layer, the header is followed by *slices* containing the picture data. The picture data itself is organized in a macroblock layer with a header, four 8×8 luma blocks that contain brightness information, and two 8×8 (subsampled) chroma blocks with colour information. A detailed description of the MPEG-1 video format is given by Mitchell *et al.* (2002).

MPEG-1 files may consist of only a video stream or both video and audio streams. In the latter case, both streams are multiplexed in MPEG system container formats like *transport streams* used for network streaming, or *program streams* used for storage purposes. The multiplexing is described in Part 1 of the MPEG-1 standard on so-called *system streams* (ISO/IEC 11172-1:1993a).

MPEG-1 video is nowadays almost replaced by modern video codecs like H.264 (as in Blu-ray Discs or HD-DVB television) or MPEG-2 (as in DVD-Video discs or SD-DVB television). However, the fundamental principles like predictive difference coding or encapsulating and multiplexing the video data in suitable container formats like MP4 are being maintained.

6.5   File Carving of Multimedia Data

In the previous sections, we introduced the concepts of general file carving and the special characteristics of multimedia files. When combining these two areas, we enter the domain of multimedia file carving. Therefore in this section, we discuss the consequences of multimedia file characteristics for file carving in more detail. First we introduce concepts for image, audio and video file carving. Then we summarize the specific characteristics of multimedia file carving on a more abstract level.

For a forensic practitioner besides the theoretic considerations given in the following sections, available tools are also of importance. There are free solutions as well as commercial products available. To get an overview about what is available, websites like the *ForensicsWiki* (http://www.forensicswiki.org) are helpful. Most commercial solutions address digital image files, but video file formats are also supported.

6.5.1   Image File Carving

Image file carving is closely related to restoring images from damaged disk systems. Also image file carving and recovering deleted images are often used as synonyms. Simple image file carvers will not consider file fragmentation and assume that an image file consists of header, body and footer in a consecutive manner. For example, images in JPEG format would be carved by searching for file system blocks beginning with the JPEG start-of-image (SOI) marker FF D8, checking for some additional features of JPEG files, and ending the carving process when the end-of-image (EOI) marker FF D9 is detected (see Figure 6.1).

Obviously, if fragmentation or nesting of JPEG files occurs, image file carving becomes more challenging as start and end sequences are identical for all JPEG images. For example, if the beginning of JPEG file X is followed by the complete JPEG file Y, the first EOI marker after the SOI marker of X will be that of Y, resulting in a damaged file starting with the beginning of X and ending with the complete content of Y. This does not only happen due to fragmentation, but also due to preview images (thumbnails) embedded into JPEG files.

Memon and Pal (2006) proposed a technique for recovering and reassembling fragmented images using greedy heuristics. Improving on this work, the same group came up with sequential hypothesis testing as an efficient mechanism for determining fragmentation points of a file (Pal *et al.* 2008). These techniques demonstrate a high success rate for recovering and reassembling fragmented JPEG files.

6.5.2   Audio File Carving

Similar to image carving approaches, a number of authors address the carving challenge for audio data. Such audio data can be available as music or voice recordings or as the sound track of an audio-visual recording. Some of the very few audio carving approaches are explained in the following.

6.5.2.1   Matching of Fragments Based on MP3 Header Syntax and ID3 Tags

One example of research on audio carving was the Digital Forensic Research Workshop (DFRWS) 2007 Challenge (http://www.dfrws.org/2007/challenge/index.shtml). The participants were challenged to carve data of different formats from given disk images, including MP3 data. Other file types were digital pictures, videos, text documents, compressed archives, executables, etc. The challenge was organized with four different levels of difficulty, namely contiguous files, files with fragments in sequential order, files with fragments in nonsequential order and files with missing fragments. This challenge meant a significant progress to the research activities in the field of fragmented file recovery.

Among the five submissions to the challenge, only the work by Al-Dahir et al. (2007) (rated second best) addressed carving of the MP3 data in the test set. It proposed a novel idea for carving fixed bit rate MP3 files by identifying sequences of valid MP3 frames as follows:

(1) First, the disk image is searched for the presence of MP3 syncwords (see Section 6.4.2) as indicator for a new MP3 frame.
(2) If such a syncword is found, the consecutive bytes are evaluated as if they were an actual MP3 header. The values for bit rate, stereo mode and sample rate are

parsed. They allow predicting the total length of the complete MP3 frame, that is the offset to the syncword in the following frame.

(3) If the respective following syncwords can be found correctly several times in sequence, that eventually allows for separation of valid MP3 frame sequences from other data: it is very unlikely to find seemingly valid sequences in actually non-MP3 data by accident. The authors proposed to require at least 20 valid consecutive syncwords/headers.

(4) Additionally, the algorithm tries to find valid ID3 tags in the disk image data. For many files, both ID3 tag versions 1 and 2 are available at the same time. The redundant semantic information about the title, artist, album title, etc. in both tags can be used to match bi-fragmented MP3 fragments by means of header/footer carving.

Although this technique performed well for the scenarios in the challenge, this is of limited use in real-world systems. An average system today has much more songs than those in the challenge. There is a high possibility of false matches as there can be expected multiple chunks with similar sizes. Obviously, unmatched MP3 fragments will be left over if the original file is actually fragmented into more than two fragments, that is it has mid/body fragments without ID tag. This research was continued by Sajja (2010) for the more challenging application on MP3 data with variable bit rate.

For completeness, the work by Yoo *et al.* (2011) shall be mentioned. It demonstrated a simpler technique to carve MP3 and other multimedia file formats. However, his technique does not take file fragmentation into consideration.

6.5.2.2   MP3 Carving Based on Frequency Similarities

Based on the work by Al-Dahir *et al.* (2007), as explained in the previous section, we presented an extension to it for matching also the left-over mid fragments (Zmudzinski *et al.* 2012). Our overall carving strategy for MP3 is an exhaustive process, exploiting also the spectral properties of the MP3 format for correlating fragments of a *multi-fragmented* file.

A single MP3 frame granule (see Section 6.4.2) has a relatively short play-length (576 audio samples, that is $\approx$ 13 ms for a sampling rate of 44.1 kHz). Therefore the spectral properties usually feature a smooth transition of the MDCT spectrum from one granule to the subsequent granule. Thus, we expect that the similarity of MDCT coefficients between two consecutive MP3 frame granules is typically much higher than the similarity of two randomly picked frames. One example can be seen in Figure 6.3.

We define the set of indices correspondent to the N *highest* MDCT coefficients as our feature for the frame matching. It should be noted that only the N-tuple of indices is considered for similarity instead of the coefficient values themselves. Other similarity measures for spectrums, like Euclidean distance or rank correlation could be used, but were not implemented for simplicity.

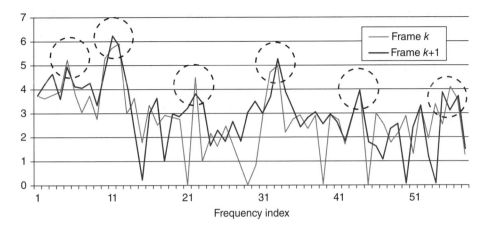

Figure 6.3   Example of MDCT spectrums in two consecutive MP3 frames in an example file. Vertical axis: Sound pressure level. Circle markers point to coincident or nearly coincident local maxima.

Before comparing and matching the frames as explained earlier, all MP3 frames to be investigated are partially decoded to the raw MDCT representation by Huffman-decoding followed by reversing the quantization. That requires fewer computational efforts than full decoding of the MP3 to raw data. The investigated MP3 frames are those at the borders of individual file fragments.

Similar to the approach by Al-Dahir *et al.* (2007), we at first identify sequences of valid MP3 frames based on MP3 headers and ID3 tags. Once we know the frame to be investigated, we decode the Huffman-encoded frequency samples into raw MDCT representations. For our research, we select the $N = 5$ most significant frequencies (only) in each granule and note the corresponding indices which are in the range from 0 to 575.

For neighbour frame matching, we compare the second granule of frame k to the first granule of frame $k + 1$, because in this constellation the compared spectrum correspond to consecutive points in time. The border frames of two fragments with most matching indices are selected as consecutive fragments and are glued together. It should be noted that our approach deals with constant bit rate as well as variable bit rate MP3 files.

We continue the same process for all the remaining fragments. In rare cases, when there is more than one candidate for being the probable neighbour, that is when we have two or more fragments with border frames having the same number of matching coefficients, human interaction is required for identifying the correct neighbour.

6.5.3   Video File Carving

Only little work has been done regarding specific carving methods for video file recovery. Yoo *et al.* (2011) proposed a strategy for AVI file carving by parsing AVI file format specific information. In their work they also proposed similar approaches for

MP3 and WAV, and NTFS compressed files. However, they assumed all media files to be sequentially stored and considered data cutoffs only, disregarding file fragmentation with two or more fragments.

The Netherlands Forensic Institute (2007) published an open-source tool for video data carving called *Defraser*. Defraser is able to carve multimedia files like MPEG-1/2, WMV and MP4. It allows extracting and viewing single frames of MPEG-1 data. Nevertheless, Defraser relies on a valid MPEG-1 structure and is not able to handle small occurrences of fragmentation or missing data in specific file areas.

In the following, a robust approach for MPEG-1 video frame recovery is described as proposed (cf. Yannikos *et al.* 2013). While MPEG-1 is rarely used today, we think that it is easy to explain and therefore provides a suitable example for the purpose of this chapter.

6.5.3.1   Robust Recovery

A robust recovery of video content should apply file carving in a more advanced way than just traditional header/footer identification. Therefore, we implemented a video frame carving component which is designed to analyze found video frame information in order to extract and decode single I-frames. If resolution information (required for decoding) is not present or cannot be found, we apply an approach for resolution identification/correction to decode all available data.

The algorithm for recovering and decoding video frames works as follows:

(1) Locate all I-frames by searching the input data for I-frame headers and by verifying the validity of the headers. If no I-frame headers are found, stop.
(2) Starting from the located I-frame headers, search the input data backwards for corresponding sequence headers within a specific threshold range.
(3) If a sequence header is found, extract the resolution information from the sequence header which is needed for a later decoding of the I-frames.
(4) Continue reading the data following the I-frame header which is typically optional data and/or I-frame slice data. Ignore optional data and verify slice data. Ignore missing or invalid slice data parts, instead read the following data within a specific threshold range until additional slice data is found.
(5) Continue reading the following data and verifying the structure until a next frame start code, a GOP start code, or a sequence end code is found.
(6) If the MPEG stream is identified as a system stream, demultiplex the found I-frame data. Otherwise, skip this step.

6.5.3.2   Robust Decoding

After the recovery of I-frames has been completed, they have to be decoded to allow further processing, for example automated content identification. Typically, decoding is possible without any problem if all required information for decoding is available.

For an I-frame a key required property for decoding is the resolution information stored in its corresponding sequence header. If that specific sequence header is not available anymore, as for example due to fragmentation or overwritten data, the correct resolution needed for decoding has to be found otherwise.

Since robust decoding should also be able to handle cases where the required information such as an I-frame resolution is missing, we propose to use resolution identification and correction by decoding frames with commonly used resolutions and subsequently measuring the pixel row difference of the decoded frames to automatically identify correct decoding results (see Figure 6.4).

All decoding results should be stored in an image file format which is widely supported, like JPEG or PNG. While using a lossless format like PNG will cause no further degradation of quality and will ensure the best possible evidence data, lossless coding will reduce the amount of data to be stored significantly. The choice therefore will depend on requirements and constrains of the examination.

The proposed algorithm for decoding can be summarized as follows:

(1) Decode all I-frames with their corresponding resolution information previously found in the sequence headers.
(2) For each I-frame where the required resolution information is missing do the following:
  • Decode the I-frame using each unique resolution previously extracted from all found sequence headers.

Figure 6.4   Determining the most likely video frame resolution. Left: Corrupted frame after carving, image data in the upper part, data junk in the lower part. Right: Corrected frame after resolution estimation by identifying the peak correlation between the video lines. Frame borders have been automatically removed. The shown scene is taken from the movie 'Big Buck Bunny' (Blender Foundation 2008).

- Decode the I-frame using the most common resolutions for this specific video format (e.g. SIF resolutions).
- Apply resolution correction.

(3) Store each decoding result as JPEG file.

Thereby, it is possible to successfully recover and decode frames of video files that suffered some fragmentation or were partially overwritten.

6.5.4   Special Considerations for Multimedia

After introducing file carving, multimedia file structures and file carving for image, audio and video files, in this section we stress the special characteristics of multimedia file carving. Many of the issues stated below have already been addressed earlier. Here they are discussed in more detail and provide the essence about what is special when carving multimedia files in forensic examinations.

6.5.4.1   File Carving vs. Stream Carving

Audio and video files are often transmitted and stored in a stream format. This means that the complete work is divided into consecutive frames consisting of header and payload sections. The header information often provides sufficient information to render the raw media content of the frame. As an example, an MP3 frame will allow creating 1152 samples of audio data, while an MPEG I-frame will allow creating a single still picture. For easy access, these headers are identified by a sync sequence, also called a *magic word*. Figure 6.5 shows an example for MP3 where header data is used to interpret the compressed audio data to render PCM audio.

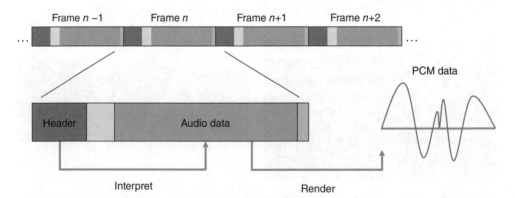

Figure 6.5   Rendering samples out of MP3 frames: All required format information to render the samples of the individual frame is stored in its header. The header of the frame is used for interpretation of the stored audio data which then is rendered to PCM data. Usage of MP3 bit reservoirs can make this more challenging.

Table 6.3 Effect of MP3 bit rate on frame size, frames per cluster and playback time of a cluster (assuming 44.1 kHz sampling rate and 4 KiB clusters)

Bit rate (kbit/s)	Frame size (B)	Frames per cluster	Playback time of cluster (ms)
320	1044	3.92	102.4
256	835	4.90	128.0
192	626	6.53	170.7
128	417	9.80	256.0
96	313	13.07	341.3
64	208	19.60	512.0
48	156	26.13	682.7
32	104	39.20	1024.0

This is a huge advantage compared to files consisting of a header/data/end-of-file structure as we can utilize even small parts of the stream without much effort for further investigation.

6.5.4.2 Cluster Sizes

The cluster size (see Section 6.2.2) has an important impact on multimedia file carving, especially stream carving. If the cluster size is larger than the smallest meaningful structure of the multimedia stream, a single cluster can allow extracting and rendering multimedia data. This is in particular the case with audio data. For example, MP3 stores 1152 audio samples in one frame. Depending on the sampling rate, a different number of frames per second of playtime is created. With the common 44.1 kHz sampling rate, roughly 38 frames per second are used. The second important factor is the MP3 bit rate. A common bit rate is 128 kbit/s. This means that 128 kbit = 16 000 B are used to encode those approximately 38 frames. A single frame at 128 kbit/s has a size of 417 bytes (if no padding byte is appended). For a cluster size of 4 KiB, at least eight complete consecutive MP3 frames are contained in one cluster at the given sampling rate and bit rate. These eight frames allow rendering more than 200 ms of audio data[11], which is in some cases enough to enable further investigation or matching as discussed below. See Table 6.3 for figures on frame length, etc. for various bit rates. As explained earlier, each MP3 frame contains two granules of 576 samples each, which supports frame and cluster analysis even further.

6.5.4.3 Multiplexed Multimedia Stream Content

Video files most often consist of multiple multiplexed streams, usually at least one video and one audio stream. These individual streams are organized by a container format that handles the individual content streams. This means that during file carving

---

[11] Usage of MP3 bit reservoirs can make the first of these eight frames undecodable.

one can expect to find an alternating sequence of audio and videos frames, but also metadata for container organization. There also can be more than one stream of the same media type, for example in movies including multiple languages. Therefore during file carving a simple approach like scanning for the next audio frame header and splicing all found content into one stream may not work. Here it is necessary to carve also for meta information like stream identifiers. Only frames of the same stream should be spliced together.

6.5.4.4   Perceived Similarity vs. Binary Identity

When dealing with forensics, it is important to distinguish between binary identical copies and copies with identical or very similar content. Similarity can be a fuzzy term. In this chapter, we define it as the state where a multimedia file consumed by a customer will be perceived as very close to the original. While a human observer/ listener can hear differences between a PCM and an MP3 copy of the same song when listening carefully in a suited environment he will usually not react differently to it. In most case he will not notice the difference. Therefore, for a forensic investigation it does not matter if an MP3, an AAC or a raw PCM copy of the song is found as evidence.

As a consequence, binary identity of copies as verified by cryptographic hash functions is a requirement most often neither necessary nor helpful in investigations. It is sufficient to verify that a piece of multimedia data is present in one of its countless potential representations. This is different to software or documents where a single bit of information can change behaviour or meaning significantly.

6.5.4.5   Role and Impact of Lossy Compression

Lossy compression is one important reason for multiple versions of one work all perceived as identical but with different binary representations. Lossy compression is a process where those parts of a multimedia file estimated to be irrelevant for an observer/listener will be removed or stored in a coarse manner to save storage space.

Simply opening a JPEG file and saving it again will produce a new, different copy of the image as lossy compression is performed during creation of the new file. As lossy compression focuses on perceptual quality, not binary fidelity, the newly saved copy will have a different cryptographic hash. Even more, an encoder can use different technical parameters which leads to different binary data and different perceptual quality. This always produces a certain amount of error which in the next iteration of decompression and compression again is seen as irrelevant and removed again. Figure 6.6 illustrates how each load/save access of a JPEG file modifies the image: Each recompression adds a certain amount of distortion. One can see that the largest difference exists between the images (a) and (b) (difference image 1). But even repeated recompression with the same parameters performed by the same software often introduces changes in the image as can be seen in the difference image 2.

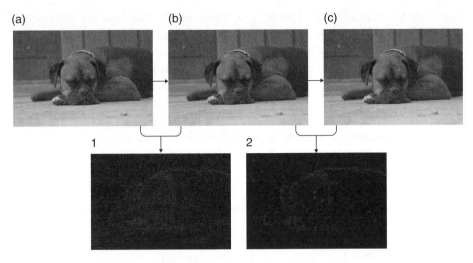

Figure 6.6 The effect of repeated JPEG compression. The original JPEG image (a) was created with JPEG quality factor 80. The images (b) and (c) were iteratively recompressed with quality factor 95. The difference between (a) and (b) is illustrated in image 1, and that between (a) and (c) in image 2. These two difference images encode the absolute differences between the respective source images in the RGB colour model, and hence areas with no difference are black, which is the colour with all channels 0. The difference images have been gamma corrected with $\gamma = 1/4$ for increasing the visibility of the pixels which are not black.

Most common multimedia file formats like MP3, MP4, OGG or JPEG and industry standards like DVD, BD or DVB are based on lossy multimedia coding standards. Smartphones and digital cameras today store recordings and photos in a lossy compression format by default. This means that during a forensic investigation it is almost inevitable to deal with such formats. It also needs to be noted that devices may not support all file formats, therefore transcoded versions of one work created by the device owners, for example from MP3 to AAC are common.

Lossless formats like RAW images created by digital cameras or FLAC audio for high-quality audio downloads gain more popularity and distribution in the recent time. The raw data stored inside them can be considered to be identical even after format changes or load/save cycles. Still, internal filters and storage formats used by multimedia software can cause differences on a LSB level, for example due to quantization algorithms when an audio software internally uses a 32-bit sample resolution for 16-bit audio samples.

6.6 Content Identification

For many binary file types, finding only a small part of the file will make it hard to identify the original file or to use the found part as evidence in a court case. But with multimedia stream files, even small segments can become helpful evidence.

For example a single I-frame of an MPEG video can show substantial content useful for investigation. This single frame is similar to a still image and is of the same value as evidence. This means that multimedia file carving can make sense even after strong fragmentation or destruction of a large amount of the original file.

For example, an MPEG-1 or MPEG-2 movie of 10 min playtime will typically feature about 1200 I-frames. Even one fragment as small as a thousandth of the original file size would likely be sufficient to reconstruct a single still image of the movie and potentially identify persons or actions shown in it. This also allows determining whether the fragment is part of a full video or a larger image evidence that has been identified as incriminating content before. Thus, the identification of known content is an important part of a forensic analysis. Especially when dealing with a large amount of data, efficient methods for content identification can significantly reduce the workload of a forensic investigator.

Practical methods for content identification employ a reference list of *hash* values of known content. Hash values are short and easily computable pieces of data representing the hashed content. Such hashes can be stored and indexed efficiently. This facilitates the compilation of large reference lists such as the National Software Reference Library (NSRL) by the NIST Information Technology Laboratory (2003–2014). Data can easily be matched against a reference list by hashing the data and looking up the hash in that reference list.

Content identification in digital forensics uses *blacklists* and *whitelists*. Whitelists are reference lists for irrelevant content like operating system binaries and other standard software. Blacklists contain relevant, incriminating material. They are crafted for certain investigation scenarios and address, for example child pornography, hacking tools[12] or documents containing corporate secrets.

Whitelisting and blacklisting demand certain properties from the applied hash function. Both strategies should avoid false matches against the reference list. More precisely, for blacklisting it suffices to avoid random, accidental false matches, but whitelisting must resist an active adversary who wants to mask illegal files as whitelisted files. Hence imitating the hash of a whitelisted file must be difficult.

A different property is the possibility to identify content even after small changes. This is particularly relevant for blacklisting as an active adversary might try to circumvent a blacklist by introducing small, but irrelevant, modifications to his illegal files. Whitelisting can benefit from robustness against small changes, too. For example, a computer program might be recognized by similarity matching if the previous version of the program is contained in the whitelist. However, a program infected by a virus will also be similar to the original version of the program. Hence similarity matching

---

[12] While in some countries the possession of hacking tools is already a crime, investigators in any country might be interested in detecting the presence of such tools in order to find evidence regarding the origin of hacking attacks.

against a whitelist does not discharge from investigating the differences of nonidentical files.

6.6.1   Cryptographic Hashing

Identification of digital content during forensic investigations is typically done by using cryptographic hash functions. They are designed to have certain security properties, which are informally defined as follows:

Preimage resistance: For a given cryptographic hash, it is computationally infeasible to find a corresponding message[13] which has the given cryptographic hash. A preimage resistant function is also called *one-way* function.

Second-preimage resistance: For a given message, it is computationally infeasible to find a different message which has the same cryptographic hash as the given message. This property is also called *weak collision resistance*.

Collision resistance: It is computationally infeasible to find two different messages which have the same cryptographic hash. This property is also called *strong collision resistance*.

These definitions have been formalized in various ways. Typically, collision resistance implies second-preimage resistance, and second preimage resistance implies preimage resistance under suitable requirements for the compression properties of the hash function[14]. Hence a collision-resistant hash function provides the highest security guarantees, and coherently, collision attacks against presumed cryptographic hash functions are usually easier than other attacks. Conversely, a preimage attack is the strongest attack. A preimage attack can also be used to find second preimages, and a second-preimage attack can also be used to find collisions.

Note that no cryptographic hash function with an unconditional security proof is known until today, and even the existence of one-way functions is not proven. Until today, security proofs in cryptography are based on the assumption that certain algebraic problems are hard, and hence cryptographic theory builds hash functions based on these assumptions. However, such hash functions are not used in practice, as they are significantly slower than other hash functions.

All so-called cryptographic hash functions used in practice satisfy the desired security properties only in a heuristic manner. Even worse, some have been assumed to satisfy these properties at the time of their publication but attacks disproving their security were developed subsequently.

---

[13] The input to a cryptographic hash function is usually called *message*.

[14] Formally, collision resistance is defined for a family of hash functions while preimage resistance and second-preimage resistance can be defined for a single function or a family of functions. Hence the view of implications presented here is actually too simplistic. Rogaway and Shrimpton (2004) give a rigorous analysis of the implications between the security properties under several definitions for these properties.

Well-known examples for cryptographic hash functions in this heuristic sense are the widely used functions MD5 and SHA-1, as well as the SHA-2 family, which has the members SHA-224, SHA-256, SHA-384, SHA-512, SHA-512/224 and SHA-512/256 named according to their output size and internal state size where needed for disambiguation. The new SHA-3 family is currently under standardization by the NIST. Currently, MD5 and SHA-1 are commonly used for forensic content identification. In particular, the NSRL (see p. 242) provides such hashes in its reference data sets.

MD5 is well known to be vulnerable to collision attacks. Wang and Yu (2005) published the first practical algorithm for producing MD5 collisions. These collisions are common-prefix collisions, that is the two colliding messages computed by the algorithm have a common prefix specified as input to the algorithm. Soon after this, algorithms for computing chosen-prefix collisions have been developed. Such collisions have independent, user-defined prefixes. A real-world attack exploiting chosen-prefix collisions has been observed in the Flame malware. Improved collision finding algorithms now allow the computation of MD5 collisions within a few seconds on standard hardware. However, until today, MD5 is resistant against practical preimage attacks. So far, the best preimage attack was published by Mao *et al.* (2009) and claims a computational complexity of $2^{123.4}$ (as opposed to a complexity of 2^{128} for a brute-force attack due to the hash size of 128 bits).

The fastest known attack against SHA-1 is a common-prefix collision attack by Stevens (2013), which claims a complexity of 2^{61}. Hence SHA-1 cannot be considered collision resistant any more although no SHA-1 collision has been published by now. However, SHA-1 can still be used for applications requiring only second-preimage resistance or preimage resistance.

For SHA-2 and SHA-3 no attack against one of the security properties is known by now. Note that all statements regarding the security of hash functions reflect the sate of the art during the time of writing this chapter in 2014. Later advancements in cryptanalysis might have changed the security assessment of the hash functions.

Forensic whitelisting is a well-suited application for cryptographic hash functions. Malicious files cannot be masked as whitelisted files with reasonable effort due to the security properties of cryptographic hash functions. More precisely, the (second-) preimage resistance prohibits the adjustment of malicious files in such a way that it has the same hash as a whitelisted file. As it is impractical for adversaries to implant benign content which collides with malicious content into the whitelist, collision resistance is not that important. Hence MD5 and SHA-1 can still be used for forensic content identification. Moreover, there are no known attacks which produce simultaneous collisions for MD5 and SHA-1.

Cryptographic hash functions cannot be used for identifying similar content. This is due to the *avalanche effect* resulting from the security properties. Any small change in the input to be hashed yields a completely different hash, which differs like a randomly chosen hash in approximately half of the bits. For example, a single additional space character in a text file yields a hash which is completely unrelated to the hash of the original content. Multimedia files may be perceived as being identical when one file resulted from another by format conversion or recompression, but they will have distinct cryptographic hashes due to the differences in binary representation.

While identification of similarities may not be demanded for whitelisting, it is a strongly desirable property for blacklisting. Hence blacklisting requires hashing approaches which can be used for similarity matching. Such approaches are described in the following sections.

6.6.2 Fuzzy Hashing

Similarity preserving hash functions map similar data to similar hash values. Similarity of data can be defined on various levels. Research in digital forensics has largely focused on similarity in terms of the raw binary data. Corresponding hash functions process their input as raw byte sequence, and hence they can be applied to any data independently from the format or media type. We discuss such functions in this section while other hashing strategies, which decode the content before deriving the hash value, are discussed later.

Until now, there is no commonly used terminology for similarity preserving hashing on binary data. *Fuzzy hashing* seems to be the most widely recognized name for this technique. This term has been coined by Kornblum (2006a) as synonym for a hashing approach he called context triggered piecewise hashing (CTPH) (Kornblum 2006b). However, it is often used with a more general meaning. In fact, the term 'fuzzy hashing' literally refers to the capability of performing fuzzy matching, that is identifying similar items as such. Hence it may even be used for referring to any similarity preserving hash function. We use fuzzy hashing as synonym for similarity hashing on the binary level in this chapter.

The most prominent fuzzy hashing tools are *ssdeep* by Kornblum (2006c) and *sdhash* by Roussev (2010b). The underlying algorithms of these tools are fundamentally different:

- The program ssdeep implements a CTPH algorithm (Kornblum 2006b). Hashes produced by ssdeep are essentially strings of base64 characters, where each character corresponds to one chunk of the hashed file. The similarity of hashes is derived from a certain edit distance applied to the base64 strings.
- The program sdhash implements the 'similarity digest' algorithm invented by Roussev (2010a). This algorithm first selects certain 'features', that is sequences of 64 bytes, and inserts these features into a sequence of Bloom filters. A similarity score of two hashes is calculated from the number of common bits between Bloom filters and the number of features inserted into these Bloom filters.

Other fuzzy hashing methods are *md5bloom* (Roussev *et al.* 2006), *mrshash* (Roussev *et al.* 2007), *SimiHash* (Sadowski and Levin 2011), *bbHash* (Breitinger and Baier

2012), *MRSH-v2* (Breitinger and Baier 2013), *mvHash-B* (Breitinger *et al.* 2013) and *saHash* (Breitinger *et al.* 2014).

Fuzzy hashing is a suitable approach where similarity of content can be recognized through similarity of the raw byte sequences contained in files. This holds especially for text files like HTML and XML files, source code files, or configuration files (.conf, .ini, etc.). Fuzzy hashing is also suitable for compiled software, for example in the form of binary executables or Java class files (bytecode). Compressed file formats like most of the commonly used multimedia formats limit the benefit of fuzzy hashing as a small change in the content of such a file leads to significant change in the compressed data stored in the file. However, there are still areas of application for fuzzy hashing on such files. Embedded objects like images in documents (PDF, Word, etc.) may be recognized without parsing the document. Likewise, identical metadata such as the same embedded thumbnail in several image files can be detected. Moreover, file carving can benefit from fuzzy hashing as fragments of files may be correlated with the original file even if decoding of the fragments is not possible.

Both blacklisting and whitelisting benefit from fuzzy hashing in the sense that such hashing approaches can recognize more content than cryptographic hashing. However, an approximate match or even a perfect match of fuzzy hashes establishes less confidence than a match of cryptographic hashes. Nevertheless, the probability for false matches – in the sense of equal or similar hashes without similarity in the binary data – is usually low enough for being manageable during a forensic analysis, for example by manually eliminating the false matches. However, there has not been much research on the security of fuzzy hashing. We should assume that masking malicious files as whitelisted files is much easier in the context of fuzzy hashing than in the context of cryptographic hashing. Hence whitelists of fuzzy hashes cannot be assumed secure against active adversaries. We should also assume that blacklisted files can be modified such that they will not be recognized any more based on their fuzzy hash. In fact, Baier and Breitinger (2011) have demonstrated strategies for anti-blacklisting on ssdeep hashes. But since anti-blacklisting on cryptographic hashes is trivial, fuzzy hashing provides more security in this aspect.

6.6.3 Perceptual Hashing

Perceptual hashing is an identification technique for digital multimedia data. Similar to cryptographic hash functions, it is a mechanism to map an input data set of arbitrary length to a compact, content-based descriptor or message digest of short length. But in the case of perceptual hashing, the hash value of a given piece of data shall be invariant or at least almost insensitive to moderate or admissible transformations of these data. Generalizing the definition by Doets and Lagendijk (2004) from audio data to arbitrary multimedia content, a perceptual hash is

> a compact representation of the perceptually relevant parts of the ... content, which can be used to identify a[] ... file, even if it is severely degraded due to compression or other types of signal processing operations. (Doets and Lagendijk 2004, p. 101)

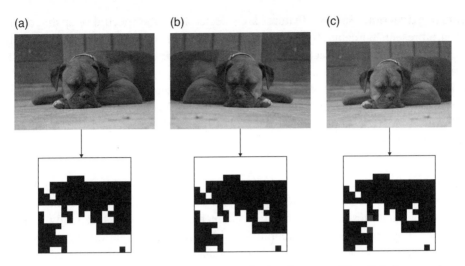

Figure 6.7 Demonstration of perceptual image hash robustness. The digital source images are perceptually identical except for allowed transformations like flipping, conversion to greyscale, resizing and JPEG compression. The robust hashes shown below the images mirror this similarity. The hashes of image (a) and (b) are identical. The hash of image (c) differs only in two bits, which are highlighted. In forensic practice, a hamming distance up to 8 bits when using a 256-bit robust hash or a BER of 0.03 can reliably be accepted.

For example, moderate JPEG, MP3 or AAC compression shall not affect the extracted hash value. In Figure 6.7 the impact of JPEG compression on a perceptual image hash is shown.

The terminology expresses that the *perceptual* hash reflects the identity of two data sets based on human perception rather than the exact binary representation. The terms *robust hashing* and *fingerprinting* are also common for these mechanisms in the literature.

As explained by Lancini *et al.* (2004), the typical perceptual hash extraction in the spectral domain includes the following processing steps:

Preprocessing: At first, the input data is prepared to improve the efficiency or the robustness, for example by down-scaling a picture or video file to a predefined size or down-mixing an audio file from stereo to mono.

Framing: The preprocessed media data stream is split in short sections, for example audio frames or picture sub blocks. The frames or blocks can overlap for overcoming the limited temporal/spacial resolution caused by framing.

Spectral transform: In many approaches, an appropriate spectral transform is applied on the frame or block data, for example DCT, Wavelets or Fourier. Other approaches analyze the data in the spatial domain in the following step which makes spectral transform redundant.

Feature extraction: Suitable features are selected from the spectral or spatial domain that represent 'perceptually relevant parts'. As pointed out by Fridrich and Goljan (2000a), the feature extraction must be *key-dependent* if a *secure* perceptual hash is required.

Output modelling: The selected features are analyzed according to a predefined criterion. Usually, the extracted robust hash is a *binary* value of length M bits. However, in some approaches it is an M-tuple with real-valued (i.e. continuous) elements instead.

Practical or even commercial applications of perceptual hashing beyond forensic inspection are broadcast monitoring, second screen services, organizing/querying multimedia databases, banning media content from a network, tampering detection or music recognition. Several perceptual hash approaches for various media types are known, which provide different levels of robustness. Some will be explained in the following sections.

As a starting point for own experiments with robust hashing, *pHash* (http://www. phash.org/) can be recommended. It provides a number of robust hash algorithms for image, video, audio and text data.

6.6.3.1   Perceptual Image Hashing

When perceptual image hashing is used in forensic applications, error rates, robustness and computational complexity are the most important aspects. Many robust image hash functions discussed in the literature apply transformations like DCT, FFT or wavelet. While there are speed-optimized libraries for these transformations, for images with sizes of 1 megapixel and more and huge quantities of images they can cause a significant delay.

While these operations help to survive many attacks and increase the robustness, they slow down the hashing process due to high computational efforts. We compared a number of hash algorithms with respect to their robustness and complexity (Zauner *et al.* 2011). We showed that the overall performance of the algorithm with the lowest complexity compares well with those of higher complexity.

The algorithm with the lowest complexity used in our aforementioned study is a block mean value based perceptual image hashing function introduced by Yang *et al.* (2006). More precisely, the algorithm is the most simple variant of four slightly different variants they proposed. This variant is described as follows:

(1) Convert the image to greyscale and normalize the original image into a preset size.
(2) Let n denote the bit length (e.g. 256 bits) of the final hash value. Divide the pixels of the image I into nonoverlapping blocks $I_1, I_2, \ldots, I_n$.
(3) *Optional*: Encrypt the indices of the block sequence $I_1, I_2, \ldots, I_n$ using a secret key K to obtain a block sequence with a new scanning order. The authors specify

no further details about what encryption method to use and what security feature to achieve with this step. For forensic content identification this encryption step is not required. Therefore we skip this step.

(4) Calculate the mean of the pixel values of each block. That is, calculate the mean value sequence $M_1, M_2, \ldots, M_n$ from the corresponding block sequence. This is a convenient way for scaling the image to n pixels. Finally obtain the median value M_d of the mean value sequence.

(5) Normalize the mean value sequence into a binary form and obtain the hash value

$$h(i) = \begin{cases} 0 & \text{if } M_i < M_d \\ 1 & \text{otherwise} \end{cases}$$

In Steinebach *et al.* (2012), the algorithm is improved by a number of additional features: To make it robust against mirroring, it automatically flips the image to be hashed in such a way that the brightest corner always is in the upper left. It calculates the hash by dividing the 16×16 area into four subareas of size 8×8 and computes the median for each of these areas to achieve a higher difference of images featuring a similar structure. For increasing the speed of the algorithm, the size normalization in Step (1) is dropped as the size is normalized to n pixels in Step (4) anyway. Figure 6.8 shows two example images and their robust hash. We call the implementation of our hashing solution *rHash*.

Many other algorithms for image hashing can be found in the literature. For example, De Roover *et al.* (2005) provided an image hash algorithm which is robust against geometrical operations like scaling and rotation. The hash draws its robustness from the use of the Radon transform. Fridrich and Goljan (2000b) proposed an approach based on random noise similarity.

Figure 6.8 Two popular images from the USC-SIPI image database (University of Southern California – Signal and Image Processing Institute 1977) and their robust hashes calculated by the improved block hash method. The hash is basically a very coarse black and white abstraction of the image with a resolution of 16×16 pixels.

6.6.3.2 Perceptual Audio Hashing

Approaches for perceptual audio hashing make use of audio features in different spectral domains:

- For example, an approach by Cheng *et al.* (2003) proposes using hidden Markov modelling (HMM) of low-level audio features like the volume (i.e. loudness), the zero-crossing-rate, the bandwidth or mel-frequency cepstrum coefficients (MFCCs).
- Ghouti and Bouridane (2006) proposed analyzing ratios of Wavelet coefficients.
- The so-called *AudioID* approach was presented by Gruhne (2009). It is based on an analysis of the spectrum flatness measure, the spectral crest factor and, again, the volume. It is included in the consumer software *mufin player* (http://www.mufin.com/) for automated reorganization of music collections.
- Another approach was presented by Mıçak and Venkatesan (2001) using the modulated complex lapped transform (MCLT).
- The work by Wang *et al.* (2003) should be mentioned as it proposes extracting audio hashes directly from spectral features in compressed bit streams without transcoding or further spectral transform.
- Two other approaches derive low-level features in the Fourier domain: the algorithm by Haitsma *et al.* (2001a,b) uses comparison of adjoining Fourier coefficients. The approach by Wang (2003) used in the commercial *Shazam* music identification service identifies dominant peaks in the Fourier spectrum as 'anchor' points.

6.6.3.3 Perceptual Video Hashing

Perceptual hashing of video files can be seen as an expansion of image hashing. One can implement a perceptual video hashing algorithm by processing all individual video frames with an image hashing algorithm. A video hash is then the sequence of the image hashes of the individual frames. In practice, this usually leads to large hash sizes and slow implementations. Therefore in the video domain often inter-frame or inter-section strategies are utilized. One intuitive strategy is to calculate a hash by measuring the frame-to-frame changes of the average frame brightness. Zhou *et al.* (2006) proposed a robust hash for video streams which is based on the similarity between spatial and temporal adjacent blocks. Alternative video hashing approaches using perceptual models have been given for example by Oostveen *et al.* (2001) and Liu *et al.* (2003).

One important design decision is how large the minimum sequence of frames needs to be for successfully identifying the video they are part of. The larger the number of frames, the more coarse the stored hash can be. Usually a few seconds of video data are sufficient for identifying a video. Typical robustness requirements for video hashes are re-encoding in lossy formats, scaling and cropping.

Common use cases for video hashing beyond forensics are broadcast monitoring, for example counting adds in a TV channel, or content blacklisting as done in YouTube.

Here identifying the content in consecutive sequences of multiple seconds is sufficient to satisfy the requirements of the applications. But when perceptual video hashing is used together with file carving, requirements change: It is likely, that only a few frames, maybe even a single one, of a video can be carved. Hash methods depending on frame-to-frame difference therefore may fail to identify the video. Therefore in file carving, despite the additional storage requirements, single-frame strategies are advised. Only then identifying a video based on one or multiple frames spread over the entire video can be successful.

As a consequence, for file carving assistance perceptual video hashing can be seen as single image perceptual hashing executed on every single frame. Optimized storage methods utilizing the often small difference between neighbour frame can help to reduce the storage load caused by this strategy.

6.6.4   Searching and Indexing of Hashes

Databases of fuzzy or robust hashes must support retrieving hashes that are similar to a query hash. However, ordinary database systems do not support similarity queries. Those systems use canonical index structures like B-trees for answering *exact* queries efficiently. These indexes cannot be employed for similarity search. Hence separate tools are needed for organizing and searching fuzzy and robust hashes.

The most simple approach for identifying similar items is a full linear search over the elements in the database. As this brute force approach is very inefficient, indexing approaches should be used for accelerating similarity search. The choice of one such approach for a particular application depends on the kind of items to be indexed (i.e. the space of items), the considered similarity function and the expected distribution of items to be indexed. Many approaches construct some kind of tree employing the metric of the underlying space (Bozkaya and Ozsoyoglu 1999; Chávez *et al.* 2001). Another family of approaches is LSH (locality-sensitive hashing) (Gionis *et al.* 1999; Paulevé *et al.* 2010). Specific approaches for strings and other symbol sequences are often based on *n*-grams or suffix trees (Boytsov 2011; Navarro *et al.* 2001). The performance and applicability of all these methods depend on the particular scenario.

6.6.4.1   *n*-Gram-Based Indexing

String-like fuzzy hashes (e.g. ssdeep hashes) can be indexed effectively and efficiently by *n*-gram indexes. An *n*-gram is a substring of length n[15]. Hashes with low distance, that is high similarity, are likely to have a common *n*-gram for a suitable value of *n*. Thus *n*-grams can be used for recognizing similar hashes. For example, our tool *F2S2* (fast forensic similarity search) (Winter *et al.* 2013) implements *n*-gram indexing for byte strings and base64 strings. Hence it is well-suited for being combined with ssdeep, and the framework in Section 6.7.3 indeed applies F2S2 to ssdeep hashes.

---

[15] For example, the 3-grams of 'ABCDE' are 'ABC', 'BCD', and 'CDE'.

When building an n-gram index, the parameter n has to be fixed first based on the application scenario. Each hash in the reference list to be indexed gets a unique ID, which can be chosen as its position in the list. The ID is used by the index for referring conveniently to individual entries of the reference list. The basic tasks of inserting entries into the index and utilizing the index during queries are performed as follows:

- For indexing a hash string, determine all n-grams of this string. Then store each n-gram together with the ID of the hash in the index structure. The n-grams serve as lookup keys for the index, that is the index can determine all IDs associated with an n-gram.
- For answering a similarity query, determine all n-grams of the query hash. The index provides the associated IDs for each n-gram. The retrieved IDs define a set of candidates for *neighbours*. Neighbours are items in the reference list which are close to the query item (typically, the similarity score shall exceed a user-defined threshold). The candidates returned by the index have to be checked with the similarity function in order to decide whether they are *true* neighbours. Hence the index serves as filter for the reference list, and the similarity function has to process only on the candidates instead of the complete reference list.

The chosen data structure in F2S2 is an array hash table, that is a hash table which uses arrays as *buckets* (see Figure 6.9). Each entry in a bucket represents an n-gram as well as the ID of an item from the reference list (see Figure 6.10). As n-grams serve as lookup keys in the index, an n-gram is used for identifying both the correct bucket of an entry and the position of this entry in the bucket. The first step is achieved with

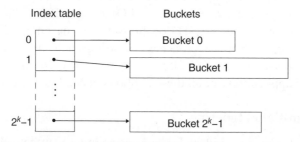

Figure 6.9 Structure of the index. Reprinted from Digital Investigation (Winter *et al.* 2013) with permission from Elsevier.

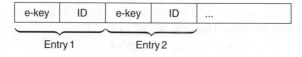

Figure 6.10 Structure of a bucket. Reprinted from Digital Investigation (Winter *et al.* 2013) with permission from Elsevier.

an address function which maps the *n*-gram to a bucket number, and the latter step by sorting the buckets according to the lexicographic order of *n*-grams during index construction.

The evaluation of F2S2 showed that it can reduce the time needed by ssdeep for performing search queries in a realistic investigation scenario from more than two weeks to several minutes (Winter *et al.* 2013) without missing any hit. This is a time reduction by a factor of more than 2000. Hence indexing strategies can be very useful or even necessary for performing similarity search in large amounts of data.

6.6.4.2   Other Indexing Approaches

Many types of similarity hashes can be indexed with *metric trees*. The basic requirement for this is a distance function between the hashes which satisfies the axioms of a metric, especially the triangle inequality[16]. Given such a distance function, a metric tree tries to put distant items into distant nodes of the tree. Hence the tree can efficiently select candidates for close items.

A different class of indexes uses *locality-sensitive hashing* (LSH). The essential building block for an LSH scheme is a family of hash functions. These functions are utilized for building hash tables which are likely to put similar items into the same bucket. As LSH is a probabilistic framework, there is some chance that the LSH index does not recognize the similarity of two neighbours. The parameters of the LSH index can be adjusted in such a way, that the chance for this error is lower than a given limit.

We have implemented and evaluated two indexing methods for rHash and related types of perceptual image hashes (Winter *et al.* 2014). The first method uses a metric tree which falls into the category of vantage point trees (*vp-trees*) and fixed-queries trees (*FQ-trees*). The second method is based on LSH using bit sampling for the family of elementary hash functions. Our evaluation showed that the LSH index is faster than the tree index. On the downside, the LSH index does not find all of the approximate matches, but as this false negative rate is low (0.23% in our evaluation), it might be acceptable in a forensic investigation.

6.7   File Carving Frameworks

As new multimedia formats are constantly being developed, a framework suitable for multimedia file carving and content identification should provide functionality which could be easily extended. In the following subsections we discuss current practice when applying the two forensic processes *multimedia file carving* and *content identification*. We define the requirements for a framework which is able to run both forensic processes and describe how to build such a framework. Therefore, we divide both processes in several tasks which we categorize in order to efficiently combine

---

[16] The triangle inequality assures $d(x, z) \leq d(x, y) + d(y, z)$ for any $x, y, z \in X$, where X is a metric space and d is the distance function on this space.

them in the framework. Furthermore, we define three processing phases which the input data passes. Finally we describe the technical aspects of the framework.

6.7.1   Current Practice and Existing Solutions

A typical digital forensic analysis of storage media comprises various processes. One important process is the recovery of deleted data which can be done with file system analysis and file carving. Another important process is the automatic identification of known content: It reduces the total workload by filtering content that is known to be harmless (whitelisted) and content known to be harmful (blacklisted). An example for the latter are pictures containing child pornography.

Current practice in a digital forensic analysis is to use one or more tools suitable for file carving and content identification. Often various tools are used and the individual results are combined in the final analysis report. Although there also exist software which combines much useful functionality, for example *EnCase Forensic* or *X-Ways Forensics*, we are not aware of any software or framework which combines the advanced approaches for multimedia file carving and content identification explained in the previous sections.

The lack of existence of such a framework is a rather interesting observation to us since we were told by many forensic investigators, for example from law enforcement agencies, that currently one of the main challenges in digital forensics is the sheer amount of data which has to be analyzed. Therefore, we think that especially standard forensic processes like file carving or content identification should be done in the most effective and efficient way possible.

Notable standalone tools for file carving are *Foremost*, *Scalpel* and *PhotoRec* (see Section 6.3). Content identification based on cryptographic hashes can be done with many tools like *md5sum* or *cfv*. Perceptual hashing tools like *rHash* (see Section 6.6.3.1) or *pHash* (see Section 6.6.3) can be used for robust identification of images or other data. The tools *ssdeep* and *sdhash* perform fuzzy hashing (see Section 6.6.2).

6.7.2   Framework Requirements

Various file carving tools exist. All of them are able to perform at least basic file carving using header and footer signatures of known file types. Some more sophisticated file carvers are able to additionally reconstruct fragmented files with a certain reliability. However, these tools do not possess the properties of a file carving *framework*, which we define as follows:

Compulsory properties
- The main functionality of the software is to recover files using file carving.
- The software should provide additional functionality for content identification.

- The functionality of the software can be easily extended using modules or plugins.

Optional properties

- The software should be platform independent/easily portable to other platforms.
- The software should be able to process multiple tasks in parallel.
- The software should require only small user intervention/should be able to run fully automated.

Some file carving tools support adding new header and footer signatures to the tool configuration for extending the number of file types which can be searched. Unfortunately, the tools provide no real API or plugin interface for implementing new functionality such as header parsing for multimedia files.

6.7.3   An Example Framework

In this section we describe the design and properties of an advanced multimedia file carving and content identification framework. The aim of the framework is to automate the process of gathering multimedia evidence as efficiently as possible by utilizing different mechanisms. The framework should use a small number of core components that provide basic functionality like multi-threading and shared data access as well as several task modules that implement specific functionality. More precisely, the framework should include all core components and task modules required to perform the following two forensic processes:

- Carving of multimedia data, for example pictures, video frames and audio files.
- Content identification, for example based on cryptographic hashes, robust hashes or fuzzy hashes.

Both forensic processes consist of several tasks. For example, the process of carving multimedia data could include the tasks JPEG picture file carving as well as MP3 audio file carving. In order to optimize the efficiency of the input data processing within the framework, it is important to understand how the data is being handled in each task.

6.7.3.1   Efficiency Optimization

Creating an efficient framework requires understanding what actually causes ineff-iciency, that is which parts of a process like multimedia file carving are the most time-consuming ones. Standalone tools for multimedia file carving or content identification obviously require input data for processing. This data is usually read by each tool individually. Therefore, processing the same input data with different tools can be very slow, especially when the same data is read multiple times from a slow storage device like a hard disk.

The first approach which comes to mind is probably just creating a large RAM disk (i.e. a virtual disk within the system memory) and placing the input data inside the

RAM disk where it can be read/processed much faster than when using a hard disk. However, since the input data could be very large, for example a hard disk image of several terabytes, this would require an amount of system memory which is typically not available.

Multi-threading can greatly improve efficiency when the same input data needs to be processed multiple times. Hence the framework should be designed so that it reads the input data once into the system memory and then runs several parallel threads which process the data currently in the system memory.

To implement the two forensic processes in our framework, that is carving of multimedia data and content identification, we need to specify how the specific tasks of each process handles input data. Therefore, we classify each task as either a *linear task* or a *nonlinear task*. The differences between linear and nonlinear tasks are explained in the following subsection.

6.7.3.2   Linear Tasks

Linear tasks are tasks which require only one single sequential read of the input data, which is efficiently implemented by using a buffer with a fixed size. Typical tasks that process input data in a linear way are calculating cryptographic hashes or simply copying a file. During the read process the buffer used to hold parts of the input data can be directly used as the input for a linear processing task. Therefore, only a single read of the input data is required while several tasks can process the data.

Several different linear tasks can be performed in parallel using multi-threading. This can significantly speed-up data processing: The read processes responsible for filling the buffer with input data must only read the input data once instead of multiple times. After the buffer is filled with input data, the data can be processed by all linear tasks in parallel, each using a separate thread. The read process only has to wait until the task with the longest running time finishes.

Our framework should be able to perform the following linear tasks:

Data duplication: In this task the available input data is written/copied to a specified target location, for example another memory area or a file on a hard drive. This is used to create a forensically sound copy (100% binary identical) of the input data.
Header-/footer-based file carving: In this task simple file carving methods are used to recover deleted (but not overwritten) files by identifying known signatures of different file types.
Cryptographic hashing: In this task common cryptographic hash functions are used to identify known files.

This list of linear tasks is not an exhaustive list of all possible linear tasks that could be performed by the framework. It should rather be seen as an example list of tasks we chose.

6.7.3.3 Nonlinear Tasks

Nonlinear tasks process input data in a nonlinear way and are typically used when interpretation of more complex data structures is required. An example for this is building the perceptual hash of a JPEG image, which requires image decoding and analysis. Interpretation of complex data structures may require navigating forwards and backwards within the data. However, the buffer used for reading the input data typically holds only a part of the input data and therefore may not provide all data required by a nonlinear task at once.

Since the buffer is always limited by the available system memory, there is a trade-off between the buffer size and the need to process large files without re-reading data: We can set the buffer size to a decent value, for example several megabytes, in order to perform specific nonlinear tasks at the same time as the linear tasks. However, since we also have to consider processing very large files partly or as a whole, it is difficult to efficiently run multiple nonlinear processing tasks by using multi-threading alone.

The framework should be able to perform the following nonlinear tasks:

Fuzzy hashing: In this task content identification based on similarity is performed. This allows identifying data which is very similar to already known data. Existing fuzzy hashing approaches may require multiple reads of the same input data thus making this task nonlinear.

Perceptual image hashing: In this task perceptual image hashing is used in order to identify similar image material regardless of the used image file format, image size or colour mode. Since perceptual image hashing requires decoding and analyzing of image files this task is nonlinear.

Audio frame carving: In this task multimedia file carving methods are used to parse supposed audio data in order to recover single audio frames. Audio frame carving may require seeking forwards and backwards in the input data which is why this task is also nonlinear.

Video frame carving: In this task multimedia file carving methods are used to parse supposed video data. Carved video frames will be recovered as still images if possible. Like with audio frame carving, this task is nonlinear since seeking backwards and forwards in the input data may be required.

Again, this list of nonlinear tasks is not a complete list but rather an example list of those tasks we want our framework to able to perform.

6.7.3.4 Data Flow

To efficiently handle linear as well as nonlinear data processing tasks as part of the framework, we divide the data flow within the framework into three successive phases: the *read phase*, the *analysis phase* and the *lookup phase*. Figure 6.11 gives an overview of the data flow in the framework.

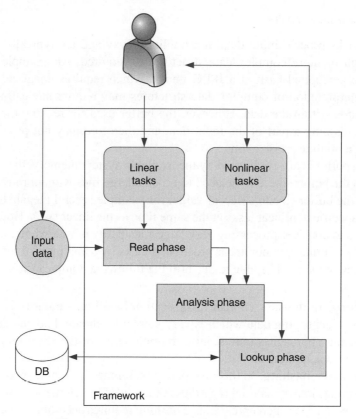

Figure 6.11   Data flow within the framework.

Read phase: The read phase denotes the first phase of the framework process in which
input data is read from a source storage, while multiple linear tasks process the input
data by directly accessing the read buffer. Preprocessing for nonlinear tasks is also
performed during this phase.

Analysis phase: After the read phase finishes the analysis phase begins. Within this
phase multiple nonlinear tasks are grouped by the type of data they analyze. Suitable
parts of the input data are assigned to the different task groups for processing. For
instance, those parts of the input data that likely contain image material are assigned
to tasks on robust image hashing. In the same way audio (video) data is assigned to
tasks which are responsible for audio (video) frame carving.

Lookup phase: After both the read phase and the analysis phase are finished the final
lookup phase starts. In this phase all relevant result data created or found in the
preceding phases are used to query a set of databases for additional information. For
instance, if a file carving task is able to recover a file and the cryptographic hashing
task builds a hash from it, that hash could be used to query a hash database in order
to find out if the recovered file is already known.

6.7.3.5   Architecture

In this section we provide a technical description of the individual framework parts. We divide the framework into *core components* consisting of core modules and *task collections* consisting of *task modules*. Core components provide basic functionality like shared input data access while the task modules implement task functionality. For example, a task module for JPEG picture file carving is a specific implementation of the linear task 'header/footer-based file carving'.

In the following we first give an overview about the core components of the framework. Then we give a more detailed description of the individual task modules implementing the linear and nonlinear tasks mentioned in the preceding section. Figure 6.12 shows the architecture of the framework.

Core components

The core components of the framework build a foundation for efficiently handling all functionality provided by the task modules. This includes providing a shared access to the buffer used for reading the input data, controlling and monitoring multiple threads which access the buffer, and a secure log of all framework activity. In the following the core components and their capabilities are described in more detail:

Input data processor: This core component is responsible for reading the input data and providing a shared buffer (i.e. a shared memory area) where the read data can be accessed by multiple threads at the same time. By sharing the same data the

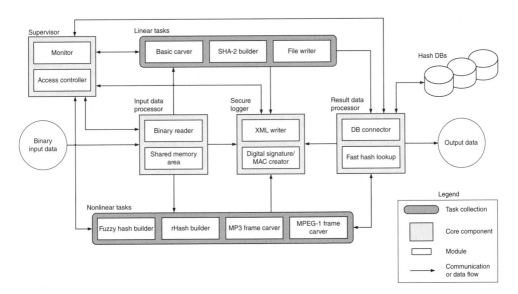

Figure 6.12   Architecture of the framework with example modules.

amount of memory required for data processing can be reduced significantly. The input data read process and the shared buffer are both implemented as an individual core module of this component.

Supervisor: This core component provides supervision over the current state of the shared memory area (shared buffer) and over any threads created by task modules. After the shared buffer is filled with input data, the Supervisor write-locks the buffer, that is denies any additional write attempts, and it unlocks the buffer for reading. Thus the task modules can access and process the data in the buffer. The Supervisor then keeps track of how many task module threads access the shared memory and when the access finishes. After all threads finished processing the shared data, the Supervisor read-locks the shared buffer and write-unlocks it again. This allows the Input Data Processor to fill the buffer with the next chunk of input data. The Supervisor is important to prevent memory corruption or deadlocks caused by multiple simultaneous read and write requests from different threads. The monitoring and the access control functionality are both implemented as individual core modules of this component.

Result data processor: This core component is responsible for transferring results created by different task modules. It provides a range of connectors to databases which provide storage for the result data or are used to retrieve additional information. The Result Data Processor can query different databases containing hashes of whitelisted and/or blacklisted files for any cryptographic, perceptual, or fuzzy hashes of the input data. Depending on the query response, the Result Data Processor then provides the framework user with information about whether or not a specific file is whitelisted or blacklisted. Indexing algorithms for supporting similarity search as described in Section 6.6.4 are part of this core component. The database connector as well as the indexing algorithms are implemented as individual core modules of this component.

Secure logger: This core component creates a secure log file that holds information about all framework activity in XML format. The log file stores information about where the input data is read from, where it is written to as well as results from linear and nonlinear tasks. To provide security against log file tampering, digital signatures or MACs are used for the whole log file and also for individual log entries. A constant communication between the Secure Logger and other core components/task modules is required to ensure a complete framework activity log. Logfile creation and digital signing are both implemented as individual core modules of this component.

Task collections

Besides the core components the framework includes task collections which again consist of multiple task modules. Each module provides very specific functionality, implementing either a linear or a nonlinear task. In the following we describe the task collections with the individual task modules we chose to include in the framework.

This list of modules is not a complete list of all possible modules but rather an example.

Collection of linear tasks: This task collection contains all those task modules that implement a linear task. The functionality of each task module is used in the read phase.

>   **File writer**: This task module is an example implementation of the linear task *data duplication* and simply writes all input data into a file on a target device.

>   **Basic carver**: This task module is an implementation of the linear task *header/footer-based file carving*. It implements file recovery based on searching header and footer signatures of known file types. As an example we chose to support signatures of JPEG, PDF, MPEG-1 and MP3 files.

>   **SHA-2 builder**: This task module is an implementation of the linear task *cryptographic hashing*. It allows building cryptographic hashes of the SHA-2 family (SHA-224, SHA-256, SHA-384, SHA-512) from the input data. Again, we use SHA-2 as an example – other cryptographic hash functions, for example from the SHA-3 family, may also be used instead or in addition.

Collection of nonlinear tasks: This task collection contains all those task modules which implement a nonlinear task. The functionality of each task module is used in the analysis phase.

>   **Fuzzy hash builder**: This task module is an implementation of the nonlinear task *fuzzy hashing*. It uses ssdeep for hashing the data. Since this fuzzy hashing approach sometimes requires multiple reads of the same input data, the task is run during the analysis phase.

>   **rHash builder**: This task module is an implementation of the nonlinear task *perceptual image hashing* and provides building robust hashes for image material using rHash (see Section 6.6.3.1). It supports different image file formats, for example JPEG, PNG, BMP, TIF and GIF.

>   **MP3 frame carver**: This task module is an implementation of the nonlinear task *audio frame carving*, that is it can recover single audio frames from complete or partial audio files found in the input data. This module supports carving for MP3 audio frames as example. Other audio file formats may also be used instead or additionally.

>   **MPEG-1 frame carver**: This task module is an implementation of the nonlinear task *video frame carving*. It provides a frame carver for video files which can recover single video frames from complete or partial video files found in the input data. The module supports carving for MPEG-1 video frames as example. Again, other video file formats may be used instead or additionally.

Since the framework is designed in a modular way, it is open to the framework user to extend it by adding desired functionality. Examples for additional modules are a task module which implements a specific algorithm for perceptual audio hashing (nonlinear task); a core module which extends the Result Data Processor by using additional remote data sources and corresponding information retrieval methods.

6.7.4   Case Study

The framework provides a straightforward workflow for processing arbitrary data. The data can in fact be of arbitrary type, since the framework can process even multimedia data that it does not recognize as such, as well as completely unknown or random data. Tasks which do not require any specific kind of input data, for example building a cryptographic hash, can still be carried out. This is especially true for all tasks which involve basic carving methods. However, when no supported multimedia files can be found in the input data, tasks which require a specific kind of input data, for example perceptual image hashing, will not be carried out at all.

In the following the workflow is described using a common scenario: A hard disk was seized from an individual which is suspected to possess illegal multimedia material. A forensic investigator analyzes the hard disk and finds one partition with a size of several gigabytes. The partition seems to contain no file system information, and therefore is not mountable or readable in the first instance. So the investigator decides to use the described framework in order to perform multimedia file carving and content identification on the partition data. The framework overview provided in Figure 6.12 aims to help following the workflow.

During the initialization of the framework, the investigator decides which tasks should be performed on the input data. He decides to enable all task modules which we described earlier and perform the corresponding tasks. More precisely, the investigator decides:

- A forensically sound copy of all partition data (the input data) should be written to a single file,
- A SHA-256 and a SHA-512 hash should be built from the complete input data,
- JPEG, PDF, MP3 and MPEG-1 files found in the input data should be recovered/extracted,
- Found MP3 files should be recovered (also single frames, if possible),
- Found MPEG-1 video files should be recovered (also single frames, if possible),
- A SHA-256 and a SHA-512 hash should be built from each recovered file,
- A robust hash should be built from each recovered JPEG file,
- A robust hash should be built from each recovered MPEG-1 video frame,
- A fuzzy hash should be built from each recovered file,
- Blacklist databases and whitelist databases should be queried for the hashes in order to find similarities to hashes in the databases.

After the initialization the framework starts. In order to set up a communication channel between the core components, one thread each is started for the Input Data Processor, the Supervisor, the Secure Logger and the Result Data Processor.

Additionally, all enabled task modules start a separate thread that waits for further instructions from the Supervisor. While the framework is running all core components and all task modules report any activity to the Supervisor. The Supervisor itself reports all activity to the Secure Logger. After the communication channels are set up, the Supervisor reports a successful initialization to the Secure Logger and instructs the Input Data Reader to start.

Next, the Input Data Processor starts the read phase by filling the shared buffer with the first chunk of input data. When the buffer is full, the Supervisor write-locks the buffer and notifies the threads of those modules which implement a linear task that new data is available in the buffer. This means that the threads of the file writer, the basic carver and the SHA-2 builder are notified and can start simultaneously processing the now available data.

During the data processing, the Supervisor communicates with each task module thread in order to exchange relevant information. For example the SHA-2 builder needs to know from the basic carver, for example whether a new file is found and, if so, at which location. Therefore, each time the basic carver recognizes a known header or a footer signature, it tells the Supervisor which in turn tells the SHA-2 builder so that it can start or finish building the corresponding SHA-256 and SHA-512 hash of the newly found file.

As soon as all data processing threads finish, the Supervisor gets notified, read-locks the buffer, releases the write lock and notifies the Input Data Processor that the buffer can be filled with the next chunk of input data. This is repeated until all input data is read and processed by the modules implementing the linear tasks. The read phase is then finished and all threads which processed any data in the read phase terminate.

Next, the Supervisor notifies the Result Data Processor that the following results are available from the read phase:

- A forensically sound copy of the complete input data (as partition image file),
- A SHA-256 and a SHA-512 hash of the complete input data,
- A number of complete JPEG, PDF, MP3, and MPEG-1 files which were recovered from the input data,
- A SHA-256 and a SHA-512 hash of each recovered JPEG, PDF, MP3 and MPEG-1 file,
- Additional temporary data used as input for some nonlinear tasks.

Now the analysis phase begins. Any results which were collected from the read phase, for example recovered files or temporary data, are now available to be processed. Therefore, the Supervisor notifies all waiting threads of those modules which implement the nonlinear tasks. In this case study the threads of the rHash builder, the fuzzy hash builder, the MP3 frame carver, and the MPEG-1 frame carver are notified that new data is ready to be processed. The threads can then start simultaneously processing the data assigned to them.

The Supervisor again communicates with all threads in order to exchange relevant information. For instance, the thread of the rHash task module, responsible for building perceptual image hashes, not only processes previously recovered JPEG files. It also needs information from the thread performing MPEG-1 frame carving about whether and where any MPEG-1 video frames have been recovered in order to build a robust hash for each found video frame. Also, the fuzzy hash builder needs information about any recovered files in order to build the corresponding fuzzy hash.

Each thread terminates after it finished processing the data. When the last thread terminated the Supervisor notifies the Result Data Processor that new results are available from the analysis phase:

- A robust hash built from each recovered JPEG file,
- A robust hash built from each recovered MPEG-1 video frame,
- A fuzzy hash for each recovered file,
- A number of recovered MP3 audio frames,
- A number of recovered MPEG-1 video frames.

The Result Data Processor can now initiate the final lookup phase where it queries a number of databases for information regarding the results collected during the read and the analysis phases. The database queries are performed in parallel by starting several subthreads where each thread handles one database connection. The Supervisor monitors all threads and notifies the Result Data Processor when all queries have finished. Then the Result Data Processor presents all results from the read phase, the analysis phase and the lookup phase to the framework user. The user can review all results as well as the individual steps of the workflow logged by the Secure Logger. The threads of the Supervisor, the Result Data Processor, and the Secure Logger continue running until the user finally exits the framework.

6.8   Conclusions

File carving multimedia data in some aspects differs significantly from file carving arbitrary binary data. The size of the data is one aspect. Especially high-quality video files should be among the largest files which forensic investigators often have to deal with. In addition, the stream-like file structure of many multimedia file formats is another important difference: Music and video files are often stored in such a way that even relatively small segments can still provide meaningful evidence. As clusters in file systems tend to be larger on larger volumes and lossy compression formats become more efficient, the trend of the recent years is that an increasing amount of meaningful information can be found in one cluster.

However, multiplexed multimedia files containing both, elementary *video and audio* streams will remain a serious challenge. Here, the multiplexing implies a large degree of *intrinsic* fragmentation of the elementary media streams inside its correspondent container format, for example MP4 or Flash Video. A framework like the one described in Section 6.7.3 can provide the synchronization of carving algorithms among the syntactic or content-based/semantic information in audio, video and container data. This facilitates analyzing other popular file formats relevant in forensic inspection, provided the framework modules are implemented that can handle the corresponding multimedia file formats.

One example is H.264 video with AAC audio contained in an MP4 file (extension .mp4). For instance, the current version of the *Defraser* multimedia file carving system actually can handle such H.264 data but not the audio sound track stored in AAC format. By combining an AAC and an H.264 carving module within the described framework, both the audio and the video track could be recovered efficiently.

The are many tools that can help a forensic examiner when dealing with multimedia files. Intelligent file carvers that use the stream structure to identify stream segments by header synchronization codes are one example. Robust hashing to match found content to blacklists independently from the coding format are another. Combining these tools in such a way that at the end a powerful framework usable to an examiner who is not a multimedia expert is an important future challenge.

Acknowledgements

This work was supported by the *CASED Center for Advanced Security Research Darmstadt*, Germany (www.cased.de), funded by the German state government of Hesse under the *LOEWE* program.

References

Air Force Office of Special Investigations 2001 Foremost, http://foremost.sourceforge.net (Accessed 21 February 2015).

Al-Dahir O, Hua J, Marziale L, Nino J, Richard III GG and Roussev V 2007 DFRWS 2007 challenge submission. Technical report, DFRWS.

Baier H and Breitinger F 2011 Security aspects of piecewise hashing in computer forensics. *IMF 2011*, Stuttgart, Germany, pp. 21–36. IEEE Computer Society.

Blender Foundation 2008 Big Buck Bunny. www.bigbuckbunny.org.

Blu-ray Disc Association 2002 *Blu-ray Disc Rewritable, Part 1: Basic Format Specifications*, 1.0 edn.

Boytsov L 2011 Indexing methods for approximate dictionary searching: Comparative analysis. *ACM Journal of Experimental Algorithmics* **16**(1), 1:1–1:91.

Bozkaya T and Ozsoyoglu M 1999 Indexing large metric spaces for similarity search queries. *ACM Transactions on Database Systems* **24**(3), 361–404.

Breitinger F and Baier H 2012 A fuzzy hashing approach based on random sequences and hamming distance. *Proceedings of the 2012 ADFSL Conference*, Richmond, VA, pp. 89–100. Association of Digital Forensics, Security and Law.

Breitinger F and Baier H 2013 Similarity preserving hashing: Eligible properties and a new algorithm MRSH-v2. *Digital Forensics and Cyber Crime*, vol. 114 of *LNICST*, pp. 167–182. Springer, Berlin Heidelberg.

Breitinger F, Åstebøl KP, Baier H and Busch C 2013 mvHash-B – A new approach for similarity preserving hashing *IMF 2013*, Nuremberg, Germany, pp. 33–44. IEEE Computer Society.

Breitinger F, Ziroff G, Lange S and Baier H 2014 Similarity hashing based on levenshtein distance *Advances in Digital Forensics X*. Vienna, Austria, 8–10 January 2014, vol. 433 of *IFIP AICT*, pp. 144–147. Springer, Berlin Heidelberg.

Carrier B 2005 *File System Forensic Analysis*. Addison Wesley, Reading, MA.

Chávez E, Navarro G, Baeza-Yates R and Marroquín JL 2001 Searching in metric spaces. *ACM Computing Surveys* **33**(3), 273–321.

Cheng WH, Chu WT and Wu JL 2003 Semantic context detection based on hierarchical audio models. *MIR '03: Proceedings of the Fifth ACM SIGMM International Workshop on Multimedia Information Retrieval*, Berkeley, CA, pp. 109–115. ACM Press.

De Roover C, De Vleeschouwer C, Lefèbvre F and Macq B 2005 Robust video hashing based on radial projections of key frames. *IEEE Transaction on Signal Processing* **53**(10), 4020–4037.

Doets PJO and Lagendijk R 2004 Theoretical modeling of a robust audio fingerprinting system. *Fourth IEEE Benelux Signal Processing Symposium*, Hilvarenbeek, the Netherlands, pp. 101–104.

DVD Consortium 1996 *DVD-ROM, Part 1: Physical Specifications*, 1.0 edn DVD Format/Logo Licensing Corporation.

ECMA 1996 Data interchange on read-only 120 mm optical data disks (CD-ROM). Standard ECMA-130, 2 edn., Ecma International. Initial specification by Philips and Sony (1985); standard also available as ISO/IEC 10149:1995.

ECMA 2001 120 mm DVD – Read-only disk. Standard ECMA-267, 3rd edn., Ecma International. Initial specification by DVD Consortium (1996); standard also available as ISO/IEC 16448:2002.

Fridrich J and Goljan M 2000a Robust hash functions for digital watermarking. *Proc. ITCC 2000*, Las Vegas, NV, 2000, pp. 173–178.

Fridrich J and Goljan M 2000b Robust hash functions for digital watermarking. *Proceeding of International Conference on Information Technology: Coding and Computing, 2000*, Las Vegoes, NV pp. 178–183. IEEE Computer Society.

Garfinkel S 2007 Carving contiguous and fragmented files with fast object validation. *Digital Investigation* **4**(Suppl.), S2–S12.

Ghouti L and Bouridane A 2006 A robust perceptual audio hashing using balanced multiwavelets. *IEEE International Conference on Acoustics, Speech and Signal Processing, ICASSP 2006*, Toulouse, France, 14–19 May 2006, vol. V, pp. 209–212.

Gionis A, Indyk P and Motwani R 1999 Similarity search in high dimensions via hashing. *Proceedings of 25th Conference on Very Large Data Bases*, Edinburgh, Scotland pp. 518–529. Morgan Kaufmann.

Grenier C 2007 Photorec. http://www.cgsecurity.org/wiki/PhotoRec (Accessed 21 February 2015).

Gruhne M 2009 Robust audio identification for commercial applications. *International Workshop for Technology, Economy, Social and Legal Aspects of Virtual Goods (VIRTUAL GOODS 2003), 22–24 May 2003*, Ilmenau, Germany.

Haitsma J, Oostveen J and Kalker A 2001a A highly robust audio fingerprinting system. *Second International Symposium of Music Information Retrieval (ISMIR 2001)*, Indiana University, Bloomington, IN, USA, October 15–17, 2001.

Haitsma J, Oostveen J and Kalker A 2001b Robust audio hashing for content identification. *Content Based Multimedia Indexing (CBMI) 2001*, Brescia Italy.

IEC 1999 Audio recording – Compact disc digital audio system. IEC 60908, 2nd edn., International Electrotechnical Commission. Initial specification by Philips and Sony (1980).

ISO 1993a Information technology – Coding of moving pictures and associated audio for digital storage media at up to about 1.5 Mbit/s – Part 1: Systems. ISO/IEC 11172-1:1993, International Organization for Standardization Geneva, Switzerland.

ISO 1993b Information technology – Coding of moving pictures and associated audio for digital storage media at up to about 1.5 Mbit/s – Part 2: Video. ISO/IEC 11172-2:1993, International Organization for Standardization Geneva, Switzerland.

ISO 1993c Information technology – Coding of moving pictures and associated audio for digital storage media at up to about 1.5 Mbit/s – Part 3: Audio. ISO/IEC 11172-3:1993, International Organization for Standardization Geneva, Switzerland.

ISO 1994 Information technology – Digital compression and coding of continuous-tone still images: Requirements and guidelines. ISO/IEC 10918-1:1994, International Organization for Standardization Geneva, Switzerland.

ISO 1998 Information technology – Generic coding of moving pictures and associated audio information – Part 3: Audio. ISO/IEC 13818-3:1998, International Organization for Standardization Geneva, Switzerland. First edition issued 1995.

ISO 2003 Information technology – Coding of audio-visual objects – Part 14: MP4 file format. ISO/IEC 14496-14:2003, International Organization for Standardization Geneva, Switzerland.

ISO 2006 Information technology – Generic coding of moving pictures and associated audio information – Part 7: Advanced Audio Coding (AAC). ISO/IEC 13818-7:2006, International Organization for Standardization Geneva, Switzerland. First edition issued 1997.

ISO 2009 Information technology – Coding of audio-visual objects – Part 3: Audio. ISO/IEC 14496-3:2009, International Organization for Standardization Geneva, Switzerland. First edition issued 1999.

ISO 2013 Information technology – Digital compression and coding of continuous-tone still images: JPEG File Interchange Format (JFIF). ISO/IEC 10918-5:2013, International Organization for Standardization Geneva, Switzerland.

JIETA 2013 Exchangeable image file format for digital still cameras: Exif version 2.3 (revised). JIETA CP-3451C, Japan Electronics and Information Technology Industries Association Tokyo Japan. Standard also available as CIPA DC-008-2012.

Kornblum J 2006a Fuzzy hashing. http://dfrws.org/2006/proceedings/12-Kornblum-pres.pdf (Accessed 21 February 2015). Presentation slides for (Kornblum 2006b).

Kornblum J 2006b Identifying almost identical files using context triggered piecewise hashing. *Digital Investigation* 3(Suppl.), S91–S97.

Kornblum J 2006c ssdeep. http://ssdeep.sourceforge.net (Accessed 21 February 2015).

Lancini R, Mapelli F and Pezzano R 2004 Audio content identification by using perceptual hashing. *2014 IEEE International Conference on Multimedia and Expo (ICME)*, Taipei, Taiwan, 27–30 June 2004.

Liu T, Zhang HJ and Qi F 2003 A novel video key-frame-extraction algorithm based on perceived motion energy model. *IEEE Transactions on Circuits and Systems for Video Technology* 13(10), 1006–1013.

Mao M, Chen S and Xu J 2009 Construction of the initial structure for preimage attack of md5. *2013 Ninth International Conference on Computational Intelligence and Security* 1, 442–445.

Memon N and Pal A 2006 Automated reassembly of file fragmented images using greedy algorithms. *IEEE Transactions on Image Processing* 15(2), 385–393.

Mıçak MK and Venkatesan R 2001 A perceptual audio hashing algorithm: A tool for robust audio identification and information hiding. In *Lecture Notes in Computer Science, 4th International Workshop Information Hiding, IH 2001*, Pittsburgh, PA, 25–27 April 2001, (ed. Moskowitz I), vol. 2137.

Microsoft 1991 *Microsoft Windows Multimedia Programmer's Reference*, Chapter 8. Microsoft Press, Redmond, WA.

Mitchell JL, Pennebaker WB, Fogg CE and LeGall DJ 2002 *MPEG Video Compression Standard*. Kluwer Academic Publishers New York.

Navarro G, Baeza-Yates R, Sutinen E and Tarhio J 2001 Indexing methods for approximate string matching. *IEEE Data Engineering Bulletin* 24(4), 19–27.

Netherlands Forensic Institute 2007 Defraser. http://sourceforge.net/projects/defraser/ (Accessed 21 February 2015).

NIST Information Technology Laboratory 2003–2014 National Software Reference Library. http://www. nsrl.nist.gov (Accessed 21 February 2015).

Oostveen JC, Kalker T and Haitsma J 2001 Visual hashing of video: Applications and techniques. *Applications of Digital Image Processing XXIV*, San Diego, CA, vol. 4472 of *Proc. SPIE*, pp. 121–131.

Pal A, Sencar HT and Memon N 2008 Detecting file fragmentation point using sequential hypothesis testing. *Digital Investigation* **5**(Suppl.), S2–S13.

Patterson DA, Gibson G and Katz RH 1988 A case for redundant arrays of inexpensive disks (RAID). *Proceedings of the 1988 ACM SIGMOD International Conference on Management of Data*, Chicago, JL pp. 109–116. ACM.

Paulevé L, Jégou H and Amsaleg L 2010 Locality sensitive hashing: A comparison of hash function types and querying mechanisms. *Pattern Recognition Letters* **31**(11), 1348–1358.

Philips and Sony 1980 *Compact Disc Digital Audio ('Red Book')*, 1st edn.

Philips and Sony 1985 *Compact Disc Read Only Memory ('Yellow Book')* 1st edn.

Richard III G and Roussev V 2005 Scalpel: A frugal, high performance file carver. *Proceedings of the 2005 digital Forensics Research Workshop (DFRWS 2005)*, New Orleans, LA.

Rogaway P and Shrimpton T 2004 Cryptographic hash-function basics: Definitions, implications, and separations for preimage resistance, second-preimage resistance, and collision resistance. *Fast Software Encryption*, Delhi, India, 5–7 February 2004, vol. 3017 of *LNCS*, pp. 371–388. Springer Berlin Heidelberg.

Roussev V 2010a Data fingerprinting with similarity digests *Advances in Digital Forensics VI*, Hong Kong, China, 4–6 January 2010, vol. 337 of *IFIP AICT*, pp. 207–226. Springer, Berlin Heidelberg.

Roussev V 2010b sdhash. http://roussev.net/sdhash (Accessed 21 February 2015).

Roussev V, Chen Y, Bourg T and Richard III GG 2006 md5bloom: Forensic filesystem hashing revisited. *Digital Investigation* **3**(Suppl.), S82–S90.

Roussev V, Richard III GG and Marziale L 2007 Multi-resolution similarity hashing. *Digital Investigation* **4**(Suppl.), S105–S113.

Sadowski C and Levin G 2011 SimiHash: Hash-based similarity detection. Technical Report UCSC-SOE-11-07, University of California, Santa Cruz CA. http://www.soe.ucsc.edu/research/technical-reports/ UCSC-SOE-11-07 (Accessed 21 February 2015).

Sajja A 2010 Forensic reconstruction of fragmented variable bitrate MP3 files. Master's thesis, University of New Orleans, New Orleans, LA. http://scholarworks.uno.edu/td/1258 (Accessed 21 February 2015).

Steinebach M, Liu H and Yannikos Y 2012 ForBild: Efficient robust image hashing. In *Media Watermarking, Security, and Forensics 2012* (ed. Memon ND, Alattar AM and Delp III EJ), Burlingame, CA, 23–25 January 2012, vol. 8303 of *Proc. SPIE*, pp. 8303 0O–1–8. SPIE, Bellingham, WA.

Stevens M 2013 New collision attacks on SHA-1 based on optimal joint local-collision analysis *Advances in Cryptology – EUROCRYPT 2013*, Athens, Greece, vol. 7881 of *LNCS*, pp. 245–261. Springer, Berlin Heidelberg.

University of Southern California – Signal and Image Processing Institute 1977 The USC-SIPI image database http://sipi.usc.edu/database/database.php (Accessed 21 February 2015).

Wang ALC 2003 An industrial strength audio search algorithm *Fourth International Conference on Music Information Retrieval (ISMIR)*, Baltimore, MD, 27–30 October 2003.

Wang X and Yu H 2005 How to break md5 and other hash functions. *Advances in Cryptology – EUROCRYPT 2005*, Aaehus, Denmark, pp. 561–561.

Wang H, Divakaran A, Vetro A, Chang SF and Sun H 2003 Survey of compressed-domain features used in audio-visual indexing and analysis. *Journal of Visual Communication and Image Representation* **14**(2), 150–183.

Winter C, Schneider M and Yannikos Y 2013 F2S2: Fast forensic similarity search through indexing piecewise hash signatures. *Digital Investigation* **10**(4), 361–371.

Winter C, Steinebach M and Yannikos Y 2014 Fast indexing strategies for robust image hashes. *Digital Investigation* **11**(Suppl. 1), S27–S35.

Yang B, Gu F and Niu X 2006 Block mean value based image perceptual hashing *International Conference on Intelligent Information Hiding and Multimedia Signal Processing, 2006. (IIH-MSP '06)*, Pasadena, CA, 18–20 December 2006, pp. 167–172. IEEE Computer Society, Los Alamitos, CA.

Yannikos Y, Ashraf MN, Steinebach M and Winter C 2013 Automation of video File carving and illegal content identification *Ninth IFIP WG 11.9 International Conference on Digital Forensics*, Orlando, FL, 28–30 January 2013, vol. 410 of *IFIP AICT*, pp. 195–212. Springer, Berlin Heidelberg.

Yoo B, Park J, Lim S, Bang J and Lee S 2011 A study on multimedia file carving method. *Multimedia Tools and Applications* **1**, 1–19.

Zauner C, Steinebach M and Hermann E 2011 Rihamark: perceptual image hash benchmarking. *Media Watermarking, Security, and Forensics III*, San Francisco, CA vol. 7880 of *Proc. SPIE*, pp. 7880 0X–1–15. International Society for Optics and Photonics.

Zhou X, Schmucker M and Brown C 2006 Video perceptual hashing using interframe similarity. *GI Sicherheit 2006*, Magadeburg, Germany, 20–22 February 2006, vol. P-77 of *LNI*, pp. 107–110.

Zmudzinski S, Taneja A and Steinebach M 2012 Carving and reorganizing fragmented *MP3 files using syntactic and* spectral information. *Proceeding of AES 46th Conference on Audio Forensics 2012, 14–16 June 2012, Denver, CO*. Audio Engineering Society.

7

On Forensic Use of Biometrics

Banafshe Arbab-Zavar[1], Xingjie Wei[2], John D. Bustard[1], Mark S. Nixon[1]
and Chang-Tsun Li[2]

[1]*School of Electronics and Computer Science, University of Southampton, Southampton, UK*
[2]*Department of Computer Science, University of Warwick, Coventry, UK*

7.1 Introduction

Forensic science largely concerns the analysis of crime: its existence, the perpetrator(s) and the modus operandi. The science of biometrics has been developing approaches that can be used to automatically identify individuals by personal characteristics. The relationship of biometrics and forensics centers primarily on identifying people: the central question is whether a perpetrator can reliably be identified from scene-of-crime data or can reliably be excluded, wherein the reliability concerns reasonable doubt. The personal characteristics which can be used as biometrics include face, finger, iris, gait, ear, electroencephalogram (EEG), handwriting, voice and palm. Those which are suited to forensic use concern traces left at a scene of crime, such as latent fingerprints, palmprints or earprints, or traces which have been recorded, such as face, gait or ear in surveillance video.

Biometrics is generally concerned with the recognition of individuals based on their physical or behavioural attributes. So far, biometric techniques have primarily been used to assure identity (in immigration and commerce, etc.). These techniques are largely automatic or semi-automatic approaches steeped in pattern recognition and computer vision. The main steps of a biometric recognition approach are, (i) acquisition of the biometric data, (ii) localization and alignment of the data, (iii) feature extraction, and (iv) matching. Feature extraction is often the pivotal part of this workflow. The biometric studies are concerned with finding a set of features, which provides the

Handbook of Digital Forensics of Multimedia Data and Devices, First Edition.
Edited by Anthony T.S. Ho and Shujun Li.
© 2015 John Wiley & Sons, Ltd. Published 2015 by John Wiley & Sons, Ltd.
Companion Website: www.wiley.com/go/digitalforensics

least deviation between the different samples of one individual and most separability between the samples of one individual and the rest of the population. Such a feature set will provide the best chance for individualization. In fingerprint recognition, the most popular and widely used features are the minutiae-based feature. Such level of consensus, however, has not been reached for most of the biometrics traits and the best set of features is subject to constant examination.

One of the earliest attempts to use biometric data for identification dates back to the 1880s when the French criminologist Alphonse Bertillon proposed a method based on anthropometric measurements. Bertillon suggested this method as a means for classification and sorting of the records of individuals and searching among them (Bertillon 1893). In 1890, Bertillon set forth a set of standards for forensic photography. He also developed a taxonomy to describe some of the physiological features of the head, including nose, forehead and ear. He called this *portrait parlé* or *spoken portrait* (Bertillon 1890). The combination of the anthropometric measurements and the *spoken portrait* developed by Bertillon is called *Bertillonage* and was fast adopted by the police and the judicial systems. Around the same time, Hendry Faulds proposed the use of fingerprints for identification (Faulds 1880). Although fingerprints were first considered with scepticism, they gradually replaced Bertillonage as the main method of forensic identification, especially after the *West v. West* (1903) case concerning a pair of suspects who could not be disambiguated by the Bertillon's methods. Among the advantages of fingerprints over Bertillonage was their relative ease of use and that one could not find traces of Bertillonage's anthropometric measurements at the scene of crime while fingerprints were in abundance. The later developments in biometrics largely followed the development of computer vision techniques, enabling identification by other bodily attributes.

In *Frye v. United States* 1923, a federal court was faced with the question of expert evidence admissibility. The court concluded that the expert evidence could be admitted to court only if this expertise had gained *general acceptance* in the field in which it belongs. In 1993, in *Daubert v. Merrell Dow Pharmaceuticals* a new standard for expert evidence admissibility was introduced by the US supreme court. In this, the proffered expert testimony must be shown to be based on reliable foundations. To show this, it is required to determine if the proffered science has been tested, if this testing was based on a sound methodology and also to take into account the results of this testing. This new standard was considered as a paradigm shift (Saks and Koehler 2005), and it was suggested that fingerprints could be one of the first forensic identification methods to make this transition since the required large databases already exist in this field. In fact, the use of handwriting and fingerprint evidence has been challenged for use in court procedure in 1999, leading to a study of whether fingerprints are permanent and unique (Pankanti *et al.* 2002). This raised concerns in the fallibility of fingerprint evidence, the performance in degraded imagery, the performance of available techniques and the need for its improvement. Such debate is not new in science since the development of any new technique must be justified in terms of societal use. Further, when it is to be deployed in serious crime investigations where punishment can be severe, then error cannot be tolerated. Indeed, the need for individualisation as a forensic paradigm was later to be questioned (Cole 2009). The current state of the art of biometrics in forensics is more nascent than established. The first *IEEE/IAPR International Workshop on Biometrics and Forensics (IWBF)* was held only recently in early 2013, arising from the EU-sponsored ICT COST Action IC1106 on Integrating Biometrics and Forensics for the Digital Age (http://www.cost.eu/COST_Actions/ict/Actions/IC1106). Just earlier the first *Workshop on Databases in Biometrics, Forensics and Security Applications (DBforBFS)* was held as a satellite workshop of the *2013 BTW Conference* (on database systems in Business, Technology and Web). The technical programs for these workshops considered face, hand-based, behavioural and other biometrics and forensics together with considerations of performance and database construction, especially for forensic deployment and analysis. There have been other previous conference sessions, and the successful emergence of conferences in new specialist topics generally underlines not only their contemporary nature, but also the importance of an emerging new subject.

When fingerprints were suggested in 1880, little investigation had been performed over their individuality, and there was no mention of the error rates for the identification predictions. In courts, other expertise were also being offered and admitted which seriously lacked the backing of proper scientific testing and statistical measures of performance. In this respect, many mistakes were made and are still being made. Saks and Koehler (2005) reported that in 86 DNA exoneration cases the error due to forensic science testing errors is ranked very high at 63% and that it is second only to the eyewitness errors with 71%. In terms of performance, the main aim of biometrics is to verify if a person has a claimed identity (a so-called *one-to-one matching*) and identification (*one-to-many matching* where a subject is compared with a database). In forensics, the conclusion concerns likelihood between a suspect and evidence. In fingerprints evidence can lead to three conclusions: individualisation, exclusion or inconclusiveness (Champod 2000). The probability of matching can also be graded as impossible, possible, probable or very likely. In DNA analysis, the potential error rate is usually couched in terms of the likelihood of mismatch, which is another representation of probability.

In terms of the literature, the majority of approaches describe analysis of latent fingerprints. However, there is also use of voice for speaker identification, face identification, dental biometrics, DNA and handwriting, which are all established biometrics in their own right (Dessimoz and Champod 2008). In terms of emerging biometrics, so far there has been one deployment of gait biometrics for identification (Bouchrika *et al.* 2011; Guan and Li 2013; Guan *et al.* 2013), and there is now a system aimed at such use (Iwama *et al.* 2012). Soft biometrics is a more recent interest and can handle low-quality surveillance data (Park and Jain 2010). Ears were considered in Bertillon's pioneering early study where the ear was described as the most identifying part of an individual and proposed a method for ear classification, and the length of the ear was one of the 11 measures that were used. One early forensics study (Spaun 2007) described interest in facial and ear individualization, adding the possibility of exploring additional biometrics including hands and gait and observing that additional ear analyses are needed, instead of databases of hundreds of ears, thousands of ears or more.

In the remainder of this chapter, we will concentrate on two case studies discussing the forensic possibilities of face and ear as biometrics. Face is the natural means for human beings to recognize each other. However, currently no fully automatic face recognition system is accepted by the judicial system. Section 7.3 introduces the manual and computer-aided forensic face recognition, discusses the disparities between the behaviour of the current automatic face recognition systems and that which is needed for forensic application and outlines the current progress towards addressing the challenges existing in face recognition. Section 7.4 examines an emerging biometric ear. The detailed examination shows the challenges that exist in introducing a new biometric feature into forensics. Ear biometrics has been chosen as the second case study as it is a potentially important biometric feature, yet its use is still under question. The current state of formal validation of ears as a forensic tool is discussed, and a set of morphological features along with an analysis of their discriminatory powers are presented. These features are important in deciding whether there is enough information available for identification in case of missing features. The terminology associated with these features may also assist with communicating ear comparison results to juries, an important step in making such evidence effective at trial. But first, in Section 7.2, we will give an overview of the general biometric system operation modes and performance metrics.

7.2   Biometrics Performance Metrics

A biometric system can be used as an assistant tool in the forensic scenarios for helping on queries against a large enrolled database. The query can be a *one-to-many search* to determine potential matches to a probe from the gallery, or a *one-to-one check* to verify the identity of an individual. These two tasks are referred to as *identification* and *verification* in the biometrics research community.[1]

In identification, the biometric system searches an enrolled database for a gallery sample matching the probe sample. An ordered list of top *n* matches may be returned as the possible identities of the probe. The performance of the system in the identification task is measured in terms of *rank-n recognition rate* which is the rate at which the true association has been included in the top *n* matches to the probe. Recognition rate is the simplified term for rank-1 recognition rate where the system returns a single match, the best match, as the most probable association for the probe sample.

On the other hand, verification is the task where the biometric system attempts to confirm an individual's claimed identity by comparing the probe sample to the

---

[1] Biometrics Glossary by National Science & Technology Council (NSTC) Subcommittee on Biometrics, 2013.

individual's previously enrolled sample. Verification is based on a decision threshold. This threshold is set by comparing all sample pairs in the gallery. The threshold is chosen to separate the genuine scores distribution from the impostor scores distribution and give the best performance based on one of the following metrics:

- *False acceptance rate* (FAR) is the rate at which the comparison between two different individuals' samples is erroneously accepted by the system as the true match. In other words, FAR is the percentage of the impostor scores which are higher than the decision threshold.
- *False rejection rate* (FRR) is the percentage of times when an individual is not matched to his/her own existing template. In other words, FRR is the percentage of the genuine scores which are lower than the decision threshold.
- *Equal error rate* (EER) is the rate at which both acceptance and rejection errors are equal (i.e. FAR = FRR). Generally, the lower the EER value, the higher the accuracy of the biometric system.

Automated biometric techniques can be used to analyze and interpret biometric traces in the forensics scenarios such as in investigation of a criminal offence and the demonstration of the existence of an offence (Meuwly 2012). These tasks are usually interrelated with each other. Biometric techniques are used to help in the three main ways:

1. *Decision.* In identity verification and identification, a decision needs to be made. Such applications include criminal ID management, suspect or victim identification, etc.
2. *Selection.* In forensics intelligence and investigation, biometrics techniques are used to link cases from biometric traces and generate short lists of candidates.
3. *Description.* In forensic evaluation, biometrics are used to describe the evidential value of the biometric evidence.

7.3 Face: The Natural Means for Human Recognition

Since the advent of photography, both government agencies and private organizations have kept face photo collections of people (e.g. personal identification documents, passports, membership cards, etc.). With the wide use of digital cameras, smart phones and CCTVs, face images can be easily generated every day. In addition, nowadays these images can be rapidly transmitted and shared through the highly developed social network such as Facebook. So face is almost the most common and familiar biometric trait in our daily lives. There are more opportunities to acquire and analyze face images of a questioned person (e.g. suspect, witness or victim) for forensic investigation purposes.

Face recognition has a long history and receives research interests from neuro-scientists, psychologists and computer scientists (Sinha *et al.* 2006). Compared with other biometric traits, face is not *perfect*. For example, it is generally less accurate than other forms of biometrics such as fingerprint and can potentially be affected by cosmetics more easily. However, face has its own advantages that make it one of the most preferred biometric traits for human recognition:

- *Biological nature*: Face is a very convenient biometric characteristic used by humans in the recognition of people, which makes it probably the most common biometric trait for authentication and authorization purposes. For example, in access control, it is easy for administrators to track and analyze the authorized person from his/her face data after authentication. The help from ordinary users (e.g. administrators in this case) can improve the reliability and applicability of the recognition systems. Whereas fingerprint or iris recognition systems require an expert with professional skills to provide reliable confirmation.
- *Non-intrusion*: Different from fingerprint and iris collections, facial images can be easily acquired from a distance without physical contact. People feel more comfortable for using face as identifier in daily lives. A face recognition system can collect biometric data in a user-friendly way, which is easily accepted by the public.
- *Less cooperation*: Compared with iris and fingerprint, face recognition has a lower requirement of user cooperation. In some particular applications such as surveillance, a face recognition system can identify a person without active participation from the subjects.

The first attempts we are aware of to identify a subject by comparing a pair of facial photographs was reported in a British court in 1871 (Porter and Doran 2000). Face recognition is one of the most important tasks in forensic investigations if there is any video or image material available from a crime scene. Forensic experts perform manual examination of facial images to match the images of a suspect's face. The use of automated facial recognition systems will not only improve the efficiency of forensic work performed but also standardize the comparison process.

7.3.1 Forensic Face Recognition

In the past, before the use of computers, face recognition was already widely used in forensics. The work of Bertillon (1893) was one of the first systematic approaches for face recognition in forensics as we mentioned in Section 7.1. Currently forensic face recognition is mainly performed manually by humans. In a typical forensic face recognition scenario, a forensic expert is given face images from a suspect (e.g. mugshot images) and a questioned person (i.e. the perpetrator). The forensic expert will give a value which represents the degree to which the these images appear to come from the same person.

There are four main categories of approaches in forensic face recognition (Ali *et al.* 2010; Dessimoz and Champod 2008): holistic comparison, morphological analysis, anthropometry and superimposition.

Table 7.1   Example of facial features examined.

Feature	Characteristic
Face	Shape, proportions, hairline
Forehead	Shape, bumps, horizontal creases, eyebrows
Eyes	Distance, angle fissure, colour, eye slit shape, creases, bags, wrinkles
Nose	Length, width, prominence, symmetry, shape of tip and nostrils, septum
Mid part of face	Cheekbones, cheek line, cheek-eye groove, cheek-nose groove
Ear	Size, protrusion, shape of helix and antihelix, darwin's tubercle, earlobe
Mouth	Size, shape, upper lip, lower lip
Mouth area	Shape of philtrum, moustache and shadow, beard and shadow
Chin	Shape, groove between mouth and chin, dimple, double chin
Low jaw	Shape
Throat	Adam's apple
Distinctive feature	Skin marks, scars, creases and wrinkles

1. *Holistic comparison.* In holistic comparison, faces are visually compared as a whole by the forensic experts. This is the simplest way and can be performed as a pre-step for other methods. Automatic face recognition systems can be designed to help for this not only on one-to-one comparison (i.e. verification) but also on one image compared to a large-scale gallery database (i.e. identification).

2. *Morphological analysis.* In morphological analysis, the local features of the face will be analyzed and compared by the forensic experts who are trained in that discipline. They carry out an exhaustive analysis on the similarities and differences in observed faces, trait by trait on the nose, mouth, eyebrows, etc., even the soft traits such as marks, moles, wrinkles, etc. The location and distribution of local facial features are considered but not explicitly measured compared with anthropometry based approaches. One example of the examined facial features currently used by the Netherland's Forensic Institute[2] is summarized in Table 7.1 (Meuwly 2012). It can be seen from the table that both internal and external features of the face are considered. These features are usually fall into two categories (Spaun 2011): (1) *class characteristics* which can place an individual within a group (e.g. facial shape, shape of the nose and freckles) and (2) *individual characteristics* which are unique to distinguish the individual (e.g. skin marks, scars, creases and wrinkles). Generally, the forensic experts need to make the conclusion based on the following comparison criteria for these local features: (i) *Similar*: imaging conditions are not optimal, in a sense that differences might be invisible. (ii) *No observation*: observation is not possible due to circumstances. (iii) *Different*: observed differences may be explained by differences in the imaging conditions.

---

[2] http://www.forensicinstitute.nl/.

3. *Anthropometry*. Anthropometry refers to the measurement of the human individual, which can be used for human recognition. Different from morphological analysis, in face anthropometry, the quantification measurements (e.g. spatial distance and angles) between specific facial landmarks (e.g. the mid-line point between the eyebrows, the lowest point on the free margin of the ear lobe, the midpoint of the vermilion border of the lower lip and the most anterior midpoint of the chin) are used for comparison. However, usually blemishes on the face such as scars are not considered. When anthropometric measurements are taken from photographs rather than from the face of a living person, it is called *photo-anthropometry*. The face images being compared should be taken from the same angle and direction and has a high quality to be able to detect the facial landmarks. These requirements limit the use of anthropometry approaches in uncontrolled scenarios (e.g. surveillance situations). At present, anthropometry-based methods are suitable to be used to exclude the questioned person rather than to make a positive identification.

4. *Superimposition*. In superimposition, one face image is overlaid onto another, and the forensic experts need to determine whether there is an alignment and correspondence of the facial features. These images should be captured under the same pose and be processed to the same scale. This category of approaches is not accurate due to their high requirement that the compared images should be taken under the same conditions. Generally, in forensics, superimposition can be performed not only between two face images but also between a face and a skull (Ibañez *et al.* 2011). In addition, superimposition is also widely used in forensic facial reconstruction (Aulsebrook *et al.* 1995), which aims to recreate the face of an individual (whose identity is often not known) for recognition purpose. Automatic face recognition system can be developed in the direction of modelling a 3D face/head model to compare with a 2D query image. In this way, the pose, angle and orientation of the face can be adjusted using the 3D models.

In holistic comparison, conclusions are generated by visually comparing images as a whole. Morphological analysis is the most applicable in modern forensics. Anthropometry and superimposition are practised by jurisdictions, but the outcomes are highly sensitive to the subject's pose and thus may easily produce inaccurate results. The choice of a specific approach depends on the face images to be compared and generally a fusion of these methods is applied in the real case analysis scenarios.

Currently there is no standard procedure and agreed upon guideline among forensic researchers. Some working groups such as the Facial Identification Scientific Working Group (FISWG[3]) of FBI, the International Association for Identification (IAI[4]) and the European Network of Forensic Science Institutes (ENFSI[5]), as well as several

---

[3] https://www.fiswg.org/.

[4] http://www.theiai.org/.

[5] http://www.enfsi.eu/.

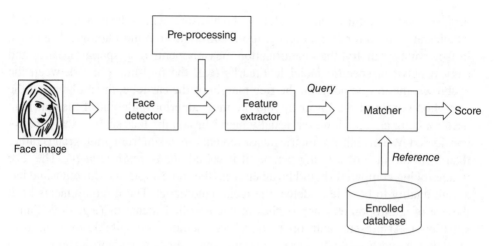

Figure 7.1 Framework of an automatic face recognition system.

international agencies such as the London Metropolitan Police are devoting to develop standards and guidelines for forensic face comparison.

Notice that in the forensic face recognition methods, ears are also considered as an external feature of the face. The ear is an important emerging biometric trait and is stable throughout adulthood. We will discuss the use of ear biometrics in the forensic tasks later.

7.3.2 Automatic Face Recognition Techniques

A general automatic face recognition system usually consists of the following modules: a face detector, a feature extractor and a matcher (Figure 7.1). The face detector crops the face area from the background of an image. The feature extractor then extracts effective information from face images for distinguishing different individuals. Usually pre-processing such as face alignment by the facial landmarks and face normalization (e.g. scale and illumination condition) will be performed before feature extraction. Then the matcher will compare two faces (e.g. one is from query and one is from the enrolled database) by the extracted features then a similarity score is calculated. Face recognition is based on the similarity scores, and its performance highly relies on the extracted features and classification algorithms used to distinguish faces.

In the early times, the main recognition approaches are geometric feature-based methods which rely on measurements between specific facial landmarks. This is similar to the anthropometry-based methods in the forensic face recognition. The first attempt to automatic face recognition started by Chan and Bledsoe (1965) in a semi-automated mode where a set of facial features were extracted from the photographs by humans. The first fully automatic face recognition system was presented by Kanade (1973), which was a milestone at that time. In 1990s, the linear subspace analysis approaches and statistical models became the mainstream. Turk and Pentland (1991) applied principal component analysis (PCA) on face images, which was referred to as *Eigenface*. These eigenfaces were the eigenvectors associated to the largest eigenvalues of the covariance matrix of the training samples, which ensured the data variance was maintained while eliminating unnecessary existing correlations among the original features (i.e. dimensions). PCA-based approaches greatly reduced the computational cost for high-dimensional data and inspired more active research in face recognition. Fisherface (Belhumeur *et al.* 1997), which was based on the liner discriminant analysis (LDA), also performed dimensionality reduction while preserving as much of the class discriminatory information as possible. Other popular methods included local feature analysis (LFA) (Penev and Atick 1996), elastic graph matching (EGM) (Wiskott *et al.* 1997), etc. From the late 1990s to present, the research of face recognition has focused on the uncontrolled and uncooperative scenarios (e.g. large pose changes, illumination and expression variations, low resolution and partially occluded faces). Locally linear embedding (LLE) (Roweis and Saul 2000), illumination core model (Georghiades *et al.* 2001), 3D Morphable Model (Romdhani *et al.* 2002), Local binary pattern (LBP) (Ahonen *et al.* 2006) and sparse representation-based classification (SRC) (Wright *et al.* 2009) are the representative methods in this period. A systematic survey of automatic face recognition can be found in the work of Zhao *et al.* (2003).

The performance of automatic face recognition techniques has been evaluated in a series of large-scale tests conducted by the National Institute of Standards and Technology (NIST[6]), such as the Facial Recognition Technology evaluation (FERET) (Phillips *et al.* 2000), the Face Recognition Vendor Test (FRVT) (Phillips *et al.* 2010) and the Face Recognition Grand Challenge (FRGC) (Phillips *et al.* 2005). Over the past decades, major advancements occurred in automatic face recognition. The false reject rate (FRR) of the best performing face recognition algorithm has decreased from 79% in 1993 to 0.3% in 2010 at a false accept rate (FAR) of 0.1% (Phillips 2012). The automatic face recognition has been successfully used in the field of security (e.g. access control and video surveillance), but the performance in unconstrained environment is still unsatisfactory. A full and systematic assessment of the automatic face recognition technology must be conducted under realistic conditions before it can be utilized for forensic applications.

7.3.3   Challenges and Trends of Face Recognition

Like in many biometric applications, the appearance variations caused by the unconstrained conditions are still challenges for face recognition in the context of forensic scenarios. Currently automatic face recognition system is only regarded as an assistant tool in forensic tasks. This section will discuss several specific face

---

[6] http://www.nist.gov/.

recognition problems which may also be difficult even for forensic experts. These challenges should be addressed in the future research (Jain *et al.* 2011).

7.3.3.1   Partial/Occluded Face Recognition

In the real-world environment, a face may be captured in arbitrary pose without the user's cooperation so it's very likely that the image only contains a partial face. Faces are easily occluded by facial accessories (e.g. sunglasses, scarf, hat and veil), objects in front of the face (e.g. hand, food and mobile phone), extreme illumination (e.g. shadow), self-occlusion (e.g. non-frontal pose) or poor image quality (e.g. blurring). In forensic face recognition, for example, it is needed to find a suspect in the crowd by matching a partially occluded face with enrolled database. The difficulty of occluded face recognition is twofold. Firstly, occlusion distorts the discriminative facial features and increases the distance between two face images of the same subject in the feature space. The intra-class variations are larger than the inter-class variations, which results in poorer recognition performance. Secondly, when facial landmarks are occluded, large alignment errors usually occur and degrade the recognition rate (Ekenel and Stiefelhagen 2009).

An intuitive idea for handling occlusion in automatic face recognition is to detect the occluded region first and then perform recognition using only the unoccluded parts. However, the types of occlusions are unpredictable in practical scenarios. The location, size and shape of occlusion are unknown, hence increasing the difficulty in segmenting the occluded region from the face images. A more practical way is to perform recognition with the presence of occlusion. There are two main categories of approaches in this direction.

The first is the *reconstruction-based approaches* which treat occluded face recognition as a reconstruction problem (He *et al.* 2011; Jia and Martínez 2008; Naseem *et al.* 2010; Wagner *et al.* 2012; Wei *et al.* 2012; Wright *et al.* 2009; Yang and Zhang 2010; Zhang *et al.* 2011). The sparse representation-based classification (SRC) proposed by Wright *et al.* (2009) is a representative example. A clean image is reconstructed from an occluded probe image by a linear combination of gallery images and the basis vectors of an occlusion dictionary. Then the occluded image is assigned to the class with the minimal reconstruction error.

The second category is the *local matching-based approaches*. Facial features are extracted from local areas of a face, for example, overlapping or non-overlapping patches of an image, so the affected and unaffected parts of the face can be analyzed in isolation. In order to minimize matching errors due to occluded parts, different strategies such as weighting (Tan *et al.* 2009), warping (Wei *et al.* 2013a,b), voting (Wei and Li 2013), local space learning (Martínez 2002; Tan *et al.* 2005) or multi-task sparse representation learning (Liao *et al.* 2013) are performed.

Klontz and Jain (2013) conducted a case study that used the photographs of the two suspects in the Boston Marathon bombings to match against a background set of mug-shots. The suspects' photographs released by the FBI were captured under uncontrolled environment and their faces were partially occluded by sunglasses and hats (Comcowich 2013). The study showed that current commercial automatic face recognition system had the notable potential to assist law enforcement. But the matching accuracy was not high enough, and more progress must be made to increase the utility in unconstrained face images.

7.3.3.2   Heterogeneous Face Recognition

Heterogeneous face recognition involves matching two face images from alternate imaging modalities. This is very practical in forensic scenarios. For instance, in the London riots in 2011, the police used face recognition system to help find the riot suspects involved in the unrest. The images of suspects are came from various sources, for example, still images captured from closed-circuit cameras, pictures gathered by officers, footage taken by the police helicopters or images snapped by members of the public. These images are usually from various sources from different modalities. In addition, in some extreme situations, only a particular modality of a face image is available. For example, in night-time environments, infrared imaging may be the only modality for acquiring a useful face image of a suspect. But the stored mug-shots by the police are visible band images. Another example is the sketch-photograph matching. When no photograph of a suspect is available, a forensic sketch is often generated according to the description of an eye-witness. Matching sketches against face photographs is very important for forensic investigation.

There are three categories of approaches in current heterogeneous face recognition. The first one is the *feature-based method* (Klare and Jain 2010; Klare *et al.* 2011; Lei and Li 2009) which represents face images with discriminative features that are invariant in different imaging modalities. The second one is the *synthesis-based method* (Tang and Wang 2004; Wang and Tang 2009; Zhang *et al.* 2010) which converts a face image in one modality (e.g. sketch) into another (e.g. photograph). And the third one is the *prototype-based method* (Klare and Jain 2013) which reduces the gap between two modalities by using a prototype as a bridge. 2D–3D face matching is a future research direction since face can be represented by heterogeneous features in the 3D and 2D modalities in the real-world cases.

7.3.3.3   Face Recognition across Aging

Facial ageing is a complex process that affects both the shape and texture (e.g. skin tone or wrinkles) of a face. The typical application scenario of face recognition systems against aging effect is to detect if a particular person is present in a previous recorded database (e.g. missing children identification and suspect watch-list check). As the age between a query and a reference image of the same subject increases, the accuracy of recognition system generally decreases.

Figure 7.2   Face samples of the same individual across ageing.

In automatic face recognition, ageing effect in human faces has been studied in two directions: (1) developing *age estimation techniques* to classify face images based on age (Geng *et al.* 2007; Guo *et al.* 2008) and (2) developing *ageing robust systems* to perform recognition. In the early time, researchers tried to simulate the ageing effects by developing the ageing function and then performing automatic age estimation based on that (Lanitis *et al.* 2002). But modelling the complex shape or texture variations of a face across ageing is a very challenging task. Nowadays, researchers propose the generative ageing model (Li *et al.* 2011) which learns a parametric ageing model in the 3D domain to generate synthetic images and reduce the age gap between query and reference images. One most challenging aspect of face recognition across ageing is that it must address all other unconstrained variations as well. Figure 7.2 shows the face samples of the same individual across aging. Pose, expression, illumination changes and occlusions can occur when images are taken years apart.

7.3.4   Summary

Face is the most natural way of recognition for human beings. A rich variety of approaches for face biometrics have been proposed, and its basic patterns are well understood over past several decades. Face recognition technology has been considered as the next-generation tool for human recognition.[7] Automatic face recognition is becoming an indispensable tool for modern forensic investigations.

---

[7] see FBI's the Next Generation Identification (NGI) program, http://www.fbi.gov/about-us/cjis/fingerprints_biometrics/ngi.

However, currently there is no generally accepted standard for forensic face comparison. Many challenging problems related to forensic face recognition still exist. A full and systematic assessment of the automatic face recognition technology must be conducted under realistic conditions before it can be utilized for forensic applications.

Up to now, we have introduced the forensic use of face recognition and discussed some challenges needed to be addressed in the future. In the real forensic scenarios, usually a combination of information from different biometric traits is applied for case analysis. In the following sections, we will introduce one emerging biometrics – ear which is highly related to face but has its own advantages.

7.4   Ears as a Means of Forensic Identification

Although ears are an external part of the head, and are often visible they do not tend to attract human attention and a vocabulary to describe them is lacking. As for the latent prints, the common ones to be found in crime scenes are of fingertips, palms and feet. Although earprints may also be found in crime scenes fingerprints are much more abundant. The fact that the forensic use of ears and some of the other biometric traits were halted by the advent of fingerprints is partly due to this practical advantage. Dutch courts have admitted numerous cases of earprint-related evidence (Van der Lugt C 2001). Earprints have also been used as a means of personal identification in other countries, such as the United States, the United Kingdom, Germany and Switzerland. In Germany both earprints and ear images have been used for identification (Champod *et al.* 2001). In Switzerland, latent earprints have been used to assist in the early stages of investigation in burglary cases (*R. v. Mark Dallagher* 2002). While in a number of higher profile cases the reliability of earprint evidence has been challenged, been refused admittance or caused erroneous convictions. The evidence regarding earprints is mainly contested due to three main factors: (i) pressure deformation, (ii) the lack of generally accepted methodologies for comparison and (iii) the lack of large-scale testing.

A study of potential identification capabilities of ears was performed by Alfred Iannarelli who examined over 10 000 ear samples over 38 years (Iannarelli 1989). He developed the Iannarelli System of Ear Identification. His system essentially consists of taking a number of measurements from a set of landmark points on the ear. He concluded:

> Through 38 years of research and application in earology, the author has found that in literally thousands of ears that were examined by visual means, photographs, ear prints, and latent ear print impressions, no two ears were found to be identical.

Despite his extensive experience with different forms of ear representation in forensics, in 1985 the Florida trial court of *State v. Polite* 1985 did not recognize him as an expert on earprint identification on the grounds that his ear identification method was not generally accepted in the scientific community. The court also raised concerns over the effects of pressure deformation on the appearance of earprints and also over the lack of studies concerning the comparison of earprints and refused to accept the earprint identification evidence altogether. The later development of ears as a biometric was to rely on the pioneering work of Iannarelli.

Ear biometric recognition has primarily been focused on automatic or semi-automatic methods for human identification or verification using 2D or 3D ear images. In comparison to the forensic references to the usage of ear morphology for recognition, the automated recognition of ear images in the context of machine vision is a recent development. Burge and Burger (1998) were amongst the first to investigate automated recognition of ears. Inspired by the earlier work of Iannarelli, they conducted a proof of concept study where the viability of the ear as a biometric was discussed theoretically, in terms of the uniqueness and measurability over time, and examined in practice through the implementation of a computer vision algorithm. Since then, there have been many ear biometric methods looking at 2D and 3D images of the ear while also attempting to overcome challenges such as occlusion, pose variation and illumination conditions.

The advantages that ear biometric studies can offer the field of ear forensic identification are twofold. Firstly, to advance and inform earprint recognition methods, and secondly, to introduce and facilitate the new emerging application of identification at a distance from surveillance footage. Pressure deformation is one of the main reasons why earprint evidence is contested. Being composed of malleable parts, the appearance of an earprint can be much influenced by the amount of pressure which is applied in making the print. A 3D model of the ear, as offered by 3D ear biometrics methods, may be useful in predicting the appearance of its earprint under different amounts of pressure. Another hindering factor for the application of earprints for identification is the missing features of the ear in an earprint. Due to the different elevations of the external ear parts, some of the ear components are commonly missing in earprints. Owing to the missing information, it can be debated that earprints present less variability than ear images. We will show that the insights offered by the ear biometric studies as to the degree of discrimination provided by different ear features can be used to evaluate the information content of earprints for identification.

Ear images from surveillance cameras have also been considered for forensic identification. Although this is considered a new development in the field of ear forensic identification, it is the core problem in ear biometrics. Thus the methodologies developed in ear biometrics may be readily transferable for use in this application. Automatic biometrics methods can also offer desirable properties for efficient processing of large datasets and attribute the performance and error rate directly to specific methodologies. Using automatic biometric methods can also provide reproducible results and eliminate operator bias.

Next, we will review earprint identification, its role as forensic evidence, its shortcomings and possible improvements. Since the forensic use of ear as a biometric is in a different stage of its life cycle compared to face, as well as looking at the

Figure 7.3 Sample earprints (from Meijerman (2006)).

methods of comparison, we will discuss the earlier question of admissibility in court. We will then look at specific automatic biometric methods and how they can be used for forensic identification from surveillance capturing. Finally, we will review the discriminant capabilities of individual ear features and how they can be used to infer the level of confidence in predictions from data which are prone to having missing features.

7.4.1 Earprints in Forensics

Earprints, which may be found in up to 15% of crime scenes (Rutty *et al.* 2005), are latent prints left behind as a result of the ear touching a surface, for example while listening at a door. In a legal context, the evidence regarding earprints could be utilized for various purposes including: dismissing a suspect, increasing evidence against a suspect or identifying possible suspects (Meijerman *et al.* 2004). Earprints have been used as a means of personal identification in different countries; however, in a number of cases the reliability of earprint evidence has been challenged. Figure 7.3 shows some sample earprints.

7.4.1.1 Earprint – A Challenged Forensic Evidence

In the cases involving earprint evidence for positive identification, two issues have been the main source of dispute. One is the admissibility of this evidence and the other is its reliability. In the United States and under the Daubert standard, all forensic expertise is subjected to a scientific scrutiny over its reliability and accuracy. In this setting, the judge acts as a *gatekeeper* and determines whether the proffered forensic evidence accords to that standard. The forensic science in question does not need to be error free to be admissible; indeed there is always a level of error involved. However, a measure of this error should be made available through rigorous testing. This, however, is not a straightforward task while the question regarding the size of the dataset, which is needed to obtain the required reliability and the statistical evaluation of performance, has not been addressed.

The admissibility of earprint evidence was a key issue in the case of *State v. David Wayne Kunze* 1999. In Washington State in 1996, David Wayne Kunze was charged with aggravated murder amongst other charges. The key evidence against Kunze was a latent earprint found at the scene. Prior to the trial, Kunze moved for excluding any evidence of earprint identification. Subsequently, the trial court convened a Frye hearing on the matter and many ear experts and latent print experts were called. The hearing concluded that earprint identification has indeed gained general acceptance and thus the earprint evidence was admitted. However, later at the appeal court, after reviewing the evidence given at this pre-trial hearing, the appeal court concluded that general acceptance was not obtained 'if there is a significant dispute between qualified experts as to the validity of scientific evidence', and since the hearing clearly showed such dispute, the appeal court ruled that the trial court erred by allowing the expert witnesses to testify and that a new trial was required. In the case of *State v. Polite* (US, Florida trial court) 1985, the court also refused to admit the earprint evidence. In excluding the earprint evidence the judge raised concerns over the unknown effect of pressure deformation and insufficient scientific background to establish reliability and validity of earprint identification.

Relevancy is another guideline for admissibility under Daubert. Relevancy is defined as (Rule 401, Federal Rules of Evidence): 'Evidence is relevant if: (1) it has any tendency to make a fact more or less probable than it would be without the evidence; and (2) the fact is of consequence in determining the action.'

In the United Kingdom, in the appeal court of Mark Dallagher 2002, the court examined the question of expert evidence admissibility. In 1998, in the Crown Court at Leeds, Mark Dallagher was convicted of murder and sentenced to life imprisonment. In this trial an earprint discovered at the scene of crime was one of the main pieces of the evidence against the defendant. Two expert witnesses testified that the defendant was the certain or highly likely maker of the latent earprints. No expert evidence was called on behalf of the defendant and the defence did not seek to exclude the evidence of the prosecution experts. Fresh evidence against the use of earprints for positive identification was offered as grounds for appeal. The appeal court subsequently refused the appellant's argument that if this expert evidence was available at the trial the prosecution's expert evidence should have been excluded. For this, references were made to other cases, such as *R. v. Clarke* 1995 on facial mapping expert evidence:

> It is essential that our criminal justice system should take into account modern methods of crime detection. It is no surprise, therefore, that tape recordings, photographs and films are regularly placed before juries. Sometimes that is done without expert evidence, but, of course, if that real evidence is not sufficiently intelligible to the jury without expert evidence, it has always been accepted that it is possible to place before the jury the opinion of an expert in order to assist them in their interpretation of the real evidence.

And continuing:

> We are far from saying that such evidence may not be flawed. It is, of course, essential that expert evidence, going to issues of identity, should be carefully scrutinized. Such evidence could be flawed. It could be flawed just as much as the evidence of a fingerprint expert could be flawed. But it does not seem to us that there is any objection in principle.

The appeal court concluded that the expert evidence could not possibly be considered irrelevant, or so unreliable that it should be excluded. Albeit, the appeal court eventually quashed the conviction and ordered a retrial on the grounds that it seemed that if the fresh evidence was given at the trial it might have affected the jury's approach toward the crucial earprint identification evidence.

In the appeal court of *R. v. Mark Kempster* 2008, the admissibility of the ear evidence was also a cause of debate. In 2001, Mark Kempster was convicted of multiple counts of burglary and attempted burglary at Southampton Crown Court. One of the main pieces of evidence against him was a positive identification of an earprint which was recovered from the scene of crime as his earprint. He appealed against the conviction twice, and in 2008 the appeal was brought on the ground that relevant fresh evidence might have undermined the expert prosecution evidence, of positive earprint identification. In the court of appeal, the defence argued against the admissibility of earprint evidence. The defence also argued that while earprint evidence may be used for excluding a suspect, a positive identification cannot be obtained using earprint evidence. Both the prosecution and defence experts agreed that this area of science was in its infancy. However, they disagreed on the results of comparing the earprint found at the scene and the prints of the appellant. The appeal court eventually concluded that the earprint evidence was admissible, and could be used by the jury to decide if it was indeed the appellant who left the mark at the scene. The judge, thus, directed the jury:

> First of all consider the evidence of the earprint. Are you sure that the earprint was Mr Kempster's? If you are not sure then you must acquit Mr Kempster on Count 1.

And the jury subsequently quashed the conviction on count 1 burglary. Thus, again, although the earprint evidence was admitted its reliability was challenged. Whether the earprint evidence is blocked out as an inadmissible expertise or is challenged on its reliability, it is apparent that it does not hold an assured status as a forensic method for positive identification. Next, we will look into the reasons for this and discuss as to how a more reliable earprint evidence maybe obtained.

7.4.1.2   Pressure Deformation

Due to their different elevation and flexibility, ear ridges react differently to the changes in pressure and cause large intra-individual variations. The unknown effects of pressure deformation is one of the main reasons why earprint evidence is contested.

To overcome this problem, it has been suggested that for each ear the control prints can be captured using different amounts of pressure and when comparing these control prints to a latent print only the best match would be considered. Junod *et al.* (2012) also proposed to combine the different earprint samples of an ear to build an earprint model. Hypothesizing that in practise a perpetrator will be listening for a sound, Alberink and Ruifrok (2007) proposed that a more realistic dataset of control prints can be acquired by applying a *functional force*. In this, the donors were instructed to listen for a sound behind a glass surface.

A different and perhaps a more comprehensive approach may be offered using a 3D model or a 3D image of the donor ear. Combined with a model of external ear part-wise elasticity, the 3D model can be used to synthesize a set of possible earprints that can be generated by an individual ear. A 3D model of the ear can be acquired using a range scanner (Chen and Bhanu 2007; Yan and Bowyer 2007). There are also methods which use 2D ear images to infer the 3D model of the ear (Bustard and Nixon 2010a; Cadavid and Abdel-Mottaleb 2008).

7.4.1.3   Variability and Missing Features

The evidence regarding the variability of ear morphology is regarded as relevant but not directly usable in the field of earprint identification, since not all the parts of the ear leave a mark in an earprint. Due to the different elevations of the external ear parts, some of the ear components are commonly missing in earprints. The parts of the ear which are frequently seen in earprints are helix, anti-helix, tragus and anti-tragus, while lobe and crus of helix are not so common (Meijerman *et al.* 2004). Owing to the missing information, it can be debated that earprints present less variability than ear images (Dessimoz and Champod 2008). Also, the amount of pressure can affect the amount of information which has been left behind in the print.

Dessimoz and Champod (2008) hypothesizes over the discrimination power of the features in different representations of ear morphology data due to the varying quality of the data. They referred to this discrimination power of the data as *selectivity* and discussed that the data with highest selectivity is of ear images captured under controlled conditions. The rest of the source data in order of diminishing selectivity are ear image occluded by hair, reference earprint, ear image taken with a surveillance camera at a distance and finally an earprint obtained from a crime scene. Note that the traditional biometrics and forensic applications of the ear morphology are at the either end of this selectivity spectrum. Dessimoz *et al.* did not explain how they arrived at this selectivity ranking. However, we suspect that, in this, the missing parts as well as the pressure deformation are the main reasons for the low selectivity of the earprints. Indeed, there is a concern that not all potentially discriminant parts of the ear are present in an earprint. This leads to the question of what features there are in an ear shape and just how discriminant they are. The findings of ear biometrics studies where occlusion and therefore missing part have been investigated may be useful to this discussion. Arbab-Zavar and Nixon (2011) have also investigated as to the origin of each part of the ear morphology and its discrimination powers. Ear parts are further discussed in Section 7.4.3.

7.4.1.4 Statistical Analysis of Performance

So far there has been relatively little analysis of earprint performance for forensic identification. The statistical analysis of performance and error rates corresponding to earprint identification was the focus of the EU-funded project Forensic Ear Identification (FearID) in 2002–2005. In this project, an earprint dataset with 7364 prints from 1229 donors from three counties was acquired (Alberink and Ruifrok 2007). For this three left and three right earprints were gathered for each donor. Also, one or two simulated crime scene prints were taken for one out of 10 donors. A semi-automatic classification method was proposed to compare the prints, and each pair of prints was classified as matching or non-matching. In this, after the earprint was lifted from the surface, first a polyline is drawn manually following the earprint contour. This polyline gives a skeleton-like representation of the earprint. A set of features are then extracted for each earprint. These features are the width and the curvature of the print along this polyline. These features are each represented as a one-dimensional signal where the horizontal axis is the position along the polyline and the vertical axis is the width or the curvature at that position respectively. A third feature vector is also extracted. This is a point pattern representing the distribution of specific anatomical locations which are manually marked by an expert. The comparison between each pair of prints is then performed by comparing the corresponding features in the two prints. A score is computed showing the similarity between the features of the two prints. An equal error rate (EER) of 3.9% for comparison of reference prints (per side) and an EER of 9.3% for the comparison of simulated crime scene prints with the reference prints are obtained using this method. Junod *et al.* (2012) have also experimented with this data. In this, the ear prints are manually segmented and pre-aligned. An earprint model is then computed for each ear (per side) by further aligning the input earprint images of that ear using a multi-resolution registration algorithm and obtaining the superposition of the aligned prints. The same alignment method is then used in testing to compute the similarity between a given earprint and a model earprint. Junod *et al.* report a 2.3% EER for the comparison of simulated crime scene prints with the reference prints and a 0.5% EER for the comparison of reference earprints. They also report hitlist results. For reference print comparisons, in over 99% of cases the true match is in the top three positions of the list and for the comparisons of simulated crime scene prints with the reference prints 88% of cases have the true match in the top three positions. In assessing the reproducibility of these results for real cases one should keep in mind that the FearID database is collected by applying a *functional force* by the print donor simulating a listening effort of a burglar. This minimizes the variability of pressure deformation in the different prints from the same donor. However, in real cases it is not practical to expect of a non-cooperative suspect to apply a *functional force*. Further evaluation of real case samples is required.

7.4.2   From Earprints to Ear Images

The effects of deformation due to pressure and the fact that some components are missing, potentially, causes large intra-individual variation in earprints, resulting in a more challenging recognition problem than ear image recognition. In biometrics, 2D or 3D images of the ear are commonly used. These images are traditionally captured in controlled environments. More recent methods have looked into improving the robustness of the algorithms and easing the controls over the image capture procedures. With rapid deployment of surveillance cameras, the number of crimes recorded on surveillance footage is also growing fast. These footage are often characterized by poor quality while effects such as occlusion, shadows and noise are commonplace. With further development of biometric approaches towards more robust methods on the one hand and the increase of crime scene surveillance footage, which calls for methods of recognition at a distance, on the other, it appears that the two fields are rapidly moving towards each other.

Compared to earprints, the use of ear images for identification has been explored and examined more frequently. Abaza *et al.* (2013) provides a list of available ear image databases which can be used for ear biometric studies. Some of the most commonly used among these databases are the UND database (Yan and Bowyer 2005) which includes 2D and 3D images of 415 individuals, XM2VTS database (Messer *et al.* 1999) comprising of 2D ear images of 295 subjects taken in four time-lapsed sessions and USTB database (UST 2005) with 500 subjects and with pose variation and partial occlusion.

The automatic recognition of ear images removes the operator bias, and so long as the probe images are comparable to the training and validation images in terms of overall quality, resolution, occlusion, illumination and pose variations the error rates reported for an algorithm are a good estimate of the reliability of the algorithm's predictions for new data. In this, the size of the validation set compared to the size of potential candidate set is also a factor which needs to be considered. However, determining the required size of the training and validation sets for each recognition problem is an open question. It should also be noted that these methods are often complex and unintuitive. Often it is not possible to point out the differences and similarities between two ear images explicitly. This is unfortunate as such descriptions can be useful for the jury.

7.4.2.1   Ear Biometrics Methods

Iannarelli (1989) proposed a method based on 12 measurements taken between a number of landmark points on an ear image. These landmark points were determined manually. An automated method based on similar measurements would primarily rely on accurate positioning and segmentation of the landmarks. This is a challenging task to perform automatically. On the other hand, an automatic analysis of samples can capture a more detailed signature, describing the sample, one which may not be viable to obtain manually. Also, there is the obvious benefit of being able to automatically search within a large dataset of samples. It is worth noting here that even the same ear would appear different, albeit slightly, in different images. Identification is possible when the intra-individual variations are smaller than the inter-individual variations. In other words, identification is possible when the samples from the same individual are more similar to each other than to the samples from other individuals. Also note that in biometrics, the focus is to design the most effective and robust algorithms to perform identification. The experimental evaluation of a biometrics technique offers the error rates associated with that specific algorithm performing identification based on a particular biometric trait. Notice that this is not the same as the error rates pertaining to a biometric trait. In biometrics, the error rates are always associated with the algorithms and no upper limit is envisaged for the recognition performance of a specific biometric trait.

One of the first automatic ear biometric algorithms was introduced by Burge and Burger (1998). They modelled each individual ear with an adjacency graph which was calculated from a Voronoi diagram of the ear curves. However, they did not provide an analysis of biometric potential. Hurley *et al.* (2005) used force field feature extraction to map the ear to an energy field which highlights *potential wells* and *potential channels* as features achieving a recognition rate of 99.2% on a dataset of 252 images from 63 subjects. Naseem *et al.* (2008) have proposed the use of sparse representation, following its successful application in face recognition. Arbab-Zavar and Nixon (2011) proposed a parts-based model approach which was guided by the biological cues as to the independent parts of the external ear morphology. The ear model was derived by a stochastic clustering on a set of scale-invariant features of a training set. The model description was extended by a wavelet-based analysis with a specific aim of capturing information in the ear's boundary structures. A recognition rate of 97.4% was achieved using this method on a dataset of 458 images from 150 individuals. Statistical methods such as PCA, independent component analysis (ICA) and LDA have also been used in ear biometrics (Chang *et al.* 2003; Hurley *et al.* 2005; Zhang *et al.* 2005; Zhang and Jia 2007). These statistical methods can obtain satisfactory results in controlled environments. However they have almost no invariance properties, thus they rely on the acquisition and pre-processing stages to window and align the data.

The 3D structure of the ear has also been exploited, and good results have been obtained (Chen and Bhanu 2007; Passalis *et al.* 2007; Yan and Bowyer 2007). Yan and Bowyer (2007) captured and segmented the 3D ear images and used Iterative Closest Point (ICP) registration to achieve a 97.8% recognition rate on a database of 415 individuals. Chen and Bhanu (2007) proposed a 3D ear detection and recognition system. Using a local surface descriptor and ICP for recognition, they reported recognition rates of 96.8% and 96.4% on two different data sets. Although using 3D can improve the performance, using 2D images is consistent with deployment

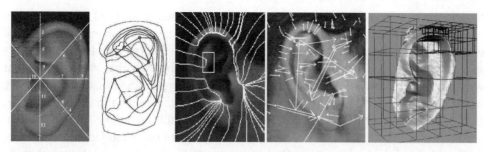

Figure 7.4 From left to right: the Iannarelli's manual measurement system; Burge and Burgers' adjacency graph; Hurley *et al.*s' force field; Arbab-Zavar and Nixons' keypoints; and Yan and Bowyers' 3D model.

in surveillance or other planar image scenarios. Figure 7.4 shows the Iannarelli's manual measurements as well as Burge and Burgers' adjacency graph, Hurley *et al.*s' force field, Arbab-Zavar and Nixons' keypoints and Yan and Bowyers' 3D model. Hurley *et al.* (2008) described the steps required to implement a simple PCA-based ear biometric algorithm. A survey of ear biometrics has been recently provided by Abaza *et al.* (2013).

7.4.2.2 Are Ear Biometrics Methods Robust?

One of the main advantages of ear biometrics is that recognition may be made at a distance, such as in surveillance videos. The images captured by a surveillance system are generally of poor quality; they might be partially occluded; the pose might not be the most desired one for identification, while poor illumination and shadows may also deter the image quality. Therefore, the automatic processing of such images requires the use of robust methods.

Bustard and Nixon (2010b) pointed out that, presently, in order to obtain good recognition rates in the area of ear biometrics it is required that the samples be captured under controlled conditions. Moving towards an unconstrained ear recognition method was the main goal of (Bustard and Nixon 2010b). The proposed method includes a registration which computes a homography transform between the probe and gallery images using scale-invariant feature transform (SIFT) point matches. In recognition, a pixel-based distance measure is used to compare the registered and normalized images. The robustness of the method is then tested in presence of pose variation, occlusion, background clutter, resolution and noise. It has been shown that this method can handle pose variations of up to $\pm 13°$ and occlusions of up to 18%, while also showing good robustness properties in the other tested cases.

Model-based methods are generally more equipped to handle noise and occlusion. Using a localized description is another way of increasing robustness to occlusion. Inevitably, some of the information will be lost as a result of occlusion. However, other measurements can also be affected by the change in the overall appearance. It is these

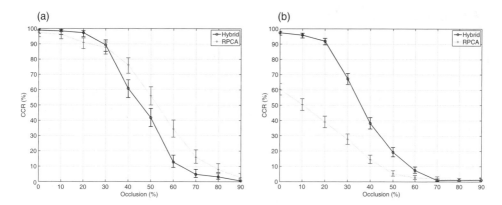

Figure 7.5 The hybrid classifier versus RPCA in occlusion on two testsets. (a) Testset A and (b) Testset C (from Arbab-Zavar and Nixon (2011)).

measurements which a localized approach can keep from being spoiled. Arbab-Zavar and Nixon (2011) demonstrated the performance advantages of their hybrid method, including a parts-based model extended by a wavelet-based analysis capturing information in the ears boundary structures, in occlusion. In this, they have compared the performance of their method with a robust PCA (RPCA) as a representative of holistic methods. Figure 7.5 shows the results of this comparison. On test set A, the hybrid method performs better than RPCA for as much as 30% of occlusion. The results on test set C exhibit the degrading effect of less accurate registration, which is obtained automatically, on RPCA. By contrast, the hybrid classification maintains good performance, and clearly outperforms RPCA on test set C. Test set C is also more challenging than test set A in terms of number of individuals and overall image quality. Yuan *et al.* (2010) proposed a localized approach with high redundancy between the local descriptions. They generated a set of 28 overlapping sub-windows for each image and used neighbourhood-preserving embedding to extract the features for each sub-window. In recognition, a weighted majority voting is used for fusion at decision level.

3D ear images have been used to overcome the difficulties encountered with variations in pose and lighting. Various recognition algorithms have been proposed (Chen and Bhanu 2007; Yan and Bowyer 2007) demonstrating high recognition performances. However, range scanners are required to capture the 3D images. Other methods have been proposed to extract the 3D information of the ear for recognition using a set of 2D images from different poses. Shape from shading (Cadavid and Abdel-Mottaleb 2008) and a B-spline pose manifold (Zhang and Liu 2008) are two examples of these methods. Similar to the methods which work with range data, the data requirements of these methods also restricts their viability for surveillance scenarios. In a more promising approach, Bustard and Nixon (2010a) proposed a new technique for constructing a 3D morphable model of the face profile and the ear using a single image.

7.4.3   Ear Morphology Features

Perhaps one of the main questions which is encountered in the forensic identification is this question: 'is there enough information available to make a positive identification?'. Hoogstrate *et al.* (2001) asked a similar question from two groups of operators: forensic experts and laymen. For this, multiple video captures using standard surveillance equipment were made from 22 subjects under different conditions. A mask was overlaid on the video showing only the ear part. Each participant was presented with 40 sets of paired videos and for each pair they were asked: (i) Is there enough information in the video for individualization or exclusion? (ii) Are the individuals in the two videos the same person? Hoogstrate *et al.* derived two main conclusions from their experiments: (i) the quality of the video influences the participant's decision of whether they have enough information; and (ii) the forensically trained persons were able to determine if they had sufficient information. Note that the dataset for this study was small and the experiment was conducted under closed set assumption. Albeit, this raises an important question of which are the ear's discriminating features and how the performance accuracy and confidence levels are affected when different parts of the ear are not visible.

The significance of various parts of the ear for identification has been rarely studied in the field of ear biometrics. In our earlier work (Arbab-Zavar and Nixon 2011), we have looked into identifying the various parts of the ear morphology and investigate as to their discriminatory powers. This study was guided by accounts of embryonic development of the external ear. The study of ear embryology reveals that the external ear is the result of six nodules whose unequal growth and coalescence give the final shape of the external ear. A conclusion was drawn that there should be a limited information redundancy between these parts since they are thought to be developed independently. Therefore missing the information of any of these parts could not be fully recovered by the other visible parts. Some of the information content is thereby lost and our capability to perform identification using this data is weakened.

The separability and variability of ear parts was investigated via two very different approaches. The first approach was based on statistical analysis of shape within a dataset of ears. In this, a parts-based model was learned from a dataset of ear images. The parts are arbitrary scaled circular neighbourhoods, called the keypoints, which are detected and described via the SIFT descriptor (Lowe 2004). A parts-based model was then built via clustering of the detected keypoints in different images. The second approach was based on embryonic development of the external ear and the embryological understanding of ear abnormalities. In this, cues from the ear abnormalities were used to hypothesize as to the independent parts of the ear. It was considered that the most variable parts of the ear are those which are most frequently the site of ear abnormalities. These parts also provide a valuable set of features that can be used to communicate with juries.

The initial appearance of the external ear in the human embryo is in the shape of six individual hillocks occurring in the 5th week of embryonic life (Streeter 1922).

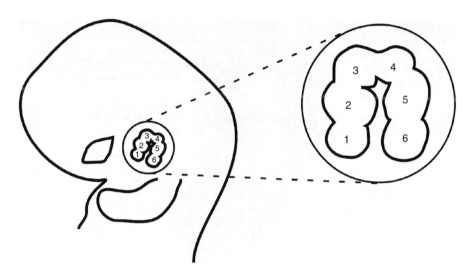

Figure 7.6   The six auricular hillocks and their location in a human embryo.

Figure 7.6 shows a drawing of an embryo with its auricular hillocks numbered. It is the unequal growth and coalescence of these six hillocks that gives the shape of the definitive auricle in a newborn baby. This is the reason for our interest in ear embryology – the premise of local and independent structures within the auricle is appealing to the classification purpose.

Streeter (1922), who provided one of the most extensive accounts of external ear embryology, argued against the individual development of the auricular hillocks and suggested that the external ear comes into existence as an intact and continuous structure which elaborates into its final form. However there is a wide range of defects which disturb the smooth continuity of the auricle. These can be best described as the failure of fusion or the lack of correct alignment of the various hillocks, which further insists on the role of separate structures in the formation of the definitive auricle (Davis 1987; Hunter and Yotsuyanagi 2005). Some other malformations can be described as excessive growth beyond, or, underdevelopment beneath the thresholds of normality. Thereby the site of such anomaly is also where a considerable variation is introduced; it is unlikely that an abnormality will be observed in locations of more constant structures.

The findings of Arbab-Zavar and Nixon (2011) have been revisited here with an eye towards earprint identification and the missing parts. Figure 7.7 shows the common terminology of the external ear.

7.4.3.1   Inferior Crus of Antihelix and the Crus of Helix

According to Streeter, the inferior crus of antihelix and the crus of helix are the least variant parts of the ear. Seemingly contradictory, these two parts are detected as the

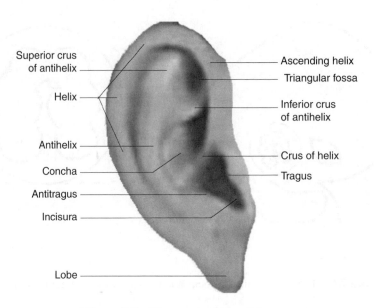

Superior crus
of antihelix

Helix

Antihelix

Concha

Antitragus

Incisura

Lobe

Ascending helix

Triangular fossa

Inferior crus
of antihelix

Crus of helix

Tragus

Figure 7.7   The terminology of the ear.

most significant parts of the parts-based model. It was discussed in Arbab-Zavar and Nixon (2011) that this is caused by the models varying capability in detecting the parts. Automatic detection of parts is a task which precedes the comparison of parts and indeed not all the parts are detected in every ear image. The inferior crus of antihelix and the crus of helix are the most frequently and accurately detected parts in different ears, so much so that they become the most significant parts of the model for recognition. It is hypothesized that the comparative consistency of these parts helps with learning them via a clustering method. We suspect that a manual labelling of the parts, presuming that such labelling can be achieved accurately and consistently, would reveal a slightly different ranking of important parts. In the next section, we will describe how the inadequate representation of helix and antihelix in the parts model have motivated Arbab-Zavar and Nixon (2011) to extend the initial parts-based model and build a hybrid model. This emphasizes the importance of choosing the right algorithm for ear image, ear print and in fact any biometric comparison. Considering the earprints, it can perhaps be considered fortunate that two parts which often do not leave a print are actually the least varying parts of the ear. Thus their absence is least significant.

7.4.3.2   Helix, Antihelix and Scapha

The outer ear rim, the helix, may be attributed to as many as three out of six embryonic hillocks. Ascending helix, the portion of the helix immediately attached to the crus of helix, is assigned to an individual hillock by Streeter (1922) and Sulik (1995). An anomaly called the *lop ear* is the product of the absence of the ascending helix, while the rest of the parts have their normal shape (Hunter and Yotsuyanagi 2005). Two other defects exhibit conspicuous clefts separating the ascending helix from the rest of the helical components on either side (Davis 1987; Park and Roh 1999). The ascending helix is also detected by the parts model and is the third most significant part of the model. As for the rest of the helix, there are two major hypotheses regarding its formation: suggested by His, the upper and lower helical components, including the helix and antihelix, are derived from hillocks 4 and 5 respectively; while Streeter believes that a single hillock (5th) gave rise to the helix and the antihelix is the product of hillock 4. In accordance with the first hypothesis, the upper helical region appears to be subject to considerable growth variations. Cryptotia and Satyr ear are two anomalies exhibiting underdevelopment of this region (Hunter and Yotsuyanagi 2005). The upper and lower helical regions have been detected as separate parts in the parts model and are both among the seven most significant parts of the model. On the other hand, the emergence of the scapha, the concave surface of free portion lying between the antihelix and the helix, provides a margin and allows the helix and antihelix to have some degree of independent development which is better described by Streeter's hypothesis. The antihelix, as mentioned earlier is subject to variations of the upper helical region, while the lower parts are more constant. Due to the limitation of the circular image descriptor which was the basis local area unit of the model parts, the elongated parts such as the helix and anti-helix were not captured adequately. A specialized representation and method was then applied to capture the variations of the two elongated structures of the helix and anti-helix separately. A recognition rate of 91.9% is achieved with helix and antihelix dominant representation on a dataset with 458 images from 150 individuals (Arbab-Zavar and Nixon 2011). In this the part model obtains an 89.1% recognition rate. The combination of these two methods, which is called the hybrid model, yields a significant improvement with a 97.4% recognition rate and further suggests that independent information content have been captured by these two methods. Also note that the helix and antihelix dominate the earprint mark. However, the upper antihelix region of the superior and the inferior cruses of antihelix are commonly missing in these prints.

7.4.3.3 The Lobe

Lobe is one of the only parts of the ear which lends itself to categorical classification. Three types of lobe are well-formed, attached and no lobe. In forensics, ear lobes are used in international standards for identification in Disaster Victim Identification (DVI) 2008. Note also that ear piercing, which is a semi-permanent body modification, was reported by Abbas et al. to occur in 46% of their population sample of 400 adults (Abbas and Rutty 2005). They reported that, in about 95% of the cases with ear piercing, the piercing occurs on the lobe. They noted that the presence or absence of such piercing itself is a useful attribute for forensic identification. The ear lobe is the only part of the ear which comprises fat rather than cartilage. This part continues to grow and change shape as the person grows older (Meijerman 2006). Albeit, it could exhibit various shapes, and in a database with a small time lapse between the captured samples it can be comparatively discriminant.

7.4.3.4 Tragus and Antitragus

In Otocephaly, which is a syndrome accompanied by an anomaly of the auricle, the tragus is missing. Other tragal anomalies may exhibit extensions or duplications of the tragus flesh (Hunter and Yotsuyanagi 2005), indicating a rich variation in the shape of this component. In contrast, antitragus has been little discussed in the analyses of ear anomalies. Tragus and antitragus are also commonly found on earprints.

7.4.3.5 Concha

Concha is the part of the external ear which will almost certainly be missing from the earprint. The depth of this cavity is the main feature of this component. The Mozart ear is characterized by its shallow concha, and it was also discussed that there is a correlation between the depth of the concha and the sensitivity of the ear to hearing sounds. However, this feature is also absent in 2D ear images.

7.4.4 Summary

In the second case study, we have examined the application of the emerging field of ear biometrics for forensic identification. Human ear is an ideal contender for such a study since it is available both in images at a distance and in latent prints. Earprints and ear images are considered separately as two different representations of ear. The less familiar features of ear, along with their correlations and variability are also discussed. We have also addressed the question of admissibility in court.

Ear is an important emerging biometric. There is a clear expectation that ears do not change in structure from cradle to grave, only in size (except with surgical intervention). There is a known taxonomy for their description, and it is accepted that people can be identified by their ears. There is a rich variety of approaches for ear biometrics and these are steeped in pattern recognition and computer vision. These show that using ears has similar performance to other biometrics, using similar methods, though the research is as yet not so deep or popular as that for the more established biometrics. As such, ears can then be deployed in scene of crime analysis where images of an ear are available, and the ear has actually already been deployed in this way. The notion that people can be recognized from a latent earprint has a more chequered history. This arises when a subject's ear makes contact with a surface, say for listening in purposes. Naturally, there are problems with image quality as there are for latent fingerprints and the problem is confounded by the absence of modelling of the change in print with ear deformation, though a combination of 3D shape analysis (which exists for ears) with a 3D plastic membrane could offer understanding in this direction. As it stands, the ear clearly has the potential to be one of the stock of biometrics in digital forensics both for imagery and for recorded prints – and given its proven identification capability it appears well worthy of future study in this respect.

7.5 Conclusions

Given that biometrics concerns automatically establishing identity and forensics requires confirmation of identity, it is perhaps surprising that the inclusion of biometrics is not further advanced within the forensic community. For this to be achieved, agreement standards for acceptability needed to be reached, and these are relatively new to biometrics. Given its long history, it is no surprise there is a richer literature in identifying subjects from fingerprints and fingerprint biometrics is becoming well established for latent fingerprint recognition. The translation of other biometrics (i.e. face, gait, ear and voice) is considerably less advanced.

This chapter has outlined the historical connections between biometrics and forensics and examined the application of face and ear biometrics for forensic identification in detail. Given that face and ear are in different stages of deployment in forensics, various aspects of this deployment were discussed. The examination of ear forensic possibilities gave rise to the early questions of admissibility, mainly regarding the use of earprints. The morphological features of the ear were also examined in detail. Such insights are essential for evaluation of partially occluded data and further influential when communicating the findings of biometric comparisons to juries. More advanced in the forensic field, manual forensic face recognition methods and the deployment of automatic techniques were discussed. The challenges looming over both face and ear applications in forensics, mainly due to poor data quality which is common of forensic data, and the current state of automatic recognition performance and robustness were examined. This chapter has aimed at closing the gap between the forensics and biometrics experts understandings of the identification task. Although further study is needed within various fields of biometrics so that they are equipped for inclusion within forensics, given the prospects they offer this appears well worthy of the effort.

References

Abaza A, Ross A, Hebert C, Harrison MAF and Nixon MS 2013 A survey on ear biometrics. *ACM Computing Surveys* **45**(2), 22:1–22:35.

Abbas A and Rutty GN 2005 Ear piercing affects earprints: The role of ear piercing in human identification. *Journal of Forensic Sciences* **50**(2), 386–392.

Ahonen T, Hadid A and Pietikainen M 2006 Face description with local binary patterns: Application to face recognition. *IEEE Transactions on Pattern Analysis and Machine Intelligence* **28**(12), 2037–2041.

Alberink I and Ruifrok A 2007 Performance of the fearID earprint identification system. *Forensic Science International* **166**, 145–154.

Ali T, Veldhuis RNJ and Spreeuwers LJ 2010 Forensic face recognition: A survey. Technical Report TR-CTIT-10-40.

Arbab-Zavar B and Nixon MS 2011 On guided model-based analysis for ear biometrics. *Computer Vision and Image Understanding* **115**(4), 487–502.

Aulsebrook W, Iscan M, Slabbert J and Becker P 1995 Superimposition and reconstruction in forensic facial identification: A survey. *Forensic Science International* **75**(2-3), 101–120.

Belhumeur P, Hespanha J and Kriegman D 1997 Eigenfaces vs. fisherfaces: recognition using class specific linear projection. *IEEE Transactions on Pattern Analysis and Machine Intelligence* **19**(7), 711–720.

Bertillon A 1890 *La photographie judiciaire, avec un appendice sur la classification et l'identification anthropometriques*. Gauthier-Villars, Paris.

Bertillon A 1893 *Identification anthropométrique; instructions signalétiques*. Impr. Administrative, Melun.

Bouchrika I, Goffredo M, Carter J and Nixon M 2011 On using gait in forensic biometrics. *Journal of Forensic Sciences* **56**(4), 882–889.

Burge M and Burger W 1998 Ear biometrics. In *Biometrics: Personal Identification in a Networked Society* (ed. Jain A, Bolle R and Pankanti S). Kluwer Academic, Dordrecht, the Netherlands, pp. 273–286.

Bustard J and Nixon M 2010a 3D morphable model construction for robust ear and face recognition *IEEE Conference Computer Vision and Pattern Recognition (CVPR)*, San Diego, CA, pp. 2582–2589.

Bustard J and Nixon M 2010b Toward unconstrained ear recognition from two-dimensional images. *IEEE Transactions on Systems, Man, and Cybernetics—Part A: Systems and Humans* **40**(3), 486–494.

Cadavid S and Abdel-Mottaleb M 2008 3-D ear modeling and recognition from video sequences using shape from shading. *IEEE Transactions on Information Forensics and Security* **3**(4), 709–718.

Champod C 2000 Encyclopedia of forensic sciences. In *Identification/Individualization: Overview and Meaning of ID* (ed. Siegel J, Saukko P and Knupfer G). Academic Press, London, U.K..

Champod C, Evett IW and Kuchler B 2001 Earmarks as evidence: a critical review. *Journal of Forensic Sciences* **46**(6), 1275–1284.

Chan H and Bledsoe W 1965 A man-machine facial recognition system: Some preliminary results. Technical Report.

Chang K, Bowyer K, Sarkar S and Victor B 2003 Comparison and combination of ear and face images in appearance-based biometrics. *IEEE Transactions on Pattern Analysis and Machine Intelligence* **25**(9), 1160–1165.

Chen H and Bhanu B 2007 Human ear recognition in 3D. *IEEE Transactions on Pattern Analysis and Machine Intelligence* **29**(4), 718–737.

Cole SA 2009 Forensics without uniqueness, conclusions without individualization: the new epistemology of forensic identification. *Law, Probability and Risk* **8**(3), 233–255.

Comcowich G 2013 Remarks of special agent in charge richard deslauriers at press conference on bombing investigation. http://www.fbi.gov/boston/news-and-outreach/press-room/2013-press-releases/ (Accessed 21 February 2015).

Daubert v. Merrell Dow Pharmaceuticals 1993 (92–102), 509 U.S. 579. http://supreme.justia.com/cases/federal/us/509/579/case.html (Accessed 21 February 2015).

Davis J (ed.) 1987 Surgical embryology. In *Aesthetic and Reconstructive Otoplasty*. Springer Verlag, New York, pp. 93–125.

Dessimoz D and Champod C 2008 Linkages between biometrics and forensic science. In *Handbook of Biometrics* (ed. Jain A, Flynn P and Ross A). Springer, New York, pp. 425–459.

Ekenel HK and Stiefelhagen R 2009 Why is facial occlusion a challenging problem? *IAPR International Conference Biometrics (ICB)*, Alghero, Italy pp. 299–308.

Faulds H 1880 On the skin-furrows of the hand. *Nature* **22**(574), 605.

Frye v. United States 1923 293 F. 1013 (D.C. Cir 1923). http://en.wikisource.org/wiki/Frye_v._United_States (Accessed 21 February 2015).

Geng X, Zhou ZH and Smith-Miles K 2007 Automatic age estimation based on facial aging patterns. *IEEE Transactions on Pattern Analysis and Machine Intelligence* **29**(12), 2234–2240.

Georghiades AS, Belhumeur PN and Kriegman DJ 2001 From few to many: Illumination cone models for face recognition under variable lighting and pose. *IEEE Transactions on Pattern Analysis and Machine Intelligence* **23**(6), 643–660.

Guan Y and Li CT 2013 A robust speed-invariant gait recognition system for walker and runner identification. *IAPR International Conference Biometrics (ICB)*, Madrid, Spain pp. 1–8.

Guan Y, Wei X, Li CT, Marcialis G, Roli F and Tistarelli M 2013 Combining gait and face for tackling the elapsed time challenges. *IEEE International Conference on Biometrics: Theory, Applications, and Systems (BTAS)*, Washington, DC, pp. 1–8.

Guo G, Fu Y, Dyer C and Huang T 2008 Image-based human age estimation by manifold learning and locally adjusted robust regression. *IEEE Transactions on Image Processing* **17**(7), 1178–1188.

He R, Zheng WS and Hu BG 2011 Maximum correntropy criterion for robust face recognition. *IEEE Transactions on Pattern Analysis and Machine Intelligence* **33**(8), 1561–1576.

Hoogstrate AJ, den Heuvel HV and Huyben E 2001 Ear identification based on surveillance camera images. *Science & Justice* **41**(3), 167–172.

Hunter AGW and Yotsuyanagi T 2005 The external ear: More attention to detail may aid syndrome diagnosis and contribute answers to embryological questions. *American journal of medical genetics* **135A**, 237–250.

Hurley DJ, Nixon MS and Carter JN 2005 Force field feature extraction for ear biometrics. *Computer Vision and Image Understanding* **98**, 491–512.

Hurley DJ, Arbab-Zavar B and Nixon MS 2008 The ear as a biometric. In *Handbook of Biometrics* (ed. Jain A, Flynn P and Ross A), Springer, NewYork, pp. 131–150.

Iannarelli A 1989 *Ear Identification*. Paramount Publishing Company, Freemont, CA.

Ibañez O, Cordon O, Damas S and Santamaria J 2011 Modeling the skull-face overlay uncertainty using fuzzy sets. *IEEE Transactions on Fuzzy Systems* **19**(5), 946–959.

International Criminal Police Organization INTERPOL 2008 Disaster victim identification. http://www.interpol.int/Public/DisasterVictim/Default.asp (Accessed 21 February 2015).

Iwama H, Muramatsu D, Makihara Y and Yagi Y 2012 Gait-based person-verification system for forensics. *IEEE International Conference on Biometrics: Theory, Applications, and Systems (BTAS)*, Washingdon, DC, pp. 113–120.

Jain A, Klare B and Park U 2011 Face recognition: Some challenges in forensics. *IEEE International Conference on Automatic Face and Gesture Recognition (FG)*, Santa Barbara, CA, pp. 726–733.

Jia H and Martínez AM 2008 Face recognition with occlusions in the training and testing sets. *IEEE International Conference on Automatic Face and Gesture Recognition (FG)*, Ansterdam, the Netherlands, pp. 1–6.

Junod S, Pasquier J and Champod C 2012 The development of an automatic recognition system for earmark and earprint comparisons. *Forensic Science International* **222**, 170–178.

Kanade T 1973 Picture processing system by computer complex and recognition of human faces. Doctoral dissertation, Kyoto University, Kyoto, Japan.

Klare B and Jain A 2010 Heterogeneous face recognition: Matching NIR to visible light images. *International Conference on Pattern Recognition (ICPR)*, Istanbul, Turkey, pp. 1513–1516.

Klare BF and Jain AK 2013 Heterogeneous face recognition using kernel prototype similarities. *IEEE Transactions on Pattern Analysis and Machine Intelligence* **35**(6), 1410–1422.

Klare B, Li Z and Jain A 2011 Matching forensic sketches to mug shot photos. *IEEE Transactions on Pattern Analysis and Machine Intelligence* **33**(3), 639–646.

Klontz JC and Jain AK 2013 A case study on unconstrained facial recognition using the boston marathon bombings suspects. Technical Report MSU-CSE-13-4.

Lanitis A, Taylor C and Cootes T 2002 Toward automatic simulation of aging effects on face images. *IEEE Transactions on Pattern Analysis and Machine Intelligence* **24**(4), 442–455.

Lei Z and Li S 2009 Coupled spectral regression for matching heterogeneous faces. *IEEE Conference Computer Vision and Pattern Recognition (CVPR)*, pp. 1123–1128.

Li Z, Park U and Jain A 2011 A discriminative model for age invariant face recognition. *IEEE Transactions on Information Forensics and Security* **6**(3), 1028–1037.

Liao S, Jain A and Li S 2013 Partial face recognition: Alignment-free approach. *IEEE Transactions on Pattern Analysis and Machine Intelligence* **35**(5), 1193–1205.

Lowe DG 2004 Distinctive image features from scale-invariant keypoints. *International Journal of Computer Vision* **60**(2), 91–110.

Martínez AM 2002 Recognizing imprecisely localized, partially occluded, and expression variant faces from a single sample per class. *IEEE Transactions on Pattern Analysis and Machine Intelligence* **24**(6), 748–763.

Meijerman L 2006 Inter- and intra individual variation in earprints. PhD thesis, Department of Anatomy and Embryology, Leiden University, Leiden, the Netherlands.

Meijerman L, Sholl S, Conti FD, Giacon M, van der Lugt C, Drusini A, Vanezis P and Maat G 2004 Exploratory study on classification and individualisation of earprints. *Forensic Science International* **140**, 91–99.

Messer K, Matas J, Kittler J, Luettin J and Maitre G 1999 XM2VTSDB: The extended M2VTS database. *Audio- and Video-Based Biometric Person Authentication (AVBPA)*, Washington, DC.

Meuwly D 2012 The use of biometric information in forensic practice.

Naseem I, Togneri R and Bennamoun M 2008 Sparse representation for ear biometrics. In *Advances in Visual Computing* (ed. Bebis G, Boyle R, Parvin B, Koracin D, Remagnino P, Porikli F, Peters J, Klosowski J, Arns L, Chun Y, Rhyne TM and Monroe L), vol. 5359 of *Lecture Notes in Computer Science*. Springer, Berlin, Germany, pp. 336–345.

Naseem AI, Togneri BR and Bennamoun CM 2010 Linear regression for face recognition. *IEEE Transactions on Pattern Analysis and Machine Intelligence* **32**(11), 2106–2112.

Pankanti S, Prabhakar S and Jain A 2002 On the individuality of fingerprints. *IEEE Transactions on Pattern Analysis and Machine Intelligence* **24**(8), 1010–1025.

Park U and Jain A 2010 Face matching and retrieval using soft biometrics. *IEEE Transactions on Information Forensics and Security* **5**(3), 406–415.

Park C and Roh TS 1999 Congenital upper auricular detachment. *Plastic and Reconstructive Surgery* **104**(2), 488–490.

Passalis G, Kakadiaris IA, Theoharis T, Toderici G and Papaioannou T 2007 Towards fast 3D ear recognition for real-life biometric applications. *IEEE International Conference Advanced Video and Signal-Based Surveillance (AVSS)*, London, U.K. pp. 39–44.

Penev PS and Atick JJ 1996 Local feature analysis: A general statistical theory for object representation. *Network: Computation in Neural Systems* **7**(3), 477–500.

Phillips P 2012 The next face challenge: Achieving robust human level performance. *International Biometric Performance Conference 2012*, Gaitherburg, MD.

Phillips P, Moon H, Rizvi S and Rauss P 2000 The FERET evaluation methodology for face-recognition algorithms. *IEEE Transactions on Pattern Analysis and Machine Intelligence* **22**(10), 1090–1104.

Phillips P, Flynn P, Scruggs T, Bowyer K, Chang J, Hoffman K, Marques J, Min J and Worek W 2005 Overview of the face recognition grand challenge. *IEEE Conference Computer Vision and Pattern Recognition (CVPR)*, San Diego, CA, vol. 1, pp. 947–954.

Phillips P, Scruggs W, O'Toole A, Flynn P, Bowyer K, Schott C and Sharpe M 2010 FRVT 2006 and ICE 2006 large-scale experimental results. *IEEE Transactions on Pattern Analysis and Machine Intelligence* **32**(5), 831–846.

Porter G and Doran G 2000 An anatomical and photographic technique for forensic facial identification. *Forensic Science International* **114**(2), 97–105.

R. v. Clarke 1995 2 England and Wales criminal appeal reports 425.

R. v. Mark Kempster 2008 England and Wales supreme court of judicature, court of appeal (criminal division), EWCA Crim 975. http://www.bailii.org/ew/cases/EWCA/Crim/2008/975.html (Accessed 21 February 2015).

R. v. Mark Dallagher 2002 England and Wales supreme court of judicature, court of appeal (criminal division), EWCA Crim 1903. http://www.bailii.org/ew/cases/EWCA/Crim/2002/1903.html (Accessed 21 February 2015).

Romdhani S, Blanz V and Vetter T 2002 Face identification by fitting a 3d morphable model using linear shape and texture error functions. In *European Conference Computer Vision (ECCV)* (ed. Heyden A, Sparr G, Nielsen M and Johansen P), vol. 2353 of *Lecture Notes in Computer Science*. Springer, Berlin Germany, pp. 3–19.

Roweis ST and Saul LK 2000 Nonlinear dimensionality reduction by locally linear embedding. *Science* **290**(5500), 2323–2326.

Rutty GN, Abbas A and Crossling D 2005 Could earprint identification be computerised? An illustrated proof of concept paper. *International Journal of Legal Medicine* **119**(6), 335–343.

Saks MJ and Koehler JJ 2005 The coming paradigm shift in forensic identification science. *Science* **309**(5736), 892–895.

Sinha P, Balas B, Ostrovsky Y and Russell R 2006 Face recognition by humans: Nineteen results all computer vision researchers should know about. *Proceedings of the IEEE* **94**(11), 1948–1962.

Spaun NA 2007 Forensic biometrics from images and video at the federal bureau of investigation. *IEEE International Conference on Biometrics: Theory, Applications, and Systems (BTAS)*, Weshington, DC, pp. 1–3.

Spaun N 2011 Face recognition in forensic science. In *Handbook of Face Recognition* (ed. Li SZ and Jain AK). Springer, London, pp. 655–670.

State v. David Wayne Kunze 1999 Court of Appeals of Washington, Division 2. 97 Wash.App. 832, 988 P.2d 977.

State v. Polite 1985 No. 84-525 (14th Judicial Circuit, Fla. Jun. 10, 1985).

Streeter GL 1922 Development of the auricle in the human embryo. *Contribution to Embryology* (69), 111–139.

Sulik KK 1995 Embryology of the ear. In *Hereditary Hearing loss and Its Syndromes* (ed. Gorlin RJ, Toriello HV and Cohen MM). Oxford University Press, Oxford, U.K., pp. 22–42.

Tan X, Chen S, Zhou ZH and Zhang F 2005 Recognizing partially occluded, expression variant faces from single training image per person with SOM and soft k-NN ensemble. *IEEE Transactions on Neural Networks* **16**(4), 875–886.

Tan X, Chen S, Zhou ZH and Liu J 2009 Face recognition under occlusions and variant expressions with partial similarity. *IEEE Transactions on Information Forensics and Security* **4**(2), 217–230.

Tang X and Wang X 2004 Face sketch recognition. *IEEE Transactions on Circuits and Systems for Video Technology* **14**(1), 50–57.

Turk M and Pentland A 1991 Face recognition using eigenfaces. *IEEE Conference Computer Vision and Pattern Recognition (CVPR)*, Maui, HI, pp. 586–591.

USA v. Byron Mitchell 1999 Criminal Action No. 96-407, US District Court for the Eastern District of Pennsylvania.

UST 2005 University of science and technology beijing USTB database. http://www1.ustb.edu.cn/resb/en/index.htm (Accessed 21 February 2015).

Van der Lugt C 2001 *Earprint Identification*. Elsevier Bedrijfsinformatiem, Gravenhage, the Netherlands.

Wagner A, Wright J, Ganesh A, Zhou Z, Mobahi H and Ma Y 2012 Toward a practical face recognition system: Robust alignment and illumination by sparse representation. *IEEE Transactions on Pattern Analysis and Machine Intelligence* **34**(2), 372–386.

Wang X and Tang X 2009 Face photo-sketch synthesis and recognition. *IEEE Transactions on Pattern Analysis and Machine Intelligence* **31**(11), 1955–1967.

Wei X and Li CT 2013 Fixation and saccade based face recognition from single image per person with various occlusions and expressions. *IEEE Conference Computer Vision and Pattern Recognition Workshops (CVPRW)*, Portland, OR pp. 70–75.

Wei X, Li CT and Hu Y 2012 Robust face recognition under varying illumination and occlusion considering structured sparsity. *International Conference Digital Image Computing Techniques and Applications (DICTA)*, Fremantle, Western Australia, pp. 1–7.

Wei X, Li CT and Hu Y 2013a Face recognition with occlusion using dynamic image-to-class warping (DICW). *IEEE International Conference Automatic Face and Gesture Recognition (FG)*, Shanghai, China, pp. 1–6.

Wei X, Li CT and Hu Y 2013b Robust face recognition with occlusions in both reference and query images *International Workshop Biometrics and Forensics*, Lisbon, Portugal, pp. 1–4.

West v. West 1903. History of the 'West Brothers' Identification, circa 1960.2008.40.26. Collection of the National Law Enforcement Museum, Washington, DC.

Wiskott L, Fellous JM, Kuiger N and Von der Malsburg C 1997 Face recognition by elastic bunch graph matching. *IEEE Transactions on Pattern Analysis and Machine Intelligence* **19**(7), 775–779.

Wright J, Yang A, Ganesh A, Sastry S and Ma Y 2009 Robust face recognition via sparse representation. *IEEE Transactions on Pattern Analysis and Machine Intelligence* **31**(2), 210–227.

Yan P and Bowyer K 2005 A fast algorithm for ICP-based 3D shape biometrics *IEEE Workshop on Automatic Identification Advanced Technologies (Auto ID)*, Buffalo, NY, pp. 213–218.

Yan P and Bowyer K 2007 Biometric recognition using 3D ear shape. *IEEE Transactions on Pattern Analysis and Machine Intelligence* **29**(8), 1297–1308.

Yang M and Zhang L 2010 Gabor feature based sparse representation for face recognition with gabor occlusion dictionary. *European Conference Computer Vision (ECCV)*, vol. 6316, Crete, Greece, pp. 448–461.

Yuan L, Wang Z and Mu Z 2010 Ear recognition under partial occlusion based on neighborhood preserving embedding *SPIE, Biometric Technology for Human Identification VII*, Orlando, FL, pp. 76670Y-1-76670Y-7.

Zhang X and Jia Y 2007 Symmetrical null space LDA for face and ear recognition. *Neurocomputing* **70**, 842–848.

Zhang Z and Liu H 2008 Multi-view ear recognition based on B-Spline pose manifold construction. *World Congress on Intelligent Control and Automation (WCICA)*, Chongqing, China, pp. 2416–2421.

Zhao W, Chellappa R, Phillips PJ and Rosenfeld A 2003 Face recognition: A literature survey. *ACM Computing Surveys* **35**(4), 399–458.

Zhang HJ, Mu ZC, Qu W, Liu LM and Zhang CY 2005 A novel approach for ear recognition based on ICA and RBF network. *International Conference on Machine Learning and Cybernetics (ICMLC)*, vol. 7, pp. 4511–4515.

Zhang L, Yang M and Feng X 2011 Sparse representation or collaborative representation: Which helps face recognition? *IEEE International Conference Computer Vision (ICCV)*, Barcelona, Spain, pp. 471–478.

Zhang W, Wang X and Tang X 2010 Lighting and pose robust face sketch synthesis. In *European Conference Computer Vision (ECCV)* (ed. Daniilidis K, Maragos P and Paragios N). vol. 6316 of *Lecture Notes in Computer Science*. Springer, Berlin, Germany, pp. 420–433.

8

Multimedia Analytics for Image Collection Forensics

Marcel Worring

Informatics Institute, University of Amsterdam, Amsterdam, the Netherlands

8.1 Introduction

Forensics are rapidly moving from an analogue to a digital world. Evidence in complex criminal cases now comprises mainly of digital data. With the omnipresence of social media as a medium for communication for the general public and criminals alike we see a shift of type of data to encounter in an investigation. In modern cases most of the data are not structured, but are in the form of text, documents, images, video, and audio recordings, and their associated metadata. In this chapter we particularly focus on techniques suited for forensic analysis of large image collections.

In digital forensics an investigator is often tasked to analyze a data source containing large numbers of images and their metadata sometimes reaching into the millions. On such large scales, the only well-established automated techniques for the analysis phase of an investigation involve filtering techniques to limit the amount of data to consider and techniques for summarizing the data. In order to filter out files, the cryptographic hashes of all files are compared with large databases of known incriminating material as well as benign files. The most famous reference set is the National Software Reference Library (NSRL) which contains hashes for over 31 million files (Richard and Roussev 2006). With such an approach we can make a significant reduction in the number of images to consider for example, used in XIRAF (Alink *et al.* 2006). However, apart from the obvious problem that such a library can never be complete, the more cautious offenders can easily circumvent such

Handbook of Digital Forensics of Multimedia Data and Devices, First Edition.
Edited by Anthony T.S. Ho and Shujun Li.
© 2015 John Wiley & Sons, Ltd. Published 2015 by John Wiley & Sons, Ltd.
Companion Website: www.wiley.com/go/digitalforensics

forensic analysis as changing even just a single bit will radically alter the hash of a file. Perceptual hashing, in which images are considered equal if they appear to be equal, is robust against such small changes (Monga and Evans 2004). Being able to match material found in a case with existing databases of known material yields significant reductions in the amount of images to be considered.

Identifying known illicit material on a captured device is an important step in identifying someone as a suspect. The true interest of the forensics analysis is finding previously unknown material of interest. With so much data automatic processing of the image collection is essential to aid the investigator in the process.

Apart from the content of the images there is a lot of information present in the metadata. For example, downloading of illicit material and its subsequent sharing by a suspect are not random processes. They are governed by usage patterns with varying complexity, and we can use the MAC (last modified, last accessed and created) times that are present in most common file system formats to understand patterns of use. Similarly, the file path can be used to extract hierarchical information that can shed light on patterns such as naming and other organizational schemes. Richard *et al.* argued that digital forensics 'not only requires better acquisition tools, but also better analysis tools', as 'investigators must be relieved of manual, time-consuming tasks' (Richard and Roussev 2006). Metadata is a great aid in finding patterns, but it alone can never be the answer as the actual content determines the meaning of the images.

Whereas the computer is the key to bulk processing of the data, it is limited in performance when dealing with the complexity in a heterogeneous dataset. Forensic experts are highly skilled analysts who are able to make the delicate decisions when judging digital evidence. Humans also have excellent skills in finding subtle patterns in data. The skills of the expert are complementary to what automatic processing tools can offer.

Along the dimension of varying degrees of automation there are essentially four different categories that arise. Most traditional designs in forensics do not go beyond presenting images as a sequential list of files that have to be inspected one by one, thus 'manual browsing' is still the dominant category.

To support browsing, methods in the second category focus on the metadata alone. CyberForensics Timelab (Olsson and Boldt 2009) is an example of a system which is focused on time as metadata. Standard file explorers and text search systems can find relevant filenames.

The third-category methods aim to arrange the images in a more meaningful way than the simple list. By doing so, images of a target class are closer together, thus finding at least one image of that class will make the process of finding-related images easier. At the same time, images that are very different from the target class will have a low probability of relevant images in its vicinity, thus the surrounding region can be browsed with less attention and, consequently, faster. Quadrianto did so using a grid-based approach (Quadrianto *et al.* 2010), whereas Nguyen uses a similarity-preserving projection (Nguyen and Worring 2008) coupled with an approach to minimize the overlap between images.

The fourth and last category is content-based image retrieval (CBIR), and there is a wide variety of different approaches and features (see Datta *et al.* 2008 for a recent survey). In theory this allows the highest possible degree of automation: given a perfect classifier for our target class we do not have to look at a single image anymore. In practice, however, humans still outperform machines in a classification task in terms of accuracy, and frequent misclassifications are not uncommon.

It should be clear that for forensic purposes using traditional techniques on the problem of browsing large image collections is insufficient. Manual browsing is clearly too time-consuming, but each of the three categories by themselves does not provide a solution either. Visualization-based browsing and content-based retrieval discard important meta-information, while retrieval purely based on meta-data doesn't consider the actual contents of images. Even in state-of-the-art research, full automation has not been achieved with any of the approaches. Interactive browsing techniques such as (Wei and Yang 2011) are promising avenues to improve the search performance, but the techniques focus on the underlying search algorithm only and ignore advanced visualization of the intermediate search results.

All of the three categories of semi-automated techniques have their merits; none of them in its own right yields an answer to the complex problem of forensic investigation. The overarching aim of this chapter is to combine the strength of each method in order to overcome the individual weaknesses. This means that we need to communicate the result of content-based retrieval methods to a human operator who uses a visualization-based browsing interface to make sense of them. Further, all of the meta-information available in the dataset has to be reflected by the design of the visualizations and accessible to the operator. Then, and only then, will we be able to take full advantage of the available data as well as computational and human resources. And even more important will be able to provide the system with feedback to improve its results. Research argues that a suitable framework for such an undertaking is visual analytics (Keim *et al.* 2010).

Visual analytics is a relatively young field, and even the term has only been in use since *Illuminating the Path* (Thomas *et al.* 2005) was published in 2005 (Andrienko *et al.* 2010). In Keim *et al.* (2010) it is defined as 'combining automated analysis techniques with interactive visualizations for an effective understanding, reasoning and decision making on the basis of very large and complex datasets'. From this definition it becomes clear that visual analytics puts a stronger emphasis on the human component than its sibling disciplines. It strives to combine the strengths of human and that of electronic data processing – a semi-automated analytical process. Even more recently visual analytics has been combined with multimedia analysis. This combined field has been coined as multimedia analytics (Chinchor *et al.* 2010; de Rooij *et al.* 2010). In all aspects it has the same characteristics as visual analytics. There is a very specific distinction between the two, though for structured data single items in the collection are of limited value; but in image collections, any picture can be of importance and by itself be an important piece of evidence. We focus on these interesting aspects of image collections within a multimedia analytics context.

This chapter is organized as follows. We first consider the specifics of datasets and tasks in the forensic field in Section 8.2. From there, we consider the two fields underlying multimedia analytics namely multimedia analysis (Section 8.3) and visual analytics (Section 8.4). The principles from those fields provide the basis for describing two-example multimedia analytics systems in Sections 8.5 and 8.6. We end with an example forensic scenario, highlighting how this could be supported by the techniques described in this paper.

8.2   Data and Tasks

We will now take a closer look at the data and tasks that play a role in forensic analysis of a large image collection.

For this chapter we consider an image collection to be a set of images with all their associated metadata such as MAC times, filenames, tags and geo-locations. Searching for images containing evidence in large collections is difficult as the information of interest might be captured in any of the various modalities associated with the images. The determining clues for the search might be in the content of the image, but might also be in its metadata. Furthermore, the best way to search is highly dependent on the task, ranging from finding an abundance of results to little nuggets of information.

To get an understanding of what these different tasks comprise, we consider it as an information retrieval task. The fundamental task in information retrieval is, given a query, to rank all elements in the dataset in such a way that the top results are the ones best corresponding to the query. It is this operation on which all common search engines are based. In practice, only a limited part of the total collection will be shown on the screen. There are two competing measures for evaluating the quality of the result on display. Let the number of results displayed, out of the total of n correct elements in the dataset, be m. Further, let m' be the number of elements in the result set which are actually correct given the query. The precision p is given by

$$p = \frac{m'}{m}.$$

It is a measure of the quality of what you see on your display.

The recall r is given by

$$r = \frac{m'}{n}.$$

Hence, this gives the proportion of elements in the dataset we have actually found when we only consider this small result set.

Different applications impose different constraints on precision and recall. Web search engines are optimized for high precision. General users seldom go beyond

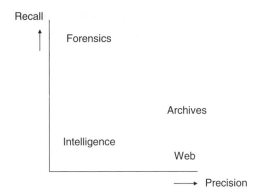

Figure 8.1   Different tasks, different ways of measuring succes.

the first or second page of results and in most cases have a limited interest in having high recall when they found some good results. When posing a query to professional archives you also expect high precision of your results. Most likely you also want to have a reasonable recall while at the same time people are not willing to go through several pages of results. Within law enforcement things are quite different. In intelligence agencies they are willing to go through large collections of data to search for any lead which might be used to go to the next step in the investigation. Recall is of a lesser concern as being able to act right away is of prime importance. Forensics has a different characteristic. There investigators spend even more time in analyzing the full collection as completeness of the result is a prerequisite for valid judgements. So where investigators are still willing to accept limited precision, they aim for the highest recall within the given time constraints. The different requirements of the applications are exemplified in Figure 8.1.

The different characteristics of the tasks in forensics mean that we have to reconsider many aspects of the standard information retrieval techniques, even more so taking into account that we are dealing with image collections. In the following sections we will therefore look at multimedia analysis, in particular analysis of visual material, the visual analytics processes that can be used to model the search process and from there we consider two example systems.

8.3   Multimedia Analysis

To derive a ranking of an image collection we need ways to query the collection. To that end we can distinguish a number of different possibilities. The first option is posing a query using any of the metadata associated with the images. Metadata is useful for filtering when the decision on every single item is binary. In that case the image either conforms to the query or not. Examples could be a specific time interval, a specific file type, or a specific filename. This does not lead to a ranking. But ranking on metadata is still very useful. For instance we can use a specific time stamp to get a temporal ordering of the dataset based on their differences in time to the reference point. Another example is using any text string that might occur in the filename of the image as query so that we can get the exact matches on top and from there a ranking in terms of how similar the filename is to the query string. Querying the metadata is relatively easy and is an easy way to filter the dataset, yet we need to know what to look for before we can pose a successful query.

Instead of using metadata for querying, we might have an example image available for instance a photo of the crime scene or a piece of evidence that we can use as query. Rather than selecting part of the dataset, the system computes the similarity to the query for every image in the collection. Based on the results the whole collection is ranked with respect to their similarity to the query.

Without a query the first possibility is to automatically cluster the images based on their visual appearance. For visual clustering a good measure for the similarity among the images is crucial. From a conceptual point of view, it is convenient to consider Euclidean distance. In reality there are various ways of measuring similarity each of which is geared towards a specific set of features describing the images. Once we are able to determine how similar images are, the actual clustering starts. There are many methods for doing so (Witten *et al.* 2011). The easiest and probably most often used method is k-Means (also called k-Centres). In this algorithm you fix the number of groups you want to have in the data and you randomly pick an image as representative for each group. From there you assign every image to the closest representative. This leads to a first grouping. For each group the algorithm determines the image which is at the center of the group and uses this as the new group representative. Now we repeat the process until either there are no more changes of group membership or when we reach the maximum number of iterations. The result is a grouping of the images into a number of sets where ideally the images in a group have a similar appearance, whereas between groups visual similarity is limited.

Are there any other ways to get to the data? In fact in the past decade a new paradigm of getting access to the content of images has appeared. This is the field known as concept based retrieval (Snoek and Worring 2009). To do so a large set of concept detectors are trained that are able to give an indication of the likely presence of the concept in the image. The result of concept detection allows to query the image collection on the basis of semantic concepts present in the image. We could, for example aim to find all images containing a creditcard logo, guns or signs of child abuse. Querying using semantic concepts is an advanced method for image collection forensics.

Deriving the concepts present in an image is a difficult task (Smeulders *et al.* 2000) for which many methods have been defined (Snoek and Worring 2009). In the early days specific models were developed for different concepts. Such methods do not scale to the large lexicons of concepts required for understanding image content. State-of-the-art techniques are based on supervised learning iven negative and positive examples for each concept. For all examples a large number of features are calculated, where commonly they are some variants of the Scale-Invariant Feature Transform

(SIFT) (Lowe 2004). SIFT is a feature describing the distribution of orientations around specific keypoints in the image. It does so at multiple scales yielding a description of both details in the image as well as global characteristics. Often salient points in the image are taken as keypoints. Alternatively dense sampling puts a regular grid of points over the image and use these as keypoints. SIFT has also been extended to colour images. A good overview of different features and their merits is (van de Sande *et al.* 2010). Based on the examples a model is learned. Any supervised learner could be employed here, but Support Vector Machines (SVM) are the method employed most in current processing pipelines, for example (van de Sande *et al.* 2010). The SVM models define a decision boundary in the feature space. When a new image is given to the model, for which the concepts are not yet known, a measure of likelihood of the concept being present is derived as function of the distance to the decision boundary. The likelihood values derived in the concept detection process form the basis for further retrieval. The whole process of concept detection is illustrated in Figure 8.2. Recently deep learning has started to appear as method for detecting concepts (Kriszhevsky *et al.* 2012). Instead of pre-defined features these methods learn the features from the data (Bengio *et al.* 2013). These are promising directions, yet they require much more training samples to be successful.

The complexity of the process of concept detection introduces an inherent uncertainty in the result. This has a number of consequences in terms of how to interpret and use the results and what subsequent operations we can perform on the values in the result.

The result of a concept detector is a score for the presence of the concept with significant uncertainty. It depends on many factors including the actual presence, but also on priors and data quality. Consequently, the value in its absolute form has limited meaning. They should mainly be used for ranking the images in the collection

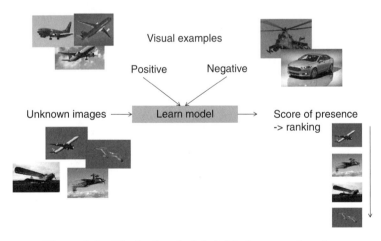

Figure 8.2   The basic principle behind concept learning.

with respect to the presence of the particular concept. Especially in forensics with its high recall requirements, thresholding the list to filter out images not containing the concept should be done with care and preferably by visual inspection of the image list. Exhaustively going through the list is prohibitively ineffecient. Semi-interactive approaches are most appropriate.

8.4 Visual Analytics Processes

As indicated before standard retrieval systems are not targeting at high recall which is crucial for forensics. Therefore, we need to consider the paradigm used to get access to the data. Rather than a 'one query to result' methodology, the investigator is interactively exploring the dataset in search of relevant images. Synergy between machine and human are needed and visual analytics has been defined with precisely such targets in mind (Keim *et al.* 2010). In the reference an iterative process for exploring data is presented considering four interrelated processes: filter, analyze and show the important, zoom, and details on demand (see Figure 8.3):

- *Analyze and show the important*: This is the starting point of access. The system has performed an analysis of the data and on the basis of this decides on which elements (images and/or metadata) to show to the investigator. In the first iteration this is based on the data alone, whereas in subsequent iterations it can include any statistics on interactions of the user.
- *Filter*: With large datasets filtering is crucial to limit the dataset to the potentially relevant items as soon as possible. This is an operation which can be initiated by the investigator by selecting a specific descriptor of the data or by the system based on data statistics.

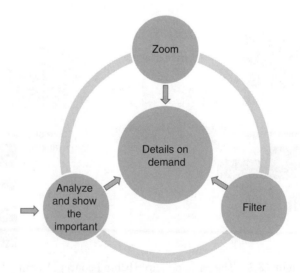

Figure 8.3   The operations involved in analyzing a dataset using a visual ananlytics approach.

- *Zoom*: When filtering the data, less elements need to be displayed. So the investigator can zoom in on specific areas of the collection to see specific relations between images or analyze patterns in metadata or content.
- *Details on demand*: True understanding of the image content can only come about when the investigator can see the content of the image. Details on demand, including metadata, give access at the most detailed level.

To illustrate how visual analytics combined with multimedia analysis leads to advanced systems for accessing large visual collections, we elaborate on two systems: ChronoBrowser (Worring *et al.* 2012) and Mediatable (de Rooij *et al.* 2010). The first system is a cluster based system whereas the second one is based on semantic concepts. Together they give a good indication of the usefulness of multimedia analytics.

8.5   ChronoBrowser

As a first example of a multimedia analytics system developed for forensic analysis we consider the ChronoBrowser (Worring *et al.* 2012) (see Figure 8.4). ChronoBrowser is a cluster based system, so it follows a data-driven approach. Next to image content it considers time and the file hierarchy as the two additional channels for forensic evidence. In the following, we first look at different visualizations that play a role and from there go through different visual analytics processes and show how they are instantiated within ChronoBrowser.

8.5.1   *Visualizations*

8.5.1.1   Timeline Visualization

The timeline aims at revealing trends that are either self-evident by looking at the temporal information itself, such as an increased amount of activity during a certain

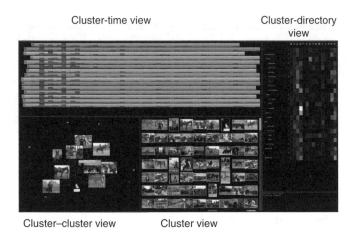

Figure 8.4   A cluster-based forensics system.

period of time, or are revealed in conjunction with other dimensions, such as a directory
that has seen many changes recently.

To support the above, primitives are structured and arranged. In addition to bars that
encode the temporal dimension on the horizontal axis, each cluster or event contains
a histogram. A histogram bin shows the activity for a fixed time segment. The icon
that represents the cluster is superimposed on the bar to aid visual search of a cluster
in related visualizations. The visualization bears similarity to Gantt diagrams and
histograms. Thus it is easy to spot trends, outliers, activity spikes or other emerging
patterns. Timelines for a set of selected set of clusters at various zoom levels are shown
in Figure 8.5.

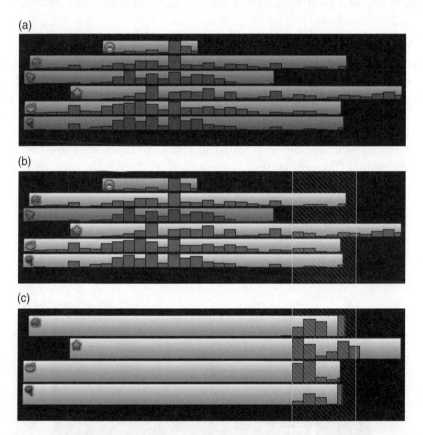

Figure 8.5 Timeline visualization in ChronoBrowser. (a) Shows the basic visualization in
which six horizontal bars representing image clusters are shown. The start- and endpoint of
each bar reflect the range in time for all images in the cluster. Within each horizontal bar,
small vertical bars represent the amount of images having a timestamp falling in the timerange
corresponding to the bar; (b) shows the effect of selecting a specific time interval. After turning
the selection into a filter; and (c) shows the result of re-evaluating the histograms stripping any
empty clusters.

8.5.1.2 Graph Visualization

The graph visualization shows a network of image clusters, whose topology is constructed using Pathfinder networks (Chen *et al.* 2000), whereas the layout is calculated using a force-based algorithm e.g. using the algorithm in Fruchterman and Reingold (1991). Each node in the network represents an image cluster. The edges between clusters encode their similarity.

The aim of the graph is to provide an overview over a part of the spatial dimension of our image collection and to support reasoning at the level of clustering. This is accomplished at several levels. After the initial layout is calculated, it is purposely held fixed throughout the remaining interaction steps (unless the user explicitly wishes to re-arrange the nodes). This helps the user to form a spatial memory of the image clusters. It will reveal how diverse the image collection is. If for example the collection consists of only holiday photos and most of these are taken outdoors, it will result in a more densely connected network. Figure 8.6 shows an example of a Pathfinder network.

8.5.1.3 Hierarchy Visualization

The purpose of the hierarchy visualization is to connect the clusters and individual images back to their physical locations on the disk, and thus to complement the spatial dimension. The major difference to the spatial information provided in the graph visualization is that the file structure has a natural hierarchy. A heatmap using grey values, as these are the best in conveying different values, provides a visual representation of the connection among the different directories and the derived clusters. This is illustrated in Figure 8.7.

Directories within the visualization itself can be highlighted as well. This will cause all other visualizations to both highlight the images that are contained in this folder, as well as the clusters that contain one or more of these images.

8.5.1.4 Detail Visualization

The detail visualization is responsible for allowing the user to browse the actual images exhaustively. As indicated the essential ingredient making multimedia analytics different from standard visual analytics solutions is the ability to visually assess the content of individual images. For the sake of simplicity and familiarity the detail visualization is a simple grid of images (see Figure 8.8).

To ensure a connection between the detail view and the clusters as well as providing some limited sense of overview without having to resort to other visualizations, the detail visualization introduces an enhanced version of a regular scrollbar. Just like a traditional scrollbar it shows the fraction and position of the actual visible region in relation to the entire set of selected images. It improves this concept by additionally plotting the different clusters as coloured regions – the order of the images as shown in the grid is partitioned into clusters – which can be clicked in order to jump between clusters. The icon representation of each cluster is embedded into the region as well. Refer to Figure 8.8 for an example of the scrollbar.

(a)

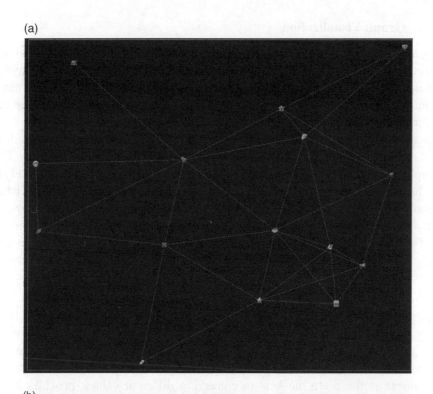

(b)

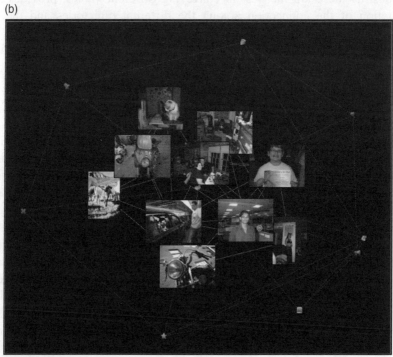

Figure 8.6 A pathfinder network based on visual similarity (a) and illustration of a zoom-in on a cluster showing representative images (b).

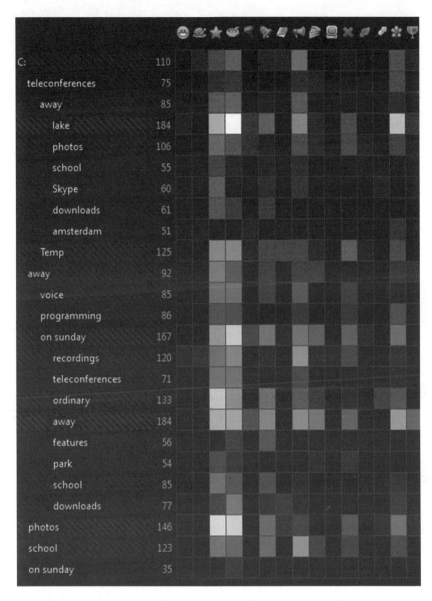

Figure 8.7 A heatmap showing the distribution of images in directories over the different clusters to aid the process of finding the directories with most candidate matches or suspicious distribution.

8.5.1.5 Cluster Representation

Different visualizations often require different ways of depicting primitives. In a timeline where time is mapped to the Y-axis and the primitives encode duration, they need to be stretched horizontally to indicate the start and the end. In the graph visualization, we can arbitrarily assign a representation for the node, but using the same

Figure 8.8   A selection of three clusters shown in the detail visualization. Icons on top of the images identify the cluster each image is part of.

bars as in the timeline hardly makes sense as the X-axis encodes spatial information. Yet the primitives in either visualization are not only of the same type, but they are two different views on the same data.

Ideally we want to be able to link the spatial (similarity) information to the temporal given those two visualizations. If we can design the primitive representation in such a way that every single primitive in the visualization is discernible along at least one of the visual channels, the user can simply single out any primitive by searching for the same properties of the channel in the other visualization. As long as the representation is identical we do not even have to know the exact channel that distinguishes an item, and our bias in the primary visual cortex is adjusted automatically along all the channels.

First, however, we need to solve the problem of primitives having different representations in the timeline and graph visualization. This turns out to be reasonably straightforward: since primitive representations in the graph are arbitrary – vertices have no visual encoding themselves – we can simply choose them to have icon representations, where each image cluster is assigned a unique icon. Regardless of what we pick for the timeline primitives (e.g. lines or bars), we can associate them by visually linking them with the same icons. Figures 8.5, 8.6 and 8.7 show how the cluster icons are used in each of the visualizations with the aim to visually link them.

8.5.2 Visual Analytics Processes

8.5.2.1 Analyze and Show the Important

The primary analysis mechanism in ChronoBrowser is the clustering of the image collection based on its visual content. In principle, any combination of features and clustering methods can be employed. In our implementation we used colour-SIFT descriptors and k-means to group the images. As the clusters are explicitly visualized in each of the individual representations k is typically set to a value in the range of 10–20.

8.5.2.2 Filter

It was Shneiderman (1996) who noted that '[users] have highly varied needs for filtering'. We use his principle of 'dynamic queries', or 'direct-manipulation queries'. Graphically manipulating objects on the screen, for example by dragging a rectangle in the timeline, is linked to a certain filter criterion (see Figure 8.5). Using 'rapid, incremental and reversible actions' along with immediate feedback, users can build up complex filtering conditions comprising temporal and spatial conditions without ever having to explicitly formulate a query. Note that selection and filtering conditions do not have to be formulated in one visualization alone – they can be freely combined across the system. After filtering all visualizations are automatically updated to reflect the new distributions.

8.5.2.3 Zoom

In ChronoBrowser zooming is especially relevant for the timeline and for the graph visualization.

For the timeline zooming is a simple operation. Given the current fixation point in time we can simply zoom in on the interval surrounding this time stamp to get a better view on the temporal distribution of images around this time point.

The zooming in the graph visualization is more difficult and uses various techniques. Fixed-shape zooming is used to retain the size of the clusters. Once zoomed in, the focused cluster is semantically enhanced by showing its representatives, an operation known as 'semantic zooming'. Clusters connected by at least one edge are projected onto a circle around the zoomed cluster, and thus the distance is no longer correct, but the angle is retained. This provides a sense of context to the zoomed view. The illusion of a geometric zoom is accomplished by moving unconnected clusters out of the view. The movement vector is calculated by taking the vector from the centre of the visualization to each cluster, normalizing it and multiplying it by a constant. Refer to Figure 8.6 for an example of using zooming in the graph.

8.5.2.4 Detail-on-Demand

Detail-on-demand is provided at two different levels. When a set of clusters is selected in any of the visualizations it will be shown in the cluster view. The timeline and directory visualization allow other interesting methods to reveal detailed information on a small selection of images.

In the timeline, histogram bins can be hovered to highlight contained images. This highlighting is done in all other visualizations revealing patterns in time, clusters and directories.

Upon hovering an image anywhere in the hierarchy visualization, the background of the corresponding folder is coloured. If a cluster is highlighted instead, the colour spreads across all the directories that contain images that are contained in this cluster. An exact number is shown next to each folder to indicate the number of hits. Additionally, the intensities of the background colouring is scaled by the number of hits over the maximum number of hits (meaning that the folder has the highest number, not the sum of all of them; see Figure 8.7). This redundant encoding helps the user to quickly determine the dominant folders for a given cluster.

8.6 MediaTable

As an example of a concept based multimedia analytics system we consider MediaTable (de Rooij *et al.* 2010).

8.6.1 Visualizations

8.6.1.1 Table Visualization

Traditional ranking-based image search engines show the result as either a list or a grid of images. In such a system every new query yields a completely new list, visually not connected to the previous one. Such a setting makes it difficult to use multiple concepts as queries and show their correlations. MediaTable takes a different approach. All rankings for a large set of concepts are displayed at the same time sorted according to the currently selected concept as a large table. This table per row shows an image and all its associated metadata. A heatmap, with black denoting maximal expected presence of the concept in the image, gives an overview of a large set of concept rankings. By scrolling horizontally and by hiding irrelevant columns the investigator can look for correlation among a set of concepts of interest (Figure 8.9).

When interacting with the collection investigators can select items of interest and put them in a category. In MediaTable the elements in a category are put in buckets denoting a specific category. These buckets need not be disjoint, images can be put into several buckets depending on different aspects of their content. Each bucket has a designated colour which is consistently used in all visualizations. In this way it is possible to observe patterns in the data based on category membership.

Interactive   Large set of concepts
categorization   as heatmap   Image scatterplot

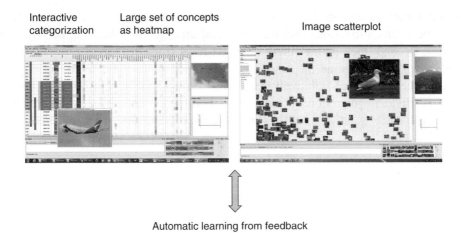

Automatic learning from feedback

Figure 8.9   Mediatable system for exploring large image collections.

8.6.1.2   Image Scatterplot Visualization

In the scatterplot two concepts are selected for the axes, and the images are placed at
the corresponding positions. Depending on the scale these elements are either shown as
individual dots allowing to get insight in overall correlations and category membership
or shown as actual images revealing their content.

The scatterplot allows to identify correlations between two concepts in a much
more detailed manner than the table visualization. It might reveal linear relationships
between different concepts, but also more subtle non-linear relationships. Clusters of
images with similar content for these two concepts are also easy to spot.

8.6.2   Visual Analytics Processes

8.6.2.1   Analyze and Show the Important

When using MediaTable a large set of semantic concept detectors are run on the full
dataset. The number of concepts we currently use in the system is typically in the
order of 100–200. The table ranks the whole collection according to the sort order of
the user selected concept. Thus it reveals any correlated concepts as they will show a
similar distribution of grey values in the corresponding heatmap column. Horizontal
scrolling brings in more concepts, while vertical scrolling show images further down
in the ranking. With a high recall as target, going to lower ranked data is essential.

8.6.2.2   Filter

MediaTable provides a faceted filter, where the whole collection can be filtered
according to a conjunction of different dimensions. These dimensions might be
concepts, thresholded to only keep the top scoring elements, or any metadata associated

with the image collection. In this way the interaction becomes a set of sort, filter, select operations. First, users sort the table, and thereby the collection, on any of its columns. Second, they define filters to select rows from the sorted results. Finally, the users visually examine the associated shots, and place these in the appropriate buckets. The process of iteratively selecting, sorting and placing items in buckets yields effective categorization. As the buckets themselves can be used as filters as well, the user can easily subdivide buckets into new buckets and thus perform fine-grained categorization.

8.6.2.3   Zoom

Zooming for the table visualization is based on dynamically adjusting column width and row height. With small values for these it becomes difficult to examine the content of individual images and the names of columns might be too small to read. It does, however, give an overview of many images at the same time as thus provides ample room for seeing correlations among various concepts. Zooming back in will reveal more details and further basis for filtering, sorting, and selecting.

When the user is zooming in on specific concepts and sees high promise for identifying forensic evidence in the dataset, she can move to the image scatter plot for more detailed analysis.

Zooming in the scatterplot is a natural operation. In combination with panning, implemented by scrolling in vertical and horizontal direction, the investigator can zoom in on any part of the space where the dots show a pattern of interest. When zoomed in far enough she can see the actual image content, furthermore she can examine all closeby images in the collection which might potentially lead to new categories easily created by selection.

8.6.2.4   Details on Demand

The current image is always shown in the detail view. In this view not only the image itself is presented but also the most important metadata. Furthermore, and likely more interestingly, it will show the highest scoring concepts for this image. Based on this the investigator might go back to the table view and start filtering on those specific concepts.

8.6.2.5   Analyze and Show the Important Revisited

Up to now we have considered the analysis of the data as a pre-processing step whose result is visualized and this forms the basis for user navigation. But true multimedia analytics goes beyond such a simple approach and makes multimedia analysis an integral part of the process. In such approaches the system considers the interactions of the user and based on this performs further analysis. There are few systems doing so in combination with advanced visualizations. An exception is the active buckets-based method proposed in (de Rooij and Worring 2013) which extends the standard

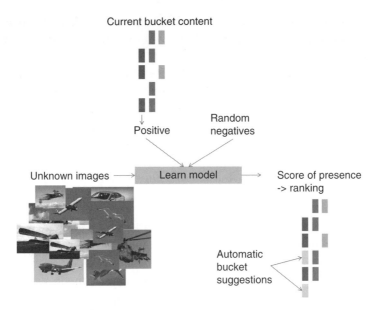

Figure 8.10 Active buckets for supporting investigators in categorizing large image collections.

MediaTable paradigm. Active buckets define a method allowing the computer to suggest new relevant items for any bucket, based on the contents of that bucket. These suggestions are based on low-level visual feature characteristics of individual items placed inside a bucket, and those that are currently not inside. The approach very much follows the same paradigm as depicted in Figure 8.2, but where explicit annotations are used for positive and negative examples active buckets rely on the current bucket content. To be precise, the active buckets unobtrusively observe changes in user selections. Based on this the active bucket engine decides which bucket is the most prominent candidate to extend with more suggestions. This bucket is then passed through to a sampling engine which selects items from the bucket as positives and some other random elements as negatives. These positive and negative examples are used dynamically to train a new model which is then applied to the whole dataset. The top scoring elements not yet in the bucket are then suggested as new bucket content to the investigator. The basics of active buckets, are illustrated in Figure 8.10. Active buckets go beyond navigation and provide real support to the investigator challenged with finding evidence in a large collection of images.

8.7   An Example Scenario

So let us now consider how multimedia analytics can aid forensic investigators. We do so by considering a potential scenario and showing how a combination of ChronoBrowser and MediaTable would be used to find evidence. In doing so we take the liberty of assuming that the systems are elements of a multimedia analytics suite where all elements are seamlessly integrated. As example domain we consider the forensic investigation of a child abuse case.

Scenario: A police team has entered the house of a man who is suspect in a child abuse case. In his bedroom they find a desktop computer with a webcam. The first thing they do is copying, in a forensically sound manner, all data on the 3 Terabyte harddisk. In addition, they take pictures of the room, with particular attention for the viewpoint of the webcam. In the forensic lab they analyze the hard disk with standard tools to get any images (and of course videos which we don't consider here) on the harddisk. In total they find over 1.2 Million images. In addition to the images, the investigators capture all metadata available on the images. This huge data set becomes the starting point for the investigation.

The first step in the analysis is filtering out any images with known MD5 hashes. These comprise both well known images such as icons used in specific programmes as well as data from international child abuse databases. Of course the latter are the ones that are important. Apart from searching exact copies, the investigator also use perceptual hashes to identify near copies of known image material. The results are overwhelming and the dataset reduces to 400K images. Clearly this is already sufficient evidence to take the suspect into custody. Yet at this point the investigators only know that he has downloaded a lot of material. They expect that there is much more interesting evidence to be found in the data.

A quick analysis reveals that for many of the images hundreds if not thousands of copies are present. ChronoBrowser shows that these copies are all downloaded in the same timeframe and that they are stored in a limited number of directories. They hypothesize that the suspect has been posing automatic queries to various websites to download the data. They don't have a clue why the suspect did so, but they will confront the suspect with these observations later.

Overnight they let the automatic concept detection system analyze the whole remaining set of images for a large set of potentially relevant concepts such as indoor, bed, child, nudity, professional porn and genitals. Viewing the result in MediaTable immediately yields an interesting result. It turns out that the professional porn detector has found many images with high scores and visual inspection shows that those are all stills from various adult porn movies. Filtering on the score, putting a high threshold to assure sufficient recall, again reduces the dataset leaving the investigators with 120K images.

The investigators now ingest the pictures taken in the bedroom of the suspect. By learning a new concept based on these examples the investigators find 20 images as top-ranked results in which a young girl is posing naked on the bed in front of the webcam. Some highly relevant evidence has been found. But who is she and is she the only one?

To continue the investigation, the dataset is ranked according to the child concept. MediaTable reveals a high correlation with the indoor as well as the nudity concept.

Ranking instead on nudity shows a great variety in images from children in bathing suits to fully naked adults. The investigator defines different buckets for those and start a manual categorization process. Quickly after the start of doing so the automatic suggestion engine starts adding relevant images to the various buckets and quickly the child nudity bucket has over 3000 images which seem to be relevant.

The 3000 images found are clustered based on their visual appearance and viewed in ChronoBrowser. Here the investigation takes a dramatic turn. Many of the clusters correspond to a number of specific time periods and inspecting them shows that two recuring locations are found. One of the locations seems to be a dressing room of a gymnastics club and another is a swimming pool. By inspection the investigators find that seven children are photographed while they are undressed. Regular tactical investigation quickly yields the conclusion that both locations are nearby the house of the suspect and investigation continues there. In addition the tactical investigation identifies the girl on the bed as his own daugther.

After a few days of intensive investigation the combination of automatic analysis and the interactive use of the system by investigators have provided the evidence to take the suspect to court. Manual investigation of the dataset with over a million images would have taken several months if not more.

8.8 Future Outlook

The scale and complexity of image collections pose a heavy burden on forensic investigators, especially because forensics poses high demands on the recall of the results. Manual inspection of all material is no longer feasible; yet current automatic techniques are not accurate enough to be employed blindly. Multimedia analytics is a new field geared towards making the best of both worlds. It lets the computer perform the laborious task of large scale simple processing, while the investigators examine the more interesting and challenging cases. In a true multimedia analytics setting the system learns from the user interactions and thus creates optimal synergy between the human and the machine.

In this chapter we have shown how the specific criteria of success for forensics require new methods. This requires advanced visualizations to bring human and machine together. The associated visual analytics processes analyze and show the important, filter, zoom, and details on demand, and provide a convenient framework for developing methods to support the investigator in understanding and navigating image collections. With two example systems, namely ChronoBrowser and MediaTable, we have shown how this framework can be instantiated in real-life systems and how it can support effective forensic investigation.

Multimedia analytics as a field has just started. It holds great promise for digital forensics of multimedia data. The two systems presented in this chapter are a major step forward. However, we expect the field to elaborate on such techniques soon with even more integration of the human cognitive processes.

Moving forward faces a number of challenges both at the multimedia analysis side as well as one the visual analytics side. Multimedia analysis will have to deal with ever more complex concepts. To reach human performance the number of detectors has to reach in to the ten thousands, each of which should have a performance which goes beyond the quality of the current detectors. Furthermore, most concept detectors are now working on the image as a whole. New methods are now looking at object-level detection. There is a lot of progess in applying detectors to video data. Yet most of them are treating the video stream as a set of independent images. Being able to reliably detect events such as sexual activity or loitering at an ATM machine has high importance for forensic investigations. Concerning visual analytics, we have to seamlessly integrate the capabilities of machine learning algorithms with the cognitive powers of the investigator. The major challenge is to exploit the power of automatic analysis to the max without the investigator taking explicit notice of it. Such an unobtrusive approach comes with an additional challenge. After evidence has been found the full process of investigation from both the human as the machine perspective should be reconstructable and the remaining uncertainty in the final result should be automatically quantified. In the years to come, the active pursuit of innovative techniques in both multimedia analysis and visual analytics and its integration in multimedia analytics will bring the quality and effectiveness of forensic investigation to the next level.

References

Alink W, Bhoedjang R, Boncz P and De Vries A 2006 XIRAF-XML-based indexing and querying for digital forensics. *Digital Investigation* **3**, 50–58.

Andrienko G, Andrienko N, Demsar U, Dransch D, Dykes J, Fabrikant S, Jern M, Kraak M, Schumann H and Tominski C 2010 Space, time and visual analytics. *International Journal of Geographical Information Science* **24**(10), 1577–1600.

Bengio Y, Courville A and Vincent P 2013 Representation learning: A review and new perspectives. *IEEE Transactions on Pattern Analysis and Machine Intelligence* **35**(8), 1798–1828.

Chen C, Gagaudakis G and Rosin P 2000 Similarity-based image browsing. *Proceedings of the 16th IFIP World Computer Congress. International Conference on Intelligent Information Processing*, Beijing, China.

Chinchor N, Thomas J, Wong P, Christel M and Ribarsky W 2010 Multimedia analysis + visual analytics = multimedia analytics. *Computer Graphics and Applications, IEEE* **30**(5), 52–60.

Datta R, Joshi D, Li J and Wang J 2008 Image retrieval: Ideas, influences, and trends of the new age. *ACM Computing Surveys (CSUR)* **40**(2), 1–60.

de Rooij O and Worring M 2013 Active bucket categorization for high recall video retrieval. *IEEE Transactions on Multimedia* **15**(4), 898–907.

de Rooij O, Worring M and van Wijk JJ 2010 der: Interactive categorization of multimedia collections. *IEEE Computer Graphics and Applications* **30**(5), 42–51.

Fruchterman T and Reingold E 1991 Graph drawing by force-directed placement. *Software: Practice and Experience* **21**(11), 1129–1164.

Keim D, Kohlhammer J, Ellis G and Mansmann F (ed.) 2010 Data mining. In *Mastering the Information Age Solving Problems with Visual Analytics*. Eurographics Association, Geneva, Switzerland, pp. 39–56.

Kriszhevsky A, Sutskever I and Hinton G 2012 Imagenet classification with deep convolutional neural networks. *Advances in Neural Information Processing Systems (NIPS)*, Lake Tahor, NV.

Lowe DG 2004 Distinctive image features from scale-invariant keypoints. *International Journal of Computer Vision* **60**(2), 91–110.

Monga V and Evans B 2004 Robust perceptual image hasing using feature points *International Conference on Image Processing (ICIP)*, Singapore.

Nguyen G and Worring M 2008 Interactive access to large image collections using similarity-based visualization. *Journal of Visual Languages & Computing* **19**(2), 203–224.

Olsson J and Boldt M 2009 Computer forensic timeline visualization tool. *Digital Investigation* **6**, 78–87.

Quadrianto N, Kersting K, Tuytelaars T and Buntine W 2010 Beyond 2d-grids: A dependence maximization view on image browsing. *ACM International Conference on Multimedia Information Retrieval*, Philadephia, PA.

Richard III G and Roussev V 2006 Next-generation digital forensics. *Communications of the ACM* **49**(2), 76–80.

Shneiderman B 1996 The eyes have it: A task by data type taxonomy for information visualizations. *Proceedings of the 1996 IEEE Symposium on Visual Languages*, pp. 336–343. IEEE Computer Society, Washington, DC.

Smeulders AWM, Worring M, Santini S, Gupta A and Jain R 2000 Content based image retrieval at the end of the early years. *IEEE Transactions on Pattern Analysis and Machine Intelligence* **22**(12), 1349–1380.

Snoek CGM and Worring M 2009 Concept-based video retrieval. *Foundations and Trends in Information Retrieval* **4**(2), 215–322.

Thomas J, Cook K, of Electrical I and Engineers E 2005 *Illuminating the Path: The Research and Development Agenda for Visual Analytics*. IEEE Computer Society, Los Alamitos, CA.

van de Sande KEA, Gevers T and Snoek CGM 2010 Evaluating color descriptors for object and scene recognition. *IEEE Transactions on Pattern Analysis and Machine Intelligence* **32**(9), 1582–1596.

Wei XY and Yang ZQ 2011 Coached active learning for interactive video search *Proceedings of ACM Multimedia*, Scattsdale, AZ.

Witten I, Eibe F and Hall M 2011 *Data Mining: Practical Machine Learning Tools and Techniques*. Morgan Kaufmann, Burlington, MA.

Worring M, Engl A and Smeria C 2012 A multimedia analytics framework for browsing image collections in digital forensics *ACM International Conference on Multimedia Information Retrieval*, Nara, Japan.

Part Three

Multimedia Device and Source Forensics

9

Forensic Camera Model Identification

Matthias Kirchner[1] and Thomas Gloe[2]

[1]*Binghamton University, Department of Electrical and Computer Engineering, Binghamton, NY, USA*
[2]*dence GmbH, Dresden, Germany*

9.1 Introduction

Camera model identification subsumes the broad class of forensic source identification techniques that aim to determine the camera model used to acquire a digital image of unknown provenance. Camera model identification seeks to answer the questions

- 'Of which model was the camera that (most likely) took this given image?', or
- 'Was this image shot with a camera of this particular make and model?',

if prior beliefs prevail. The premise of camera model identification is that images acquired with the same camera model share common characteristics, which can be exploited for inference about the source of an image. Camera manufacturers are free to fine-tune their models so as to create visually pleasing images according to their preferences. In consequence, a plethora of distinct variants of camera-internal components, processing algorithms and combinations thereof lead to images that vary across a wide range of different model-specific image characteristics. The 16 images of a selected scene from the Dresden Image Database (Gloe and Böhme 2010) in Figure 9.1 give an indicative example. Each image was acquired with a different camera model, suggesting that immense differences are often already apparent from a comparative visual inspection.

Handbook of Digital Forensics of Multimedia Data and Devices, First Edition.
Edited by Anthony T.S. Ho and Shujun Li.
© 2015 John Wiley & Sons, Ltd. Published 2015 by John Wiley & Sons, Ltd.
Companion Website: www.wiley.com/go/digitalforensics

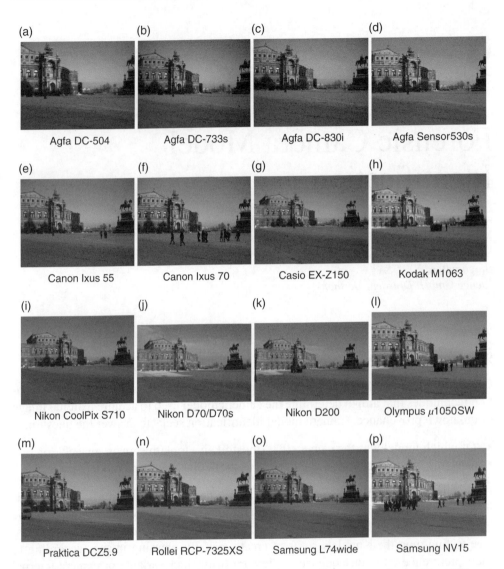

Figure 9.1 Variations across images of the same scene, each one acquired with a different camera model at the smallest available focal length, using the camera's default settings. The images are part of the Dresden Image Database (Gloe and Böhme 2010). The depicted scene is 'Theaterplatz : Semperoper'. (http://forensics.inf.tu-dresden.de/ddimgdb/locations)

This chapter gives a concise overview of the state of the art of technical means to achieve camera model identification. We put most emphasis on the general ideas, rather than on particular implementations of identification algorithms. The chapter starts with a brief introduction to forensic source identification in Section 9.2, leading up to the specific case of digital camera model identification in Section 9.3.

Section 9.4 comments on suitable image datasets for setting up practical algorithms, the foundations of which we discuss in Sections 9.5 and 9.6. Section 9.7 then points to problems that arise in identification scenarios with unknown camera models, before the penultimate Section 9.8 alludes to connections between camera model identification and device identification. Section 9.9 concludes the chapter with a special focus on practical challenges. Two accompanying case studies (in Sections 9.6.2 and 9.7.1) shed light on technical aspects of camera model identification under realistic settings.

9.2 Forensic Source Identification

Forensic source identification links multimedia content to a particular (class of) acquisition device(s). The key assumption of forensic source identification is that acquisition devices leave traces in the acquired content, and that instances of these traces are specific to the respective (class of) device(s). Source identification works by extracting *acquisition characteristics* from content of unknown provenance and comparing them to a reference database of known characteristics. Reference data is compiled from labelled content of potentially relevant acquisition devices in a training phase. This makes source identification a typical classification problem, with a class space defined to represent distinct instances of acquisition characteristics (Böhme and Kirchner 2013).

9.2.1 Identification Granularity

Depending on the specific case and the available reference information, forensic source identification may be approached at different levels of *identification granularity*. Figure 9.2 exemplarily illustrates for the analysis of digital images that we can distinguish between methods to determine the *type* of acquisition device, its *make* or *model*, and ultimately also the actual *device* itself. The vertically stacked instances

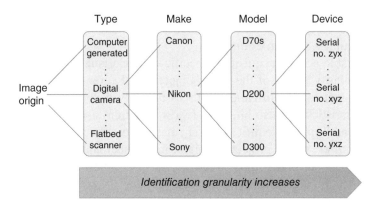

Figure 9.2 Granularity levels in forensic source identification, exemplarily illustrated for the analysis of digital images.

of (groups of) acquisition devices and/or acquisition methods allude to possible class spaces of each of the horizontally arranged granularity levels. Granularity increases from left to right: at the lowest level, we might be interested in separating between scanned and photographed material, whereas device-level identification aims to differentiate individual digital cameras (or scanners, etc.). Arbitrary intermediate levels are conceivable, for instance to distinguish consumer-grade point-and-shoot cameras from professional DSLR cameras.

Intuitively, forensic source identification should strive for the identification of the very acquisition device that was used to create the questioned content. Yet it is not always possible (or wanted in the first place) for practical reasons to work at this high level of granularity. Compiling instances of characteristics for a reference database requires controlled access to content acquired with the respective candidate devices. This is generally a resource-expensive and time-consuming task. Pre-classification with lower granularity can help to single out a relevant subset of acquisition devices for running more targeted identification algorithms in a subsequent stage. In real case-work we might also encounter situations where no additional content of the actual acquisition device is available. It is then useful to narrow down the source to a particular group of devices at least. To this end, it is usually sufficient to have access to other representatives of the same device class.

9.2.2 Intra- and Inter-Class Similarity, Feature Space Representation

Independent of the aspired or practically imposed level of identification granularity, forensically suitable acquisition characteristics ideally exhibit both a high *intra-class similarity* and a low *inter-class similarity*. This means that instances of respective characteristics should be highly similar for all content captured with the same (class of) acquisition device(s), but differ substantially across different (classes of) acquisition devices. Device-level source identification, for instance, will only work with *device* characteristics free of model-specific, 'non-unique' (Fridrich 2013), artefacts. Otherwise, arbitrary devices of the same model could be mistaken for the actual acquisition device (cf. Section 9.8 for an example).

Finding suitable sets of acquisition-specific characteristics in terms of inter- and intra-class similarity is one of the major objectives of forensic source identification research. Yet practical identification methods not only depend on the quality of the chosen characteristics but also hinge on (ways to obtain) accurate estimates thereof from the content. Typical identification schemes represent acquisition characteristics via a set of *features*, which project the high-dimensional and intractable input space to a much smaller and much more tractable subspace, the feature space. It is clear that an identification algorithm can only be as good as the set of features it employs. The better a feature space matches reality, the more confident we can base decisions on it. Figure 9.3 gives an example and illustrates how a number of content samples should typically distribute in a two-dimensional feature space for source identification

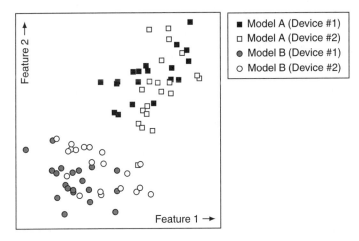

Figure 9.3 Exemplary distribution of content samples in an imaginary two-dimensional feature space for model identification. Features have a high intra- and a low inter-model similarity.

at the level of device models. Observe that content acquired with the same model clusters nicely (i.e. we observe a high intra-model similarity), while there exists no trivial decision criterion to distinguish between individual devices (i.e. the inter-model similarity is low).

9.2.3 Digital Camera Acquisition Characteristics

Good feature space representations naturally require a good model of the content acquisition process. To this end it is often useful to understand acquisition devices as a sequence of individual *components*, each of which may be specified by its input–output relation and a number of parameters.

The stylized image acquisition pipeline in Figure 9.4 gives an illustrative example and depicts the most relevant components of a digital camera (Holst and Lomheim 2007; Ramanath *et al.* 2005). Digital cameras focus the light of a scene on the *sensor* via a *system of lenses*. An interposed *optical filter* reduces undesired light components (e.g. infrared light). The sensor of most cameras comprises CCD or CMOS elements. Each element represents a pixel in the final image. Sensor elements output an electric charge proportional to the light intensity at their location on the sensor. The vast majority of designs employ a *colour filter array* (CFA) to obtain RGB colour images, so that individual sensor elements only record light of a certain range of wavelengths. Missing colour information is estimated from surrounding pixels of the raw image (possibly after pre-processing and white-balancing). This process is referred to as *CFA interpolation* or *demosaicing*. After demosaicing, the image is subject to a number of camera-internal *post-processing* steps, including for instance colour correction, edge

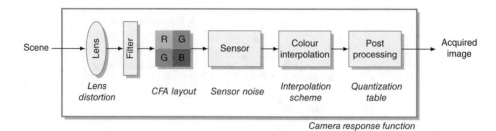

Figure 9.4 Stylized digital camera processing pipeline. Different camera components leave behind different characteristics in acquired images. Selected characteristics are printed in italics below respective camera components.

enhancement and compression, before it is eventually stored on the device, possibly in combination with metadata.

Both the very existence of certain components, as well as differences in the manufacturing, implementation and parametrization of components may result in acquisition-specific characteristics. Figure 9.4 assigns selected forensically relevant characteristics to components of our simplified camera pipeline. They are printed in italics below the respective component, where they originate. Most of the depicted components introduce model-, make- or type-specific acquisition characteristics. Different camera models employ different lens systems, yielding a variety of characteristic lens distortion artefacts in the respective images (Choi *et al.* 2006; Gloe *et al.* 2010, i. a.). CFA layout and interpolation scheme are usually model-specific (Cao and Kot 2009; Kirchner 2010; Swaminathan *et al.* 2007, i. a.), although the general presence of periodic inter-pixel dependencies due to the existence of a CFA may also be interpreted as a coarse indicator of digital camera images: computer-generated content or images obtained with a flatbed scanner will typically lack such characteristics (Gallagher and Chen 2008). Many camera models also employ customized post-processing steps and quantization tables (Deng *et al.* 2011; Kee *et al.* 2011, i. a.). The sensor stands out in this regard, as it is the most unique component of a digital camera. Variance in the light sensitivity of individual sensor elements results in camera-specific sensor noise 'fingerprints' (Fridrich 2013). Nevertheless, coarser representations of sensor characteristics, such as the general noise level, may also find applications at lower levels of identification granularity (Fang *et al.* 2009). A noteworthy characteristic at the lowest level is the use of different sensor types in digital cameras (array sensor) and flatbed scanners (line sensor), which causes periodically correlated sensor noise across multiple scan lines of scanned images (Caldelli *et al.* 2010).

Similar component-based variations can be found across different (classes of) digital video cameras and audio recording devices. Eventually, they allow forensic methods to infer the source of questioned multimedia content and to distinguish between different instantiations of acquisition processes. Yet the remainder of this chapter focuses on digital cameras, and in particular on *camera model identification*. Our emphasis on this type of acquisition devices reflects the state of the art in the forensic literature, which has largely been devoted to the examination of digital still images acquired with digital cameras. Nevertheless, many of the to-be-discussed concepts are equally applicable to other types of imaging devices and digital video cameras. Readers with interest in the examination of scanned images or audio recordings are also referred to subsequent chapters of this book.

9.3   Digital Camera Model Identification

Digital camera model identification exploits image characteristics that differ between images taken with different camera models but occur similarly in images taken by arbitrary devices of the same model. Within the general framework of forensic source identification, camera model identification is thus particularly appealing whenever

1. It is known that a questioned image was captured with a digital camera, and
2. A reference set of independent images, unequivocally shot with exactly the same camera, is unavailable.

The first point establishes the dividing line between camera model identification schemes and methods to identify computer-generated images (Dirik *et al.* 2007; Gallagher and Chen 2008; Lyu and Farid 2005; Ng *et al.* 2005), or methods to pre-classify scanner and digital camera images (Caldelli *et al.* 2010; Khanna *et al.* 2009). The second point emphasizes that camera identification at device level – typically based on sensor noise fingerprints (Fridrich 2013), possibly combined with additional clues from sensor dust (Dirik *et al.* 2008) – only works with reference data of the actual camera.[1]

On the contrary, for acquisition characteristics commensurate with the desired identification granularity, camera model identification requires only one device per model to obtain reference data. This is generally the case when instances of relevant characteristics exhibit a high intra-model similarity and a low inter-model similarity, cf. Section 9.2.2. We will discuss in detail in Section 9.5 how virtually all components of the camera processing pipeline leave behind suitable model-specific characteristics in the resulting images (cf. Figure 9.4). In Section 9.6, we will also see that it is possible to abstract from the component level and to understand camera model characteristics as properties of a black box instead.

Based on a suitable feature representation, $\mathcal{F}$, a camera model identification algorithm partitions the image space, $\mathcal{I}$, by assigning feature vectors, $\mathbf{f} \in \mathcal{F}$, to camera

---

[1] The absence of such reference data does not rule out measures to refute false hypotheses about the source device.

models, $m \in \mathcal{M}$. For a general discussion of camera model identification, we will assume that feature vectors are extracted via an *extraction function*, extract $: \mathcal{I} \rightarrow \mathcal{F}$. The partition function, assign $: \mathcal{F} \rightarrow \mathcal{M}$, is learned in the *training phase* from tuples of labelled training data, $(I^{(m)}, m) \in \{\mathcal{I}_{\text{train}} \times \mathcal{M}_{\text{train}}\} \subseteq \{\mathcal{I} \times \mathcal{M}\}$, where the notation $I^{(m)}$ indicates that image I was acquired with camera model $m \in \mathcal{M}$. A standard learning approach is to minimize the probability of mis-classification, $\text{Prob}(\text{assign}(\text{extract}(I^{(m)})) \neq m)$, or some related measure, over the *training set*, $\mathcal{I}_{\text{train}}$. Most forensic algorithms nowadays employ a support vector machine (SVM, Cortes and Vapnik 1995) for this purpose, one of the most widely used general-purpose classification algorithms in the domain of machine learning. Training data should generally comprise a reasonably large number of representative samples. Too homogeneous training data is likely to result in a partition function that does not generalize well. Before practical use, any camera model identification algorithm should additionally be tested against validation data, $\{\mathcal{I}_{\text{val}} \times \mathcal{M}_{\text{train}}\}$, and possibly refined based thereon, to rule out a mismatch with the real world. After training and validation, function assign can be used for practical camera model identification in the *test phase* by predicting labels from images $I^{(m)} \in \mathcal{I}_{\text{test}}$, $m \in \mathcal{M}_{\text{test}}$, of unknown provenance.

In the above general setting, we can distinguish between two different scenarios (Alles *et al.* 2009; Gloe 2012a). Source identification in a *closed set* of camera models assumes that all candidate models are known in the training phase, that is, $\mathcal{M}_{\text{test}} \subseteq \mathcal{M}_{\text{train}}$. Here, we assign an image of unknown provenance to the camera model that is most likely given our prior knowledge. While this scenario is appealing for testing and benchmarking camera model identification algorithms in a controlled research setting, it is decidedly more difficult in practical case work to consider *all* potential camera models that could have been used for image acquisition. It is reasonable to assume that a real-world reference database will never be comprehensive. As a consequence, unknown models in an *open set* of camera models may cause false accusations, and the codomain of function assign should be extended by a special symbol $\perp$ for unknown models, assign $: \mathcal{F} \rightarrow \{\mathcal{M}_{\text{train}} \cup \perp\}$.

We close this section with the remark that a considerable number of camera model identification schemes do not explicitly define the class space based on camera models, but rather use particular model-specific components of the camera acquisition pipeline as proxies. In practice, such methods – often subsumed under the term *component forensics* (Swaminathan *et al.* 2009) – thus assign component instances to digital images. Depending on the addressed acquisition characteristics this may imply that a number of camera models share the very same component instantiation. We call such camera models *equivalent* with respect to a acquisition characteristic (or feature representation). It is then useful to redefine the aforementioned partition function so as to map to *equivalence sets* of camera models, assign $: \mathcal{F} \rightarrow \mathcal{M}^{+}$, $\mathcal{M}^{+} = \bigcup_{n=1}^{|\mathcal{M}|} \mathcal{M}^{n}$, where operator $^{+}$ denotes the 'Kleene plus'.

9.4   Benchmarking Camera Model Identification Algorithms

Digital camera model identification algorithms need to be developed and benchmarked in a controlled research environment before they can be applied to real-world situations. The underlying evaluation process hinges on a suitable reference and benchmark image dataset. A thoroughly designed dataset is a necessary condition for a sound assessment of an algorithm's strengths and weaknesses. First and foremost, this concerns efforts to reduce the general effects of dataset bias (Torralba and Efros 2011) to a reasonable degree by employing a representative and heterogenous image sample of large size. Yet there also exist a number of peculiarities that are more specific to the domain of camera model identification, as we will discuss in the following text.

9.4.1   A Dataset Template for Camera Model Identification Research

Camera model identification requires features that capture characteristics of the camera model rather than of the device or of the image content. A good database thus comprises

1. Images $I \in \mathcal{I}$ of similar scenes, $s \in \mathcal{S}$, where
2. Each scene is captured by multiple camera models, $m \in \mathcal{M}$, and
3. Multiple devices $d \in \mathcal{D}^{(m)}$ per model m.

Here, set $\mathcal{S}$ denotes the infinite set of all conceivable scenes, and set $\mathcal{D}^{(m)}$ represents all devices of a particular camera model $m \in \mathcal{M}$. We will further use the convenience notation $I^{(d(s))}$ to indicate that image I was captured by device $d \in \mathcal{D}$ and depicts scene $s \in \mathcal{S}$.

Capturing images with multiple devices per model is necessary to account for realistic scenarios. In particular, we must expect that images under investigation were acquired with a device that we cannot access for training. Hence, it is reasonable to partition the set of available devices into a subset for training and validation, $\mathcal{D}_{\text{train}}$, and a disjoint subset for testing, $\mathcal{D}_{\text{test}}$. The respective sets are given by

$$\mathcal{D}_{\text{train}} \subseteq \bigcup_{m \in \mathcal{M}_{\text{train}}} \mathcal{D}^{(m)}, \quad \mathcal{D}_{\text{test}} \subseteq \bigcup_{m \in \mathcal{M}_{\text{test}}} \mathcal{D}^{(m)}, \quad \text{and} \quad \mathcal{D}_{\text{train}} \cap \mathcal{D}_{\text{test}} = \emptyset. \qquad (9.1)$$

The set of available scenes in the dataset should be organized similar to the set of available devices. In practical settings, it is unlikely that the depicted scene in an image under investigation can be recaptured for the sake of training and validation under exactly the same environmental and lighting conditions. In addition, test images of similar scenes across all camera models guarantee a high comparability of benchmark results, relatively independent of the image content. Ideal datasets are thus composed from images of a large number of scenes, from which we derive three mutually disjoint sets for training, validation, and testing, $\mathcal{S}_{\text{train}}$, $\mathcal{S}_{\text{val}}$ and $\mathcal{S}_{\text{test}}$, respectively. In summary, the above partition of devices and scenes results in the following image datasets for benchmarking:

$$\mathcal{I}_{\text{train}} = \left\{ I^{(d(s))} \mid d \in \mathcal{D}_{\text{train}} \land s \in \mathcal{S}_{\text{train}} \right\} \tag{9.2}$$

$$\mathcal{I}_{\text{val}} = \left\{ I^{(d(s))} \mid d \in \mathcal{D}_{\text{train}} \land s \in \mathcal{S}_{\text{val}} \right\} \tag{9.3}$$

$$\mathcal{I}_{\text{test}} = \left\{ I^{(d(s))} \mid d \in \mathcal{D}_{\text{test}} \land s \in \mathcal{S}_{\text{test}} \right\} \tag{9.4}$$

9.4.2   The Dresden Image Database

Creating a set of images based on the aforementioned recommendations is a vastly time-consuming process, which also requires physical access to a large number of different devices in the first place. At the same time, custom-made datasets make the reproduction and the assessment of published experimental results through independent researchers cumbersome and error-prone (Garfinkel *et al.* 2009; Vandewalle *et al.* 2009). Images from public photo sharing platforms are useful to a limited extent only. Neither can we control for acquisition conditions, nor can we ensure the images' authenticity in a straight-forward way.

The *Dresden Image Database* is a notable publicly available alternative,[2] compiled for the very purpose of camera model identification research (Gloe and Böhme 2010). It was designed with great care to ensure that it is most useful to researchers and forensic investigators alike. All images were captured under controlled conditions following a specific acquisition protocol. To provide images of similar contents across different models and devices, a tripod was fixed for each *motif*. A systematic variation of the cameras' focal length settings yielded at least three *scenes* per device and motif. Camera model, device identifier, camera settings and the acquired motif are stored along with each image to support analyses under specific constraints. The database currently holds approximately 17,000 full-resolution natural images stored in the JPEG format. It covers 27 different camera models, with up to 5 devices per model, and it includes both typical consumer digital camera models and semi-professional digital SLR cameras. An addition of 1500 uncompressed raw images was stored simultaneously with the JPEG images from cameras with raw-support.

The Dresden Image Database meets most of the requirements from Section 9.4.1 by design. A noteworthy (but minor) limitation is that the set of camera models had to be split into two disjoint sets for logistical reasons. Each set covers different motifs. Also, the overall number of available motifs (currently 81) may still be to small for machine learning in very high-dimensional feature spaces. Yet the modular design of the database also encourages the integration of new image data, so that we foresee a continuous increase of the dataset's value with future releases.

---

[2] The database is available online at http://forensics.inf.tu-dresden.de/ddimgdb.

9.4.3 Benchmarking Procedure

With a suitable image dataset at hand, camera model identification algorithms can be benchmarked by 'hiding' the labels $m \in \mathcal{M}_{\text{test}}$ from the trained classifier in the testing phase and then checking the results against the known labels. Note that this is in contrast to practical case work, where we cannot assume any knowledge about the camera model of a questioned image. The performance of an algorithm is often measured and reported in terms of its *classification accuracy*, that is, the percentage of correctly labelled images across all tested images. Accuracy may be reported per camera model or aggregated over all tested camera models. It is generally advised to randomly re-partition the dataset multiple times in a *cross-validation* procedure, and to report average accuracies over all partitions. This guarantees that the partition function assign is trained from and tested against different subsets each time, and thus mitigates effects of dataset bias to some degree. We finally note that the most realistic (and challenging) scenario sets $|\mathcal{D}_{\text{train}}| = |\mathcal{M}_{\text{train}}|$, that is, camera model characteristics are trained from only one device per model.

9.5 Model-Specific Characteristics of Digital Camera Components

Equipped with the basic notions and prerequisites for camera model identification, we can now turn to specific camera characteristics that have found use in the literature. The digital camera acquisition pipeline in Figure 9.4 will serve as a blueprint. All components therein may leave behind model-specific image characteristics. Forensic algorithms strive to capture them, often in an attempt to model and to infer component parameters. The following subsections focus on individual components, before Section 9.6 discusses camera model identification from a black box perspective.

9.5.1 Compression Parameters, Metadata and File Format

The entry point to the forensic examination of a digital image is naturally the *file* it is stored in. In the vast majority of cases we will deal with compressed images stored in the JPEG format (ITU 1992). JPEG files store a number of mandatory parameters with the file, most importantly *quantization tables* and *Huffman tables* for decompression. Virtually all digital cameras also attach a rich collection of *metadata* to the file to document details about the image acquisition process. The preferred method to organize and store such metadata is specified in the EXIF standard (Japan Electronics and Information Technology Industries Association 2002). EXIF data may contain information about the acquisition device, the acquisition time, camera settings, or GPS coordinates, amongst many others. Digital cameras may further store one or multiple *preview images* for fast image display along with the full resolution image.

While the JPEG and EXIF standards define frameworks how image data and metadata ought to be stored in a structured way, they leave room for interpretation.

Naturally, digital images assume different resolutions and aspect ratios, depending on the format of the sensor. Digital cameras can use customized quantization and Huffman tables based on the camera vendors' preferences. EXIF data may contain a plethora of different entries, or adopt specific naming conventions. Size and compression settings of the preview image(s) are not standardized at all (and often unrelated to the full resolution image, as Kee *et al.* (2011) pointed out).

It is not surprising that all these degrees of freedom have led to a wide range of model-specific instances in the wild. Farid (2008) found that quantization tables alone are already a good indicator of the digital camera model. Many *software* vendors employ their very own unique tables. A combination of quantization settings, image and thumbnail dimensions, and the number of EXIF entries yields an even more distinctive signature of camera configurations (Kee *et al.* 2011). At the *core file level*, also the order of so-called JPEG marker segments and EXIF data structures varies to considerable extent between different camera models (Gloe 2012b). Because standard image processing software is very likely to modify compression settings, EXIF data and file internals, feature vectors based thereon are particularly useful to verify that an image has not been manipulated or resaved after it was acquired with a specific camera model (Gloe 2012b; Kee *et al.* 2011). JPEG quantization tables and peculiarities of the file structure may also narrow down the origin of an image when no EXIF data is present, or when EXIF data is considered unreliable.[3] We further refer to Chapter 15 for a discussion on how quantization tables from previous compression steps can be estimated from double-compressed images (Pevný and Fridrich 2008, i. a.), or from bitmap images that had been JPEG compressed before (Neelamani *et al.* 2006, i. a.).

An in-depth discussion of all these characteristics is beyond the scope of this chapter. We refer interested readers to other chapters of Part IV in this book for a detailed treatment of file format and metadata forensics.

9.5.2 Lens Distortion

Each digital camera is equipped with a complex optical system to project the scene to the much smaller sensor. Such projections are in general not perfect, and plenty of inherent *lens distortion* artefacts – so-called optical aberrations – can be found in digital images. Different camera models naturally employ different optical systems, all of which have their own individual distortion profile. Some of the most relevant aberrations are *radial lens distortion*, *chromatic aberrations* and *vignetting*.[4] The camera model identification literature exploits these lens distortion artefacts preferably by means of parametric distortion models, that is respective model parameters define the feature space. Parameters of interest are then estimated from an image of unknown provenance, which is assigned to a camera model – or, more precisely, to a lens model – based on a pre-compiled reference set of characteristic parameter vectors.

---

[3] Otherwise we would readily extract the camera model from the corresponding metadata entry.

[4] We refer to Chapter 14 for a discussion of further types of lens distortion.

Before we briefly discuss specific lens distortion artefacts below, we note that most aberration models assume radial symmetry with respect to the optical axis of the lens. Hence, knowledge of the *optical center*, that is the intersection of the optical axis and the sensor plane, is crucial. Because the optical center of an image does not necessarily align with the geometric center (Willson and Shafer 1994), it generally needs to be considered to be part of the parameter estimation problem.

9.5.2.1 Radial Lens Distortion

Radial lens distortion is a non-linear geometrical aberration that lets straight lines in the real world appear curved in an image. It is caused by the lens' inability to magnify all parts of a scene with the same constant factor. The specific shape and strength of the distortion depends on the focal length and the concrete lens(es) in use. Figure 9.5 illustrates the two general forms, barrel distortion and pincushion distortion. Choi *et al.* (2006) adopted a standard polynomial model to relate undistorted coordinates $(\hat{x}, \hat{y})$ to distorted image coordinates (x, y) at radius $r = \sqrt{(x - x_o)^2 + (y - y_o)^2}$ from the optical center (x_o, y_o),

$$\hat{x} = (x - x_o)\left(1 + \kappa_1 r^2 + \kappa_2 r^4\right) + x_o ,$$
$$\hat{y} = (y - y_o)\left(1 + \kappa_1 r^2 + \kappa_2 r^4\right) + y_o . \tag{9.5}$$

Parameters (κ_1, κ_2) are understood as a feature vector for camera model identification. Following Devernay and Faugeras (2001), the parameters can be estimated from an image by minimizing the total error between curved and straight lines. Choi *et al.* (2006) obtained a 89% classification accuracy for a setup with five different camera models, operated at fixed focal lengths, and a support vector machine classifier. The same authors note that classification over a range of focal lengths is more problematic. This applies in particular to mid-range focal lengths, where distortion artefacts practically vanish (Fischer and Gloe 2013).

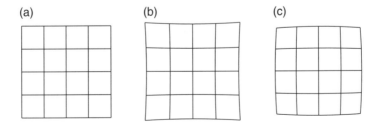

Figure 9.5 Radial lens distortion lets straight lines appear curved in the image. (a) An ideal rectangular grid, and the same grid subject to (b) barrel distortion and (c) pincushion distortion, respectively.

9.5.2.2 Chromatic Aberrations

Chromatic aberrations occur because a lens' dispersion index varies for different wavelengths. By Snell's law, this causes polychromatic light rays to be spread over the sensor plane, as Figure 9.6 depicts. The resulting displacements become visible as colour fringes in high-contrast areas. Axial chromatic aberration refers to wavelength-dependent longitudinal variations of the focal point along the optical axis, whereas lateral chromatic aberration explains off-axis displacements of different light components relative to each other. Johnson and Farid (2006) modelled the latter as linear expansion/contraction of the red and blue colour channel coordinates $(x^{(c)}, y^{(c)})$, $c \in \{R, B\}$, relative to the green channel coordinates $(x^{(G)}, y^{(G)})$,

$$x^{(c)} = \alpha^{(c)} \left(x^{(G)} - x_o^{(c)} \right) + x_o^{(c)} \,,$$
$$y^{(c)} = \alpha^{(c)} \left(y^{(G)} - y_o^{(c)} \right) + y_o^{(c)} \,, \tag{9.6}$$

where coordinate $(x_o^{(c)}, y_o^{(c)})$ denotes the optical centre. The parameters can be found by maximizing a global similarity criterion between each pair of colour channels (Johnson and Farid 2006), or via a more efficient block-based variant (Gloe *et al.* 2010). Van *et al.* (2007) constructed a six-dimensional feature vector from the two optimal parameter sets and fed it into a support vector machine for camera model identification. The authors reported a 92% classification accuracy over a small set of three cellphone camera models. However, more recent studies indicate a complex interaction between measurements of chromatic aberration and the lens' focal length and focal distance (Gloe *et al.* 2010; Yu *et al.* 2011). Observations of non-linear behaviour (Gloe *et al.* 2010) and a considerable intra-model difference between some lenses of the same type (Yu *et al.* 2011) further emphasize that chromatic aberrations are currently not fully understood and thus still subject to ongoing research efforts.

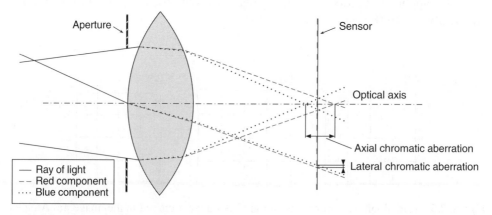

Figure 9.6 Formation of chromatic aberrations due to varying dispersion indices for light components of different wavelengths.

9.5.2.3   Vignetting

Vignetting is an aberration that describes the light intensity fall-off towards the corners of an image. It appears strongest for wide apertures, when not all light rays reach the sensor's outer parts. Figure 9.7 gives illustrative examples for an image of a homogeneously lit scene and a natural image. Lyu (2010) employed a parametric vignetting model of the form

$$I(x,y) = v(x,y,\boldsymbol{\theta})\hat{I}(x,y)\,, \tag{9.7}$$

to characterize the relation between the undistorted image $\hat{I}$ and the distorted version I in terms of a vignetting function $v(\cdot,\boldsymbol{\theta})$ with parameters $\boldsymbol{\theta}$. The vignetting function is defined to combine an illumination factor and a geometric factor. Lyu generalized the standard model by Kang and Weiss (2000) to account for the intrinsic intensity fall-off as a function of the distance from the optical centre and for non-circular vignetting artefacts, respectively. The 10-dimensional parameter vector $\boldsymbol{\theta}$ is estimated from image derivatives based on a generalized Gaussian distribution assumption. Camera model identification results from natural images are less promising than for other lens distortion artefacts. Using a nearest neighbour classifier, Lyu reported a 57% classification accuracy from an experiment with eight different lenses. By far more reliable estimates can be obtained from synthetically vignetted images (Lyu 2010) or from ideal test scenes (Fischer and Gloe 2013, see also Figure 9.7). This indicates that the low accuracy under more realistic conditions is largely caused by problems with measuring rather weak vignetting artefacts from textured image regions.

(a)   (b)

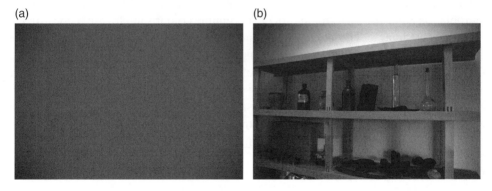

Figure 9.7   Vignetting artefacts in an image of a homogeneously lit scene (a) and a natural image (b). The images (camera: Nikon D70) were subject to non-linear colour correction for better visibility. Note the intensity fall-off towards the corners of the images. Both images are part of the Dresden Image Database (Gloe and Böhme 2010). The scene depicted on the right is 'Home V: Wall Unit'. (http://forensics.inf.tu-dresden.de/ddimgdb/locations)

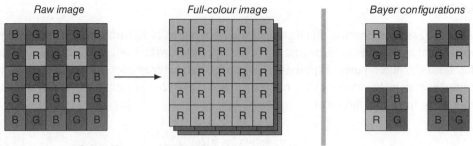

Colour filter array interpolation

Figure 9.8 Digital cameras with a colour filter array (often one of the four Bayer configurations shown on the right) acquire a full-colour images by interpolating missing colour information from surrounding pixels of the raw image.

9.5.3 CFA and Demosaicing

Most digital camera models combine their sensor with a CFA, such that individual sensor elements measure specific colour information only. This is necessary because a CCD/CMOS sensor element itself is merely sensitive to the received light intensity, not to a particular colour. Colour filters split up the light into corresponding colour components. A full-colour image is then obtained by interpolating missing colour information from surrounding samples of the raw signal in a so-called *demosaicing* procedure.

Both the *CFA configuration* and the *CFA interpolation algorithm* may vary between camera models. The former defines the specific arrangement of colour filters across the sensor plane. Although theory does not restrict the CFA to RGB components or to a particular layout (Menon and Calvagno 2011), it is often one of the four 2 × 2 Bayer patterns (Bayer 1976) that is repeated over the sensor plane, cf. Figure 9.8. The CFA interpolation algorithm determines how non-genuine colour samples are reconstructed from the raw image. The demosaicing literature is rich of proposals, making the CFA interpolation algorithm one of the most important characteristics of a camera model. Algorithms may differ in the size of their support, their way of incorporating neighbouring samples, or their adaptiveness to the image content (Gunturk *et al.* 2005; Menon and Calvagno 2011). Yet all demosaicing procedures have in common that at least two thirds of all pixels in an RGB image are interpolated,[5] introducing strong interdependencies between neighbouring pixels.

9.5.3.1 Inter-pixel Correlations

Forensic techniques have focused on modelling and measuring these inter-pixel correlations, because their specific appearance is largely a product of the camera

---

[5] We say 'at least' because demosaicing procedures may also 're-compute' samples of the raw signal.

model's CFA configuration and interpolation algorithm. Most published works assume that colour image pixels $I(x, y, c)$ at position (x, y) in colour channel $c \in \{R, G, B\}$ are linearly correlated with their neighbours,

$$I(x, y, c) = \sum_{(i,j) \in \Omega} \sum_{c' \in \{R,G,B\}} w(i, j, c' \mid x, y, c, I) \cdot I(x + i, y + j, c') + \varepsilon(x, y, c). \qquad (9.8)$$

The notion of 'neighbour' does not necessarily only refer to a suitable spatial neighbourhood, Ω, but may also include pixels in other colour channels. Weights w reflect the characteristic dependencies among pixels accordingly. The weights may vary conditional on a pixel's position and colour channel, and also depending on image characteristics in the pixel's neighbourhood. A standard approach to camera model identification is to estimate those weights from the image and to understand them as demosaicing-specific feature vector. Residual ε denotes the approximation error and may serve as an additional feature pool.

Popescu and Farid (2005) first applied this model and estimated global weights from a small intra-channel neighbourhood around each pixel, that is $w(i, j, c' \mid x, y, c, I) = w(i, j, c' \mid c)$ and $w(i, j, c' \mid c' \neq c) = 0$. The weights are found in an expectation-maximization (EM) procedure by minimizing the approximation error per colour channel.[6] Although their main focus was image manipulation detection, Popescu and Farid also observed that the estimated weights differ between demosaicing algorithms. Bayram *et al.* (2005) combined the estimated weights with frequency domain features to reflect periodic artefacts in the approximation error, caused by the periodic structure of CFA patterns. They reported a 83% classification accuracy from 5×5 neighbourhoods in a small series of tests with three camera models and a support vector machine (SVM) classifier. The results improved to 96% accuracy when smooth and textured image regions were handled differently (Bayram *et al.* 2006). The rationale behind this distinction is that demosaicing algorithms mostly differ in their treatment of high-contrast neighbourhoods. Also Long and Huang (2006) worked in the framework of global intra-channel correlations, but did not explicitly estimate the weights w from the image. Instead they proposed to compute a normalized correlation matrix over all local neighbourhoods Ω and to feed it into a neural network classifier. The authors obtained similarly high classification accuracy, yet again only for a rather small set of four camera models.

Swaminathan *et al.* (2007) deviated from the global correlation model and estimate intra-channel weights for interpolated pixels only. They took the underlying CFA C, $C(x, y) \in \{R, G, B\}$, into account by defining pixel neighbourhoods where only genuine raw pixels contribute to the sum in Equation 9.8, i.e., $w(i, j, c \mid C(x + i, y +$

---

[6] The EM algorithm is a widely used iterative maximum likelihood method to obtain parameter estimates in the presence of latent variables (Moon 1996). Popescu and Farid (2005) model Equation 9.8 as a mixture of correlated and uncorrelated pixels and consider the fact whether a pixel belongs to the first or the second group as latent variable.

$j) \neq c) = 0$. Assuming that these pixels are relatively invariant to the demosaicing procedure, the estimated weights reflect the CFA interpolation algorithm more directly than a global estimate from all pixels of a colour channel. Swaminathan *et al.* (2007) proposed to determine the weights in a total least squares procedure by minimizing the approximation error over all plausible CFA configurations. Similar to Bayram *et al.* (2006), they also distinguished between different image regions. Three independent estimation procedures for neighbourhoods with strong horizontal gradients, strong vertical gradients, and overly smooth areas lead to overall 147 weights per colour channel. The estimated $3 \times 147 = 441$ weights form a feature vector and are fed into a probabilistic SVM (Wu *et al.* 2004). A test with 19 camera models yielded a 86% classification accuracy. Mis-classifications were largely confined to disjunct camera subsets of the same make, indicating that camera vendors re-use processing units across different models.

Cao and Kot (2009) extended the work by Swaminathan *et al.* in two directions. Their state-of-the-art method transferred Equation 9.8 into the domain of partial derivatives and, for the first time, also considered inter-channel correlations. Estimating correlations from derivatives was shown to be equivalent to the original formulation, while at the same time attenuating the impact of image content on the estimation procedure. By explicitly allowing inter-channel correlations, Cao and Kot reflected that many demosaicing algorithms reconstruct samples from raw samples across all three colour channels. Instead of grouping pixels based on surrounding image content, the authors proposed 16 different pixel categories based on the pixels' positions in the CFA. Corresponding weights were again estimated by minimizing the approximation error per category. Cao and Kot did not attempt to determine the correct CFA configuration but ran their procedure for all four plausible Bayer patterns instead. The result is a 1536-dimensional feature vector, from which the 250 most relevant features were found with sequential floating forward selection (SFFS, Pudil *et al.* 1994). A probabilistic SVM achieved a 98% accuracy over a set of 14 digital camera models. Follow-up studies have supported the promising results and reported comparable accuracies for the identification of mobile phone camera models and camera models with similar processing units (Cao and Kot 2010, 2011).

9.5.3.2 Other Approaches

A number of alternative research streams have focused on CFA characteristics beyond the estimation of inter-pixel correlations. Ho *et al.* (2010) measured inter-channel correlations via spectra of green/red and green/blue colour channel differences, $I(x, y, \mathsf{G}) - I(x, y, \mathsf{R})$ and $I(x, y, \mathsf{G}) - I(x, y, \mathsf{B})$, respectively. Their motivation was that many demosaicing algorithms are based on a smooth hue assumption and thus interpolate colour differences instead of colour channels directly (Gunturk *et al.* 2005). Gao *et al.* (2011) drew on the same observation and computed variances of colour channel differences. They combined these features with a work by Gallagher and Chen

(2008), who observed differences in the variance of high-pass filtered pixels of CFA interpolated images depending on the pixels' positions' in the CFA.[7] Specifically, Gao *et al.* obtained the residual variances from raw and interpolated sites, iterating over all plausible CFA configurations. Assuming that interpolated samples exhibit a lower variance, the ratios of variances from presumably raw and interpolated samples are a good indicator of demosaicing characteristics. Overall 69 features led to a 99% SVM classification accuracy over seven camera models from the Dresden Image Database.

9.5.3.3 CFA Configuration

Most of the aforementioned methods examine demosaicing artefacts by making assumptions about the specific configuration of the CFA. It is often necessary to distinguish between interpolated and raw pixels. In the absence of concrete knowledge about a camera's CFA, a standard approach is to compute features for all plausible configurations (Cao and Kot 2009; Gao *et al.* 2011). Other works jointly estimate the CFA configuration along with correlation weights by minimizing the approximation error (Swaminathan *et al.* 2007).

For some applications, it may likewise be helpful to determine the CFA configuration first, before conducting further analyses. More generally, the CFA configuration is itself a model-specific characteristic, which can serve as a means to decrease the degrees of freedom in forensic source identification. Takamatsu *et al.* (2010), for instance, used ratios of site-specific variances to determine the most likely Bayer configuration before they estimated correlation weights. The correct Bayer configuration can also be inferred based on a CFA synthesis approach by approximating the raw signal from the CFA interpolated image (Kirchner 2010; Kirchner and Böhme 2009), or via the examination of order statistics over 2×2 pixel blocks (Choi *et al.* 2011). Determining the position of green Bayer pattern elements is typically more reliable because of their diagonal layout.

9.5.4 Camera Response Function

Different from the component-driven approaches to camera model identification we have discussed so far, it is also possible to understand the whole digital camera as one single system that maps irradiance, R, to pixel intensities, I. The mapping is assumed to follow a specific *camera-response function* (CRF), whose parameters can be estimated from questioned images. The principal camera model identification approach is then relatively closely related to the component-based ones above. Different camera models are expected to have different camera response functions, which are usually modelled

---

[7] A linearly high-pass filtered image is equivalent to the approximation error in Equation 9.8, when the weights are replaced with appropriate filter coefficients.

to be monotonically increasing and of non-linear nature. Yet CRF estimation is generally a highly non-trivial problem in the absence of reliable ground-truth data (Mitsunaga and Nayar 1999). A standard parametric CRF model with applications in forensic settings assumes a generalized gamma curve of order K (Ng *et al.* 2007),

$$I = R^{\sum_{k=0}^{K} \alpha_k R^k} . \qquad (9.9)$$

More advanced versions extend this model and include a linear term for low irradiance (Ng 2009; Ng and Tsui 2009). The parameter vector, $\alpha = (\alpha_0, \ldots, \alpha_K)$, can be estimated from edge profiles that correspond to locally linear irradiance surfaces. Experimental results by Ng (2009) suggested a promising estimation accuracy and indicated differently shaped curves for a small set of four camera models. However, most practical forensic applications of CRF estimation have focused on the detection of inconsistencies due to image manipulations (Hsu and Chang 2010; Lin *et al.* 2005). A large-scale study on CRF distinctiveness across different camera models has yet to be conducted.

9.5.5 Summary and Limitations

This section has illustrated how characteristics of virtually all components of a digital camera acquisition pipeline can be turned into ingredients for model identification schemes.

Lens distortion artefacts are one such source of acquisition characteristics. A common challenge of forensic methods based thereon is the relatively large parameter space, which has not been comprehensively explored (for forensic applications) so far. All optical aberrations strongly depend on a variety of lens settings, most prominently focal length, focal distance and aperture. Also camera orientation, image stabilization and efforts to correct for lens distortion artefacts play a role. The most promising results were thus obtained for rather small sets of camera models with fixed lens settings. Prospects of large-scale applications have yet to be determined. Along these lines, it will also be important to study interdependencies between different types of lens distortion. While the current literature is mostly confined to isolate specific distortion parameters, a more holistic approach may better account for the fact that all aberrations are always intertwined with each other (Fischer and Gloe 2013; Yu *et al.* 2011).

Algorithms that address the colour reproduction of a scene by means of the *camera response function* take a more holistic perspective by design. Future research will have to show whether the parameter space is tractable in forensic applications that go beyond camera calibration under controlled settings. Yet it is worth noting that methods with focus on more specific aspects of colour reproduction, for instance the auto white balancing, have already yielded very promising results (Deng *et al.* 2011).

Model-specific CFA and *demosaicing* characteristics are generally better understood. Although the premise of linear inter-pixel correlations may not hold under all circumstances, it has proven a robust working assumption for many algorithms. Reported classification accuracies are among the highest in the literature on camera model identification, even after a good amount of JPEG compression. However, the relatively small number of distinct instances of CFA layouts and interpolation algorithms may turn into a disadvantage in practical large-scale investigations that aim to distinguish actual camera models. Cao and Kot (2009) bypassed this aspect to some degree by computing a high-dimensional feature set over *all* possible CFA configurations. Their features thus very likely also capture a variety of characteristics that are not solely determined by CFA interpolation, in particular image noise, sharpness and texture. We will discuss in the next section how the examination of such more general characteristics can also be approached from a black box perspective, without restricting assumptions about specific components in the image acquisition pipeline.

9.6 Black Box Camera Model Identification

Although many model-specific image characteristics can be attributed to particular components of the image acquisition pipeline in Figure 9.4, the general appearance of an image is equally the result of these components' complex interplay. Strength and characteristics of image noise, for instance, are not only inherently affected by properties of the camera's sensor, but also depend on the demosaicing algorithm and internal post-processing procedures. Just in a very similar way, characteristics of colour reproduction, saturation, sharpness and other determinants of what is often subsumed under the broad term of image quality cannot easily be traced back to a single component of the acquisition pipeline.

The literature on camera model identification exploits such characteristics from a *black box* perspective, without explicitly modelling the relation between input and output of the camera (components). Instead, forensic algorithms here use *general-purpose image descriptors* so as to cover a wide range of different image characteristics. We call them 'general-purpose' because most of these features have their origin in other domains of image analysis and classification. Typical examples are texture classification, visual quality assessment or steganalysis. Their application to camera model identification is driven by the belief that they provide a good feature space representation of digital images under a variety of aspects that are at the same time inherently relevant to distinguish between particularities of different image sources.

In retrospective, it is interesting to note that the first published work on forensic camera model identification used such general-purpose features. In their seminal paper, Kharrazi *et al.* (2004) fed a combination of colour features, image quality metrics and wavelet statistics into a support vector machine to distinguish between a small set of five camera models. For an examination of 16 cellphone camera models, Çeliktutan *et al.* (2008) later added binary similarity measures to the pool of features. More recent works derived features from local binary patterns (Xu and Shi 2012) and DCT-coefficient transition probabilities (Gao *et al.* 2012; Xu *et al.* 2009).

9.6.1   General-Purpose Image Descriptors

This section gives a brief overview of relevant feature types, before we apply some of them in a large-scale case study based on the Dresden Image Database in Section 9.6.2.

9.6.1.1   Colour Features

The colour reproduction of a scene largely depends on physical properties of a camera's CFA and colour correction algorithm. Kharrazi *et al.* (2004) proposed a set of colour features, $\mathcal{F}_{col}$, in an attempt to capture these differences in the RGB colour space. The 12-dimensional feature vector characterizes the shape of individual colour channel histograms, and also considers inter-channel correlation and energy ratios, respectively. The examination of interrelations between colour channels here not only addresses effects of CFA interpolation (cf. Section 9.5.3), but is also thought to account for the so-called grey-world assumption. Many colour correction algorithm inherently draw on this assumption, which states that the average intensities of all colour channels should equal (Ramanath *et al.* 2005). A more recent extension of the original colour feature set reflects this more explicitly by adding the cumulated absolute intensity difference across all pairs of colour channels and further inter-channel features inspired by grey-world colour correction coefficients (Gloe *et al.* 2009).

9.6.1.2   Image Quality Metrics

While colour features focus on the colour reproduction of a scene, also sharpness and noise characteristics determine the visual appearance of an image. Avcibaş *et al.* (2002) proposed so-called image quality metrics (IQM) to measure these aspects of scene reproduction, which are mainly a result of different optical systems, sensor properties and internal post-processing pipelines. The basic idea of IQMs is to examine differences between an image and a denoised version, typically obtained through filtering with a Gaussian kernel. Overall 13 measures can be broadly classified into measures based on pixel differences, correlation-based measures, and measures based on spectral distances. Kharrazi *et al.* (2004) computed these features, $\mathcal{F}_{iqm}$, as average over all three colour channels, whereas Çeliktutan *et al.* (2008) considered each colour channel individually.

9.6.1.3   Wavelet Statistics

A set of more general features is based on the wavelet decomposition, which is well known for its ability to capture image and noise characteristics across different spatial positions, scales, and directions (Mallat 1989). Farid and Lyu (2003) first used statistical moments of high-frequency detail subband wavelet coefficients in a forensic context. While their focus was largely the distinction between computer-generated and natural images, Kharrazi *et al.* (2004) adopted a similar set of wavelet features,

$\mathcal{F}_{wav}$, to capture basic model-specific noise characteristics of an image in the wavelet domain. The nine-dimensional feature vector contains the average values of all detail sub-bands, obtained from one-level wavelet decompositions of all colour channels. Çeliktutan *et al.* (2008) increased the decomposition level to three and further extended this set by additional first-order statistics (standard deviation, skewness and kurtosis). A 36-dimensional grey scale version is obtained by averaging the features over all three colour channels. Following Farid and Lyu, Çeliktutan *et al.* added another 36 features to reflect the strength of inter-subband correlations. These features are obtained by predicting subband coefficients from their spatial, orientation and scale neighbourhoods. A 27-dimensional variant considers only mean, standard deviation and skewness of all three one-level decomposition detail sub-bands from all three colour channels (Gloe 2012a).

9.6.1.4 Local Binary Patterns

Local binary patterns (Ojala *et al.* 1996, 2002) are another relatively general representation of a variety of image characteristics. Primarily introduced for texture classification, this broad class of features captures inter-pixel relations by thresholding a local neighbourhood at the intensity value of the center pixel into a binary pattern. Xu and Shi (2012) combined so-called Ojala histograms of these patterns from 3×3 neighbourhoods into 59-dimensional feature vectors.[8] They computed them for each colour channel from corresponding intensity values, a linear predictor residue image, and the diagonal detail sub-band of a one-level wavelet decomposition, respectively. The final feature set, $\mathcal{F}_{lbp}$, comprises $2 \times 3 \times 59 = 354$ features from the green and the red colour channel.[9]

9.6.1.5 Binary Similarity Measures

While local binary patterns refer to bit sequences obtained by thresholding local intensity neighbourhoods, binary similarity measures subsume a set of features to analyze binarised neighbourhood characteristics in and across different bit planes of a digital image (Avcibaş *et al.* 2005). The features can be broadly classified into measures based on differences between bit planes, histogram and entropy features, and measures based on Ojala histograms. Çeliktutan *et al.* (2008) adopted these measures for camera model identification and broadened the idea of what defines a neighbourhood so that relations between colour channels are also taken into account. The complete feature set, $\mathcal{F}_{bsm}$, amounts to 480 features.

---

[8] Ojala histograms (Ojala *et al.* 1996) count the frequency of local binary patterns across the image.
[9] Xu and Shi assumed that red and blue colour channels by and large exhibit similar characteristics due to comparable demosaicing procedures.

9.6.1.6 DCT Domain Features

Most digital cameras store compressed images in the JPEG format. Hence, it is reasonable to examine model-specific image characteristics in the DCT domain directly. Xu *et al.* (2009) argued that the absolute magnitudes of the 8×8 DCT-coefficient blocks bear a distinct residual degree of inter-coefficient correlations. The authors modelled these correlations by means of Markov transition probabilities, computed from thresholded first-order differences of the DCT-coefficient array. The corresponding transition matrices then serve as a feature representation. Gao *et al.* (2012) improved upon this approach by considering difference arrays with multiple lags and averaged transition matrices. The resulting feature set, $\mathcal{F}_{tp}$, obtained from the luminance component of the image, has a dimensionality of 58. Also Wahab *et al.* (2012) considered DCT domain characteristics. However, instead of differences between the absolute values of DCT coefficients, they exploited empirical conditional probabilities of the relative order of magnitudes of three selected coefficients in the upper-left 4×4 low-frequency bands. The 72-dimensional feature space, $\mathcal{F}_{cp}$, is composed from conditional probabilities over overall eight different coefficient subsets.

9.6.1.7 Merged Feature Sets

Table 9.1 summarizes the major feature sets that have found application in the literature. Following the seminal work of Kharrazi *et al.* (2004), a number of studies have employed combinations of the individual feature sets we discussed earlier. Specifically, Kharrazi *et al.* merged colour features, image quality metrics and wavelet statistics into a 34-dimensional feature set, $\mathcal{F}_{Khar}$. Its adaption to colour images, supplemented with higher-order wavelet statistics, increased the dimensionality to 82 (Gloe 2012a). In reference to our own earlier work, we denote this feature set as $\mathcal{F}_{Gloe}$. Çeliktutan *et al.* (2008) did also employ image quality metrics and wavelet statistics,

Table 9.1 Camera model identification feature sets and their dimensionality.

Feature set	$\mathcal{F}_{col}$	$\mathcal{F}_{iqm}$	$\mathcal{F}_{wav}$	$\mathcal{F}_{lbp}$	$\mathcal{F}_{bsm}$	$\mathcal{F}_{tp}$	$\mathcal{F}_{cp}$	$\sum$	References
$\mathcal{F}_{Khar}$	12	13*	9					34	Kharrazi *et al.* (2004)
$\mathcal{F}_{Gloe}$	18	37	27					82	Gloe (2012a)
$\mathcal{F}_{Çeli}$		40	72†		480			592	Çeliktutan *et al.* (2008)
$\mathcal{F}_{Xu}$				354				354	Xu and Shi (2012)
$\mathcal{F}_{Gao}$						58		58	Gao *et al.* (2012)
$\mathcal{F}_{Wah}$							72	72	Wahab *et al.* (2012)

Feature types are aligned in columns, combined feature sets in rows. The last column denotes the overall size of the combined feature sets.
* Averaged over colour channels.
† three-level wavelet decomposition.

but replaced colour features with binary similarity measures. Their complete feature set, $\mathcal{F}_{\text{Celi}}$, thus comprised 592 features.

The promise of merging is that the resulting feature space provides a better representation of model-specific image characteristics – and thus a higher classification accuracy – than individual (smaller) feature sets. Yet this comes at the cost of an increased feature space dimensionality. It can not only increase computational complexity but also give rise to *overtraining*. The larger a feature space becomes, the more training samples are necessary for a classifier to generalize well to unknown test data. This 'curse of dimensionality' in particular also affects the widely used support vector machines with non-linear kernels (Bengio *et al.* 2005). *Feature selection* has thus become an integral part of camera model identification in high-dimensional feature spaces. Sequential forward floating search (SFFS, Pudil *et al.* 1994) is often the method of choice (Cao and Kot 2009; Gloe 2012a). It iteratively adds (removes) the most (least) informative features to (from) a subset of the complete feature set.

As far as classification performances are concerned, the literature indicates that features derived from general-purpose image descriptors are well able to reliably distinguish between a variety of different camera models. All studies report accuracies of above 90%. Xu and Shi (2012) achieved a 98% average accuracy over 18 camera models from the Dresden Image Database, yet without strictly following the rigorous methodology that we outlined in Section 9.4. It goes without saying that a direct experimental comparison of the different proposed feature sets is only viable as long as all algorithms are tested under the same controlled conditions. The most comprehensive benchmark up-to-date considered three feature sets (namely $\mathcal{F}_{\text{Khar}}$, $\mathcal{F}_{\text{Celi}}$ and $\mathcal{F}_{\text{Gloe}}$) in a close-to-real-world scenario and noted a 92% average accuracy over 26 camera models after feature selection in the $\mathcal{F}_{\text{Gloe}}$ feature space (Gloe 2012a). We will also use this benchmark as a basis for a case study on closed-set camera model identification in the next section.

9.6.2 Dresden Image Database Case Study: Closed-Set Camera Model Identification

Our case study focuses on feature sets inspired by the seminal work of Kharrazi *et al.* (2004). Specifically, we employ $\mathcal{F}_{\text{Khar}}$, $\mathcal{F}_{\text{Celi}}$ and $\mathcal{F}_{\text{Gloe}}$ for closed-set forensic camera model identification under realistic assumptions, based on all natural images in the Dresden Image Database (cf. Section 9.4.2, and also Gloe and Böhme 2010). Where previous works considered only relatively small camera model sets with one device per model, our experiments will give a good impression of how these general-purpose image descriptors can be applied to camera model identification in real-world scenarios. Furthermore, we will shed light on practically relevant aspects of creating appropriate training data sets by discussing how the number of images, devices and camera models affects classification accuracy. A brief outlook on the problem of handling unknown camera models in open sets follows in Section 9.7. Features of all images in our case study are available as supplementary material at the website of the Dresden Image Database,[10] along with a number of exemplary camera model identification scripts.

9.6.2.1 Test Setup

The core data set of our case study entails $16,956$ full-resolution natural images stored in the JPEG format. Table 9.2 gives an overview of the employed camera models,[11] along with the number of corresponding devices and images. We distinguish between two sets of camera models. Set $\mathcal{M}_{all}$ contains all 26 models listed in the table (74 devices total). Set $\mathcal{M}_{sub} \subset \mathcal{M}_{all}$ is a subset of 10 camera models (overall 39 devices). We use it for experiments with high computational complexity. Our experiments generally adopt the methodology outlined in Section 9.4. However, we use only two independent sets of scenes, namely one set for training, $\mathcal{S}_{train}$, and one disjoint set for validation and testing, $\mathcal{S}_{val,test}$, respectively. This trade-off balances the large body of available camera models and images against the still relatively low number of motifs in the database.

We apply cross-validation based on a fixed set of 100 image space partitions in all our experiments, unless otherwise stated. Each partition assigns devices and scenes randomly to training, validation and test sets. Each camera model is trained based on a single device and 26 motifs[12] of the database. For camera models where only one device is available, test results are obtained from validation images. We use support vector machines (SVMs) with a radial-based kernel function for classification, implemented by Chang and Lin (2001) and accessed via the R[13] package e1071. The library solves the multi-class problem through individual binary SVMs for all pairs of different camera models. A voting scheme (one-versus-one) determines the most likely class. SVM parameters are found in a grid search with γ in the range of $2^{3,2,\ldots,-15}$ and C in the range of $2^{-5,-4,\ldots,15}$ for each of the 100 partitions independently.

9.6.2.2 Intra- and Inter-Camera Model Similarity

The ability to separate between different camera models and not between individual devices is a crucial requirement for all camera model identification schemes. Good feature representations should thus exhibit a high intra-camera model similarity, while keeping the inter-camera model similarity low. Figure 9.9 visualizes the intra- and inter-camera model similarity between all 74 devices in the feature space $\mathcal{F}_{Gloe}$. Specifically, we computed for each device its centroid in the high-dimensional feature space and applied multi-dimensional scaling (Borg and Groenen 2005) to map these

---

[10] https://forensics.inf.tu-dresden.de/ddimgdb/publications/modelid.

[11] We consider the camera models Nikon D70 and Nikon D70s as equivalent in this study.

[12] Remember that each motif was captured multiple times with different camera settings, so that the number of actual training images is larger (cf. Section 9.4.2).

[13] http://www.r-project.org.

Table 9.2   Digital camera models in this study, number of devices per model, image size in pixels and number of available images.

Make	Model	# Devices	Image size	# Images		
				Set A	Set B	Σ
Agfa	DC-504	1	4032 × 3024	78	91	169
Agfa	DC-733s	1	3072 × 2304	150	128	278
Agfa	DC-830i	1	3264 × 248	176	187	363
Agfa	Sensor505-X	1	2592 × 1944	87	85	172
Agfa	Sensor530s	1	4032 × 3024	195	177	372
Canon	Ixus55	1	2592 × 1944	224		224
Canon	Ixus70*	3	3072 × 2304	567		567
Canon	PowerShot A640	1	3648 × 2736		188	188
Casio	EX-Z150*	5	3264 × 2448	924		924
FujiFilm	FinePix J50	3	3264 × 2448		630	630
Kodak	M1063*	5	3664 × 2748	1070	1321	2391
Nikon	CoolPix S710*	5	4352 × 3264	925		925
Nikon	D200*	2	3872 × 2592	752		752
Nikon	D70/D70s	2/2	3008 × 2000	736		736
Olympus	μ1050SW*	5	3648 × 2736	1040		1040
Panasonic	DMC-FZ50	3	3648 × 2736		931	931
Pentax	Optio A40	4	4000 × 3000		638	638
Pentax	Optio W60	1	3648 × 2736		192	192
Praktica	DCZ5.9*	5	2560 × 1920	1019		1019
Ricoh	GX100	5	3648 × 2736		854	854
Rollei	RCP-7325XS*	3	3072 × 2304	589		589
Samsung	L74wide*	3	3072 × 2304	686		686
Samsung	NV15*	3	3648 × 2736	645		645
Sony	DSC-H50	2	3456 × 2592		541	541
Sony	DSC-T77	4	3648 × 2736		725	725
Sony	DSC-W170	2	3648 × 3648		405	405
Σ		74		9863	7093	16956

* Camera model is part of the reduced set, $\mathcal{M}_{\text{red}}$.

points to a two-dimensional subspace. The distances of the mapped centroids represent the (dis)similarity of devices in the higher-dimensional feature space. The resulting graph clearly suggests that devices of the same camera model form spatial clusters. Hence, it supports the assumption that the employed features are able to separate between camera models.

Figure 9.10 takes a closer look at feature representations at the level of individual images. Here, we used a principal component analysis (PCA) to project images to the two-dimensional subspace that is spanned by the two most distinctive principal

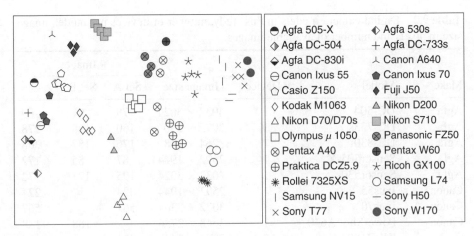

Figure 9.9 Intra- and inter-camera model similarity in the $\mathcal{F}_{\text{Gloe}}$ feature space. Shown is, for each device, the centroid of all images in the feature space, mapped to a 2D subspace using multi-dimensional scaling. Individual devices of the same camera model share the same symbol. Devices of the same camera model are close to each other, whereas devices of different models are further apart.

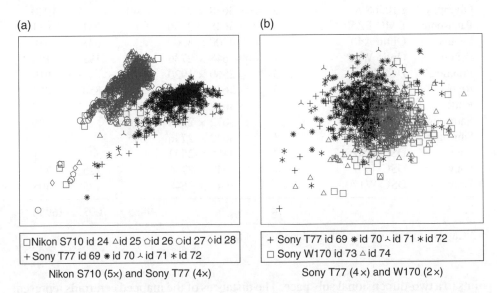

Nikon S710 (5×) and Sony T77 (4×)   Sony T77 (4×) and W170 (2×)

Figure 9.10 Similarity of images in the $\mathcal{F}_{\text{Gloe}}$ feature space for all devices of two selected pairs of camera models. (a) Five Nikon S710 vs. four Sony T77 cameras. (b) Four Sony T77 vs. two Sony W170 cameras. Shown are the projections to the corresponding two most distinctive principal components. Individual devices of one camera model share the same colour. Camera models are visually separable, whereas individual devices of the same model are not. (Also cf. Figure 9.3.)

components. Figure 9.10a illustrates the dissimilarity of individual images of two selected camera models (Nikon S710 and Sony T77) in the feature space. Figure 9.10b, on the contrary, indicates that images captured with camera models of the same manufacturer are sometimes harder to separate (Sony T77 and Sony W170 in this case), as they are likely to share some components. Observe that in both cases, the intra-model similarity is very high, that is it is not trivially possible to distinguish between individual devices.

9.6.2.3   Camera Model Identification

Forensic source identification in a closed set of models assumes that all candidate models are known in the training phase. For a realistic scenario, we train the classifier based on a single device per model. The classifier is tested against 'unknown' devices whenever possible. For the $\mathcal{F}_{\text{Çeli}}$ feature space, we restrict the classifier to the smaller $\mathcal{M}_{\text{sub}}$ subset of camera models to keep the high-dimensional problem computationally tractable. In general, it can be expected that increasing the number of training models reduces classification accuracy, as it becomes more and more likely that different models share similar characteristics. This effect is indeed evident from Table 9.3, which reports the average classification accuracies over 100 image space partitions for the three feature sets. Both the $\mathcal{F}_{\text{Khar}}$ and the $\mathcal{F}_{\text{Gloe}}$ classifier loose about 5–6 percentage points in accuracy when switching from 10 ($\mathcal{M}_{\text{sub}}$) to 26 ($\mathcal{M}_{\text{all}}$) camera models. Table 9.3 also indicates that classification over the validation set, $\mathcal{I}_{\text{val}}$ (i.e. images acquired with the same device that was used to generate the training data) always yields better results than running the same classifier over the test set, $\mathcal{I}_{\text{test}}$ (i.e. images acquired with other devices). This suggests that the features are subject to small variations across different devices of one model. There still exists a non-negligible degree of inter-model *dis*similarity. Nevertheless, all three tested feature sets generally achieve a promising classification performance, with average accuracies close to or above 90%. Table 9.4 provides a more detailed view on the performance of the best-performing feature set,

Table 9.3   Average classification accuracies (in percentage points) over camera model sets $\mathcal{M}_{\text{all}}$ and $\mathcal{M}_{\text{sub}}$ for different feature sets and 100 image space partitions.

	Feature set					
	$\mathcal{F}_{\text{Khar}}$		$\mathcal{F}_{\text{Gloe}}$		$\mathcal{F}_{\text{Çeli}}$	
Model set	$\mathcal{I}_{\text{val}}$	$\mathcal{I}_{\text{test}}$	$\mathcal{I}_{\text{val}}$	$\mathcal{I}_{\text{test}}$	$\mathcal{I}_{\text{val}}$	$\mathcal{I}_{\text{test}}$
$\mathcal{M}_{\text{sub}}$	95.62	93.14	98.06	96.36	95.32	92.80
$\mathcal{M}_{\text{all}}$	89.29	85.62	93.08	90.67	–	–

Breakdown by validation and test images, $\mathcal{I}_{\text{val}}$ and $\mathcal{I}_{\text{test}}$, respectively.

Table 9.4 Per-model average classification accuracies (in percentage points) over camera model set $\mathcal{M}_{sub}$ for feature set $\mathcal{F}_{Gloe}$ and test images ($\mathcal{I}_{test}$) from 100 image space partitions.

Device	I70	Z150	M10	S710	D200	μ	DCZ	7325	L74	NV
					Identified as					
Ixus70	**93.5**	0.6	0.1	–	0.1	–	0.3	0.3	5.1	–
Z150	0.1	**98.7**	0.3	–	0.4	–	0.2	–	0.1	0.1
M1063	–	0.1	**99.4**	0.1	0.4	–	–	–	–	–
S710	–	–	0.1	**99.6**	–	–	–	–	–	0.3
D200	0.1	1.4	2.9	–	**91.1**	2.2	–	0.4	–	1.8
μ1050	0.1	0.7	0.4	–	1.4	**93.5**	0.1	–	–	3.8
DCZ5.9	0.1	0.2	0.1	–	–	–	**99.5**	0.1	–	–
7325XS	0.1	0.1	0.2	–	–	–	0.1	**98.1**	1.2	0.1
L74	2.2	0.1	0.1	–	–	–	0.2	0.2	**97.0**	–
NV15	0.5	0.3	0.5	0.1	2.6	4.3	–	0.2	–	**91.5**

$\mathcal{F}_{Gloe}$. It reports the test set classification accuracies over $\mathcal{M}_{sub}$ per camera model. The numbers indicate that the actual accuracy greatly depends on the camera model. Images acquired with a Nikon S710 camera, for instance, can be reliably identified with only few errors. On the contrary, Samsung NV15 images yield a much lower accuracy due to a higher similarity to images acquired with other camera models. Adding the high-dimensional binary similarity measures, $\mathcal{F}_{bsm}$, interestingly does not improve the performance in our test scenario. We suspect that this is an effect of overfitting due to a higher training complexity. We will see in the following that feature selection can be one way to alleviate this 'curse of dimensionality'.

9.6.2.4 Feature Selection

We employ sequential forward floating search (SFFS, Pudil *et al.* 1994) feature selection to determine the most influential features in the set $\mathcal{F}_{Gloe}$. Because the particular choice of training motifs and devices might influence the feature selection procedure, we apply SFFS to ten out of the 100 fixed image space partitions for both sets of camera models, $\mathcal{M}_{sub}$ and $\mathcal{M}_{all}$, respectively. Figure 9.11 depicts the relation between classification accuracy and the number of selected features for the two partitions that yield the best and the worst achieved maximum accuracy over $\mathcal{I}_{val}$. The camera model set is $\mathcal{M}_{sub}$ in this example. All ten trials result in accuracies of above 98% in the validation step. Feeding test images into the reduced-set classifiers again yields lower average accuracies, although it is possible to achieve more than 98% for one out of the ten partitions.

While SFFS generally selects features based on their importance to the overall classification accuracy, we observed a surprisingly large variability in the order of specific selections across the 10 image space partitions. Consequently, we determined

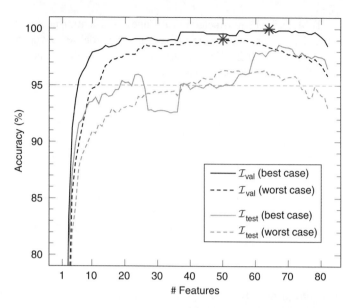

Figure 9.11 SFFS feature selection for feature set $\mathcal{F}_{\text{Gloe}}$ over camera model set $\mathcal{M}_{\text{sub}}$. Shown are the results for two specific image space partitions, which yielded the best-case and worst-case accuracies over $\mathcal{I}_{\text{val}}$. Asterisks mark the highest accuracy in both cases.

2×10 feature selections from the maximum average accuracies over the corresponding validation data and apply the resulting classifiers to all 100 image space partitions. Table 9.5 summarizes the average classification accuracies for both validation data, $\mathcal{I}_{\text{val}}$, and test data, $\mathcal{I}_{\text{test}}$, respectively. A comparison with the results in Table 9.3 indicates only relatively small differences. Yet it is worth noting that feature selection based on the complete set of camera models, $\mathcal{M}_{\text{all}}$, increases the performance by one percentage point in the most challenging case (classification over $\mathcal{M}_{\text{all}}$ and $\mathcal{I}_{\text{test}}$).

9.6.2.5 Influence of the Number of Images and Devices

Time and money are limiting factors in practical forensic investigations. It is thus helpful to know how many images and devices are sufficient in the training stage for reliable camera model identification.

As part of the design of the Dresden Image Database, individual images vary significantly more across different motifs than within subsets of scenes of the same motif. Based on the $\mathcal{F}_{\text{Gloe}}$ feature set, we investigate the influence of the amount of training images and motifs by varying both numbers per training device $d \in \mathcal{D}_{\text{train}}$. Specifically, we use a fixed set of 225 image space partitions,[14] obtained in a similar procedure as before. Firstly, one training device is randomly selected per camera model, for which, secondly, the number of training motifs is varied from 1 to 45. Both

---

[14] Based on our earlier experiments, we fix the SVM parameters as $\gamma = 2^{-9}$ and $C = 2^8$ for the $\mathcal{F}_{\text{Gloe}}$ feature set to decrease computational complexity.

Table 9.5 Feature selection average classification accuracies (in percentage points) over camera model sets $\mathcal{M}_{\mathrm{all}}$ and $\mathcal{M}_{\mathrm{sub}}$ over 20 SFFS feature selection procedures and 100 image space partitions.

| | SFFS on feature set | | | |
| | $\mathcal{F}_{\mathrm{Gloe}}\ (\mathcal{M}_{\mathrm{sub}})$ | | $\mathcal{F}_{\mathrm{Gloe}}\ (\mathcal{M}_{\mathrm{all}})$ | |
Model set	$\mathcal{I}_{\mathrm{val}}$	$\mathcal{I}_{\mathrm{test}}$	$\mathcal{I}_{\mathrm{val}}$	$\mathcal{I}_{\mathrm{test}}$
$\mathcal{M}_{\mathrm{sub}}$	97.72	96.08	97.79	95.97
$\mathcal{M}_{\mathrm{all}}$	92.89	90.77	93.59	91.68

Breakdown by validation and test images, $\mathcal{I}_{\mathrm{val}}$ and $\mathcal{I}_{\mathrm{test}}$, respectively. The camera model set used for feature selection is indicated in brackets.

steps are repeated five times. To investigate the influence of training scenes per motif, we additionally vary this parameter for each device from 1 to the maximum number of available images. Figure 9.12a illustrates how classification accuracy over camera model set $\mathcal{M}_{\mathrm{all}}$ changes dependent on the number of training motifs and scenes. The graphs depict test set ($\mathcal{I}_{\mathrm{test}}$) average accuracies over all five fixed sets of training devices. While the performance clearly increases with the number of motifs, we can also observe a saturation effect after adding more than 30 motifs. Nevertheless, training the classifier with images acquired with a broad range of different camera settings (i.e. a large number of scenes) is still important to achieve the maximum possible accuracy.

For an impression on how the number of training devices affects classification performance, we repeat the aforementioned experiments with all available scenes per motif, while randomly selecting one up to four devices for training with five-fold cross-validation. In this procedure, we make sure that there is always at least one distinct device left for testing. Figure 9.12b reports the average classification accuracies over all 15 camera models in set $\mathcal{M}_{\mathrm{all}}$ that supply three or more devices. The graphs re-emphasize that the feature set provides a high degree of inter-model similarity. Increasing the number of training devices has only a negligible impact on the overall performance. Practical investigations should thus focus available resources on the acquisition of enough images per camera model, covering a variety of motifs and camera settings. Borrowing or purchasing more than one device per camera model is less important as far as the average performance is concerned.

9.6.3 Summary

General-purpose image descriptors promise a broad and generalizable representation of model-specific image characteristics in the feature space. A considerable number of

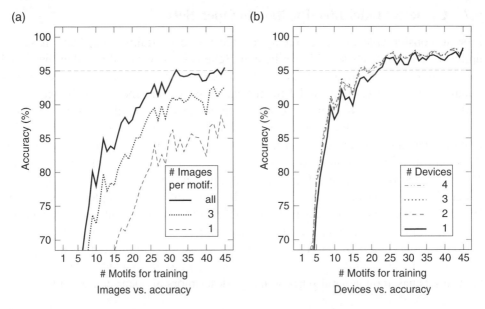

Figure 9.12 Relation of the number of training motifs and the classification accuracy over $\mathcal{I}_{test}$. (a) Influence of the number of training images per motif (and per device). Increasing the number of images improves accuracy. (b) Influence of the number of training devices per model. (The graphs are obtained from camera models with >3 available devices.) More devices per camera model yield only minor improvements.

camera model identification schemes in the literature have meanwhile extended and advanced the seminal work of Kharrazi *et al.* (2004) along this direction. The black box perspective does not restrict them to specific aspects of individual components in the image acquisition pipeline. This allows the classifiers to distinguish between actual camera models, and not just between (a typically limited number of) component instances. Yet the potentially higher dimensionality of the class space comes at a price. Reported classification accuracies tend to be lower, in particular as more and more camera models are taken into consideration (Gloe 2012a). Our case study based on the Dresden Image Database supports this impression. Increasing the dimension of the model space from 10 to 26 resulted in about 6% performance loss on average (cf. Table 9.3). Also the high dimensionality of the feature spaces can pose practical problems. While feature selection might be a way to cope with the curse of dimensionality, our experiments have demonstrated that finding a set of optimal features is difficult due to dependencies on the selected training devices and images. Even more than component-based schemes, camera model identification algorithms based on general-purpose image descriptors thus hinge on a carefully designed and comprehensive training methodology.

9.7 Camera Model Identification in Open Sets

Despite considerable efforts to create a large reference training database, it seems unlikely that this database will ever be comprehensive. A residual risk that the actual camera model of an image of unknown provenance is not in the database always remains. *Open set* camera model identification poses an inherent challenge to the widely used multi-class support vector machines (SVMs), as they always assign a test sample to one of the trained classes. Yet for critical forensic applications (i.e. in particular in relation to criminal investigations) unknown camera models must be detected with almost certainty so as to avoid false accusations under all circumstances. The relevance of the open set problem naturally grows with the dimension of the underlying class space, and thus also with the generality of the employed feature set. While there exist only four plausible Bayer CFA configurations and most likely only a few dozen different relevant demosaicing algorithms (cf. Section 9.5.3), the distinction between actual camera models has to cope with a basically unlimited number of potential classes. Consequently, it is particularly important to understand the limitations of feature sets based on general-purpose image descriptors in open set scenarios.

9.7.1 Dresden Image Database Case Study: One-Class SVM

A standard machine learning approach to deal with unknown classes is subsumed under the concept of 'novelty detection'. *One-class classification* is one way to solve the problem (Schölkopf *et al.* 2001). When applied to support vector machines, the idea of one-class classification is to transform feature vectors of one single class via a kernel, so as to determine a small region that represents most known vectors. Feature vectors of unknown classes are expected to lie outside the trained region. The fraction of outliers and support vectors in the training phase is controlled via parameter ν.

We use this concept for open-set camera model identification in the Dresden Image Database by training one one-class SVM for each known camera model, $m_{known} \in \mathcal{M}_{all}$. The same fixed 100 image space partitions as before apply (cf. Section 9.6.2). Figure 9.13a illustrates how average test data classification accuracies vary as a function of parameter ν for both classes. The graphs indicate that the absence of negative information (i.e. feature vectors of unknown camera models) comes at a price. We obtain higher accuracies either for unknown or for known camera models, depending on the choice of ν. As the fraction of training outliers increases, unknown camera models are detected more reliably. Yet this works only against the backdrop of lower classification accuracies for the known models. Setting $\nu = 0.17$ yields the maximum overall average accuracy of 76%. An alternative is to determine the optimal ν for each camera model separately. Figure 9.13b visualizes the resulting performance for a number of selected camera models by means of ROC curves. In this setting, the true positive rate reflects the percentage of assignments to the correct camera model, whereas the false positive rate indicates mis-classifications. A good classifier should exhibit very low false positive rates (i.e. a high likelihood of identifying unknown models) to avoid false accusations. Overall, the graphs suggest very mixed results,

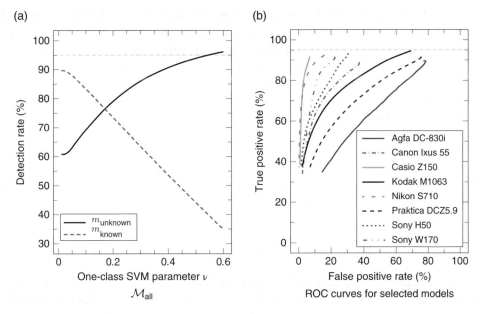

Figure 9.13 Open-set camera model identification based on one-class SVMs. (a) Average test data classification accuracies as a function of SVM parameter ν for both known and unknown camera models over 100 image space partitions. (b) Individual results for selected camera models. ROC curves were obtained by varying parameter ν and observing the percentage of correctly classified and mis-classified test images.

with a wide performance range between the best case (Casio EX-Z150) and the worst case (Agfa DC-830i).

9.7.2 Summary and Outlook

Camera model identification in realistic open-set scenarios is still subject to ongoing research. Despite its high practical relevance, it has not received much attention in the literature so far. The one-class approach can be a means to support decisions under certain circumstances. Yet a straight-forward application within the accuracy margins of highly critical investigations seems currently beyond its scope. With a modified training procedure, also traditional binary support vector machines can be employed for open-set camera model identification (Gloe 2012a). In closed sets, a multitude of binary SVMs is trained for all pairs of distinct classes (one-versus-one), and a voting scheme determines the most likely camera model. To cope with lacking training data for unknown camera models, arbitrary models in the training database can be converted into 'dummy' unknown models. As a result, a set of binary SVMs is trained for all combinations of known models, $m_{known} \in \mathcal{M}$, and remaining dummies. Specifically, images for the respective unknown class are sampled from models $m_{unknown\ (train)} \in \mathcal{M}/\{m_{known}\}$. Similar to the one-class approach, also this method has drawbacks. First, the classification accuracies for known models tend to be considerably lower than in the closed-set scenario. This is an effect of the missing voting scheme over several binary SVMs in case of multi-class classification in closed sets. Second, the use of dummy data can only attenuate the problem of missing training data, but does not solve it completely. The performance of the classification scheme will rather depend on the quality and the generality of the training dummies.

In summary, we see both approaches as very first attempts to handle unknown camera models. Further research is necessary to improve upon the prevention of false accusations. This goes naturally along with a better understanding and modelling of class likelihoods over the image space (Böhme and Kirchner 2013). Given the high dimensionality and the empirical nature of the problem, we can only surmise that classifier-based approaches will very likely never reach the confidence levels of techniques that (can) draw on rigorous hypothesis testing frameworks, such as camera identification based on sensor noise for instance (Fridrich 2013).

9.8　Model-Specific Characteristics in Device-Level Identification

Before we close this chapter with a discussion of open problems towards practical applications in Section 9.9, this penultimate section takes a brief look at the role of model-specific characteristics in device-level identification. This means that we have to shift our perspective: a major challenge for the identification of individual cameras is the *suppression* of so-called 'non-unique' artefacts (Fridrich 2013). Non-unique artefacts are more specific to camera model or make than they are specific to a particular device, that is they exhibit an undesirably high inter-class similarity in the realm of device-level identification granularity, or, equivalently, a low intra-class similarity at lower levels of identification granularity. They may thus lead to mis-classification because of an increased similarity across distinct devices.

Sensor noise, and in particular photo-response non-uniformity (PRNU), is the most widely used acquisition characteristic for digital camera identification. PRNU – commonly accepted as a unique sensor 'fingerprint' (Fridrich 2013) – is a multiplicative noise that is caused by inevitable differences in the quantum efficiency of individual sensor elements. Large-scale tests indicated a very high identification reliability. Goljan *et al.* (2009) reported almost 98% detection rate at a false positive rate as low as 3×10^{-5} even across thousands of different devices. PRNU is typically examined in the form of spatial noise patterns, which can be estimated from noise residuals of reference or questioned images. A camera's reference noise pattern is ideally estimated from a number of homogeneously lit images. PRNU-based device identification then measures the similarity between the noise pattern of a questioned image and all known reference noise patterns in terms of the peak-to-correlation energy ratio (PCE).[15] Experiments have indicated that PCE values ≥ 60 are a reliable indicator of the correct camera (Goljan *et al.* 2009).

---

[15] PCE weights the maximum squared cross-correlation between noise residual and reference pattern over all spatial shifts with respect to the average squared cross-correlation over all shifts outside a small neighbourhood around the maximum.

Relevant non-unique artefacts in PRNU-based camera identification include, for instance, JPEG blocking artefacts, or pixel interdependencies as they result from demosaicing. JPEG and CFA artefacts occur in regular (linear) periodic spatial patterns that can be corrected for relatively easy, but also non-linear model-specific artefacts due to the correction of radial lens distortion have been reported (Goljan and Fridrich 2012).

Figure 9.14 illustrates another practical example of non-unique artefacts in noise patterns of three Nikon CoolPix S710 digital cameras. Each of the three panels shows the same 256 × 256 region of the cameras' respective reference noise patterns, estimated from 50 homogeneously lit images each. Diagonal structures are visible in all three noise signals, particularly so in the two rightmost patterns ('S710 #2' and 'S710 #3'). While it is currently an open question where those diagonal artefacts originate (Gloe *et al.* 2012), it is evident that they might interfere with device-level source identification if similar structures exist in all images acquired with cameras of the same model. Figure 9.15 a indeed supports our concerns. Shown are PCE values from a typical camera identification scenario: noise residuals from images of the three aforementioned cameras were tested against the reference noise pattern of camera 'S710 #2'. Observe that hardly any of the images from the correct camera yields sufficiently large PCE values. Hence, most images would not be classified correctly. The good news for practical camera identification is that such periodic artefacts can be well reduced by Wiener filtering in the frequency domain (Fridrich 2013). Results from a second run of the camera identification procedure are depicted in Figure 9.15b, this time based on Wiener-filtered noise estimates. The removal of the periodic artefacts yields the desired outcome: all images from camera 'S710 #2' give substantially larger PCE values.

(a)   (b)   (c)

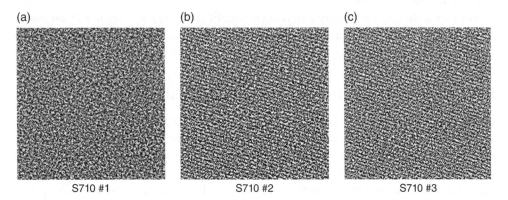

S710 #1   S710 #2   S710 #3

Figure 9.14 Diagonal non-unique artefacts in sensor noise patterns of three different Nikon CoolPix S710 digital cameras. 256 × 256 crops of reference noise patterns, estimated from 50 homogeneously lit images each. The noise patterns were subject to histogram equalization and sharpening for better visibility of the diagonal structures. All images are part of the Dresden Image Database (Gloe and Böhme 2010), http://forensics.inf.tu-dresden.de/ddimgdb/locations/frames.

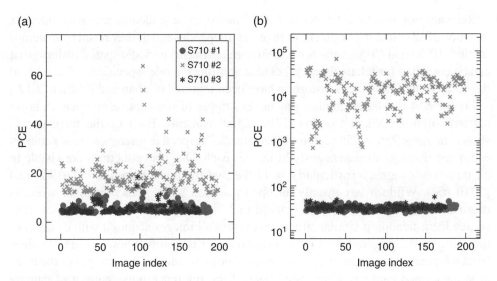

Figure 9.15 PRNU-based digital camera identification without (a) and with (b) removal of non-unique artefacts through Wiener filtering in the frequency domain. Images of three different Nikon CoolPix S710 digital cameras, tested against a reference noise pattern of camera 'S710 #2' (see also Figure 9.14). Larger PCE values indicate higher similarity. Note the different scales.

Our example emphasizes the dual role of model-specific characteristics in forensic source identification. While the observed diagonal patterns are certainly a welcome characteristic to narrow down the source of an image to a small set of camera models,[16] they are considered an undesired distortion when it comes to the identification of individual devices. This indicates that research on characteristics for a particular level of identification granularity may well benefit from findings regarding higher or lower levels. Eventually, this points to a more holistic approach to forensic source identification, and thus leads us to the final section of this chapter.

9.9 Open Challenges Towards Practical Applications

In retrospect of the large variety of different aspects of model-specific image characteristics that emanate from the overly simple digital camera acquisition pipeline in Figure 9.4 alone, it may not be surprising that camera model identification as a whole is the most actively pursued stream of forensic source identification research. However, this abundance of research directions has also led to a rather scattered discipline, which defines itself primarily through its flourishing specialized branches. Interactions

---

[16] In fact, we are currently not aware of another camera model that exhibits the same type of artefacts.

between these sub-fields barely exist in the literature. In particular, benchmarks and evaluations are typically confined to specific types of image characteristics, leaving a comprehensive comparison of strengths and weaknesses across different approaches to future work.

At the same time, it should be clear that model-specific artefacts do not occur independent of each other. A combination of different approaches is thus generally likely to contribute to an increased overall reliability. The few existing works in this direction merge general-purpose image descriptors with component-based measures (Choi *et al.* 2006; Deng *et al.* 2011; Tsai *et al.* 2012). In the worst-case, however, interferences between artefacts can also render modelling assumptions about specific characteristics obsolete. Geometrical warping due to lens distortion correction, for instance, can interfere with and desynchronize sensor noise (Gloe *et al.* 2012; Goljan and Fridrich 2012). It is likely to affect demosaicing inter-pixel characteristics as well (although this aspect has not been studied in the literature yet). In general, it can be expected that the ever-increasing sophistication and adaptability of camera-internal image processing primitives will only increase the complexity and the parameter space of camera model identification algorithms further.

But not only advanced camera-internal processing poses a challenge to existing and future forensic techniques. Digital images are often also post-processed *after* they left the camera. It is without question that practical investigations will encounter non-genuine images at times, which underwent, for instance, colour enhancement, resizing or cropping. Cropping is already known to lower the accuracy of classifiers based on general-purpose image descriptors (Fang *et al.* 2009; Gloe *et al.* 2009; Wahab *et al.* 2012). The less pixels are available for feature extraction the more noisy the resulting feature representations become. Downsizing is even more critical (Fang *et al.* 2009), as it acts as a low-pass filter that smooths out subtle inter-model differences. Typically, such post-processing also goes along with a final lossy compression step. The effects of a single JPEG compression pass have been mostly studied for CFA-based algorithms, for which a relatively high robustness under moderate JPEG qualities (down to quality factor 90) was reported (Cao and Kot 2009; Swaminathan *et al.* 2007). In general, little is known about the performance of camera model identification schemes in more complex scenarios with combined post-processing and multiple compression steps.

As forensic camera model identification finds applications in practical investigations, a more explicit treatment of these different types of artefact interdependencies will gain more and more relevance. Future work will have to show how (and which) well-understood and highly targeted component-based characteristics and high-dimensional feature spaces are traded off best to handle practical challenges like large open-set class spaces covering processed images. For the time being, camera model identification should be understood as a valuable tool for *narrowing down* the source of a questioned image, which can offer very reliable results to forensic investigators who are well aware of the underlying assumptions and inherent technical limitations.

References

Alles EJ, Geradts ZJ and Veenman CJ 2009 Source camera identification for heavily JPEG compressed low resolution still images. *Journal of Forensic Sciences* **54**(3), 628–638.

Avcibaş İ, Sankur B and Sayood K 2002 Statistical evaluation of image quality measures. *Journal of Electronic Imaging* **11**(2), 206–223.

Avcibaş İ, Kharrazi M, Memon N and Sankur B 2005 Image steganalysis with binary similarity measures. *EURASIP Journal on Applied Signal Processing* **17**, 2749–2757.

Bayer BE 1976 Color imaging array US Patent, 3971065.

Bayram S, Sencar HT, Memon N and Avcibaş İ 2005 Source camera identification based on CFA interpolation. *IEEE International Conference on Image Processing (ICIP)*, Geneva, Italy, vol. 3, pp. III–69–72.

Bayram S, Sencar HT and Memon N 2006 Improvements on source camera-model identification based on CFA. In *Advances in Digital Forensics II, IFIP International Conference on Digital Forensics* (eds. Olivier MS and Shenoi S), vol. 222 of *IFIP Advances in Information and Communication Technology*, Springer, New York. pp. 289–299.

Bengio Y, Delalleau O and Le Roux N 2005 The curse of dimensionality for local kernel machines. Technical report 1258, Département d'Informatique et Recherche Opérationnelle, Université de Montréal, Montréal, Quelec, Canada.

Böhme R and Kirchner M 2013 Counter-forensics: Attacking image forensics. In *Digital Image Forensics* (eds. Sencar HT and Memon N). Springer-Verlag, New York, pp. 327–366.

Borg I and Groenen PJF 2005 *Modern Multidimensional Scaling: Theory and Applications* 2nd edn. Springer, New York.

Caldelli R, Amerini I and Picchioni F 2010 A DFT-based analysis to discern between camera and scanned images. *International Journal of Digital Crime and Forensics* **2**(1), 21–29.

Cao H and Kot AC 2009 Accurate detection of demosaicing regularity for digital image forensics. *IEEE Transactions on Information Forensics and Security* **4**(4), 899–910.

Cao H and Kot AC 2010 Mobile camera identification using demosaicing features. *IEEE International Symposium on Circuits and Systems (ISCAS)*, Paris, France.

Cao H and Kot AC 2011 Similar DSLR processor identification using compact model templates. *Asia Pacific Signal and Information Processing Association Annual Summit and Conference (APSIPA)*, Xian, China.

Çeliktutan O, Sankur B and Avcibaş İ 2008 Blind identification of source cell-phone model. *IEEE Transactions on Information Forensics and Security* **3**(3), 553–566.

Chang CC and Lin CJ 2001 LIBSVM: A library for support vector machines. *ACM Transaction on Intelligent System and Technology* **2**, 27.1–27.27.

Choi KS, Lam EY and Wong KKY 2006 Automatic source camera identification using intrinsic lens radial distortion. *Optics Express* **14**(24), 11551–11565.

Choi CH, Choi JH and Lee HK 2011 CFA pattern identification of digital cameras using intermediate value counting *MM&Sec'11, Proceedings of the 13th ACM Multimedia and Security Workshop*, ACM Press, New York. pp. 21–26.

Cortes C and Vapnik V 1995 Support-vector networks. *Machine Learning* **20**(3), 273–297.

Deng Z, Gijsenij A and Zhang J 2011 Source camera identification using auto-white balance approximation. *IEEE International Conference on Computer Vision (ICCV)*, Barcelona, Spain, pp. 57–64.

Devernay F and Faugeras O 2001 Straight lines have to be straight. *Machine Vision and Applications* **13**(1), 14–24.

Dirik AE, Bayram S, Sencar HT and Memon ND 2007 New features to identify computer generated images *IEEE International Conference on Image Processing (ICIP)*, San Antonio, TX, vol. 4, pp. 433–436.

Dirik AE, Sencar HT and Memon ND 2008 Digital single lens reflex camera identification from traces of sensor dust. *IEEE Transactions on Information Forensics and Security* **3**(3), 539–552.

Fang Y, Dirik AE, Sun X and Memon ND 2009 Source class identification for DSLR and compact cameras. *IEEE International Workshop on Multimedia Signal Processing (MMSP)*, Rio de Janeiro, Brazil.

Farid H 2008 Digital image ballistics from JPEG quantization: A followup study. Technical report TR2008-638, Department of Computer Science, Dartmouth College, Hanover, NH.

Farid H and Lyu S 2003 Higher-order wavelet statistics and their application to digital forensics. *IEEE Workshop on Statistical Analysis in Computer Vision (in Conjunction with CVPR)*, Madison ,WI.

Fischer A and Gloe T 2013 Forensic analysis of interdependencies between vignetting and radial lens distortion. *Proceedings of SPIE: Media Watermarking, Security, and Forensics 2013* (eds. Alattar AM, Memon ND and Heitzenrater CD), Burlingame, CA, vol. 8665, 86650D.

Fridrich J 2013 Sensor defects in digital image forensics. In *Digital Image Forensics* (eds. Sencar HT and Memon N). Springer Verlag, New York pp. 179–218.

Gallagher AC and Chen T 2008 Image authentication by detecting traces of demosaicing. *IEEE Computer Society Conference on Computer Vision and Pattern Recognition Workshops (CVPRW)*, Jampa, FL.

Gao S, Xu G and Hu RM 2011 Camera model identification based on the characteristic of CFA and interpolation. *Digital Forensics and Watermarking, 10th International Workshop, IWDW 2011* (eds. Shi YQ, Kim HJ and Pérez-González F), vol. 7128 of *Lecture Notes in Computer Science*. Springer-Verlag, Berlin, Germany, pp. 268–280.

Gao S, Hu RM and Tian G 2012 Using multi-step transition matrices for camera model identification. *International Journal of Hybrid Information Technology* **5**, 275–288.

Garfinkel S, Farrell P, Roussev V and Dinolt G 2009 Bringing science to digital forensics with standardized forensic corpora. *Digital Investigation* **6**, S2–S11.

Gloe T 2012a Feature-based forensic camera model identification. *LNCS Transactions on Data Hiding and Multimedia Security VIII (DHMMS)*, vol. 7228 of *Lecture Notes in Computer Science*, Springer, Berlin, Germany, pp. 42–62.

Gloe T 2012b Forensic analysis of ordered data structures on the example of JPEG files. *IEEE International Workshop on Information Forensics and Security (WIFS)*, Jenerife, Spain, pp. 139–144.

Gloe T and Böhme R 2010 The Dresden image database for benchmarking digital image forensics. *Journal of Digital Forensic Practice* **3**, 150–159.

Gloe T, Borowka K and Winkler A 2009 Feature-based camera model identification works in practice – Results of a comprehensive evaluation study. In *11th Information Hiding, Revised Selected Papers* (eds. Katzenbeisser S and Sadeghi AR), vol. 5806 of *Lecture Notes in Computer Science*. Springer-Verlag, Berlin, Germany, pp. 262–276.

Gloe T, Borowka K and Winkler A 2010 Efficient estimation and large-scale evaluation of lateral chromatic aberration for digital image forensics. *Proceedings of SPIE: Media Forensics and Security II* (eds. Memon ND, Dittmann J, Alattar AM and Delp III EJ), San Jose, CA, vol. 7541, p. 754107.

Gloe T, Pfennig S and Kirchner M 2012 Unexpected artefacts in PRNU-based camera identification: A 'Dresden Image Database' case-study. *MM&Sec'12, Proceedings of the 14th ACM Multimedia and Security Workshop*, Coventry, UK, pp. 109–114.

Goljan M and Fridrich J 2012 Sensor-fingerprint based identification of images corrected for lens distortion. *Proceedings of SPIE: Media Watermarking, Security, and Forensics 2012* (eds. Memon ND, Alattar AM and Delp III EJ), San Francisco, CA, vol. 8303, p. 83030H.

Goljan M, Fridrich J and Filler T 2009 Large scale test of sensor fingerprint camera identification In *Proceedings of SPIE: Media Forensics and Security XI* (eds. Delp III EJ, Dittmann J, Memon N and Wong PW), San Jose, CA, vol. 7254, p. 72540I.

Gunturk BK, Glotzbach J, Altunbasak Y, Schafer RW and Mersereau RM 2005 Demosaicking: Color filter array interpolation. *IEEE Signal Processing Magazine* **22**(1), 44–54.

Ho J, Au OC and Zhou J 2010 Inter-channel demosaicking traces for digital image forensics. *IEEE International Conference on Multimedia and EXPO (ICME)*, Singapore, pp. 1475–1480.

Holst GC and Lomheim TS 2007 *CMOS/CCD Sensors and Camera Systems*. SPIE Press, Bellingham, WA.

Hsu YF and Chang SF 2010 Camera response functions for image forensics: An automatic algorithm for splicing detection. *IEEE Transactions on Information Forensics and Security* **5**(4), 816–825.

ITU 1992 *Information Technology – Digital Compression and Coding of Continuous-Tone Still Images – Requirements and Guidelines*. ITU-T Recommendation T.81, ITU, Gneneva, Switzerland. http://www.itu.int/rec/T-REC-T.81/en (Accessed 26 February 2015).

Japan Electronics and Information Technology Industries Association 2002 *JEITA CP-3451, Exchangeable Image File Format for Digital Still Cameras: Exif Version 2.2* Japan Electronics and Information Technology Industries Association, Tokio, Japan.

Johnson MK and Farid H 2006 Exposing digital forgeries through chromatic aberration *MM&Sec'06, Proceedings of the Multimedia and Security Workshop 2006* ACM, New York, pp. 48–55.

Kang SB and Weiss R 2000 Can we calibrate a camera using an image of a flat, textureless lambertian surface? In *Computer Vision – ECCV 2000, Proceedings, Part II* (ed. Vernon D), vol. 1843 of *Lecture Notes in Computer Science*. Springer Verlag, Berlin, Germany, pp. 640–653.

Kee E, Johnson MK and Farid H 2011 Digital image authentication from JPEG headers. *IEEE Transactions on Information Forensics and Security* **6**(3), 1066–1075.

Khanna N, Mikkilineni AK and Delp EJ 2009 Scanner identification using feature-based processing and analysis. *IEEE Transactions on Information Forensics and Security* **4**(1), 123–139.

Kharrazi M, Sencar HT and Memon ND 2004 Blind source camera identification. *IEEE International Conference on Image Processing (ICIP)*, Singapore, pp. 709–712.

Kirchner M 2010 Efficient estimation of CFA pattern configuration in digital camera images. *Proceedings of SPIE: Media Forensics and Security II* (eds. Memon ND, Dittmann J, Alattar AM and Delp III EJ), San jose, CA, vol. 7541, p. 754111.

Kirchner M and Böhme R 2009 Synthesis of colour filter array pattern in digital images *Proceedings of SPIE: Media Forensics and Security* (eds. Delp III EJ, Dittmann J, Memon ND and Wong PW), San jose, CA, vol. 7254, p. 72540K.

Lin Z, Wang R, Tang X and Shum HY 2005 Detecting doctored images using camera response normality and consistency. *IEEE Computer Society Conference on Computer Vision and Pattern Recognition (CVPR)*, San Diego, CA, vol. 1, pp. 1087–1092.

Long Y and Huang Y 2006 Image based source camera identification using demosaicking. *IEEE Workshop on Multimedia Signal Processing (MMSP)*, Victoria, BC, Cananda, pp. 419–424.

Lyu S 2010 Estimating vignetting function from a single image for image authentication. *MM&Sec'10, Proceedings of the Multimedia and Security Workshop 2010*. ACM, New York, pp. 3–12.

Lyu S and Farid H 2005 How realistic is photorealistic? *IEEE Transactions on Signal Processing* **53**(2), 845–850.

Mallat SG 1989 A theory for multiresolution signal decomposition: The wavelet representation. *IEEE Transactions on Pattern Analysis and Machine Intelligence* **11**(7), 674–693.

Menon D and Calvagno G 2011 Color image demosaicking: An overview. *Signal Processing: Image Communication* **26**(8–9), 518–533.

Mitsunaga T and Nayar SK 1999 Radiometric self calibration. *IEEE Computer Society Conference on Computer Vision and Pattern Recognition (CVPR)*, Fort Collins, CO, pp. 347–380.

Moon TK 1996 The expectation-maximization algorithm. *IEEE Signal Processing Magazine* **13**(6), 47–60.

Neelamani R, de Queiroz R, Fan Z, Dash S and Baraniuk RG 2006 JPEG compression history estimation for colour images. *IEEE Transactions on Image Processing* **15**(6), 1365–1378.

Ng TT 2009 Camera response function signature for digital forensics – Part II: Signature extraction. *IEEE International Workshop on Information Forensics and Security (WIFS)*, London, UK, pp. 161–165.

Ng TT and Tsui MP 2009 Camera response function signature for digital forensics – Part I: Theory and data selection *IEEE International Workshop on Information Forensics and Security (WIFS)*, London, UK, pp. 156–160.

Ng TT, Chang SF, Hsu J, Xie L and Tsui MP 2005 Physics-motivated features for distinguishing photographic images and computer graphics. *Proceedings of the 13th Annual ACM International Conference on Multimedia*. ACM Press, New York, pp. 239–248.

Ng TT, Chang SF and Tsui MP 2007 Using geometry invariants for camera response function estimation. *IEEE Conference on Computer Vision and Pattern Recognition, CVPR 2007*, Minneapolis, MN.

Ojala T, Pietikäinen M and Harwood D 1996 A comparative study of texture measures with classification based on featured distributions. *Pattern Recognition* **29**(1), 51–59.

Ojala T, Pietikäinen M and Mäenpää T 2002 Multiresolution gray-scale and rotation invariant texture classification with local binary patterns. *IEEE Transactions on Pattern Analysis and Machine Intelligence* **24**(7), 971–987.

Pevný T and Fridrich J 2008 Detection of double-compression in JPEG images for applications in steganography. *IEEE Transactions on Information Forensics and Security* **3**(2), 247–258.

Popescu AC and Farid H 2005 Exposing digital forgeries in colour filter array interpolated images. *IEEE Transactions on Signal Processing* **53**(10), 3948–3959.

Pudil P, Novovičová J and Kittler J 1994 Floating search methods in feature selection. *Pattern Recognition Letters* **15**(11), 1119–1125.

Ramanath R, Snyder W, Yoo Y and Drew M 2005 Color image processing pipeline. *IEEE Signal Processing Magazine* **22**(1), 34–43.

Schölkopf B, Platt JC, Shawe-Taylor J, Smola AJ and Williamson RC 2001 Estimating the support of a high-dimensional distribution. *Neural Computation* **13**(7), 1443–1471.

Swaminathan A, Wu M and Liu KJR 2007 Nonintrusive component forensics of visual sensors using output images. *IEEE Transactions on Information Forensics and Security* **2**(1), 91–106.

Swaminathan A, Wu M and Liu KJR 2009 Component forensics. *IEEE Signal Processing Magazine* **26**(2), 38–48.

Takamatsu J, Matsushita Y, Ogasawara T and Ikeuchi K 2010 Estimating demosaicing algorithms using image noise variance. *IEEE Computer Society Conference on Computer Vision and Pattern Recognition (CVPR)*, San Francisco, CA.

Torralba A and Efros AA 2011 Unbiased look at dataset bias. *IEEE Computer Society Conference on Computer Vision and Pattern Recognition (CVPR)*, Colorado Spring, CO, pp. 1521–1528.

Tsai MJ, Wang CS, Liu J and Yin JS 2012 Using decision fusion of feature selection in digital forensics for camera source model identification. *Computer Standards & Interfaces* **34**(3), 292–304.

Van LT, Emmanuel S and Kankanhalli MS 2007 Identifying source cell phone using chromatic aberration. *IEEE International Conference on Multimedia and EXPO (ICME)*, Beijing, China, pp. 883–886.

Vandewalle P, Kovačević J and Vetterli M 2009 Reproducible research in signal processing. *IEEE Signal Processing Magazine* **26**(3), 37–47.

Wahab AWA, Ho AT and Li S 2012 Inter-camera model image source identification with conditional probability features. *Proceedings of the IIEEJ Image Electronics and Visual Computing Workshop*, Kuching, Malaysia. Paper ID 2P-2.

Willson RG and Shafer SA 1994 What is the center of the image? *Journal of the Optical Society of America A* **11**(11), 2946–2955.

Wu TF, Lin CJ and Weng RC 2004 Probability estimates for multi-class classification by pairwise coupling. *The Journal of Machine Learning Research* **5**, 975–1005.

Xu G and Shi YQ 2012 Camera model identification using local binary patterns. *IEEE International Conference on Multimedia and EXPO (ICME)*, Melbourne, VIC, Australia, pp. 392–397.

Xu G, Gao S, Shi YQ, Hu R and Su W 2009 Camera-model identification using Markovian transition probability matrix. *Digital Watermarking, 8th International Workshop, IWDW 2009* (eds. Ho ATS, Shi YQ, Kim HJ and Barni M), vol. 5703 of *LNCS*. Springer-Verlag, Berlin, Germany, pp. 294–307.

Yu J, Craver S and Li E 2011 Toward the identification of DSLR lenses by chromatic aberration. *Proceedings of SPIE: Media Forensics and Security III* (eds. Memon ND, Dittmann J, Alattar AM and Delp III EJ), San Francise, CA, vol. 7880, p. 788010.

10

Printer and Scanner Forensics

Shize Shang and Xiangwei Kong
School of Information and Communication Engineering, Dalian University of Technology, Dalian, China

10.1 Introduction

With the development of electronics, printing and photocopying have become fast, high quality and low cost. Printed and photocopied materials are an important way by which information is exchanged by government departments, companies and individuals. Printers and photocopiers produce many files, including documents, contracts, tickets, patient records, receipts, etc. Some files are important and cannot be replaced by electronic documents. However, printed and photocopied documents can be easily altered using physical manipulation, chemical manipulation or sophisticated software. In recent years, the number of instances in which documents have been altered has increased dramatically, resulting in economic crime and disturbance of social order. Criminals change documents directly, such as by erasing, changing, replacing, or copying and pasting content on the documents. Others modify scanned images of documents, created using image processing software such as Photoshop.

In the process of criminal investigation, it is important to track the sources of documents and verify their authenticity. If textual evidence is provided in a court trial, the authenticity of the documents must be verified. Document examination is helpful for analyzing cases, narrowing the scope of investigation, tracing criminals, etc. Thus, document examination is an important part of judicial identification.

Two techniques can be used to resolve the security problems of printed and photocopied documents, and to protect document authenticity and integrity: proactive security measures and passive forensics. The aim of proactive forensics is to

Handbook of Digital Forensics of Multimedia Data and Devices, First Edition.
Edited by Anthony T.S. Ho and Shujun Li.
© 2015 John Wiley & Sons, Ltd. Published 2015 by John Wiley & Sons, Ltd.
Companion Website: www.wiley.com/go/digitalforensics

add watermarking information to digital images and documents in advance. Robust watermarking protects the copyrights of the products, and fragile watermarking prevents tampering. For example, yellow dot watermarking is added to colour printed documents in a way that records information about the source of the document, including device make, model, printing time, etc. (Foundation n.d.). Other watermarking techniques, such as special ink, special paper and holographic images, are also used in copyright protection. The aim of passive forensics is to trace the sources of documents and expose forgeries of documents by analyzing documents in bulk (rather than individual, suspicious documents). Prosecutors usually uses professional equipment to examine documents, as well as the assistance and judgement of forensic experts.

Being able to determine whether a document is original, counterfeit or forged is crucial for protecting government departments, companies and individuals from tangible and intangible losses and damages. Forensic results can be regarded as evidence in a court and can help in the detection of crimes. Kelly and Lindblom (2006) presented several questions to be answered by print forensics: Who created the document? Which device created the document? Which part has been modified in the document? Is the document as old as it purports to be? These questions can be answered by doing two things: identifying the source printer and detecting tampering. Traditional technology for inspecting documents requires not only professional theoretical knowledge and rich experience but also expensive document inspection equipment.

There are several traditional technologies for document inspection, such as file format examination, visible light inspection, infrared inspection ultraviolet-ray inspection, fluorescent inspection, microscopic examination, chemical inspection, and residual word inspection (Liu and Guo 2008; Ma 2007). These technologies require document inspection equipment and professional testing personnel. They are costly and slow. Furthermore, some technologies are lossy for documents, especially chemical forensic methods. Various methods of digital image forensics were introduced in the book *Digital Image Forensics: There is More to a Picture than Meets the Eye* edited by Sencar and Memon (2013). Although print forensics are also based on scanned images, document forensics differ greatly from digital image forensics, such as in regards to image source, image content and tampering methods being detected. New lossless methods should be developed to reduce the reliance on equipment and personnel, improve detection efficiency and reduce testing costs. Tchan (2000, 2004), Khanna *et al.* (2006) and Martone *et al.* (2006) used a scanner and computer for forensics, and achieved some satisfactory results. These new technologies work without document inspection equipment and personnel, and deal with documents in bulk. The task of passive print forensics is to provide a partial alternative to tedious file inspection. Passive print forensics can reduce reliance on professional equipment and personnel, make detection results more objective and efficient and reduce detection costs and time.

In this chapter, we exhibit the technologies on device source identification and document forgery detection, and give three examples on printer identification, scanner identification and document forgery detection for details. We expect that the readers could know the development situation on this research area, and understand how the technologies work through the examples. This chapter is organized as follows. Section 10.1 provides an overview of printer and scanner forensics and compares print forensics and digital image forensics. It also introduces the document lifecycle. Section 10.2 presents the difference between laser printers and inkjet printers, as well as a flowchart of printer forensics and forensic algorithms that are based on hardware defects and intrinsic character features. Section 10.3 shows algorithms used for scanner forensics and a flowchart of scanner forensics. Section 10.4 presents algorithms for photocopier forensics. Section 10.5 introduces some tampering methods and presents flowcharts and algorithms for detecting document forgery. These algorithms focus on forgery of documents and scanned document images. Section 10.6 provides three sample algorithms: for printer forensics, scanner forensics and tampering detection. The section also illustrates details of the algorithms that should clarify how the algorithms work. In Section 10.7, we present open problems and challenges for forensics. In Section 10.8, we draw conclusions.

10.1.1   Comparison with Digital Image Forensics

1. *Sources of document images differ*: The images in digital image forensics are captured by cameras or generated by computers. Image information stored on a hard drive cannot be damaged, except by destroying the hard drive, image content won't degrade over time. Images used for print forensics come from printed and photocopied documents, and the paper documents can be easily polluted or damaged by human. In addition, paper documents age over time. Therefore, before images of paper documents are captured, some valuable information may be lost. Finally, the influence of scanners should also be considered in print forensics, as some noise and intrinsic features are introduced into scanned images.
2. *Image contents differ*: Digital images can include landscapes, architectures, people, etc. Digital pictures have rich content, containing lots of texture and colour information. Colour information is also an aspect of research in digital forensics. Document image content usually consists of characters, and document images are usually in greyscale. So the content of text documents is monotonous and lacks textural information. The background of such scanned images is the copy paper; sometimes it is useless for print forensics, and the detection result can easily be influenced by the texture of the paper.
3. *Tampering methods differ*: Image processing software such as Photoshop, Auto Collage Studio and Graphic Workshop Pro is used to tamper with images. All tampering based on software occurs after an image is captured, so the detection of forgery involves analyzing image edge information, statistical characteristics, re-compression, re-sampling, etc. For print forensics, most detectable samples are the paper documents, and most tampering occurs before image scanning. Unlike

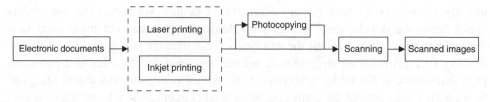

Figure 10.1 Document lifecycle.

with digital image forensics, we cannot research print forensics only based on the characteristics of scanned images; the way to solve this problem is to consider the signatures in the document introduced by hardware defects and tampering methods.

10.1.2 Document Lifecycle

The document lifecycle includes all of the processes by which an electronic document is transformed into a scanned document image, such as laser printing or inkjet printing, photocopying and scanning. The document lifecycle is illustrated in Figure 10.1. The document is printed using either laser printing or inkjet printing, and the document may then be photocopied. Because passive print forensics is based on image processing and analyzing, the printed or photocopied document must then be scanned to a digital image. The lifecycle of a document printed on a laser printer is characterized by charging, exposure, development, transference, fusing and cleaning of the laser printer (Chiang *et al.* 2009). For a document printed on an inkjet printer, it is characterized by carriage movement, ink jetting and paper delivery. Besides printing, the document lifecycle includes the scanning (or photocopying and scanning) process. The characteristics of a scanned document image reflect those of the laser or inkjet printer, the photocopier and the scanner used on it, and these characteristics overlap in the image.

In the laser printing process, the document completes six steps presented earlier, in which the mechanical defects and toner distribution will be reflected on the printed documents. An inkjet printer consists of three principal components: the printhead, carriage and paper advancement mechanism. In the inkjet printing process, the printhead jets ink, and then the carriage slides and the paper advances at the same time. In this process, also, the characteristics of the mechanisms and the ink affect the document. After the printing process, the document is scanned as a document image. The scan head, with its charge coupled device (CCD) or complementary metal oxide semiconductor (CMOS) sensor, translates linearly to capture the image using a belt and stepper motor. Sensor noise and scanner bed defects will be reflected in the document image.

At each step in the document lifecycle, characteristics are introduced into the document or document image, which result from hardware defects, intrinsic signatures, sensor noises, etc. These characteristics can be considered the fingerprints of devices; they can usually be used to identify the devices and to expose forgeries.

10.2 Printer Forensics

10.2.1 Working Principles of Laser Printers and Inkjet Printers

Because the construction and working principles of laser printers and inkjet printers differ greatly, their printouts demonstrate different character morphology and characteristics. The differences in printouts are clearly visible using a microscope or enlargements of scanned character images. Figure 10.2 shows scanned images of the English words 'Robinson Crusoe' as produced by a laser printer and an inkjet printer. The characters printed by the laser printer have clearer contours and less noise; by contrast, the edges of characters printed by the inkjet printer are rough, with periodic 'tails' or 'satellites' (Chiang *et al.* 2009), and character contours are less clear. These characteristics are closely related to the working principles. We describe the reasons why they have divergences in character morphology as follows.

The process of laser printing can be summarized in six steps: (i) charging, (ii) exposure, (iii) developing, (iv) transferring, (v) fusing and (vi) cleaning (Chiang *et al.* 2009). The architecture of a laser printer is shown in Figure 10.3a. The transmission of printing information mainly depends on the exchange of optical signals and electrical signals, printing information is transported with very little distortion. The first four steps are associated with the optical photoconductor (OPC), so the printing quality is determined by the OPC. In the fusing step, the high temperature of the fuser melts the toner on the paper, and the melting toner is pressed into the paper by the rollers.

An inkjet printer consists of three parts: printhead, carriage and paper advancement mechanism. The architecture of an inkjet printer is shown in Figure 10.3b. In the printing process, the printhead moves back and forth on the carriage and scans the paper under the carriage line by line. Simultaneously, the nozzles on the printhead eject the ink onto the paper and the paper advancement mechanism moves the paper out of

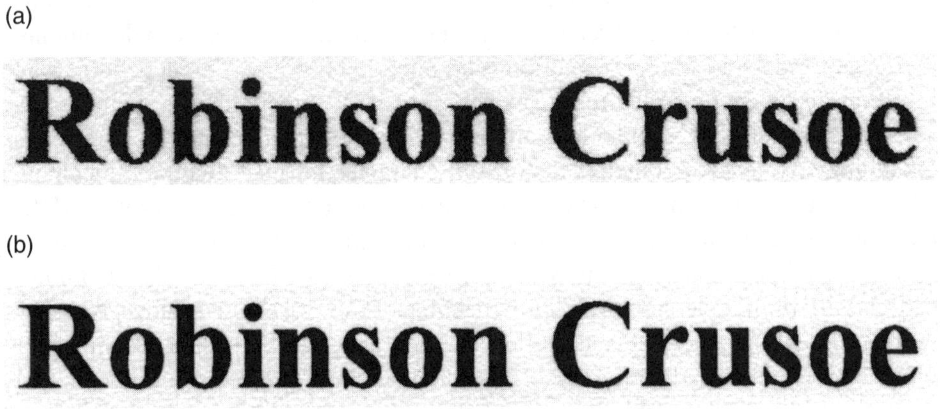

(a)

Robinson Crusoe

(b)

Robinson Crusoe

Figure 10.2   Words printed with a laser printer (a) and an inkjet printer (b).

(a)   (b)

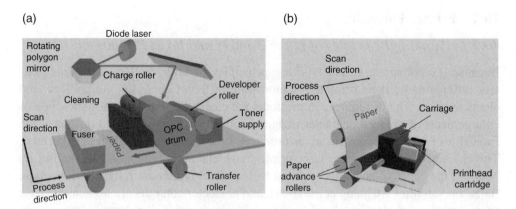

Figure 10.3   Architecture of (a) a laser printer and (b) an inkjet printer (Chiang *et al.* 2009).

the printer, and then the inkjet printing process is finished. Because the moving of the printhead and the ejecting of ink are synchronous, there will be a 'tail' at the edges of the characters. In addition, the ink in the inkjet printer is liquid and the diffusion rate of ink on paper is faster than that of toner, so the black-to-white transition at the edges of characters is wider than that of laser printing. For this reason, the contours of inkjet printed characters are not as clear as those of laser printed characters.

10.2.2   Flowchart of Printer Forensics

The aim of printer forensics is to trace the manufacturer, model and individual printer of a document. A flowchart of the printer forensics process is shown in Figure 10.4. The steps in the process are document scanning, feature extraction, classification or correlation matching, and decision making. First, the document should be scanned to a digital image, as printer forensics is based on image processing and pattern recognition. Normally, images are saved in TIFF format, because TIFF images have no quantization distortion and retain the most original information from documents of any image formats. Scanning resolution is commonly 600–1200 dpi, depending on the algorithms used. Second, features are extracted from document images. Some features focus on character images and others are from whole document pages. Features consist of banding frequency, geometric distortion, character intrinsic features, mechanical defects, gear indentation, etc. The features extracted are based on hardware defects and character signatures. Third, classifier or correlation matching is used to identify the features. Some classifiers need feature samples to train a model. Commonly used classifiers include support vector machines (SVMs), the k-Nearest Neighbour algorithm (k-NN), Fisher linear classifiers and neural networks. Last, a decision should be made based on classification or correlation matching. The decision may be a label number, a probability that the document belongs in a category, or a graph that shows, visually, which category the document likely belongs to.

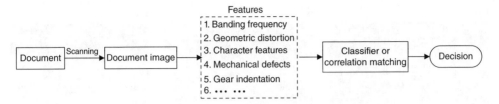

Figure 10.4   Flowchart of printer forensics.

10.2.3   *Laser Printer Forensics*

We introduce laser printer forensics from three different aspects: banding frequencies, geometric distortion and intrinsic character features. Banding frequencies and geometric distortion are hardware defects. Banding frequencies are mainly extracted from printed images, and much work must be done before they can be used in practice; geometric distortion is a feature of the whole page of a document but not of local areas or characters, which occurs for both text documents and printed images. Intrinsic character features are embodied in character morphology, noise levels, grey value textures and printing quality. These features provide more detail than the other two kinds of features, and they can be extracted from individual characters.

10.2.3.1   Banding Frequencies

Four of the six steps in the laser printing process are associated with OPC, so the OPC greatly influences document quality. When the laser printer is working, the rotational velocity of the OPC is not absolutely constant; this kind of hardware defect is reflected on the printed document and is called 'banding'. If the rotational velocity of the OPC decreases just a little, the grey level of the corresponding printing region will be dark; if the rotational velocity of the OPC increases just a little, the grey level on the corresponding printing region will be light. Either of these can appear as non-uniform light and dark lines perpendicular to the process direction (Chiang *et al.* 2009). The rotational velocity of the OPC is determined by the gears on the two sides, so the frequency of banding is also associated with the gears. Figure 10.5 shows three original images and three printed images with banding frequencies. Figure 10.5a–c are standard testing images from the Matlab toolkit and Figure 10.5d–f are their respective printed images.

Different brands of laser printers use gears of different sizes and rotational velocities, resulting in different banding frequencies. This hardware defect thus reveals information about the source printer, so some researchers want to use banding frequency to identify source printers. But in practical cases, it is difficult to extract banding features from grey or colour images. Because changes in the grey level of an image depend on image content and image content occurs randomly in the image, banding characteristic will be submerged in image content. Ali *et al.* (2009) extract

(a)   (b)   (c)

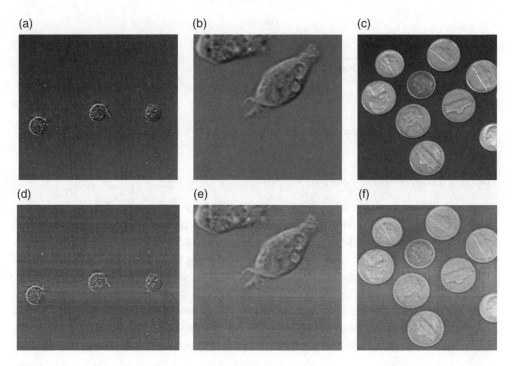

(d)   (e)   (f)

Figure 10.5 Examples of banding frequency in printed images: images a–c are standard testing images from the Matlab toolkit; images d–f are their respective printed images, which show the banding frequencies.

banding frequencies from analyzing the grey levels of images chosen in advance. After the images are scanned, a horizontal projection is applied to compute the histogram of row projection. The histogram spectrum is computed using a fast Fourier transform (FFT) and the banding frequency is the peak in the Fourier spectrum.

10.2.3.2 Geometric Distortion of Pages

In addition to the signature of banding frequency, another hardware defect can result in geometric distortion of pages, which is reflected on documents as character location distortion and line slope distortion. The distortion is caused by spinning velocity fluctuations of polygon mirrors and imperfections of the paper feeding mechanism (Bulan *et al.* 2009; Wu *et al.* 2009). When the laser printer works, the polygon mirror rotates constantly and reflects the laser beam, which is the same as the electronic signal to print on the surface of the OPC. But in practice, the rotational velocity of the polygon mirror is not absolutely constant, and tiny differences in velocity can shift character location. In addition, there may be tiny differences in the distances between pairs of rollers in the paper feeding mechanism, which can cause line slope distortion across the whole page. Different brands of printers produce different types

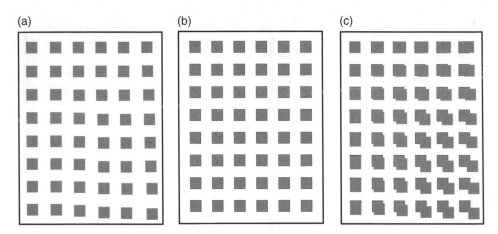

Figure 10.6 Illustration for document image matching to measure geometric distortion: (a) real document image, (b) ideal document image and (c) result for image matching.

of geometric distortion, even individual printers of the same brand and model may have different distortion distributions. Because the signature of distortion remains for a long time and differs for individual printers, geometric distortion can be used to help trace the source printer of a document. Figure 10.6 shows the model of image matching between a real document image and an ideal document image. Each colour block denotes one character in the document. The characters in the real document are subject to geometric distortion; the ideal document image is a perfectly geometrical image estimated based on the real document. The character in the upper-left corner is used as the basis matching character. The image matching result on the right of Figure 10.6. Any character except the basis matching character may have geometric distortion. Figure 10.7 shows two geometric distortion graphs: one each for two laser printers of different make and model (scanning resolution is 1200 dpi and the unit is one pixel). Each dot denotes the original location of a character and each small arrow denotes the orientation and strength of the geometric distortion for that character. The colour curves label the isolines of distortion strength.

Wu *et al.* (2009) created a model of geometric distortion and estimated the model parameters for identifying source printers. The ideal image for a document can be obtained using optical character recognition (OCR) on the scanned image, and then the distortion can be computed at any location on the document through by matching the scanned image and ideal image. Supposing that the projection relationship between the two images will correspond to projection transformation, the parameters of projection transformation can be estimated by the least squares method and SVD. Then, the estimated parameters can be used as features for classifying the source printer based on SVM. The signature of geometric distortion is caused by hardware defects in the laser printer and has nothing to do with the toner. So the performance of this technique is not influenced by the changing of toner or by toner

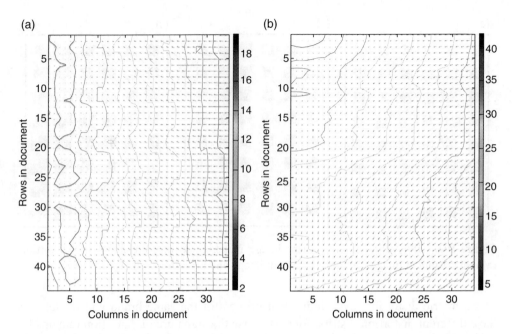

Figure 10.7 Graphs of geometric distortion of document pages for (a) Canon 3010 and (b) Samsung ML3471ND laser printers.

density. Bulan *et al.* (2009) also developed a method to identify the source printer based on geometric distortion of a scanned image in the document. Printed images comprise halftone dots and permutations of the dots follow certain rules. In real printed images, halftone dots are offset from their original locations by geometric distortion. The actual locations of halftone dots can be estimated using dot centroid computation. The source printer of the printed image can be determined by computing the correlation of distortion parameters with a training set.

10.2.3.3 Intrinsic Character Features

Because different types of printers have different hardware and software structures, printed characters differ in toner density distribution, morphological characteristics and the roughness of edge profiles. Some characteristics are peculiar to specific printer brands or types, while some features are specific to certain printers.

The grey-level co-occurrence matrix (GLCM) can describe the spatial relationships between pixels, and this kind of feature is sensitive to image texture. Because the letter 'e' is the most frequently used English letter, Bulan *et al.* (2011) identified printers by analyzing changes in the grey level of the letter 'e'. GLCM features are extracted from eight qualities of a letter, including means, variances, entropy measures, maximum entry, etc. Sequential forward selection (SFS) is applied to select features that contribute greatly to classification. The aim of using SFS is to reduce computational complexity. The dimensions of features are reduced to 3D by linear discriminant analysis (LDA) and SVM is used as the classifier.

Shen *et al.* (2007) identified printers based on the features of quality measurements. Because laser printed documents have quality defects, such banding, it is possible to identify printers based on character quality. Each character image is de-noised using a Gaussian filter. Image quality measure (IQM) features are extracted from the differences between the original image and the Gaussian filtered image. Seven dimensions of features are extracted, such as grey-level features, correlation features and spectrum features. This approach can use arbitrary characters for detection. Based on the IQM features, Kong *et al.* (2010) improved the performance by increasing the number of measured features to 24. Two levels of wavelet transformation are conducted on characters, and four features are extracted from the six sub-images in wavelet transformation, except for the low-frequency component in the first level. The features extracted are mean, variance, skew and kurtosis. The SVM classifier is applied to classify the characters and the decision regarding the document is based on a majority voting.

The approaches discussed earlier can identify the source printers of documents based on characters. If the accuracy is high enough, these approaches can help expose forgeries and locate characters that have been tampered with. Features extracted using these approaches are based on statistical characteristics, so they are sensitive to changes in toner density. For example, if we use a model trained using characters with toner density of 90% to test documents with toner density of 40%, the classification results will include a large deviations.

10.2.4   Inkjet Printer Forensics

Inkjet printers are another common kind of printer. Although they provide less printing quality for text documents, they provide better performance for picture printing. We introduce inkjet printer forensics in regards to mechanical defects and gear indentation. Mechanical defects result in character deformation, but the defects are inconsistent across printers so analysis of mechanical defects cannot be performed in exclusion. Gear sets are the essential components of inkjet printers, they leave indentation information on paper when printing. However, profession equipment is required to extract indentation information.

10.2.4.1   Mechanical Defects

Because laser printers and inkjet printers have different structures with different hardware defects, manifestations of the defects on documents also differ. Inkjet printheads have many nozzles, and ink is ejected from these nozzles onto the paper. If some nozzles are blocked (which often happens), blank lines will reoccur throughout the document. The carriage carries the printhead back and forth in the scanning direction during printing. If tracks or gears are misaligned, some characters are distorted or the positions of adjacent lines are obviously distorted. For colour inkjet

Figure 10.8 The paper conveyance unit of an inkjet printer.

printing, colour distortion and colour misalignment on documents are other kinds of defects. These signatures remain stable on the document. Wang (1998) identified source inkjet printers based on these defects. These signatures can be observed under a microscope and source printers can be identified by comparison.

10.2.4.2   Gear Indentation

The paper conveyance gears are one of the important parts of an inkjet printer; they can fix the position of the paper when the paper is ejected. After inkjet printing, the sharp spurs of the gears leave indentations on the document. For different brands and models, the distances between gears and the distances between spurs on the same gear differ, so gear indentation can be used as a signature for identifying source inkjet printers. Because indentations are too small to see with the naked eye, (Yoshinori *et al.* 2002, 2009) used high power microscopic scanning equipment to obtain high-resolution sample images, then used infrared oblique lighting and gradient image processing to capture indentation images. Figure 10.8 shows an graph of paper conveyance gears. There are two rows of gears that fix the paper when documents are being printed. Source inkjet printers can be identified by computing the distances between gears and the distances between spurs on the same gear. Because the technique uses infrared rays, the extraction of indentations is independent of image content.

10.3   Scanner Forensics

A flatbed scanner comprises a scanner bed, motor, scan head, lamp, mirror lens and imaging sensor, as shown in Figure 10.9. Most flatbed scanners use CCD/CMOS imaging sensors, these sensors are 1D. The document is put on the scanner bed, the

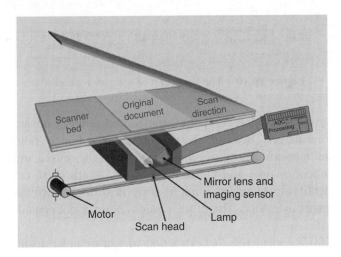

Figure 10.9   The paper conveyance unit of inkjet printer (Chiang *et al.* 2009).

lamp illuminates the document through the scanner bed and the light is reflected onto the 1D imaging sensor. The motor moves the imaging sensor and the sensor scans information line by line. When the scan head with the imaging sensor moves from one end to the other, the scanner finishes scanning work and records the document image on the computer.

The scanning process is essential in printer and scanner forensics, and the scanner leaves its signature on the image just like with a digital camera. We introduce two aspects of scanner forensics: one is sensor noise, the other is dust and scratches. Sensor noise is the fingerprint for the imaging sensors; dust and scratches come from the scanner bed and also can be remained for a long time.

10.3.1   Flowchart of Scanner Forensics

Scanner forensics are used when the original document is unavailable and only the scanned document image is supplied. The aim of scanner forensics is to identify the make, model and individual scanner used to scan an image. Because the sources of scanned images vary, image formats may be JPEG, TIFF or BMP, and the resolution of the scanned image may be lower than 600 dpi. Scanner forensics is hard work in practice, because the scanned image may not be high definition or the scanning process may have introduced quantization error. A flowchart of scanner forensics is shown in Figure 10.10, it is similar to the flowchart of printer forensics and only lacks the process of scanning. The features are extracted from the whole document page normally.

10.3.2   Sensor Noise

The sensors in flatbed scanner are usually CCD and CMOS imaging sensors. Both sensors introduce noise into the scanned images. There are three types of noise

Figure 10.10   The flowchart of scanner forensics.

from sensor: sensor array defects, pattern noise and random noise. Array defects consist of hotpoint defects, dead pixels, pixel traps, etc.; Pattern noise is caused by dark currents and non-uniformity of photo response in the sensor array, this type of noise is independent of time and image content, so pattern noise can be used as a signature for identifying scanners; random noise is strongly correlated with image content and changes over time. Therefore, random noise is not considered for source identification.

Gou *et al.* (2007) considered the extraction of sensor noise from three aspects. (i) First they estimated the noise of an image using different de-noising algorithms, such as an averaging filter, Gaussian filter, median filtering and Wiener adaptive image de-noising algorithm, and computed the mean and the standard deviation of the log 2 transformed absolute values of the noise as the features. (ii) Then, they analyzed the noise in wavelet coefficients and computed the mean and the standard deviation of high-frequency coefficients as the second part of noise features. (iii) Finally, they characterized neighbourhood prediction error in the smooth regions as the statistical noise features. In total, 60 dimensions of features were extracted from the scanned images. Khanna *et al.* (2009) identified the scanners based on the pattern noise, because CCD/CMOS sensors in scanners are 1D, unlike sensors in cameras. A 2D wavelet transformation was applied to de-noise the scanned image, and the 1D pattern noise is obtained by averaging the 2D noise image in rows and columns. The 51 dimensions of features included standard deviation, skew, kurtosis, etc.

10.3.3   Dust and Scratches

In the scanning process, the laser beam irradiates the document through the plate glass on the scanner, and the laser is reflected on the CCD/CMOS sensor. If there is dust or scratches on the plate glass, the noise caused by the dust and scratches is stored on the scanned image. So not only the pattern noise in the CCD/CMOS sensor is introduced into the image but also the noise from the plate glass. Usually, the dust is on the inner surface of the plate glass and the scratches are on the glass. This kind of signature does not change for a long time, so it can be considered the scanner's fingerprint. So the dusts and scratches on the plate glass are applied to identify the scanner. Figure 10.11 is an image block in a scanned picture. Dust is labelled with rectangles. Dirik *et al.* (2009) used blemish detection methods to capture image dust. After the scanned image is filtered with a high-pass filter, texture information, such as edge dust and scratches, is left on the image. A dust and scratch model established in advance is applied to extract the dust and scratch information from the filtered image. The dust and scratch

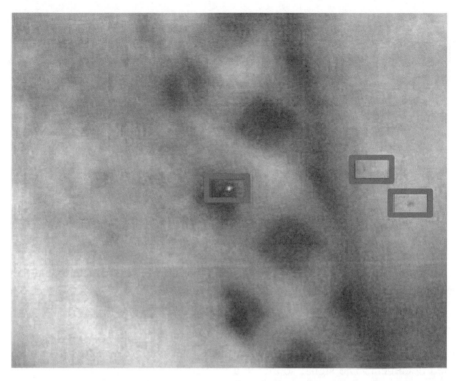

Figure 10.11   A printed graph with the signs of dust marked in the printing (Dirik *et al.* 2009).

positions can be determined by computing normalized cross correlation (NCC) of the model and filtered image. Several scanners are modelled on this method and store dust and scratch positions in a data set. If an unknown scanned image matches well with one of the scanned models in the data set, the source of the unknown image can be identified.

10.4   Photocopier Identification

10.4.1   Contact between Printer and photocopier

Most common photocopiers are xerographic copiers; the working principle of these photocopiers is similar to that of laser printing. Besides the six steps in laser printing, the process in photocopying consists of scanning step before the printing process. In other words, the source signals for laser printing and photocopying differ. Because paper documents are distorted relative to electronic documents and a lot of noise is introduced during the scanning processing, the character quality is not as good as that of laser printed characters. Figure 10.12 compares laser printed words to photocopied words. Because a lot of noise is introduced into document, the roughness

(a)

Robinson Crusoe

(b)

Robinson Crusoe

Figure 10.12   (a) is laser printed words and (b) is photocopied words.

of photocopied character edges is greater than that of printed characters, and there is more impulsive noise in the smooth character regions.

10.4.2   Character Signature

The acquisition process of a photocopied document is more complex than that of a printed document. The signal source is a printed document, and the signatures and noise in the printed document are introduced into the photocopied document. Also, the document is scanned into the photocopier, which also introduces noise such as AWGN and impulsive noise. The two issues greatly influence document quality, so the quality of photocopied documents is not as great as that of laser printed documents. Figure 10.13 shows the letter 'e' from different photocopiers, each of which has different noise and character morphology.

The traditional forensic methods for photocopied documents usually involve main component analysis of photocopy toner, which requires document examination equipment and testing personnel, and they can only identify the brands and styles of photocopiers. For some applications, the manufacturer, model or even individual photocopier for a given document must be known in advance. Wang *et al.* (2013) used texture features to identify source photocopiers automatically, and this approach is suitable for arbitrary character identification. After the characters are segmented from the document image, three sets of features are extracted from each individual character, including the grey-level features, the gradient differential matrix features and the grey-level gradient co-occurrence matrix features. The characters in a document are classified using a Fisher classifier and a majority voting is performed on the character classification results to identify the source photocopier. Before feature extraction, in order to minimize the influence of paper texture on the background of the scanned image, the paper texture image on the background is removed by binary segmentation, so that only the character portion remains for feature extraction.

(a)

(b)

(c)

(d)

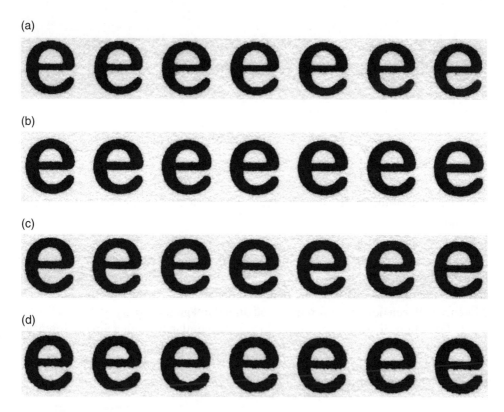

Figure 10.13   The letter 'e' as photocopied by (a) MF3010, (b) Lanier LD360, (c) Lanier LD528 and (d) Toshiba 4540C photocopier.

10.5   Forgery Detection for Printed and Scanned Documents

In addition to the source of a document, its authenticity should also be determined. There are four main tampering methods:

1. *Collage-and-paste tampering*: Two parts of a document are spliced and the combined document is photocopied to create an integral forgery. After photocopying, the edges of the document pieces are removed by the photocopier.
2. *Adding of extra information to blank spaces*: If a document includes blank space in the document, a forger could re-print the document to add extra information there.
3. *Changing of document content*: If a forger wants to change the signature or some other important information on the document, physical and chemical processing may be conducted on the document. Some information is deleted from the document or covered by the new information.
4. *Modification of scanned images*: Image processing software, such as Photoshop, can be used to modify the content of scanned images. This tampering method is the

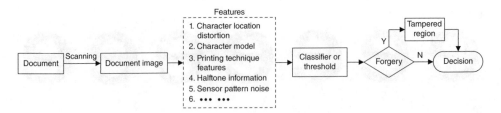

Figure 10.14   Flowchart of document forgery detection.

same as that of digital image tampering, the physical structure of the document is not modified.

In this section, we introduce some algorithms for exposing forgeries. Different characteristics are introduced to forgeries for different tampering methods, so the selection of forensic approaches depends on specific tampering methods. The difference between document forensics and digital image forensics is shown in Section 10.3.1. Detection of document image forgery is similar to digital image forensics, which mainly focuses on statistical properties and noise in the scanned image. Detection of printed document forgery mainly focuses on statistical properties of characters and character position distortion.

10.5.1   Flowchart of Forgery Detection

Forgery detection can be divided into four steps: scanning, feature extraction, classification and decision making. A flowchart of forgery detection is shown in Figure 10.14. Forgery includes both document forgery and image forgery; there is no scanning process for detection of image forgery. Features that reveal forgery include character location distortion, character models, printing technique features, halftone information, sensor pattern noise, etc. Some features focus on the detection of document tampering, others focus on the detection of image tampering. In the process of classification, the commonly used classifiers are the same as those of printer forensics. The difference is that classification in forgery detection includes two classification categories: REAL and FAKE. Because some classifiers need to train a model, some of the algorithms use only a threshold to identify forgeries. In the decision-making step, the algorithms not only return a decision label but also need to return a document graph which labels the tampered region. For example, if the document has not been tampered with, the algorithm returns only the decision label REAL; if the document has been tampered with, the algorithm returns the decision label FAKE and a document graph in which the tampered regions are labelled with different colours or special symbols.

Ascender line
X-height line
Base line
Descender line

Figure 10.15 Text lines for English words.

10.5.2 *Forgery Detection for Printed Documents*

10.5.2.1 Character Location Distortion

For pasting, reprinting and page changing forgeries, character location distortion is introduced into tampered regions compared to the genuine document. Distortion parameters can be obtained by detecting character locations, the tampered regions can then be identified.

The positions of English letters are determined by certain rules. Figure 10.15 flowcharts the text lines of English words and their position distortion. The text lines consist of four lines which are ascender line, x-height line, base line and descender line. All English letters are aligned using these four lines; some letters require three lines for alignment and some letters require only two lines. The special structure of English letters allows us to use character position distortion to expose forgery. In Figure 10.15, the text line position for word 'English' and 'text-line' have a distortion d, that means one of the words is forged. In addition, the widths of margin for all the paragraphs and all rows should be consistent for the whole page. For some forgeries, some paragraphs or sentences are spliced into the document, which may be misaligned relative to other paragraphs and sentences. (Beusekom *et al.* 2010, 2013) presented the extraction of text lines and alignment lines for document inspection. The text lines of English words are extracted and shifts in the positions and angles of characters are computed based on the alignment of text lines. If the position shift or angle shift is bigger than that of other characters, there is a high probability that the characters have been tampered with. Beusekom and Shafait (2011) considered the distortion problem of whole page from another perspective. Suppose the scene includes multiple documents with the same or similar content, and forgeries account for only a small portion of the documents. Pairs of documents are matched, as because the documents generated from different printers have different geometric distortion on the whole page, if one of the document in the data set have great distortion with other documents, that means this document is generated by a different printer and determined as a forgery. This approach does not need to extract text lines from the document, so it is also suitable for documents in other languages. However, this method requires many genuine pages for matching purposes and needs the document have similar content.

10.5.2.2 Character Characteristics Model

Because documents can be tampered with using different printers, in addition to exposing forgeries by detecting distortion of character position, we can also locate tampered characters by detecting inconsistencies in the statistical properties of document text. Kee and Farid (2008) have established printer models for each English letter, and have identified the positions of characters that have been tampered with by matching printer models with the unknown characters. This image matching method is used to compare the same letter (i.e. 'e' or 'a') in different places in a document. SVD is performed on the letters to obtain the eigenvalues, and the printer profile is calculated based on the eigenvalues and the mean grey value of the letter. A cost function is identified to judge the similarity of the profiles to known printer profiles. In addition, a printer model for each letter is established, and the model can re-construct letters based on the printer profiles. The re-construction error is computed between the re-constructed letter and the original letter. The model of the source printer is then identified as the printer model that minimizes the re-construction error. This approach can expose forgeries produced by different brands of printers, and is sensitive to the changing of toner density. It is not suitable for detecting forgeries produced by printers of the same brand or model either.

10.5.2.3 Printing Technique Detection

Frequently used device types include laser printers, inkjet printers and xerographic copiers. The identification of printing techniques and photocopying narrows the forensic region. If the device type of an unknown document is identified correctly, then the document can be identified by source forensics to detect the brand and model of the device. If the accuracy of device-type identification is high enough to detect documents based on characters, this technology can also be used to expose forgeries. Because tampering sometimes occurs on different types of devices, if the device types for each character is classified correctly, the characters that have been tampered with will be located.

The contour regions of inkjet printed characters are rougher than those of laser printed characters. Lampert *et al.* (2006) proposed a method to distinguish the device types based on the features in character contour region. For each character, 15 dimensions of features are extracted, which include roughness of character edge, correlation coefficient features, texture features, etc. SVM is used to classify the features. Umadevi *et al.* (2011) considered device type identification from another perspective, the text word image is divided into three regions: foreground text, noise and background. An expectation maximization (EM) algorithm is utilized to derive features of models. One of the indices, called the print index (PI), is used for print technology discrimination. The training words and testing words should have the same number of letters. Schulze *et al.* (2009) described a frequency domain method for printing techniques and photocopying detection. The mean and standard deviation were extracted from frequency sub-band boxes. This method also depends on SVM, and results in an average accuracy of up to 99.08% for full-page detection.

10.5.3   Forgery Detection for Scanned Documents

10.5.3.1   Halftone Information

For a laser printer, the printing models to describe the images are different from the document text. Because the grey level of an image changes continuously, halftone dots are applied to describe the grey values of image pixels. Sparse halftone dots denote the light levels and dense halftone dots denote the dark levels in grey values. When a piece of a text document is scanned as a digital image and is printed again, the characters are constituted of halftone dots. A forger may use image processing software to splice two scanned image and print it again. Figure 10.16a shows the laser printed letter 'e' and Figure 10.16b shows the twice printed letter 'e', the halftone phenomenon is obviously in Figure 10.16b. So forgery can be exposed through the detection of the halftone phenomenon. Kong and Wu (2012) researched halftone images based on analysis of image spectra. The halftone phenomenon, which is introduced into document text image by double printing can be considered cosine noise, and the frequency and amplitude of the cosine noise are stable, so there will be peaks in the spectrum associated with the halftone. Figure 10.16c and d show the logarithms of the Fourier spectra of Figure 10.16a and b, respectively. There are four peaks in the four quadrants in Figure 10.16d, associated with the halftone image. A band pass filter is established to detect the peaks. If the the peaks in the spectra are detected, this character could be determined as twice printed.

10.5.3.2   Sensor Pattern Noise

As we have noted, the pattern noise introduced by CCD/CMOS sensors can be used to identify source scanners. If the pattern noise can be extracted from image blocks and the image blocks can be identified correctly, the pattern noise can also be used to expose

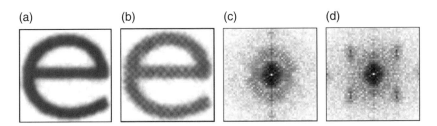

(a)   (b)   (c)   (d)

Figure 10.16   Halftone graphs of (a) the scanned image of the letter 'e', (b) the scanned image of a twice printed letter 'e', (c) the logarithmic Fourier amplitude spectrum of (a) and (d) the logarithmic Fourier amplitude spectrum of (b).

the forgery. A digital image can be tampered with using image processing software to splice two images together. If the two images were scanned by different scanners, they will have different pattern noise. Image processing software can only modify image content to make it more visually clean, but not to remove pattern noise. Based on this signature, Khanna *et al.* (2008) divided scanned images into several blocks, pattern noise is extracted from each image block and SVM classifies the sub-images of the spliced image. Different regions of a spliced image with different types of pattern noise can be labelled. The effectiveness of pattern noise identification is based on statistics, so the size of the image block influences the actual performance of identification.

10.6 Sample Algorithms with Case Studies

In this section, we introduce several algorithms for document forensics in detail. Document forensics includes printer identification, document forgery detection and scanner identification. In each algorithm, the flowchart, calculation and experiments are illustrated step by step. Examples clarify how the forensic methods work and how the algorithms are implemented.

10.6.1 Printer Identification

In this part, we introduce the method proposed by Wu *et al.* (2009), which is based on the geometric distortion of document pages. In an ideal document, such as a PDF document, the rows on one page of the document are strictly parallel. In practice, the slopes of rows in printed documents change regularly along the printing direction. The changes may result from imperfections and differences in paper feeding mechanism of different printers. Distortion is caused by fluctuations in the spinning velocity of polygon mirrors and imperfections of paper feeding mechanisms (Bulan *et al.* 2009; Wu *et al.* 2009). The geometric distortion model is established by projective transformation and model parameters can be used for printer identification.

The main steps of this method are summarized as follows. First, the document image and ideal image are preprocessed to obtain the feature point set. Second, a projective transformation model is established based on the document geometric distortion. Third, singular value decomposition (SVD) and the least squares method are applied to solve the equation and compute the model parameters. Finally, SVM classification is used for printer identification based on model parameters.

10.6.1.1 Pre-processing

When a printed document is being identified, the ideal image and scanned image should be obtained first. OCR software should be applied on the scanned image to obtain the ideal image, if the original PDF or Word document is not available. The resolution of the image should be 600 dpi.

The grey-level document image should be converted to a binary image based on the threshold, and the characters should be segmented from the binary image. Because the

(a)   (b)   (c)

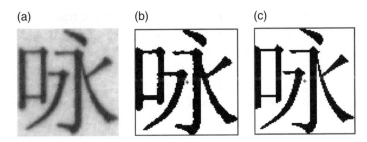

Figure 10.17   The center determination for a character image, including (a) a character in a document image, (b) a binary image of the character with its center and boundary marker and (c) the character in an ideal image at the corresponding location.

characters are listed in the document row by row and the values of character pixels are different from those of blank areas, each row can be segmented from the image by horizontal projection. The grey-value accumulation is computed by the projection. For each row, the column projection is applied to segment each character. The aim of character segmentation is to extract the center coordinate of each character, and the distortion of displacement can be estimated by comparing the center coordinates of characters in the document image and the ideal image.

Extraction of the center coordinate of a character is illustrated as Figure 10.17. The total character area is outlined with a rectangle. The center of the rectangle is the center of the character. Figure 10.17a shows a scanned image of a Chinese character. Figure 10.17b shows a binary image segmented from document image and its center (marked with a dot). Figure 10.17c shows a binary image segmented from the ideal image and its center. After all of the coordinates of character centers are computed from the document image and the ideal image, the distortion can be computed by comparing the two images. The character in the upper-left corner is used as the reference point, it has the same coordinate in both images. After coordinate adjustment, the character center coordinates of the two images are saved in a feature point set. A distortion graph is shown in Figure 10.7. Because the character in the upper-left corner is the reference point, the distortion of the reference point is 0. The distortion value increases when the location is far away from the reference point.

10.6.1.2   Projective Transformation Model

The projective transformation model describes the geometric distortion of a document page. Equation 10.1 describes the form of two-dimensional projective transformation mapping coordinates of point (x_1, y_1) to coordinates of point (x_2, y_2):

$$\begin{cases} x_2 = \dfrac{m_0 x_1 + m_1 y_1 + m_2}{m_6 x_1 + m_7 y_1 + 1} + e_x \\[2mm] y_2 = \dfrac{m_3 x_1 + m_4 y_1 + m_5}{m_6 x_1 + m_7 y_1 + 1} + e_y \end{cases}, \tag{10.1}$$

where (m_0, m_4) are the scaling coefficients, (m_2, m_5) are the translation coefficients, (m_6, m_7) are the coefficients that represent the degree to which parallel lines are distorted into intersecting lines in the x and y directions respectively, (m_1, m_3) are the rotation coefficients and (e_x, e_y) are the residual error and direction, respectively.

The aim of solving Equation 10.1 is to estimate the parameters m_0 to m_7. These parameters can be considered the model of the projective transformation. If the values of the parameters are estimated, the printer can be identified. In order to compute the parameters conveniently, Equation 10.1 should be transferred to the following matrix operations:

$$
\begin{bmatrix} x_2 \\ y_2 \end{bmatrix} = \begin{bmatrix} x_1 & y_1 & 1 & 0 & 0 & 0 & -x_1 x_2 & -y_1 x_2 \\ 0 & 0 & 0 & x_1 & y_1 & 1 & -x_1 y_2 & -y_1 y_2 \end{bmatrix} \mathbf{M} + \begin{bmatrix} e_x \\ e_y \end{bmatrix}, \tag{10.2}
$$

where $\mathbf{M} = \begin{bmatrix} m_0 & m_1 & m_2 & m_3 & m_4 & m_5 & m_6 & m_7 \end{bmatrix}^{\mathrm{T}}$. This is the projective transformation equation for the one character coordinate. Of course, the document page is constructed of many characters. Assuming that there are n characters in one document page, then the equation can be described as follows:

$$
\begin{bmatrix} x_{21} \\ y_{21} \\ \cdots \\ x_{2n} \\ y_{2n} \end{bmatrix} = \begin{bmatrix} x_{11} & y_{11} & 1 & 0 & 0 & 0 & -x_{11}x_{21} & -y_{11}x_{21} \\ 0 & 0 & 0 & x_{11} & y_{11} & 1 & -x_{11}y_{21} & -y_{11}y_{21} \\ \cdots & \cdots & \cdots & \cdots & \cdots & \cdots & \cdots & \cdots \\ x_{1n} & y_{1n} & 1 & 0 & 0 & 0 & -x_{1n}x_{2n} & -y_{1n}x_{2n} \\ 0 & 0 & 0 & x_{1n} & y_{1n} & 1 & -x_{1n}y_{2n} & -y_{1n}y_{2n} \end{bmatrix} \mathbf{M} + \begin{bmatrix} e_{x1} \\ e_{y1} \\ \cdots \\ e_{xn} \\ e_{yn} \end{bmatrix}. \tag{10.3}
$$

Equation 10.3 is an over-determined equation set, which means that the number of the equations is greater than the number of unknown variables (the number of equations is $2n$ and the number of unknown variables is 8 in $\mathbf{M}$), the vector or matrix in Equation 10.3 is denoted by the bold letters and the equation can be expressed as follows:

$$
\mathbf{b} = \mathbf{AM} + \phi. \tag{10.4}
$$

Except for the eight unknown variables in $\mathbf{M}$, there are also some unknown variables in ϕ. If the parameters in $\mathbf{M}$ can be estimated when ϕ has the minimum value, the estimated parameters are closest to the actual situation. The over-determined equation can be solved in the least squares sense shown in Equation 10.4. The aim is to find the $\hat{\mathbf{M}}$ that minimizes $\mathbf{AM} + \phi$. $\hat{\mathbf{M}}$ is the model of geometric distortion.

$$
\hat{\mathbf{M}} = \arg \min_{\mathbf{M}} ||\mathbf{AM} - \mathbf{b}||. \tag{10.5}
$$

Because the minimal value of the norm of $\mathbf{AM} + \phi$ can be 0, the ideal solution to the equation is

$$
\mathbf{AM} - \mathbf{b} = 0. \tag{10.6}
$$

Table 10.1   The list of printers used
in experiments.

Printer model	Label
Hp 1000	1, 3, 4, 6
Hp 1020	5, 9
Hp 1320n	2, 10
Lenovo 2312P	7
Samsung ML 1510	8

$\mathbf{M}$ can be estimated by computing the inverse of $\mathbf{A}$. Because $\mathbf{A}$ may not have an inverse matrix, SVD is applied to solve the least squares problem to ensure the stability of computing. The matrix $\mathbf{A}$ can be decomposed by SVD as follows

$$\mathbf{A} = \mathbf{U\Sigma V}^{-1}, \tag{10.7}$$

where the orthogonal matrices $\mathbf{U} \in \mathbb{C}^{2n \times 8}$ and $\mathbf{V} \in \mathbb{C}^{8 \times 8}$. $\mathbf{\Sigma}$ is a diagonal matrix, so the solution of $\hat{\mathbf{M}}$ can be obtained by

$$\hat{\mathbf{M}} - \mathbf{V\Sigma}^{-1}\mathbf{U}^T\mathbf{b}. \tag{10.8}$$

10.6.1.3   Experimental Results

The printer models used in the experiment are listed in Table 10.1. Ten laser printers were used in experiments for five models. For some printers, several individual printers with different label numbers were selected. Each printer printed 12 pages and each page consisted of 1496 Chinese characters, selected at random from among frequently used Chinese characters. The font used was 'small IV Song'. After preprocessing, 1496 character centers were extracted and a feature point set was constructed.

The distribution of parameters m_4 and m_7 is shown in Figure 10.18. In the figure, the clustering and the interclass separability are obvious. Parameter m_7 is close to 0, and m_4 is close to 1, which demonstrates that the geometric distortion is minimal.

Because translation and rotation are inevitable during the process of printing and scanning, the parameters m_1, m_2, m_3, m_5 cannot represent the intrinsic features of the printer. However, the scaling and the distortion by which parallel lines degrade into intersecting lines are stable, so the four parameters m_0, m_4, m_6, m_7 are selected to represent the intrinsic features of the printer. The selected features are classified by SVM for printer forensics. The 12 document images for each printer were divided into two groups of six. The first six were used to train the SVM model, and the second six were used for testing. Test results show that the correct classification rate of the 10 printers is 100%.

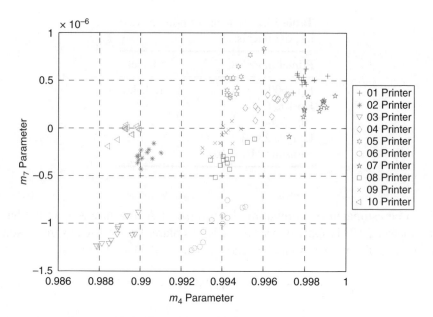

Figure 10.18 Distribution of the m_4 and m_7 values from 10 printers (Wu *et al.* 2009).

10.6.2 Scanner Identification

In this part, we introduce an approach of scanner identification based on the method of detecting dust and scratches proposed by Dirik *et al.* (2009). For a flatbed scanner, the flat glass is used to hold the document while the laser goes through the glass unhindered. The quality of the flat glass greatly influences the scanned image. Some dust and scratches are on the flat glass, and they remain for long periods of time. In Professor Memon's method, the dust and scratches are considered the fingerprint of a scanner and can be used to identify the scanner.

10.6.2.1 Detection of Dusts and Scratches

Dust and scratches are very small and 3×3 to 9×9 pixels at a resolution of 300 dpi. But they differ greatly from the surrounding pixels, with much faster grey value changes than for other regions of the image. Blemishes of scanned images are shown in Figure 10.19. Figure 10.19a and b are the dust images for a Canon scanner and Epson scanner, respectively. The images are 15 × 15 pixels at a resolution of 300 dpi.

Before the positions of dust and scratches are identified, the scanned image should be preprocessed. Because dusts and scratches can be seen as high-frequency noise in the scanned image, a high-pass filter is applied to remove the low frequency information. The remaining high frequency information consists of dust, scratches, edges and texture. As with previous work in image identification (Dirik *et al.* 2008) based on the dusts information, the authors provide a model for the dust spots. All the dust

(a)   (b)

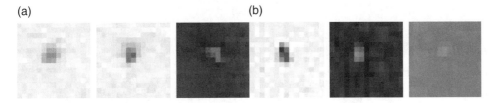

Figure 10.19 Blemishes on scanned images from (a) a Canon scanner and (b) an Epson scanner (Dirik *et al.* 2009).

Figure 10.20 Dust and scratch model for a high-pass filtered scanned image (Dirik *et al.* 2009).

spots can be simulated by a Gaussian model with a fixed pattern. The simple absolute values filtered image model in Figure 10.20 is used to detect the dust and scratches. The dimension of the white region is 5×5 pixels.

To separate the dust and scratches from other detail information, such texture and edges, an NCC is applied to detect the positions of the dust and scratches. The computation of NCC is

$$\mathrm{NCC}(x,y) = \frac{\sum_{(i,j)} \mathbf{M}(i,j) \cdot \mathbf{I}(x+i, y+j)}{\sqrt{\sum_{(i,j)} \mathbf{M}^2(i,j) \cdot \sum_{(i,j)} \mathbf{I}^2(x+i, y+j)}}, \qquad (10.9)$$

where $\mathbf{M}$ is the model of dust and scratches shown in Figure 10.20, $\mathbf{I}$ is the scanned image (which is high pass filtered), (i,j) is the pixel coordinate in the $\mathbf{M}$ model and (x,y) is the pixel coordinate in scanned image $\mathbf{I}$. From this formula, we can see that, when the model slides on the scanned image, one NCC value is returned to describe the correlation between the region around (x,y) and model $\mathbf{M}$. The closer NCC is to 1, the more similar for the two images. Figure 10.21 shows examples in which model $\mathbf{M}$ and the scanned image block (a) match well and (b) do not match well. In Figure 10.21a the correlation value has a single high peak surrounded by much lower values. In Figure 10.21b the correlation value is low and randomly distributed. After

(a)   (b)

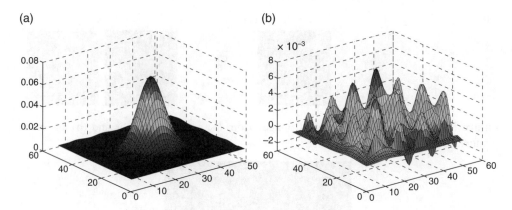

Figure 10.21 Computation of correlation in scanner identification: (a) the model matches the scanned image block well and (b) the model does not match the scanned image block well (Dirik *et al.* 2009).

(a)   (b)

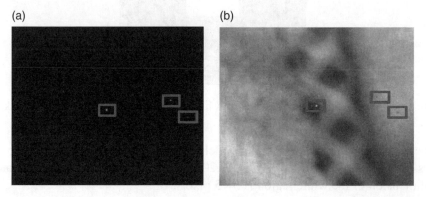

Figure 10.22 (a) A scanner template with a black background and (b) a scanned image block from the same location as the template; for both images, dust specks are labelled with rectangles (Dirik *et al.* 2009).

the correlation between the model and the scanned image is computed, a threshold is used to locate the positions of dust and scratches.

10.6.2.2 Scanner Identification

Templates of scanners should be generated to represent the fingerprints of the scanners. This method considers only a black background as a template. Because dust and scratches are white compared to a black background, they are easy to distinguish from a black background image. Such a template is also easy to generate. No black paper or other black baffle is required to generate the template, the scanner lid is simply left open while the scanner runs. Figure 10.22 shows the template of a scanner and a scanned image with dust. Figure 10.22a and b show the same part of the scanner

(a)   (b)

Figure 10.23   Images scanned by (a) an Epson Perfection 1250 and (b) a Canon CanoScan LiDE90.

plate. Dust specks are labelled with rectangles. The scanner template is generated for each scanner during an identification step, and the correlation computation introduced earlier is also used to identify the scanner. Given a detectable scanned image, the correlation with all templates in the database is computed. A correlation value is chosen that is greater than the threshold set previously, and decide the scanner corresponding to the greater correlation as the scanner source.

10.6.2.3   Experimental Results

In order to demonstrate the performance of the method, we selected two scanners for an experiment: Epson Perfection 1250 and Canon CanoScan LiDE90. The Epson scanner is an older model and the Canon scanner is a new model. We generated two scanner templates and several test scanned images at 300 dpi. By cross computation for the two scanners, we chose an NCC threshold of 0.02. Figure 10.23 shows two scanned images made by the two brands of scanners and their correlation values. Figure 10.23a shows the image scanned by the Canon scanner and Figure 10.23b shows the image scanned by the Epson scanner.

Figure 10.24 shows the experimental results for the two scanners. Figure 10.24a shows the cross correlation for the Canon scanner: the circles show correlations between the Canon scanner template and the images scanned by the Canon scanner and the crosses show correlations between the Canon scanner template and the images scanned by the Epson scanner. Figure 10.24b shows similar results for the Epson. With a threshold of 0.02, the scanners could be identified correctly. This method is also demonstrated to robust for JPEG compression.

10.6.3   Document Forgery Detection

Here we introduce the method proposed by Beusekom *et al.* (2010). This method is focused on detection of document forgery based on the alignment of text lines. When

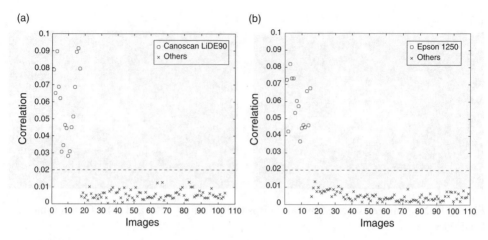

Figure 10.24 Cross correlation between the scanner template and scanned images for (a) the Canon template, (b) the Epson template (Dirik *et al.* 2009).

a document is forged by the printing of characters in blank areas, or by collaging two parts of a document and photocopying, the forged part may be misaligned with the original parts of the document, the misalignment results from the translation and angles of text lines. English has a special layout mode, each letter and each word in English should be aligned to the ascender line, x-height line, base line and descender line. Except for the text lines, there are three alignment lines for document pages: the left alignment line, center alignment line and right alignment line. When some characters or paragraphs are forged, the text lines and the alignment lines may be distorted.

10.6.3.1 Text Line Finding

Before extracting the text lines for each word, the location of each letter should be founded. First the connected components in the document should be identified. The segmentation threshold is determined by the detectable document with the largest connected component making up an alphanumeric character, such as the letter 'i'. English words are aligned by the text lines shown in Figure 10.15. Some letters are aligned using the ascender line, x-height line, and base line, such as the letters 'T', 'k' and 'b'. Some letters are aligned using the x-height line, base line, and descender line, such as the letters 'q' and 'g'. Some letters are aligned using only the x-height line and the base line, such as the letters 'e', 'o' and 'w'. There is no unified template to extract text lines for all letters, and the bounding box is first applied to identify the top and bottom of a letter. Rectangles identify the bounding boxes and bottom center of the bounding box is identified as the reference points.

When the words are forged, there may be an angle distortion. The centres of the bottom and top sides of the bounding box for each letter are used as the reference

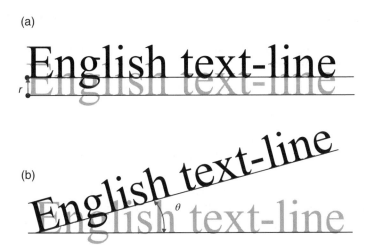

Figure 10.25   Two parameters (r, θ) for a text line (a) the parameter r for character translation, (b) the parameter θ for character rotation.

points. Reference points at the bottom are on either the base line or the decender line, and reference points at the top are on either the ascender line or the x-height line. When all of the reference points have been extracted for a line, the four text lines can be identified. Two parameters (r, θ) are extracted from the text lines in Figure 10.25, where r is the distance of the words from the original location, θ is the angle at which the text is offset from the image. Because there are many reference points on one text line, the two parameters are estimated by the least squares method.

10.6.3.2   Alignment-Line Finding

On a document page, each paragraph and text line should be aligned to the left alignment line, centre alignment line and right alignment line. The left alignment line is a vertical passing through the starting points of the text lines, the right alignment line is a vertical line passing through the ending points of the text lines, and the centre alignment line is a line directly in the centre between the left and right alignment lines.

Alignment lines can be extracted based on the reference points of bounding boxes. The first and last bounding boxes for each text line should be found. The centre point on the left side of the first bounding box defines the left alignment line and the centre point on the right side of the last bounding box defines the right alignment line. The centre alignment line can then be identified based on the left and right alignment lines. The alignment line consists of two parameters, (r, θ), and they are computed similarly to text lines, the least squares method is applied to compute the parameters based on the reference points on the alignment lines. In addition to this method for extracting alignment lines, the RAST line finding method (Breuel 2003) can also be used.

10.6.3.3 Decision Making

In the decision-making step, Bayes theorem is applied to estimate the probability of forgery. The Bayes formula is

$$P(f|d) = \frac{P(d|f) \times P(f)}{p(d)}, \tag{10.10}$$

where $P(f|d)$ is the probability of the text being forged for distance d between reference points at the start, center or end of the text line relative to the left, center or right alignment lines. $P(d|f)$ is a distribution of distance d when the text is forged. The distribution is assumed to be a Gaussian probability distribution and the mean value and standard deviation can be obtained by training the database. $P(f)$ denotes the prior probability that the text is forged. The probability that a text line is the original for distance d is

$$P(\neg f|d) = \frac{P(d|\neg f) \times P(\neg f)}{p(d)}, \tag{10.11}$$

where $\neg f$ denotes the negation operation of f, $P(\neg f) = 1 - P(f)$ and $P(\neg f)$ is computed by training in the database. The parameters $P(f|d)$ and $P(\neg f|d)$ are computed for each text line. If $P(f|d) > P(\neg f|d)$, the text line is considered an original line. If $P(f|d) \leq P(\neg f|d)$, the text line is considered a forged line. In this step, the unknown parameter $p(d)$ can be ignored.

Two decisions are made for each line of text line, corresponding to the two parameters (r, θ). One decision is made for each alignment line. If at least one of the decisions is that the text line is forged, this text line is considered forged. An example of inspection of the alignment lines is shown in Figure 10.26. Figure 10.26a shows the full page of a forged document, b and c show the left and right alignment lines and the margins. The left and right margins of original text are d_1 and d_3 respectively, and the margins for the forged text are d_2 and d_4 which have large deviation with d_1 and d_3. The forged text could be detected by the margin deviation.

10.7 Open Problems and Challenges

Compared to traditional document examination, passive print forensics provides great advantages in terms of forensic performance. Passive print forensics uses scanners and computers for forensic activities, and thus does not damage the documents. The new technology can reduce the detection cost and time required for print forensics, and can also reduce the need for professional personnel. The technology can also scan documents in bulk and provide objective decisions made by computers. Nevertheless, passive print forensics must also overcome problems and challenges. Significant advancement must be made before passive print forensics can be applied in practice. We list the problems and challenges in the hope that they will be resolved as the technology develops.

(a)   (b)   (c)

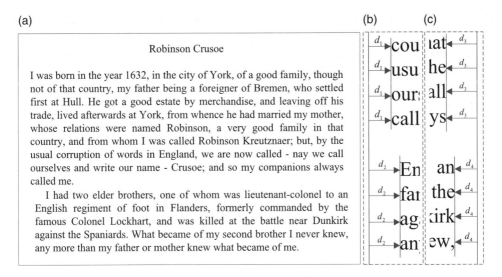

Figure 10.26 Inspection of the alignment lines of a document, including (a) the full page of a forged document, (b) the left alignment line and margins for each text line and (c) the right alignment line and margins for each text line.

1. Print forensics are influenced greatly by the objective environment, including the source of paper, document formats and tampering methods. Documents such as receipts, contacts, tickets, bills and certificates all use different types of paper and document formats. The texture of paper and the material of which have great influence on document examination. Forgers are also using ever more sophisticated methods that can cheat our eyes. Compared to digital image forensics, we face a complex objective environment that makes document forensics more difficult.

2. Printing equipment has unstable and overlapped characteristics. The characteristics of documents are unstable, because as the printer or the toner being used changes, so does the signature of the document. If the type of toner is changed, that can also modify the signature of the printer. Document images also come from many sources, such as laser printers, inkjet printers, photocopiers and scanners. Because the characteristics of these devices overlap, it is hard to separate out the characteristics and consider them individually.

3. Paper documents age and get damaged. Unlike digital images, documents age after they are printed. Detecting when a document was created is challenging, and the characteristics of paper may degrade over long periods of time. Documents can also be damaged either inadvertently or due to vandalism, valuable information lost in damaged documents can be hard to recover.

In the future, documents should be examined by people and computers together. Humans and machines use their respective advantages to reduce dependence on professional equipment and personnel. The extraction of features, analysis and primary decision making should be done by computers, which have the advantage of being able to perform efficient calculation. Based on the calculation results, testing personnel should provide a final decision based on their experience. The forensics process can thus become semi-automatic. Decision results will become more objective.

10.8   Conclusions

As document images are printed and scanned, different characteristics are introduced into the images during different processes. Forensics algorithms are summarized and classified based on these characteristics. In recent years, passive print forensics, which uses only scanners and computers to analyze documents and document images, have received greater attention and have developed rapidly. However, some problems remain unresolved, such as the instability of equipment, the influence of the objective environment, document ageing and damage, and the overlapping of multiple characteristics. The challenges and opportunities for passive print forensics coexist. We hope that new technologies will reduce the cost and time required to detect forgery, improve the efficiency of forensics and ensure the integrity of documents.

Acknowledgements

This work is supported by the National Natural Science Foundation of China under Grant Number 61172109.

References

Ali GN, Mikkilineni AK, Chiang PG, Allebach JP, Chiu GT and Delp EJ 2009 Intrinsic and extrinsic signatures for information hiding and secure printing with electrophotographic devices. *NIP19 International Conference on Digital Printing Technologies*, September 2009; New Orleans, LA pp. 511–514.

Beusekom JV and Shafait F 2011 Distortion measurement for automatic document verification *Proceedings of the International Conference on Document Analysis and Recognition* 18–21 September, 2011; Beijing, China, pp. 289–293.

Beusekom JV, Shafait F and Breuel TM 2010 Document inspection using text-line alignment, China *Proceedings of the 9th International Workshop on Document Analysis Systems*, 9–11 June, 2010; Boston, MA, pp. 263–270.

Beusekom JV, Shafait F and Breuel TM 2013 Text-line examination for document forgery detection. *International Journal on Document Analysis and Recognition* **16**, 189–207.

Breuel TM 2003 On the use of interval arithmetic in geometric branch-and-bound algorithms. *Pattern Recognition Letters* **24**, 1375–1384.

Bulan O, Mao J and Sharma G 2009 Geometric distortion signatures for printer identification. *Proceedings of the International Conference on Acoustics, Speech and Signal Processing* 19–24 April 2009; Taipei, April, Taiwan, pp. 1401–1404.

Bulan O, Mao J and Sharma G 2011 Forensic printer detection using intrinsic signatures. *Proceedings of the SPIE International Conference on Media Watermarking, Security and Forensics*, 23–27 January 2011; San Francisco, CA, pp. 78800R-1–78800R-11.

Chiang PG, Khanna N, Mikkilineni AK, Segovia MVO, Suh S, Allebach JP, Chiu GTC and Delp EJ 2009 Printer and scanner forensics. *IEEE Signal Processing Magazine* **26**, 72–83.

Dirik AE, Sencar HT and Memon N 2008 Digital single lens reflex camera identification from traces of sensor dust. *IEEE Transactions on Information Forensics and Security* **3**, 539–552.

Dirik AE, Sencar HT and Memon N 2009 Flatbed scanner identification based on dust and scratches over scanner platen. *IEEE International Conference on Acoustics, Speech and Signal Processing*, 19–24 April 2009; Taipei, Taiwan, pp. 1385–1388.

Foundation EF n.d. Docucolour tracking dot decoding guide. https://w2.eff.org/Privacy/printers/docucolour/ (Accessed 21 February 2015).

Gou H, Swaminathan A and Wu M 2007 Robust scanner identification based on noise features. *Proceedings of SPIE-IS and T Electronic Imaging – Security, Steganography, and Watermarking of Multimedia Contents IX*, 29 January-1 February 2007; San Jose, CA, pp. S5050–S5060.

Kee E and Farid H 2008 Printer profiling for forensics and ballistics *Proceedings of the 10th ACM Workshop on Multimedia and Security Workshop*, 22-23 September 2008; Oxford, U.K., pp. 3–10.

Kelly JS and Lindblom BS 2006 *Scientific Examination of Questioned Documents*. CRC Press, Boca Ration, FL.

Khanna N, Mikkilineni AK, Martone AF, Ali GN, Chiu GTC, Allebach JP and Delp EJ 2006 A survey of forensic characterization methods for physical devices. *Digital Investigation* **3**, 17–28.

Khanna N, Chiu GTC, Allebach JP and Delp EJ 2008 Scanner identification with extension to forgery detection *Security, Forensics, Steganography, and Watermarking of Multimedia Contents* X 28–30 January 2008; San Jose, CA, pp. 6819:68190G.

Khanna N, Mikkilineni AK and Delp EJ 2009 Scanner identification using feature-based processing and analysis. *IEEE Tranactions on Information Forensics and Security* **4**, 123–139.

Kong X and Wu Y 2012 Document authenticity detection based on halftone information. Chinese patent ZL 201010154461.6.

Kong X, You X, Wang B, Shang S and Shen L 2010 Laser printer source forensics for arbitrary chinese characters. *Proceedings of International Conference on Security and Management*, 12–15 July 2010; Las Vegas, NV, pp. 356–360.

Lampert CH, Mei L and Breuel TM 2006 Printing technique classification for document counterfeit detection. *2006 International Conference on Computational Intelligence and Security*, 3–6 November 2006; Guangzhou, pp. 639–644.

Liu X and Guo T 2008 The test on the forged document made by printer printing method. *Journal of Guizhou Police Officer Vocational College* **4**, 53–55.

Ma J 2007 Frontiers of present questioned document examination. *Chinese Journal of Forensic Sciences* **1**, 47–50.

Martone AF, Mikkilineni AK and Delp EJ 2006 Forensics of things. *Proceedings of the 2006 IEEE Southwest Symposium on Image Analysis and Interpretation*, 26–28 March 2006; Washington DC, pp. 149–152.

Schulze C, Schreyer M, Stahl A and Breuel T 2009 Using dct features for printing technique and copy detection In *Advances in Digital Forensics V* (ed. Peterson G and Shenoi S) vol. 306. Springer, Berlin, Germany, pp. 95–106.

Sencar HT and Memon N 2013 *Digital Image Forensics: There Is More to a Picture than Meets the Eye.* Springer, New York.

Shen L, Kong X and You X 2007 Printer forensics based on character image quality measures. *Journal of Southeast University (Natural Science Edition)* **37**, 92–95.

Tchan J 2000 Classifying digital prints according to their production process using image analysis and artificial neural networks. *Proceeding of the SPIE International Conference on Optical Security and Counterfeit Deterrence Techniques*, 27–28 January 2000; San Jose, CA, pp. 105–116.

Tchan J 2004 The development of an image analysis system that can detect fraudulent alterations made to printed images. *Proceeding of the 5th International Conference on Optical Security and Counterfeit Deterrence Techniques*, 20–22 January 2004; San Jose, CA, pp. 151–159.

Umadevi M, Agarwal A and Rao R 2011 Printed text characterization for identifying print technology using expectation maximization algorithm. *Multi-disciplinary Trends in Artificial Intelligence. Proceedings 5th International Workshop*, 7–9 Deccember 2011; Hyderabad, India, pp. 201–212.

Wang J 1998 Inkjet printed documents identification. *Forensic Science and Technology* **4**, 21–23.

Wang C, Kong X, Shang S and You X 2013 Photocopier forensics based on arbitrary text characters. *Proceedings of SPIE-IS and T Electronic Imaging–Media Watermarking, Security, and Forensics*, 5–7 February 2013; Burlingame, CA, p. 86650G (11pp.).

Wu Y, Kong X, You X and Guo Y 2009 Printer forensics based on page document's geometric distortion. *Proceedings of International Conference on Image Processing*, 7–10 November 2009; Cairo, Egypt, pp. 2909–2912.

Yoshinori A, Kazuhiko K, Shigeru S and Yoko S 2002 Discrimination of inkjet printed counterfeits by spur marks and feature extraction by spatial frequency analysis. *Proceedings of the Conference on Optical Security and Counterfeit Deterrence Techniques IV*, 23–25 January 2002; San Jose, CA, pp. 129–137.

Yoshinori A, Atsushi Y and Yoshiyasu H 2009 Estimation of inkjet printer spur gear teeth number from pitch data string of limited length. *3rd International Workshop on Computational Forensics*, 13–14 Auguest 2009; The Hague, the Netherlands, pp. 25–32.

11

Microphone Forensics

Christian Kraetzer and Jana Dittmann
Department of Computer Science, Otto-von-Guericke-University, Magdeburg, Germany

11.1 Introduction

Forensic investigations often have the goal of finding the ground truth for authenticity, integrity or non-repudiation of entities and/or data (see, e.g. related, commonly known security aspects in Dittmann *et al.* (2001)). In respect to audio-based investigations, the (usually) recorded signals might contain a variety of individual audio influences, such as those from subjects and objects in a given environment that might underlie changing environmental conditions (wind, rain, etc.). Subjects and objects have the ability to produce sound phenomena and/or are able to influence the overall appearance of the recorded sound. Examples for such influences are changes imposed to the reflection or absorption characteristics of the recording environment.

In respect to the overall sound source, audio signals can therefore be seen as having a patchwork signal characteristic fused from multiple individual sound source characteristics containing original or replayed sources. For individual source investigations, a separation (or fission) of signal components might be of interest to study the ground truth more accurately. However, due to the manifold physical signal interplay (interference as well as additional constant and dynamic environmental effects) the separation task is difficult and might cause errors in the forensic investigation or information loss and/or might not be done transparently as required in court applications (see, e.g. the considerations on usability in court, that is compliance to the Daubert standard, in Kraetzer 2013). Furthermore, microphones (as the recording devices) display frequency- and content-depending intrinsic transfer functions (see Pawera 2003). Eliminating individual identified microphone characteristics might therefore affect the original sound source signal due to the nature of the microphone's intrinsic trans-fer functions and might potentially produce distortions unacceptable for a forensic investigation (for a discussion on acceptable audio signal manipulations in forensic investigations (see, e.g. Koenig *et al.* 2007).

An universal method of *microphone forensic analysis*, which directly analyses the originally fused individual signals during the recording process, can represent the patchwork audio signal and keep all individual signal properties in their *acquisition context*. In this chapter, we therefore focus on a universal microphone forensic method by analyzing the original recorded signal to estimate ground truth for different investigation goals. Since this analysis is by its nature a typical *pattern recognition problem*, the main issues to be considered are the *feature space* used to represent the signal, the *matching or classification mechanism* used to implement the recognition process and the precise definition of the *patterns* that have first to be segmented from background noise and second to be classified.

In particular, we consider the *original sound sources* as overall complex acoustic phenomena in an environment formed by subjects and objects within complex environmental-dependent interactions. The *microphone* (the acquisition sensor) and the *context* (describing in detail microphone characteristics used and the actual situation (physical properties as well as static and dynamic attributes) of the original sound sources) are forming the overall sampled *content*.

For example, a particular scenario of a sound recording could consist of two subjects (one singing, one playing piano) with a CD player for backing track (as replay of a recorded audio) in front of an audience of 20 people (giving further individual interference to the sound characteristics) in a small theatre with a particular room acoustics, and with a frontally positioned microphone imposing its own individualizing effects to the recording. This scenario, although rather lengthy in its textual description, has still to be considered a rather simplistic scenario in terms of audio signal recording contexts. Figure 11.1 summarizes the acquisition context in microphone forensics in

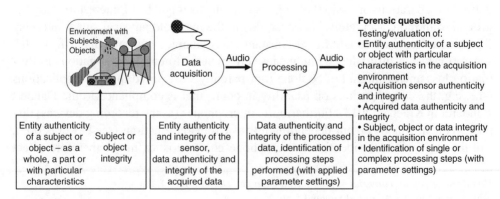

Figure 11.1   The acquisition context in microphone forensics and corresponding forensic questions.

terms of authenticity or integrity issues and maps selected security aspects to the signal generation, data acquisition and following processing operations.

In this chapter, *microphone forensic investigations* specify selected components of (complex) audio signal composition as the patterns to be detected, identified or verified by forensic methods. Such patterns are primarily the characteristics of the sensor (i.e. the microphone), hence the name *microphone forensics*, but also include characteristics of the environment (original sound sources, the room, etc.) or processing operations used to modify the recordings. In this chapter, we just focus on the signal acquisition context–for a complete mapping of security aspects to an entire pattern recognition system architecture, see for example Vielhauer *et al.* (2013).

In the context of our work in this chapter, we restrict the selection of the patterns to be recognized mainly to the microphone's intrinsic transfer functions, and therefore the recording characteristics imposed to the audio signal by the recording with a specific microphone. Thereby, our approach to microphone forensics is intended to address:

- Authenticity:
  - Authentication (entity authenticity) of the microphone as the acquisition sensor. Here, the question is whether a recording has been made by using exactly one specific microphone.
  - Authentication (entity authenticity) of other selected contextual components, such as the recording location (room characteristics), distinction between original sound sources and/or playback/backing track signals, a search for specific background noises, etc.
- Integrity:
  - Detection of discontinuities in the recorded signals as indicators for deletions,
  - Detection of specific attacks that violate the source characteristics, such as filter and compression operations, the insertion or mixing with audio material from other sources, etc.

The rest of this chapter is organized as follows: In the following section we discuss the general approach of pattern recognition for microphone forensics by briefly addressing pattern recognition and its sub-disciplines as well as the state-of-the-art in pattern recognition-based microphone forensics. In Section 11.3 we give general guidelines for microphone registration (i.e. the training of a model description required as a prerequisite for the pattern recognition approach). Section 11.4 contains descriptions on experiments performed for this chapter. For four specific empirical analyses the investigation tasks are specified. Furthermore, an implementation of the statistical pattern recognition pipeline relying on the established data mining suite WEKA (Hall *et al.* 2009) and the author's own general-purpose audio feature extractor are presented and applied to the four investigation tasks specified. Finally, we summarize some challenges in microphone forensics identified by the investigations described here as well as general challenges to this field.

11.2 Pattern Recognition for Microphone Forensics

As summarized Section 11.1, microphone forensics, as considered here, is by its nature a typical pattern recognition problem. All approaches addressing this problem currently found in the literature implement one of the various kinds of pattern recognition. Section 11.2.1 briefly summarizes the main concepts of pattern recognition and the two sub-disciplines of *template matching* and *statistical pattern recognition*, which are the most relevant pattern recognition sub-disciplines for this work. Section 11.2.2 performs a comprehensive review of the current state-of-the-art in pattern recognition based microphone forensics.

11.2.1 Pattern Recognition and Its Sub-Disciplines

According to Bishop (2007) pattern recognition is the act of taking in raw data, projecting it onto a feature space and taking an automated action based on the 'category' (or class) of the pattern – a type of theme of recurring events, objects or characteristics.

In general it can be distinguished between supervised pattern recognition (also known as *classification*) on the one hand and unsupervised statistical pattern recognition (*clustering*) on the other hand. While in classification the existing classes are known and used to solve a recognition problem, in clustering the classes are unknown prior to the application of the clustering algorithm which is used to deduce the classes (their exact number as well as their descriptive models). In other words, classification can be seen as the task of recovering the model that generated the patterns or at least a suitable representation of impact of the generation process (in this chapter as representation of the reference required for microphone forensics – see Figure 11.2). Clustering is the technique to deduce classes represented in a dataset.

There are different approaches (or sub-disciplines) for implementing supervised pattern recognition. The two very prominent ones in the context of forensic analyses are template matching and statistical pattern recognition (see, e.g. Duda *et al.* 2001). The first approach creates the references used in the classification step (as shown in Figure 11.2) directly from template (or candidate) representations of registered representative training samples; the latter builds statistical models representing the distinguishing class characteristics derived from the training samples. Usually, the statistical approaches are more flexible and at the same time have a higher computational complexity attached.

In practice there exists a large variety of pattern recognition architectures (or pipelines), depending on the practical application scenario, the context (the characteristics of the data, including the question of whether the signal to be analyzed originates in the analogue or digital domain) and content or focus (i.e. the patterns and their classes) of the pattern recognition problem as well as the pattern recognition sub-discipline used. Furthermore, the pattern recognition architecture could be designed in a way

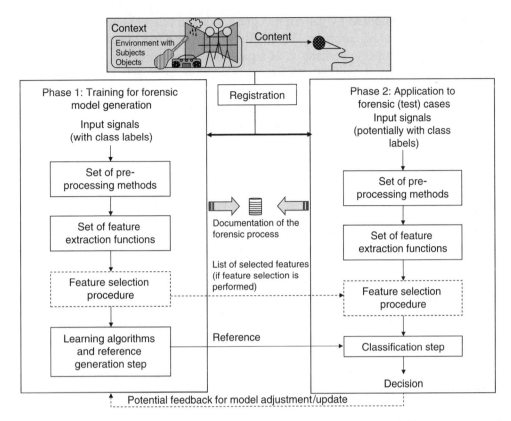

Figure 11.2 Pattern recognition driven microphone forensics – acquisition context and process phases.

that the model can be adaptive or can evolve incrementally. Figure 11.2 visualizes the whole process of *pattern recognition-driven microphone forensics* as described in Section 11.1 and the corresponding audio source forensic analysis pipeline.

Besides the signal generation in a certain recording context (i.e. the sensing by a microphone) the registration of the recording and the two-phase pattern recognition process, Figure 11.2 puts an emphasis on the documentation of the whole investigation process, particularly the case data collected and information derived thereof. This documentation component, symbolized in the figure by a central data storage unit which collects data and information generated in every step of the process and is slowly filled during the investigation, has to include case related meta-data, organizational and technical details describing every performed investigation step as well as data and information generated. It is obvious that the data to be stored includes a close description of the recording context, the originally recorded audio content, the parameterization and output of every processing step as well as any deviation from the normal procedures encountered.

A requirement imposed by pattern recognition driven microphone forensics is the need for *microphone registration*. This registration process (represented as phase 1 in Figure 11.2) performs the generation of a pattern representation (or *model*) for each of the classes to be distinguished in a pattern recognition approach. Depending on the used classification or matching algorithm either explicit representations (*templates*) or implicit representations (*statistical models*) are generated to describe the microphone's intrinsic transfer functions and therefore the recording characteristics imposed to the audio signal by the recording with a specific microphone.

For further details on pattern recognition and its sub-disciplines, the reader is referred to a text book on this topic, for example Duda *et al.* (2001) or Bishop (2007).

For details on practical applications of pattern recognition techniques using open-source data-mining software like the extremely powerful and widely used suite Waikato Environment for Knowledge Analysis (WEKA) provided by the Machine Learning Group at the University of Waikato, New Zealand, the interested reader is referred to examples in Witten *et al.* (2011) and Hall *et al.* (2009).

11.2.2 State-of-the-Art in Pattern Recognition-Based Microphone Forensics

There are mainly three classes of existing approaches in this field of recording source authentication using context information for digital audio signals (Kraetzer 2013):

- Side-information based authenticity and integrity analyses using the electric network frequency (ENF) of a recording setup,
- Content-based consistency analyses on local phenomena found in the time-domain representation of a recording
- Microphone response-based pattern recognition using acquisition setup nonlinearities

Regarding the questions for forensics summarized in Figure 11.1, the first approach focuses primarily on acquired data authenticity and integrity – it is capable of verifying the authenticity of a claimed date for a recording and provides a rather imprecise location determination (in terms of identifying the power grid a recording setup is attached to). This approach cannot be used to verify that a specific recording was made with a given microphone, but it allows for the detection of inconsistencies in the recording.

The second approach uses local phenomena specific to a certain recording location (precisely the reflection and absorption properties of the room the recording is made in) to analyse the acquisition sensor authenticity and the integrity of the data generated.

The third approach performs primarily an acquisition sensor authenticity analysis based on microphone (or more precisely the acquisition setup) intrinsic nonlinearities imposing characteristic patterns to each recording. This allows for subsequent analyses on the authenticity and integrity of the acquired data, as well as considerations on the authentication of other recording context components (i.e. the sound source for the recorded signal, the recording environment and its reflection, and absorption and convolution effects).

11.2.2.1 Side-Information-Based Authenticity and Integrity Analyses Using ENF

From the three classes of approaches, the currently most widely studied class is the *electric network frequency (ENF)-based approach* introduced by Catalin Grigoras in 2003 (Grigoras 2003). It's based on the fact that if a recording setup or device with an alternating current (AC) power supply (or a direct current (DC) powered device influenced by electromagnetic fields) is used to record an audio signal, the 50/60 Hz[1] ENF as well as its harmonics become part of the recorded signal. The reason for this is that the used equipment usually lacks ideal voltage regulators and perfect shielding.

The work in Grigoras (2003) shows that the ENF is a continuously differentiable function which displays variations in the form of fluctuations of a small value (usually less then ± 1 Hz). It proposes to constantly measure the ENF within a power grid and store a description of this continuous signal in a reference database. This registered ENF is then used as side information on the influence imposed to recording devices and therefore for assessing the authenticity (in the context of this paper the authenticity of a claimed date for an audio recording) and integrity (by means of detecting signal inconsistencies in the harmonic signals recorded) of digital audio/video evidence.

Early state-of-the-art publications on ENF-based microphone (and/or video) forensics are, for example Cooper (2008), Grigoras (2005, 2007), Brixen (2008b) and Nicolalde and Apolinario (2009); more recent work is for example Gerazov *et al.* (2012) and Garg *et al.* (2013). The complete electro-physical requirements for this approach are summarized in a European Network of Forensic Science Institutes (ENFSI) standard published in 2009 (Grigoras *et al.* 2009). The core functionalities of this approach include a pre-registration (template generation) step for the ENF at a recording locationin the power grid of interest, extraction of ENF-related features from the recorded signal and correlation-based template matching of the feature vector against the pre-registered ENF side-information.

The acquisition context considered in ENF analyses is the influence of the electrical power grid to the attached recording device. As long as certain strict requirements are met (see the list of limitations given in the text), no subject, object or other environmental factors are assumed to have an impact on the results of this method. The ENF approach unfortunately has a number of limitations:

* It is of limited use for properly shielded devices using direct current (DC) power supply.

---

[1] In most parts of the world either 50 or 60 Hz electric network frequencies are used, depending on the national regulations.

- The approach works for devices powered by an observed AC power supply (e.g. public power grids). It will fail if the recording location uses its own AC power source not monitored by sensors connected to the forensic database, for example an electric generator or an uninterruptible power supply.
- Certain types of microphones are immune to electromagnetic fields (and their changes) (Brixen 2008a).
- As described in Grigoras *et al.* (2009), the standard approach requires a continuously updated database documenting ENF features for all possible recording locations, that is it does not allow for the authentication of legacy content for which no ENF template exists.
- It is highlighted in Grigoras *et al.* (2009) that the approach does not work well with audio files after lossy encoding, such as GSM or MP3 encoding.
- The approach is assumed to work well with common speech recordings as the ENF is usually outside of the bandwidth in the signal spectrum occupied by human speech. However, if the signal has strong frequency components in the frequency band covering the ENF (e.g. background music), it is highly probable that the approach will fail.

To draw a summarizing statement on these limitations, we would like to quote what Brixen said about the usability of the ENF features (in Brixen 2008b): 'In praxis, approximately 40–60% of the digital recordings in question contain traceable ENF.'

For a more complete discussion on the forensic usages of the ENF approach for (including its usages for source and data authentication), the reader is referred to Grigoras *et al.* (2009), Brixen (2008a) or REWIND (2011).

11.2.2.2   Content-Based Consistency Analyses on Time-Domain Local Phenomena

The second class of approaches is the *time-domain and local phenomena-based evaluations*. Malik and Farid (2010) described a technique for estimating the amount of reverberation in an audio recording by correlation based template matching. Because reverberation depends on the recording context, especially the shape and composition of a room and the subjects and objects located within, differences in the estimated reverberation can be used in a forensic setting for context-based authentication (and/or integrity verification against composition of audio material from multiple source recordings). The computed consistency of the reverberation behaviour can be considered as a special kind of global feature, even though it is based on temporal phenomena found in time domain. However, it might not be extractable in each recording, since it is strongly dependent on the recording context and the content: There exist a wide range of environments which do not display the required constant reverberation behaviour (e.g. any outdoor recording location as well as crowded places) and certain recording contents will not display any useful reverberations.

Thus, the application of this approach is seriously limited as summarized in Gupta *et al.* (2012): 'Currently, this measure has been successfully applied to synthesized audio with assumptions that cannot be fulfiled by most real-world signals. Thus, it needs to be generalized for a wider range of applications.' Its application is further hindered by signal post-processing operations (e.g. blind de-reverberation) automatically performed in many application scenarios such as audio and video conferencing, hands-free telephony, etc. (see e.g. REWIND (2011) for more examples).

For more details on the prospects and limitations of the recording context authentication using time-domain and local phenomena-based evaluations, especially acoustic reverberation, the reader is referred to Chaudhary and Malik (2010), Ikram and Malik (2010), Malik and Miller (2012), Zhao and Malik (2012), Peters *et al.* (2012), which extend the original idea and report empirical investigations to show its capabilities.

11.2.2.3 Microphone-Response-Based Pattern Recognition

The third alternative approach is the *microphone response based pattern recognition approach*. In 2005, Oermann *et al.* (2005) indicated that identification of a microphone as source of a recording might be possible, based on the observation that two different microphones cause noticeable differences in the recorded spectra of the same content recorded in a given recording context.

The first class of practical investigations on this approach are presented in Kraetzer *et al.* (2007), where a statistical pattern recognition-based approach is introduced to classify four different microphones operating in 10 different recording contexts (covering a wide range of different room types). The original work on this approach was further extended in Kraetzer *et al.* (2009, 2011, 2012), Buchholz *et al.* (2009), and a very recent summary can be found in Kraetzer (2013). The research works summarized in Kraetzer (2013) include the following:

- The introduction of a general-purpose audio feature extractor called the Advanced Multimedia and Security Lab (AMSL) – an institution at the department of computer science, Otto-von-Guericke-University Magdeburg, Germany) Audio Feature Extractor (AAFE) as the basis for the implementation of this microphone forensics approach;
- Considerations on empirical ground truth for the microphone forensics approach, including intra-microphone class classifications, influence of number and dimensionality of feature vectors in training, considerations on how to select classifiers and features based on experiments;
- Evaluations on different recording context influences: like the influence of the recording environment (especially the room), microphone orientation, microphone mounting settings, content selection and content dependence.

Besides those considerations on the authentication of microphones in specific recording contexts, Kraetzer (2013) covers investigations on:

- The persistence of the pattern in the feature space extracted from recordings of the same microphone under selected post-processing operations (normalization, MP3 conversion and de-noising) and playback recording;
- Composition detection.

Furthermore, a reflection of the achieved results against the Daubert standard is performed, evaluating the degree of maturity that is currently achieved by this approach – as one of the requirements imposed by Daubert.

The second class of practical realizations and evaluations of the idea of microphone response-based pattern recognition can be found in the work of Garcia-Romero and Espy-Wilson (2010), where the authors developed a Gaussian mixture model (GMM)-based template matching to implement automatic acquisition device identification on speech recordings. The main motivation behind their work is that determining the microphone would improve the performance of speaker recognition approaches. In their evaluations with two sets of microphones (a low-quality set of eight telephone handsets and a normal quality set of eight microphones) the authors reported classification accuracies higher than 90%. If these results would be generalizable, this would allow for a reliable selection of microphone-specific speaker recognition models and thereby would assumedly result in an increase of the performance of the subsequent speaker recognition process. A very interesting part of their approach is the generation of the template for each microphone, which allows for an extremely compact and recording length-independent representation of the microphone-response-based recording influences. The main drawback of this approach is that, while it assumedly works quite well on speech signals, it is determined to fail on other more complex audio signals. Other researchers' work like (Dufaux 2001) and (Moncrieff et al. 2006) showed that, for complex (both foreground and background) audio signals, even the complex GMMs (with a high number of Gaussian components) could not succeed in adequately solving the high diversity of potential recording contexts.

The third class of practical realizations and evaluations of the microphone-response-based pattern recognition approach is the work of Malik and Miller (2012). In their paper, the authors report a threshold based template matching method using first- and higher-order statistics of estimated Hu's moments (see Hu (1962) for the explanation of the original concept of transform-invariant moments and Malik and Miller (2012) for the corresponding application to audio signals) to implement microphone authentication, and they tested their method with a set of eight microphones in a fixed context.

The fourth and rather specialized instantiation of the basic concept introduced in Oermann et al. (2005) is an authentication of mobile phone recording contexts as presented for example in Hanilçi et al. (2012) and Hanilçi and Ertas (2013) (source cell-phone authentication using speech signals) or Panagakis and Kotropoulos (2012) (telephone handset identification). Besides the focus on specific types of microphones, they follow the same basis principles as the approaches described earlier.

For all the four above-mentioned instantiations of the basic concept of microphone-response-based pattern recognition, the considered recoding contexts are limited to small-scale setups. Until the time of this writing usually up to 8 (for normal microphones) and 14 (for the mobile phones in Hanilçi and Ertas (2013)) recording devices, recording a small number of audio contents in a small number of environments (up to 10 in Kraetzer (2013)), are used. With two sets of four microphones recording multi-genre audio signals in 10 rooms, (Kraetzer 2013) displays the most complex setup among all the work we are aware of. There are two main reasons behind the small scale of all instantiations reported so far. On the one hand, it is a very laborious, hardware-intensive and expensive task to create and operate large-scale audio recording setups, and on the other hand, the computational effort for training and applying pattern recognition algorithms to the large amounts of data generated is huge. The second reason, which is in our opinion more challenging than the first, can only be addressed by finding new, low-dimensional feature spaces that are capable of describing the patterns imposed by the intrinsic transfer functions of individual microphones in varying contexts. The discovery of such features would significantly reduce the complexity of the classification problem and thereby allow for implementations of the forensic analysis approach to be tested on large-scale setups.

11.2.2.4 Other Related Work

Besides the work described earlier, there exists scientific work in closely related fields. Some examples identified include benchmarking and quality assurance in microphone impulse response and distortion measurement (see, e.g. Farina 2000), audio signal copy detection (see, e.g. Cooper 2006), room acoustics (see, e.g. Kuttruff 2009), double recording detection for analogue recordings (see, e.g. Koenig 1990) or manual audio forensic analysis of digitally recorded signals by forensic experts (see, e.g. Koenig and Lacey 2009). In the near future, such approaches might be transferred into approaches for recording context or microphone authentication.

For further details on audio signal authentication, the reader is referred to some survey papers such as Owen (1988), Bijhold *et al.* (2007) and Maher (2010) (which focus on forensic analysis of analogue audio signals), standard-like documents (SWGDE 2014) as well as Brixen (2007), Rumsey (2008), Koenig and Lacey (2009), Tibbitts and Lu (2009), Maher (2009, 2010) and, Gupta *et al.* (2012) (which focus on forensic analysis of digital audio signals).

11.3 Guidelines for Microphone Registration

The principal requirement for the registration of microphones for forensic authentication is the necessity to represent the recording context of the recording sample to be authenticated as precisely as possible. This means that the original recording context has to be estimated (or reconstructed) in terms of generating suitable training data for the microphone to be authenticated (i.e. the null-hypothesis in the authentication) as well as similar microphones (acting as representatives for the alternative hypothesis in the authentication).

This importance of fulfiling this requirement is demonstradnt in the investigations performed in Kraetzer (2013) on the impact of context influences (e.g. room characteristics, mounting influences to the microphone and content influences) on the performance of the scheme. Only when there is a strong contextual similarity between the original recording setup and the forensic authentication setup, the classification model will be good enough for the classification task at hand.

If such a strong similarity between original recording setup and the forensic setup is ensured, the complexity of the pattern recognition problem might be reduced to a level where it can be successfully solved by template matching approaches (e.g. the works of Garcia-Romero and Espy-Wilson (2010) and Malik and Miller (2012)), instead of having to rely on the more flexible but usually computationally more complex statistical pattern recognition approaches.

The above-mentioned requirement for microphone registration is addressed in practical investigations performed in the examples described in this chapter by using stratified cross-validation in closed set setups as the primary evaluation method.

A second requirement for microphone registration is imposed by the computational complexity associated with solutions for statistical pattern recognition problems. The statistical models for problems with a strong overlap between the patterns to be recognized and the background pattern, microphone forensics being a typical problem of this kind, tend to be rather complex, thus leading to large run-times in the training phase as well as in the application phase. The derived requirement is the need to find features that show a strong distinguishing power between the background pattern and microphone-induced patterns, therefore simplifying the classification task significantly.

The second requirement is met in the examples described in this chapter by using an audio feature set that has already been proved to be able of supporting implementations of microphone forensics (see, e.g. Kraetzer 2013). This set of features is used here to establish, amongst other investigation tasks, how much recorded content is required to achieve a suitable detection performance for a given context.

Currently, all research work on microphone forensics is still in its early prototype stage. When the approaches reach a higher degree of maturity, a new requirement will emerge: the need to establish a clear picture of the scalability of the approaches and the associated error rates. The second part of this new requirement is derived from the Daubert standard (see, e.g. Fradella et al. (2004) or Hildebrandt et al. (2011)), which are being accepted in an increasing number of judicial systems as the most important set of criteria for handling forensic evidence in courtrooms (i.e. for the admissibility of an expert testimony based on such evidence).

11.4   Case Studies

To demonstrate how a typical microphone forensic analysis can be done using statistical pattern recognition, new practical investigations are performed in this chapter to further advance the microphone-response-based pattern recognition approach introduced in Kraetzer *et al.* (2007) and in its various extensions summarized in Kraetzer (2013).

Section 11.4.1 defines the investigation tasks for the intended extensions. Section 11.4.2 briefly summarizes the implementation of the statistical pattern recognition pipeline used for our evaluations and Section 11.4.3 describes the evaluation setups. Finally, Section 11.4.4 summarizes the evaluation results.

11.4.1   Investigation Tasks

For the investigations described here, the following four tasks are defined:

Task A: Benchmarking different feature extractors by asking 'Is the selected audio feature extractor performing as good as its predecessor and competitors?'

To address this task it is important to determine whether the results of performance evaluations published in previous work are still valid for the new updated version of the AMSL Audio Feature Extractor (AAFE; see Sections 11.2.2 and 11.4.2), which is first used in this book chapter.

Task B: Impact of size and composition of the training and test data by asking 'How much audio content is sufficient to generate a suitable context representation for recording setup authentication in the training phase?'

The size of the evaluation data has a tremendous influence on the runtime of a practical classification setting. The size and composition of the training and test data generate an optimization problem: the training data have to be representative for the application domain and all possible contexts as well as patterns under investigation; while the amount of training data used should be as small as possible to keep the time required for training at a minimum and the model size as small as possible. In testing (or application) phase the performance of the classifier and a specifically trained model are supposed to be good if (and only if) there exists a strong correlation between the training and test data (i.e. if the training data contains representative samples for the tested cases). In the field of pattern recognition this problem is termed 'fitting' and is part of the considerations on the generalizability of an approach.

Here, stratified cross-validation is used as the performance evaluation method, which naturally solves the problem of the required correlation between and training and test data for closed-set evaluations.[2] We need to evaluate how much recorded audio content

---

[2] That is, all potentially existing classes for the problem are known, and therefore no unknown data will have to be classified in future which is not present in the given training data.

is sufficient to generate a suitable context representation using AAFE (in its current version v.3.0.1) for specific recording setup authentication attempts.

Task C: Discriminating the classifier's distinguishing power by asking 'Is the inter-class variability in contrast to the intra-class variability large enough to allow for microphone forensics?'

The statistical properties of different classes (i.e. their distributions of the extracted audio features and their intra- and inter-class variances) are the most important influencing factors for any pattern recognition-based authentication attempt. With high intra-class and low inter-class variances, any classification will be difficult; the smaller the intra-class variance and the higher the inter-class variance, the simpler the classification problem will (theoretically) be. Here, this problem is strictly addressed by usage of pre-existing statistical pattern recognition methods provided with by the open source data mining suite WEKA (Hall *et al.* 2009). Alternative approaches (e.g. the usage of inferential statistics to actually analyze the intra- and inter-class variances) are beyond the scope of the work described here.

Task D: Impact of classifier's parameter settings by asking 'Which influence do the classifier's parameters have on the classification process?'

The work described in the state of the art (e.g. Kraetzer 2013) often relies on the usage of existing classification algorithms operating in their default parameterization. Since the authors of classification algorithms in many cases have a specific application scenario in mind while implementing their classification algorithms and in most cases the original scenario is different from latter application scenarios of these tools, it is not surprising that the default parameterization of a classification algorithm does not show an optimal detection performance in many application scenarios. Here, it is important in the scope of forensic investigations to determine the actual influence the classifier's parameters have on the actual classification process. A first study is performed, and it shows some tendencies by investigating alternative classification parameter settings, for the WEKA classifier *weka.classifiers.functions.Logistic* in a grid search through a large proportion of the parameter space.

Figure 11.3 illustrates a brief summary of the evaluated context and statistical pattern recognition pipeline setup. This figure projects core components of the acquisition context for pattern recognition-driven microphone forensics, as shown in Figure 11.2, onto a simplified statistical pattern recognition pipeline. Furthermore, the documentation of the forensic investigation and the case data collected and generated during the whole investigation process are not displayed in Figure 11.3. For all of our investigations, we use a high-quality monitor loudspeaker for playing our pre-recorded audio reference set, minimizing the impact on the re-recording with the microphones under investigation. In contrast to the examples on potentially very complex recording contexts given in Section 11.1 of this chapter, here thereby a rather simplistic, yet better controllable and easier reproducible recording context is created.

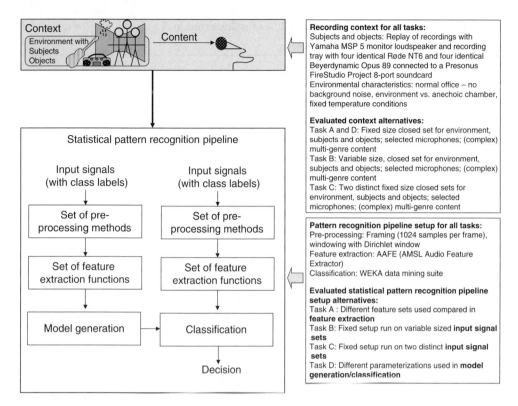

Figure 11.3 Summary of the evaluated context and statistical pattern recognition pipeline setups evaluated.

The precise evaluation setups for the practical investigations described in this work are presented in Sections 11.4.2 and 11.4.3 below.

11.4.2 Implementation of the Statistical Pattern Recognition Pipeline

The number of **pre-processing** operations is restricted within the evaluations performed to the framing and windowing (the splitting into consecutive, fixed sized windows of audio samples and the application of a windowing function – here Dirichlet function, see for example Kumar *et al.* (2010) required for the window-based processing of audio material in feature extraction.

The software used for **feature extraction** in our implementation of the statistical pattern recognition pipeline is the AMSL Audio Feature Extractor (AAFE) developed by the Advanced Multimedia and Security Lab (AMSL) at the Otto-von-Guericke-University of Magdeburg since 2006. Initially it was intended to be used within an audio steganalysis framework, but it has grown into a general-purpose audio feature extractor successfully applied to different application fields including microphone forensics.

Table 11.1 A summary of the audio features implemented in AAFE v.3.0.1; for details on the feature implementations see Kraetzer (2013).

Intra-window features	
Time-domain (9 features)	zero cross rate, entropy, LSB ratio, LSB flip rate, median, mean, RMS amplitude, short time energy, short time average magnitude
Frequency-domain ($19 + n/2$)	pitch, spectral entropy, spectral centroid, spectral bandwidth, spectral rolloff, spectral irregularity, spectral variation, spectral smoothness, 11 formants (the first two for each vowel and the first for singer), $n/2$ frequency coefficients
Cepstral-domain(52 features)	13 Mel-frequency cepstal coefficients (MFCCs), 13 filtered Mel-frequency cepstal coefficients (FMFCCs), 13 second-order derivative MFCCs (based on Liu *et al.* (2009)) and 13 second-order derivative FMFCCs
Inter-window features	
Totaling features	total zero cross rate
Averaging features	average entropy, average LSB ratio, average LSB flip rate, average median, average mean, average RMS amplitude, average short time energy, average short time average magnitude, average pitch, average spectral entropy, average spectral centroid, average spectral bandwidth, average spectral roll off, average spectral irregularity, average spectral variation, average spectral smoothness

In Kraetzer (2013) the feature space of AAFE version 2.0.5 is described in detail. Here, with version 3.0.1 an updated version of the AAFE is used for the first time. The main difference to its predecessor version (2.0.5) is the porting of the tool from MATLAB to Java, which, together with an optimization of the routines used for the computation of the windowing functions (Dirichlet (Kumar *et al.* (2010), a.k.a. rectangular) and Hamming windows are supported) and individual features, reduces the time for the feature extraction by approximately a factor of 10. Besides this run-time improvement, a GUI is also provided (earlier versions are command line tools) and slight modifications to the feature space have been madeto add one new time-domain feature (short time average magnitude) and one new frequency-domain feature (spectral variation). The feature space for AAFE version 3.0.1 now includes 80 plus $n/2$ intra-window features (a.k.a. segment-wise computed features), where n is the window size used (by default $n = 1024$), and 17 inter-window features (a.k.a. global features). Table 11.1 briefly summarizes the computed features.

With the default parameterization for AAFE (Dirchletwindows with size $n = 1024$) two outputs are generated per audio file: a set of intra-window features of dimensionality 592, and a set of inter-window features (one per channel in the audio file) of dimensionality 17.

Feature selection is an optional component in the pattern recognition pipeline. It enables the empirical identification of relevant features for a given decision or classification problem, which, on the one hand, potentially reduces the dimensionality of the feature space (and thereby speeds up the classification process) and on the other hand, generates domain-knowledge for the problem to be solved.

Witten *et al.* proposed (2011) a methodology for feature selection, which identifies the best way to select features as a manual selection process based on a deep understanding of the decision or classification problem and the actual meaning of the features. Since the knowledge required is not always available, different approaches for automatic feature selection have been proposed. Witten *et al.* identify two fundamentally different approaches for scheme-independent feature evaluation: The usage of filters on the feature set, to find the most promising features and the usage of classifiers in a wrapper to identify the significant features. The work discussed here relies on the filter- and wrapper-based feature selection strategies implemented in the WEKA data-mining suite (Hall *et al.* 2009).

For details on basic feature selection strategies, the interested reader is referred to Witten *et al.* (2011). More advanced methods for feature evaluation (e.g. symmetric uncertainty, principal component analysis (PCA) or random projection based methods) are discussed for example in Witten *et al.* (2011), Lu *et al.* (2007) and Blum (2006).

For the practical investigations described here, the considerations on the assignment of the class labels to the candidate signals are limited on supervised statistical pattern recognition (i.e. **classification** – see Section 11.2.1 of this chapter). Other pattern recognition sub-disciplines as well as unsupervised statistical pattern recognition (clustering) are omitted in the investigations performed.

The work discussed here uses the large portfolio of **classification** algorithms already implemented in the well-known open source data-mining suite WEKA (Hall *et al.* 2009).

11.4.3   Evaluation Setups

A fixed set of two different recording locations (labelled *R01* and *R06*) is used to provide a controlled set of different*context influences* on the environmental shaping for the investigations performed. This set is a sub-set of the larger set of recording locations introduced and described in detail in Kraetzer *et al.* (2007). The original set consists of eight rooms and two outside locations of the main building of the Faculty of Computer Science, Otto-von-Guericke University of Magdeburg.

The recording locations *R01* and *R06* are a large office and an anechoic chamber, respectively. The first location provided us with a realistic recording context while the second provided a (nearly) perfect recording context, minimizing subject, object and environmental influences to the recorded content.

To provide further control over the recorded content, a fixed set of reference signals (labelled *ref10*) is used. This set contains 10 files from the audio test set described in Kraetzer *et al.* (2006), which represent 10 different classes of audio material. All material is provided in CD-quality (44.1 kHz sampling frequency, 16-bit quantization, stereo and PCM coded). The reference files (and the recordings based on these references) have a duration of 30 s each.For the playback of the reference files at the recording locations described earlier, exactly one sound source (a very precise Yamaha MSP 5 high-quality monitor loudspeaker) is used to provide a very good transmission function and a minimal noise component for constant playback influences.

Only two sets of microphone and pre-amplifier combinations are used to provide a controlled set of different recording influences that is the intrinsic characteristics producing the patterns distinguished by the microphone forensics approach considered in this chapter. These sets are homogeneous sets of four microphones each. The set *RS4_Beyer* consists of four Beyer dynamic Opus 89 dynamic microphones, and the set *RS4_Rode* consists of four Røde NT6 condenser microphones. Both sets were recorded in parallel using a Presonus Fire Studio Project 8-port high-quality sound card. The audio signal was recorded with 44.1 kHz sampling frequency, 16-bit quantization, mono and PCM coded.

For evaluation 10-fold stratified cross-validation was used with separate training and test sets. More details of the evaluation strategy actually applied are specified in the setup summary for each test described in Section 11.4.4. The metric used to evaluate the classifier performance is the *Kappa statistics*. It measures the agreement of prediction with the true class (i.e. the agreement normalized for chance agreement) in the range $[-1; 1]$. A value of Kappa $= 1$ indicates perfect agreement and Kappa $= 0$ indicates chance agreement for the overall classification. Negative values for Kappa imply the choice of a classifier model trained for a different classification problem. To be able to use Kappa statistics, the classes are equally distributed in training as well as test sets.

For the implementation of the pattern recognition pipeline, pre-processing was restricted to windowing with 1024 samples per non-overlapping, consecutive frame, using Dirichlet windows (see Section 11.4.2). For feature extraction, AAFE version 3.0.1 was used consistently for all tests. For one evaluation AAFE version 2.0.5 was used additionally. The investigations performed were limited to the intra-window features computed by the feature extractor, due to the low performance shown for the global features in Kraetzer (2013).

For classification, some supervised classification algorithms implemented in WEKA v.3.6.8 (see Section 11.4.2) were used in their default parameterizations. The platform for the implementation of the statistical pattern recognition solution for the evaluations performed is an array of workstations with Intel Core 2 Duo E8400 CPUs 3 GHz and 4 GB RAM, running Microsoft Windows XP, WEKA v.3.6.8 on Java SE 6 (32-bit Windows version) with 1.6 GB allocated RAM for each WEKA instance (i.e. classifier).

Table 11.2 Comparison of the detection performance (expressed in Kappa statistics) of AAFE versions 2.0.5 (from Kraetzer 2013) and 3.0.1 (used here).

Classifier	100 frames per reference file (4000 frames, ca. 90s audio)		200 frames per reference file (8000 frames, ca. 180s audio)	
	AAFE v.2.0.5	AAFE v.3.0.1	AAFE v.2.0.5	AAFE v.3.0.1
functions.Logistic	0.612	0.624	0.589	0.709
meta.MultiClassClassifier	0.572	0.581	0.603	0.712
meta.RandomSubSpace	0.641	0.612	0.586	0.689
meta.RotationForest	0.552	0.705	0.637	0.767

Parameterization of the pattern recognition pipeline: Training & Test set: *RS4_Beyer* in *R01*; recorded content: *ref10* playback by an Yamaha MSP 5 high-quality monitor loudspeaker; evaluation mode: 10-fold stratified cross-validation; pre-processing: framing (1024 samples per frame), windowing with Dirichlet window; feature extraction: AAFE v.2.0.5 and AAFE v.3.0.1; feature selection: none (590 vs. 592 intra-window features plus class); classification: best four classifiers from Kraetzer (2013) in default parameterization (originally best five but WEKA v.3.6.8 lacks *weka.classifiers.meta.EnsembleSelection*).

11.4.4 Evaluation Results

One major issue regarding microphone forensics is the question on the scalability of the approach in terms of the number of different microphone (or recording setups) that can be distinguished. So far this has not yet been answered in the research community. The evaluation tasks defined for the examples described in this chapter in Section 11.4.1 cover important intermediate steps to extend the state of the art in this field by new investigations, which can further contribute to the maturity of the research field and in the near future can hopefully lead to the first estimation of the distinguishing power of the microphone forensic approach

11.4.4.1 Evaluation Task A

To address evaluation task A defined in Section 11.4.1, the detection performance of feature extractor versions AAFE v.2.0.5 (used extensively, e.g. in Kraetzer (2013)) and v.3.0.1 (new version, first used here) in a representative setup are compared in Table 11.2. The four chosen classifiers are the best four classifiers from Kraetzer (2013) in default parameterization. For detailed descriptions on those classifiers see Witten *et al.* (2011).

The results show that the new version 3.0.1 of AAFE is not only about 10 times faster in the feature extraction than version 2.0.5 but also achieved a slightly higher detection performance in seven out of the eight tested cases. Even though this small experiment is not an absolute proof, but only an indication, it can still be assumed that the new version can achieve detection performances in microphone recording set authentication that are equivalent to (or even slightly better than) those already reported in the literature. Therefore, generalizing statements on the system performance made in previous work should still be valid for the new updated version of AAFE, which is first used here.

11.4.4.2 Evaluation Task B

To address the question 'How much audio content is sufficient to generate a suitable context representation for recording setup authentication in the training phase?' (evaluation task B defined in Section 11.4.1), four empirical investigations were performed using two sets of microphones each containing four microphones in two different recording environments.

Figures 11.4 and 11.5 show the development of Kappa values with increasing evaluation set sizes for a set of four Røde NT6 condenser microphones.

As can be seen for both evaluations the detection performance increases rapidly for small numbers of frames per reference file. In case of the recording environment room *R01* the achieved increase levels for the slowest levelling classifiers in the set at about 75 frames per reference file.

For the recording environment room *R06* better detection performances were achieved for all four tested classification algorithms. Furthermore, the achieved increase levels earlier (at about 35 frames per reference file for the classifiers *meta.Rotation Forest* and *meta.Random SubSpace* and at about 70 for the other two classifiers). This can be attributed to the fact that *R06*, as an anechoic chamber, can produce a less noisy representation of the reference signals than a normal office space like *R01*.

Another set of two evaluations (the same two rooms but a different set of microphones – Beyer dynamic Opus 89 dynamic microphones instead of Røde NT6 condenser microphones) showed exactly the same behaviour in terms of the achieved detection performances. Therefore the details of these results are omitted here.

The results of this investigation led to two main insights:

1. The context representation requires only a very small amount of audio content: the detection performance achieved for the ten different reference signals in the reference set *ref10* already levels out at about 75 frames (about 2 s of audio material with the used parameterization) per reference file for normal recording environment and much earlier for less noisy recording environments.
2. The modelling of the context for the forensic setup should be made as accurate as possible (including the expected audio source and environment influences) to achieve optimal performances for the forensic approach.

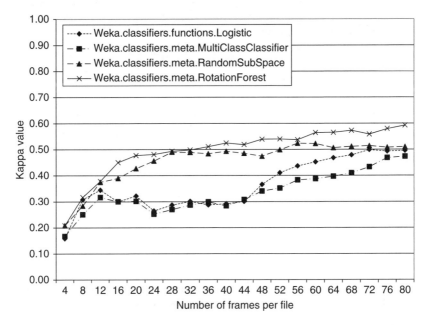

Figure 11.4 Kappa values development with increasing evaluation set size for the recording environment room *R01* (an office of size 4 × 5 m). Note: Parameterization of the pattern recognition pipeline: Training and test set: *RS4_Rode* in *R01*; recorded content: *ref10* playback by an Yamaha MSP 5 high-quality monitor loudspeaker; evaluation mode: 10-fold stratified cross-validation; pre-processing: framing (1024 samples per frame), windowing with Dirichlet window; feature extraction: AAFE v.3.0.1; feature selection: none (592 intra-window features plus class); classification: best four classifiers from Kraetzer (2013) in default parameterization (originally best five but WEKA v.3.6.8 lacks *weka.classifiers.meta. EnsembleSelection*).

11.4.4.3   Evaluation Task C

In respect to inter-class variability in contrast to the intra-class variability, the performance of classification methods can be evaluated by looking only at the achieved detection performance (e.g. as accuracy or Kappa statistics). In forensic setups, a precise analysis of the associated error behaviour is also required. Table 11.3 shows the distribution of errors for the classification of four Beyer dynamic Opus 89 microphones (M20 to M23) and a model trained for eight different microphones (the four Røde NT6 microphones (M16 to M19) and the four Beyer dynamic Opus 89 microphones (M20 to M23)). Each line in the table represents 500 feature vectors originating from one of the microphones in the test set. Each column indicates the number of class assignments to the eight classes present in the training phase.

As can be seen Table 11.3, the true classifications (marked in bold) dominate this confusion matrix with 1266 out of 2000 (63.3%) individual class assignments.

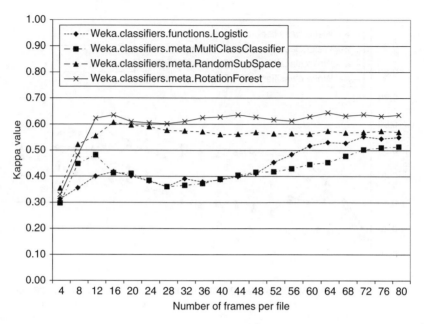

Figure 11.5   Kappa values development with increasing evaluation set size for the recording environment room *R06* (an anechoic chamber). Note: Parameterization of the pattern recognition pipeline: Training and test set: *RS4_Rode* in *R06*; recorded content: *ref10* playback by an Yamaha MSP 5 high-quality monitor loudspeaker; evaluation mode: 10-fold stratified cross-validation; pre-processing: framing (1024 samples per frame), windowing with Dirichlet window; feature extraction: AAFE v.3.0.1; feature selection: none (592 intra-window features plus class); classification: best four classifiers from Kraetzer (2013) in default parameterization (originally best five but WEKA v.3.6.8 lacks *weka.classifiers.meta. EnsembleSelection*).

Besides the true classifications, a larger number of errors were made in assignments to microphones of the same model (407 out of 734 misclassifications in total, with an average of 33.92) than to microphones of a different model (327 out of 734 misclassifications in total, with an average of 20.44).

Table 11.4 presents similar result to those in Table 11.3 for the distribution of errors for the classification of four Røde NT6 microphones (M16 to M19) and a model trained for eight different microphones (the four Røde NT6 microphones (M16 to M19) and the four Beyer dynamic Opus 89 microphones (M20 to M23)). Here, also the true classifications (marked in bold in the table) dominate this confusion matrix with 1112 out of 2000 (55.6%) individual class assignments. Besides the true classifications, the absolute number of errors made in assignments to other Røde microphones and that to the Beyerdynamics are very close to each other (436 vs. 452), but the average number of errors for the other Røde instances is much higher than that for the Beyer dynamics microphones (36.33 vs. 28.25).

Table 11.3 Distribution of correct matches (in bold) and errors in the classification results of four Beyer dynamic Opus 89 microphones (M20 to M23) using a model trained for eight different microphones.

	M16	M17	M18	M19	M20	M21	M22	M23
M20	18	24	18	11	**310**	28	20	71
M21	15	48	17	9	25	**319**	48	19
M22	21	32	16	12	17	65	**318**	19
M23	23	29	19	15	48	27	20	**319**

Table 11.4 Distribution of correct matches (in bold) and errors in the classification of four Røde NT6 microphones (M16 to M19) using a model trained for eight different microphones.

	M16	M17	M18	M19	M20	M21	M22	M23
M16	**249**	21	35	100	21	21	17	36
M17	30	**312**	19	24	31	26	24	34
M18	23	9	**318**	31	28	57	18	16
M19	73	32	39	**233**	46	20	20	37

Parameterization of the pattern recognition pipeline: Training set: *RS4_Beyer&RS4_Rode*; Test set: *RS4_Rode* in *R01*; recorded content: *ref10* playback by an Yamaha MSP 5 high-quality monitor loudspeaker; 50 frames per reference file; evaluation mode: separate training and test sets; pre-processing: framing (1024 samples per frame), windowing with Dirichlet window; feature extraction: AAFE v.3.0.1; feature selection: none (592 intra-window features plus class); classification: *weka.classifiers.meta.RotationForest*.

In summary, for the investigation performed here it can be argued that the used approach not only achieved detection performances better than the performance of randomly guessing (i.e. Kappa = 0 or an average of 25% true classifications per class in the experiments summarized in Tables 11.3 and 11.4), but also shows a plausible misclassification behaviour: On average, more feature vectors are wrongly assigned to microphones of the same model than to microphones of a different model. The detection performance and distribution of errors indicate that the evaluation sets in the used feature space display an intra-class and inter-class variance relationship that helps to distinguish between different types of microphones as well as different individual microphones.

11.4.4.4 Evaluation Task D

To address evaluation task D for the impact of alternative classification parameter settings, for the one exemplarily selected classifier from the set of four used for the evaluation described above (here the classifier *weka.classifiers.functions.Logistic*, chosen for its comparably short run-time) a grid search through a large proportion

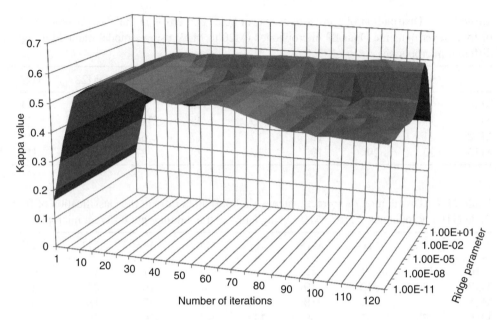

Figure 11.6   Kappa values achieved by a (partial) grid search through the two-dimensional parameter space of the WEKA classifier *functions.Logistic*. Note: Parameterizsation of the pattern recognition pipeline: Training & Test set: *RS4_Beyer* in *R01*; recorded content: *ref10* playback by an Yamaha MSP 5 high-quality monitor loudspeaker; evaluation mode: 10-fold stratified cross-validation; pre-processing: framing (1024 samples per frame), windowing with Dirichlet window; feature extraction: AAFE v.3.0.1; feature selection: none (592 intra-window features plus class); classification: *weka.classifiers.functions.Logistic* with grid search through the parameter space.

of the parameter space was performed. Figure 11.6 visualizes the result of this grid search.

The Kappa value archived for the default parameterization[3] of this classifier is Kappa = 0.496. The best detection performance[4] achieved in the performed grid search is much higher: Kappa = 0.645.

Even though this evaluation shows the benefit of parameter optimization, it also highlights the immense cost (it terms of amount of computation) attached. In the tested example, one classification took about 17 min on our test machine. For reasons of visual clarity, Figure 11.6 shows only 390 out of 765 performed classifications – the whole evaluation ran 216.75 h on the test machine.

---

[3] Default parameters (see Witten *et al.* (2011) for the description of the classifier's parameters): ridge (−R) 1.0E-8; max number of iterations (−M) − 1 (until convergence).

[4] For parameters: ridge (−R) 1.0E-1; max number of iterations (−M) − 1 (until convergence).

In summary, the example given earlier shows that the classifier's parameters can have a significant influence on the classification performance achieved. Therefore, an application of classifiers without optimized parameterscould be far away from being a fair benchmark of the performance of different individual classifiers.

11.5   Chapter Summary

The context of forensic investigations plays an important role as it is the groundtruth for the findings that we can derive from experiments. For microphone forensics to become an accepted tool for handling audio evidence used by forensic experts, it needs to be compliant to the Daubert standard as mentioned earlier in this chapter. Here the error rates of microphone forensics need to be presented within the context in which the experiments are performed, so to determine whether a particular microphone forensic investigation is appropriate for a particular court case. The precise description of the acquisition context is one of the steps required for the systematic evaluation as well as the comparison (bench marking) of forensic techniques.

Furthermore, such a precise context description allows the answering of questions such as: Is the recording context in an audio signal consistent with the claimed context in verification? Is the recording environment really an office space or do the context characteristics point towards a different kind of recording environment? Is the recording context detected consistently, or does it give indications that might lead to assumptions regarding integrity violations?

In summary, such context model would allow for extensions of the considerations and move the field away from the rather narrow focus of an authentication of the microphone as the acquisition sensor to a much wider context analyses. Therefore, recording context modelling will become increasingly important in future work. Simple models like the one proposed in Gaubitch *et al.* (2011) (modelling an audio recording as a simple convolution between the content to be recorded, the total environmental effects and an additive observation noise) should be enhanced with more complex context models. Of course, the interplay between authentication on the one hand and acquisition context estimation raises further issues that need to be researched in more detail. One such issue is the need to face challenges imposed by counter-forensics (see e.g. Barni 2012).

In general, questions of generalizability of approaches, performance (detection performance and errors, throughput as well as scalability) and compliance to forensic requirements (e.g. legal criteria such as the Daubert standard) are challenging tasks that remain to be answered.

As summarized in Section 11.3, currently all pattern recognition-based approaches for microphone forensics are still in their infancy. To meet the challenge of achieving acceptance under the ultimate benchmark for forensic methods (i.e. the admissibility in court) good features have to be found first that could distinguish between microphone induced recording patterns and other context influences to the recorded content. Then, based on such features, a template matching or statistical pattern recognition-based approach would be applied to obtain appropriate scalability (in terms of distinguishing power and collusion rates). This approach would have to be evaluated in large-scale experiments to investigate its precise capabilities and limitations, including the associated error rates. To achieve the necessary acceptance it has to be openly discussed in the corresponding scientific community.

Initial considerations and steps to achieve these challenges would be the following:

- Perform investigations into the exact scaling behaviour of current existing approaches for microphone authentication.
- Perform feature selection and feature space fusion operations on existing approaches to search for better feature spaces to represent the problem.
- Model and quantify context influences – some context models for the recording process already exist, ranging from simplistic models like the one proposed in Gaubitch *et al.* (2011) to complex context models (see, e.g. Kraetzer (2013) and Kraetzer *et al.* (2012)).
- Introduction of fair bench marking strategies for pattern recognition schemes, including means for parameter optimization as discussed in Section 11.4.4.4 of this chapter.
- Discuss all approaches in terms of compliance to forensic legal requirements.

If the microphone authentication problem could be solved, the proposed approaches could be further extended to other applications such as the following:

- Room authentication, for example is the recording environment really an office space or do the context characteristics point towards a different kind of recording environment?
- Integrity verification, that is the recording context detected consistent or does it give indications that might lead to assumptions regarding integrity violations? First considerations on such an extension of the authenticity of the recordings towards integrity analyses using the same approach have already been discussed in Kraetzer (2013) and Kraetzer *et al.* (2012).

Acknowledgements

The work on audio feature extraction in this chapter has been supported in part by the European Commission through the IST Programme under Contract IST-2002-507932 ECRYPT. The work on recording source authentication has been supported in part by the European Commission through the FP7 ICT Programme under Contract FP7-ICT-216736 SHAMAN. The information in this document is provided as is, and no guarantee or warranty is given or implied that the information is fit for any particular purpose. The user thereof uses the information at its sole risk and liability.

Christian Kraetzer and Jana Dittmann wish to express thanks to all experts for our joint work and/or discussion on our ideas on microphone forensics during the past few years. As researchers who have published upon this topic in IT-security-related workshop and conference proceedings, we would like to express our gratitude to all reviewers involved in the publication process. They mostly remained anonymous, but nevertheless their comments helped to shape our research presented within this chapter.

The context visualization in Figures 11.1, 11.2 and 11.3 was sketched by Jana Dittmann and is/will be part of different figures of several publications to visualize context aspects in different research fields of her research group. The copyrights for these figures are held by the authors.

References

Barni M, 2012, A game theoretic approach to source identification with known statistics. In *Proceedings of the IEEE International Conference on Acoustics, Speech and Signal Processing* (ICASSP) pp. 1745–1748. http://dx.doi.org/10.1109/ICASSP.2012.6288236 (Accessed 5 February 2015).

Bijhold J, Ruifrok A, Jessen M, Geradts Z, Ehrhardt S and Alberink I, 2007, Forensic audio and visual evidence 2004–2007: a review. In *15th INTERPOL Forensic Science Symposium*. Lyon, France.

Bishop CM, 2007, Pattern Recognition and Machine Learning.Springer-Verlag, New York.

Blum A, 2006, Random projection, margins, kernels, and feature-selection. In Proceedingsof the *2005 International conference on Subspace, Latent Structure and Feature Selection*, SLSFS'05. Springer Verlag, Berlin, Heidelberg, pp. 52–68. http://dx.doi.org/10.1007/11752790_3. (Accessed 5 February 2015).

Brixen EB, 2007, Techniques for the authentication of digital audio recordings. In *Proceedings of the 123rd Audio Engineering Society Convention*, New York. Audio Engineering Society (AES), Inc., New York.

Brixen EB, 2008a, ENF – quantification of the magnetic field. In *Proceedings of the Audio Engineering Society Conference: 33rd International Conference: Audio Forensics-Theory and Practice*, Denver, CO. Audio Engineering Society (AES), Inc., New York.

Brixen EB, 2008b, How to extract the ENF from digital audio recordings. In *Proceedings of the American College of Forensic Examiners Institute (ACFEI) 2008 National Conference*, San Diego, CA. American College of Forensic Examiners Institute (ACFEI), Springfield, MO.

Buchholz R, Kraetzer C and Dittmann J, 2009, Microphone classification using Fourier coefficients. In *Proceedings of the 11th Information Hiding*. Springer-Verlag, Berlin, Heidelberg, pp. 235–246.

Chaudhary U and Malik H, 2010, Automatic recording environment classification using acoustic signature. In *Proceedings of the 129th Audio Engineering Society Convention*, San Francisco, CA. Audio Engineering Society (AES), Inc., New York.

Cooper AJ, 2006, Detection of copies of digital audio recordings for forensic purposes. PhD Thesis, Department of Information and Communication Technology, The Open University, Milton Keynes, UK.

Cooper AJ, 2008, The electric network frequency (ENF) as an aid to authenticating forensic digital audio recordings an automated approach. In *Proceedings of the Audio Engineering Society Conference, 33rd International Conference, Audio Forensics-Theory and Practice*, Denver, CO. Audio Engineering Society (AES), Inc., New York.

Dittmann J, Wohlmacher P and Nahrstedt K, 2001, Multimedia and security–using crypto-graphic and watermarking algorithms. *IEEE Multimedia* **8**(4), pp. 54–65.

Duda R, Hart P and Stork D, 2001, Pattern Classification and Scene Analysis: Pattern Classification, 2nd edn. Wiley Interscience, New York.

Dufaux A, 2001, Detection and recognition of impulsive sound signals. Ph.D. Thesis, Institute of Microtechnology, University of Neuchatel, Switzerland.

Farina A, 2000, Simultaneous measurement of impulse response and distortion with a swept-sine technique. In *Proceedings of the 108th. Audio Engineering Society Convention*, Paris, France. Audio Engineering Society (AES), Inc., New York.

Fradella HF, O'Neill L and Fogarty A, 2004, The impact of Daubert on forensic science. *Pepperdine Law Review* **31**, pp. 323–325.

Garcia-Romero D and Espy-Wilson CY, 2010, Automatic acquisition device identification from speech recordings. In *Proceedings of the IEEE International Conference on Acoustics Speech and Signal Processing* (ICASSP), pp. 1806–1809. http://ieeexplore.ieee.org/xpl/articleDetails.jsp?tp=& arnumber=5495407 (Accessed 5 February 2015).

Garg R, Varna AL, Hajj-Ahmad A and Wu M, 2013, Seeing ENF: power-signature-based timestamp for digital multimedia via optical sensing and signal processing. *IEEE Transactions on Information Forensics and Security* **8**(9), pp. 1417–1432.

Gaubitch ND, Brookes M, Naylor PA and Sharma D, 2011, Single-microphone blind channel identification in speech using spectrum classification. In *Proceedings of the 19th European Signal Processing Conference* (EUSIPCO 2011), Barcelona, Spain.

Gerazov B, Kokolanski Z, Arsov G and Dimcev V, 2012, Tracking of electrical network frequency for the purpose of forensic audio authentication. In *13th International Conference on Optimization of Electrical and Electronic Equipment* (OPTIM), pp. 1164–1169. http://ieeexplore. ieee.org/stamp/stamp.jsp?tp=&arnumber=6231908&isnumber=6231751 (Accessed 5 February 2015).

Grigoras C, 2003, Digital audio recording analysis: the electric network frequency criterion. Technical Report. Diamond Cut Productions, Inc., Hibernia, NJ.

Grigoras C, 2005, Digital audio recording analysis: the electric network frequency (ENF) criterion. *Speech, Language and the Law* **12**(1), pp. 63–76.

Grigoras C, 2007, Application of ENF criterion in forensic audio, video, computer and telecommunication analysis. *Forensic Science International* **167**, pp. 136–145.

Grigoras C, Cooper A and Michalek M, 2009, Forensic speech and audio analysis working group: best practice guidelines for ENF analysis in forensic authentication of digital evidence. Technical Report. European Network of Forensic Science Institutes (ENFSI): Forensic Speech and Audio Analysis Working Group (FSAAWG). The Hague, Netherlands. http://www.enfsi.eu/sites/default/files/documents/ forensic_speech_and_audio_analysis_wg_-_best_practice_guidelines_for_enf_analysis_in_forensic_ authentication_of_digital_evidence_0.pdf

Gupta S, Cho S and Kuo CCJ, 2012, Current Developments and Future Trends in Audio Authentication. *IEEE Multi Media* **19**(1), pp. 50–59.

Hall M, Frank E, Holmes G, Pfahringer B, Reutemann P and Witten IH, 2009, The WEKA data-mining software: an update. *SIGKDD Explorations Newsletter*, **11**(1), pp. 10–18.

Hanilçi C and Ertas F, 2013, Optimizing acoustic features for source cell-phone recognition using speech signals. In *Proceedings of the 1st ACM Workshop on Information Hiding and Multimedia Security*, IH&MMSec'13. ACM, New York, pp. 141–148. http://dl.acm.org/citation.cfm?id=2482513.2482520. (Accessed 5 February 2015).

Hanilçi C, Ertas F, Ertas T and Eskidere Ö, 2012, Recognition of brand and model of cell-phones from recorded speech signals. *IEEE Transactions on Information Forensics and Security* **7**(2), pp. 625–634.

Hildebrandt M, Kiltz S and Dittmann J, 2011, A common scheme for evaluation of forensic software. In *Proceedings of the 6th International Conference on IT Security Incident Management and IT Forensics* (IMF2011). Stuttgart, Germany, pp. 92–106. http://ieeexplore.ieee.org/xpl/articleDetails.jsp?tp=&arnumber=5931115 (Accessed 5 February 2015).

Hu MK, 1962, Visual pattern recognition by moment invariants. *IRE Transactions on Information Theory* 8(2). pp. 179–187. http://ieeexplore.ieee.org/xpl/articleDetails.jsp?tp=&arnumber=1057692 (Accessed 5 February 2015).

Ikram S and Malik H, 2010, Digital audio forensics using background noise. In *Proceedings of the IEEE International Conference on Multimedia and Expo 2010* (ICME), pp. 106–110. http://ieeexplore.ieee.org/xpl/articleDetails.jsp?tp=&arnumber=5582981 (Accessed 5 February 2015).

Koenig BE, 1990, Authentication of forensic audio recordings. *Journal of Audio Engineering Society*, **38**(1–2), pp. 3–33.

Koenig BE and Lacey DS, 2009, Forensic authentication of digital audio recordings. *Journal of Audio Engineering Society*, **57**(9), pp. 662–695.

Koenig BE, Lacey D and Killion S, 2007, Forensic enhancement of digital audio recordings. *Journal of Audio Engineering Society* **55**(5), pp. 352 – 371.

Kraetzer C, 2013, Statistical pattern recognition for audio forensics – empirical investigations on the application scenarios audio steganalysis and microphone forensics. PhD Thesis, Faculty of Computer Science, Otto-von-Guericke University Magdeburg, Germany.

Kraetzer C, Dittmann J and Lang A, 2006, Transparency benchmarking on audio watermarks and steganography. In *Security, Steganography, and Watermarking of Multimedia Contents VIII* (Ed. Delp EJand Wong PW), Vol. 6072 of Society of Photo-Optical Instrumentation Engineers (SPIE) Electronic Imaging Conference Series, San Jose, CA.

Kraetzer C, Oermann A, Dittmann J and Lang A, 2007, Digital audio forensics: a first practical evaluation on microphone and environment classification. In *Proceedings of the 9th ACM Workshop on Multimedia & Security*, MM&Sec'07. ACM, New York, pp. 63–74. http://dl.acm.org/citation.cfm?id=1288869.1288879 (Accessed 5 February 2015).

Kraetzer C, Schott M and Dittmann J, 2009, Unweighted fusion in microphone forensics using a decision tree and linear logistic regression models. In *Proceedings of the 11th ACM Workshop on Multimedia and Security*, MM&Sec'09. ACM, New York, pp. 49–56. http://dl.acm.org/citation.cfm?id=1597817.1597827 (Accessed 5 February 2015).

Kraetzer C, Qian K, Schott M and Dittmann J, 2011, A context model for microphone forensics and its application in evaluations. In *Media Watermarking, Security, and Forensics XIII* (Ed. Memon ND, Dittmann J, Alattar AM and Delp EJ), Vol. 7880 of Society of Photo-Optical Instrumentation Engineers (SPIE) Electronic Imaging Conference Series, San Francisco, CA.

Kraetzer C, Qian K and Dittmann J, 2012, Extending a context model for microphone forensics. In *Media Watermarking, Security, and Forensics XIV* (Ed. Memon ND, Alattar AM and Delp EJ), Vol. 8303 of Society of Photo-Optical Instrumentation Engineers (SPIE) Electronic Imaging Conference Series, San Francisco, CA.

Kumar S, Singh K and Saxena R, 2010, Analysis of Dirichlet and Generalized Hamming-window functions in the fractional Fourier transform domains. *Signal Processing* **91**(3), pp. 600–606. http://dl.acm.org/citation.cfm?id=1879620. (Accessed 5 February 2015).

Kuttruff H, 2009, Room Acoustics, 5th edn. Taylor & Francis, New York.

Liu Q, Sung AH and Qiao M, 2009, Novel stream mining for audio steganalysis. In *Proceedings of the 17th ACM International Conference on Multimedia*, MM'09. ACM, New York, pp. 95–104. http://dl.acm.org/citation.cfm?doid=1631272.1631288 (Accessed 5 February 2015).

Lu Y, Cohen I, Zhou XS and Tian Q, 2007, Feature selection using principalfeature analysis. In *Proceedings of the 15th International Conference on Multimedia*, MULTIMEDIA'07. ACM, New York, pp. 301–304. http://dl.acm.org/citation.cfm?doid=1291233.1291297 (Accessed 5 February 2015).

Maher RC, 2009, Audio forensic examination: authenticity, enhancement, and interpretation. *IEEE Signal Processing Magazine* **26**(2), pp. 84–94.

Maher RC, 2010, Overview of audio forensics. In *Intelligent Multimedia Analysis for Security Applications* (Ed. Sencar H, Velastin S, Nikolaidis N and Lian S), Vol. 282 of Studies in Computational Intelligence, Springer Berlin Heidelberg, pp. 127–144.

Malik H and Farid H, 2010, Audio forensics from acoustic reverberation. In *Proceedings of the IEEE International Conference on Acoustics, Speech, and Signal Processing*, ICASSP'10. Dallas, TX, pp. 1710–1713. http://ieeexplore.ieee.org/xpl/articleDetails.jsp?tp=&arnumber=5495479 (Accessed 5 February 2015).

Malik H and Miller JW, 2012, Microphone identification using higher-order statistics. In *Proceedings of the Audio Engineering Society Conference: 46th International Conference*, Denver, CO. Audio Engineering Society (AES), Inc., New York.

Moncrieff S, Venkatesh S and West G, 2006, Unifying background models over complex audio using entropy. In *Proceedings of the 18th International Conference on Pattern Recognition*, Vol. 04, ICPR'06. IEEE Computer Society, Washington, DC, pp. 249–253. http://ieeexplore.ieee. org/xpl/articleDetails.jsp?tp=&arnumber=1699827 (Accessed 5 February 2015).

Nicolalde DP and Apolinario JA, 2009, Evaluating digital audio authenticity with spectral distances and ENF phase change. In *Proceedings of the 2009 IEEE International Conference on Acoustics, Speech and Signal Processing*, ICASSP'09. IEEE Computer Society, Washington, DC, pp. 1417–1420. http://ieeexplore.ieee.org/xpl/articleDetails.jsp?tp=&arnumber=4959859 (Accessed 5 February 2015).

Oermann A, Lang A and Dittmann J, 2005, Verifier-tuple for audio-forensic to determine speaker environment. In *ACM Proceedings of the 7th Workshop on Multimedia and Security*, MM&Sec'07 (Ed. Eskicioglu AM, Fridrich J and Dittmann J), pp. 57–62. http://dl.acm.org/ citation.cfm?id=1073170.1073181 (Accessed 5 February 2015).

Owen T, 1988, Forensic audio and video-theory and applications. *Journal of Audio Engineering Society* **36**(1/2), pp. 34–41.

Panagakis Y and Kotropoulos C, 2012, Automatic telephone handset identification by sparse representation of random spectral features. In *Proceedings of the ACM Multimedia and Security Workshop*. ACM, New York, pp. 91–96. http://dl.acm.org/citation.cfm?id=2361407.2361422 (Accessed 5 February 2015).

Pawera N, 2003, Microphone Practice, 4th edn. PPV Medien, Bergkirchen.

Peters N, Lei H and Friedland G, 2012, Name that room: room identification using acoustic features in a recording. In *Proceedings of the 20th ACM International Conference on Multimedia*, MM'12. ACM, New York, pp. 841–844. http://dl.acm.org/citation.cfm?id=2393347.2396326 (Accessed 5 February 2015).

REVerse engineering of audio-VIsual coNtent Data (REWIND), 2011, Deliverable D3.1 State-of-the-art on multimedia footprint detection, FP7-ICT project REWIND, contract number 268478. Technical Report http://cordis.europa.eu/docs/projects/cnect/8/268478/080/deliverables/001-REWINDD31 final.pdf (Accessed 27 February 2015).

Rumsey F, 2008, Forensic audio analysis. *Journal of Audio Engineering Society* **56**(3), pp. 211–217.

Special Working Group on Digital Evidence (SWGDE), 2014, SWGDE best practices for forensic audio, version 2. Technical Report SWGDE. https://www.swgde.org/documents/Current%20Documents/ 2014-09-08%20SWGDE%20Best%20Practices%20for%20Forensic%20Audio%20V2 (Accessed 27 February 2015).

Tibbitts J and Lu Y, 2009, Forensic applications of signal processing. *IEEE Signal Processing Magazine* **26**, pp. 104–111.

Vielhauer C, Dittmann J and Katzenbeisser S, 2013, Design aspects of secure biometric systems and biometrics in the encrypted domain. In *Security and Privacy in Biometrics* (Ed. Campisi P). Springer-Verlag, London/New York.

Witten IH, Frank E and Hall MA, 2011, Data-Mining: Practical Machine Learning Tools and Techniques, 3rd edn. Morgan Kaufmann, Amsterdam/Boston, MA.

Zhao H and Malik H, 2012, Audio forensics using acoustic environment traces. In *Proceedings of the IEEE Statistical Signal Processing Workshop*, SSP'12. Ann Arbor, MI, pp. 373–376. http://ieeexplore.ieee.org/xpl/articleDetails.jsp?tp=&arnumber=6319707 (Accessed 5 February 2015).

12

Forensic Identification of Printed Documents

Stephen Pollard[1], Guy Adams[1] and Steven Simske[2]
[1]*Hewlett Packard Research Laboratories, Bristol, UK*
[2]*Hewlett Packard Printing and Content Delivery Lab (PCDL), Ft. Collins, USA*

12.1 Introduction

It is desirable to be able to authenticate individual documents to a high level of statistical confidence. Where the term *document* refers to a wide range of printed materials starting from traditional legal documents, identification papers, birth and marriage certificates, currency and other financial negotiables, etc., extending far and wide to also include tickets, receipts, labels and all forms of printed packaging. In fact almost any printed material has the potential to be exploited as a means of authentication in order to protect itself or an associated item from forgery. Hence, unlike the related and partially overlapping areas of printer forensics (where the goal is to identify the printer or printer family used to create a document) and paper/substrate classification (where the goal is to identify the specific paper type or printing material that the document is to be or has been printed on), the goal of print forensics is to uniquely identify individual printed items. Thus, it shares many similarities with human biometrics where individual fingerprints, iris patterns, face features, etc., are used to authenticate an individual. Such biometric systems involve two phases: an enrollment phase where the biometric information is stored in a database along with other personal information, and an identification phase wherein the biometric of the individual is tested against the stored version. The process for printed documents is very similar. When a document is selected for inclusion a forensic signature is extracted

Handbook of Digital Forensics of Multimedia Data and Devices, First Edition.
Edited by Anthony T.S. Ho and Shujun Li.

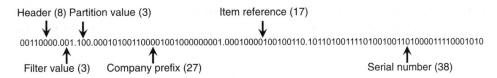

Figure 12.1 Format of the 96-bit SGTIN including a final randomly generated 38-bit serial number able to uniquely identify individual product samples. Note that the partition value (in this case 4) determines the partition of the next 44 bits into company prefix (in this case 27 bits) and item reference (in this case 17).

from the document and then either stored in a database or alternatively included in some form on, or with, the document itself (e.g. electronically using radio-frequency identification/RFID as is the case with electronic passports or alternatively printed on the document as a barcode). As with human biometrics, it is important that the chosen forensic signature uniquely identifies the individual document with a high degree of statistical accuracy and that it is extremely difficult, if not impossible, to replicate.

Traditional forms of document security begin by uniquely identifying individual items using unique serial numbers as is the case for banknotes, birth certificates and the identification documents including passports. A modern-day counterpart is the Serialized Global Trade Item Number (SGTIN) which is an extension to the traditional GTINs (used in 1D barcodes for product identification) to include a unique serial number for every item (GS1 2014). This can be stored in electronic form (e.g. in an RFID tag) or printed on a product label as a 2D barcode (i.e. a QR code (BS ISO 2007) or a DataMatrix (GS1 2011)). The format of the 96 bit SGTIN is shown in Figure 12.1 (there are in fact three variants with either 64, 96 or 198 bits).

While unique codes are useful for identifying individual documents or products, they are easy to replicate or spoof once read, and so do not provide a forensic level of authentication. Instead, it is necessary to associate some form of physical authentication with the document or label that is difficult to replicate or copy. These have traditionally included finely detailed lithographic printing techniques and the addition of security features such as watermarks, holograms, security inks, and special coatings or laminations (see Pizzanelli (2009) for an in depth review of these techniques). While these approaches raise the bar for counterfeiters and help identify large-scale criminal activities, they still do not provide authentication of individual documents to the forensic levels (here we adopt a statistical interpretation where the chance of a false positive by chance is less than one in a billion, alternatives are discussed in Chapter 1) of evidence that we desire here. Instead, we will concentrate on those techniques that aim to extract a unique and reproducible forensic signature from a printed document or the substrate on which the document is printed. These techniques are able to provide high levels of statistical robustness sufficient to uniquely identify printed documents.

Such techniques can broadly be split into two categories: those that require the addition of special materials and processes to provide a random forensic signature that is very hard to copy, and those that make use of the natural randomness of the print process and the underlying substrate to generate the forensic signature. We will introduce examples of the former approach but concentrate for the majority of this chapter on the latter. The reason why we focus more on the latter category is because it is universally applicable and potentially low cost with a modest investment in the infrastructure and no added cost to the printed items themselves.

The chapter is organized as follows. First we discuss some general hardware (HW) requirements and introduce some special purpose and potentially low-cost HW we have developed for the extraction of forensic signatures from printed documents. In Section 12.2 we briefly outline the use of special materials and additives in paper and print to support print forensics. In Section 12.3 we review a number of methods to recover forensic signatures from substrate materials alone. In Section 12.4 we present methods, mainly our own, that use forensic printed marks, character glyphs and halftone images as the basis for forensic signature recovery. In that section we also show that using a model to locate the printed forensic mark and measuring the signature relative to it allows a much more robust signature to be extracted. In Section 12.5 we illustrate the potential use of forensic authentication in the context of currency protection. Finally, in Section 12.6, we provide some concluding comments and a broad discussion on the deployment of print based forensics.

12.1.1   Hardware Considerations

As will become apparent, quite an array of imaging technology has been used for extracting forensic signatures from substrate and print material, with image capture devices varying in resolution by a factor of 30 (microns) to 1. The requirements for print forensics constrain the problem to be able to capture a reasonably large region of print or substrate at a resolution such that the random perturbations of the printing process or the substrate upon which it is printed are evident and the deliberate counterfeiting of the forensic signature is very difficult if not impossible. While a variety of off the-shelf digital microscopes and high-quality document scanners have been used to develop methods for the recovery of forensic signatures we have developed a special purpose high-resolution low-cost capture device able to image a reasonably large area of print at a high resolution. The only other dedicated hardware device of which we are aware is the Laser Surface Authentication device discussed in Section 12.3.2 which images speckle patterns caused by laser interference.

Effective hardware must be able to resolve features smaller than the minimum addressable mark size printable. Typically, this is measured using modulation transfer function (MTF) testing (which measures the highest frequency sine wave reliably resolved by the imaging device). In order to withstand a print and scan (or 'copy') attack the MTF of the hardware must be sufficiently high as to capture the microscopic variations of the print with sufficient resolution for reliable authentication and of course significantly higher than the resolution the print hardware can reproduce intentionally. As the random print parasitics typically range in size from 5 to 20 μm, the target MTF needs to accommodate the lower end of this range.

A resolution of 5 μm equates to 5080 ppi or lines per inch in MTF parlance. A key realization is the fact that this sub-5 μm dimension is similar to the size of pixels found in many image sensors. For example, the Aptina 3MP CMOS image sensor (MT9T001) used in our prototypes has 3.2×3.2 μm pixels. Thus, the magnification of the lens system could be 1:1. This requirement means that a Dyson Relay catadioptric lens (which are more commonly seen in semiconductor stepper optics) can be used in place of a more conventional multi-element lens. This design has the combined advantages of low cost, high resolution, fixed magnification, low chromatic aberration and low distortion.

Figure 12.2 shows both the standard Dyson Relay lens design and a modified version that was introduced by Adams, (2010). In the standard design (at the top) rays from the object surface pass through the lens element, are refracted on exit, travel to a concave mirror and are directed back through the refractive surface. The rays then travel to the image surface, which is symmetrical below the mid line of the optical system. The overall dimensions are 95 mm long and 16 mm diameter. In order to incorporate this basic design for imaging a surface into a hand-held device, the optical path to the image sensor needs to be moved to a different physical plane by using a mirror as can be seen at the bottom of Figure 12.2. This modified design includes

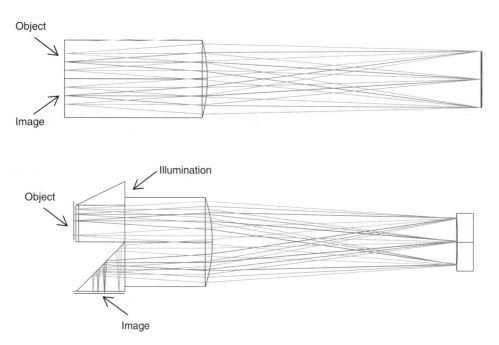

Figure 12.2   Standard Dyson Relay design (top) and modified version for print forensics (bottom).

an additional prism used to direct light from an LED (i.e. aligned in the notch at the top and points towards the center of the object field) onto the object surface and also includes the cover glass for the sensor and the associated air gap to the die as well as a small stand-off air gap at the object surface. The latter air gap ensures the glass does not actually contact the printed surface, which could result in abrasion and contamination. The parameters of the illumination system using a side prism were chosen to ensure total internal reflection of the light from the LED.

We have built a number of prototype devices based on this design which we call Dyson Relay CMOS Imaging Device (DrCID). The glass elements and mirror of the design were sourced, assembled and built into a suitable rigid tube housing which included a fine-threaded mount for the mirror for fractional adjustment. This was then assembled into a custom built, 3D printed frame that locates the lens (with the object surface stand-off gap), sensor and LED illumination in a small handheld package as shown in Figure 12.3. The prototypes built thus far all deliver close to the design performance, which was an MTF of 0.4 at the 3.2 μm resolution of the Aptina sensor (i.e. 312 lines or 156 line pairs per mm). As can be seen from the image regions extracted from DrCID in Figure 12.4 this resolution is sufficient to expose the random print variations in both InkJet and Laser Jet samples.

12.1.2 Performance Characterization

When seeking to authenticate a document, we want to differentiate the genuine item from the pool of alternative documents and attempts to generate forged documents

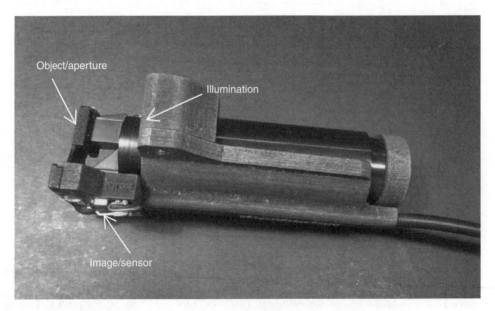

Figure 12.3   First prototype of DrCID with a total length of 10 cm.

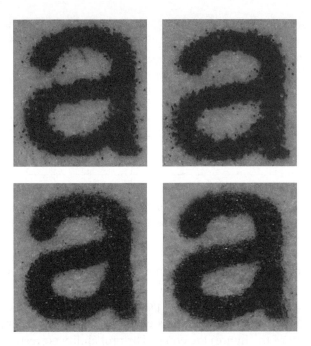

Figure 12.4   Pairs of InkJet (HP K5400), top, and LaserJet (HP CP6015xh), bottom, six-point Arial characters printed successively on the same printer and imaged with DrCID.

through some form of copy or replication process. As with biometrics, two scenarios present themselves: identification, in which the entire database is searched for a document with a matching forensic signature; and verification, in which a one-to-one comparison is made for a presented document against a specific stored document. For the latter scenario it is useful to use a serial number associated with the document to identify it so that the associated forensic signature can be quickly identified in the database.

When making comparisons it is important to understand the probability density functions (PDFs) of similarity scores for both genuine and erroneous matches. For example, consider the PDFs of signature difference values (many possible distance metrics could be used to discriminate signatures, including sum of absolute difference, sum of squared distance, Hamming distance, etc.) for the populations of genuine and erroneous comparisons shown on the left of Figure 12.5. The related false-positive and false-negative rates (FPRs and FNRs, respectively), which are simply the area under the curve of the respective erroneous and genuine PDFs, are shown on the right. As a difference threshold is varied, to make a binary identification decision, the FPR and FNR indicate the cumulative probability of false positives and false negatives occurring, respectively. Sweeping the threshold value and plotting the cumulative probability of correct positives (1-FNR) against FPR gives the receiver operating characteristic, or simply the ROC curve which characterizes the overall performance

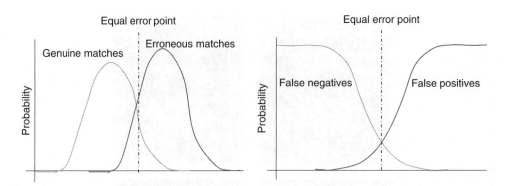

Figure 12.5 Illustrations of probability density distributions for genuine and erroneous matches on the left and probabilities of false positives and negatives for each possible selection of the decision threshold on the right.

of the system taking both types of errors into account. The point where the probability of false negatives is equal to the to the probability of false positives (as shown on the right of the figure) is called the equal error point (EEP) and coincides by definition with the point at which the area under the tails of the PDFs of the genuine and erroneous matches (in the left of the figure) are equal. The probability corresponding to the EEP is called the equal error rate (EER), and it provides a useful shorthand of the system performance; however, as we will see later, this is not always the best measure.

In order to estimate the threshold corresponding to the EEP and the associated probability value (EER) it is necessary to better understand the details of these distributions. The situation is particularly straightforward if the distributions are both Gaussian. In this case the EEP can be characterized in terms of the standard scores (commonly referred to as z-scores or z-values) of the distributions. A standard score simply refers to the number of standard deviations a point is from the mean of a Gaussian distribution and is directly related to the cumulative probability of the normalized Gaussian distribution. Hence the EEP between a pair of Gaussian distributions occurs at the point where the z-score with respect to each is equal and has a value

$$Z_E = \frac{|\mu_2 - \mu_1|}{\sigma_1 + \sigma_2} \tag{12.1}$$

where μ_1, μ_2 and σ_1, σ_2 are, respectively, the means and standard deviations of the two Gaussian distributions under consideration. In this case the EER is given by the error function, erf, of the normal distribution

$$\text{EER} = \frac{1}{2} - \frac{1}{2}\text{erf}\left(\frac{Z_E}{\sqrt{2}}\right) \tag{12.2}$$

which is the equivalent to looking up the cumulative probability corresponding to the single tailed z-score in a set of standard statistical tables. The relationship between

Table 12.1 Comparison of 1 sided z-scores and corresponding cumulative probabilities from the Gaussian distribution of the z-score being exceeded by chance.

z-score	2	6	10	14	18	22
Probability	0.023	9.9×10^{-10}	7.6×10^{-24}	7.8×10^{-45}	9.7×10^{-73}	1.4×10^{-107}

z-scores and probability is highly non-linear with example values shown in Table 12.1. While a z-score of 2 is equivalent to around a 2% (1 in 50) probability of the violation by chance, a z-score of 6 relates to a cumulative probability of around 1×10^{-9} (1 chance in a billion). We shall use this latter value as a working definition of forensic levels of authentication.

While the EER is a useful description of the statistical separations between two distributions, often in print forensics we are more interested in preventing false positives than allowing the occasional false negative. The former is an indication that a system has been spoofed and a counterfeit has been accepted as a valid example. False negatives on the other hand can be a result of user or equipment error. As print forensics is largely based on the analysis of image data, an out of focus or motion blurred image can result in a false negative even though the printed data itself is perfectly valid. Thus the mean z-score of all positive examples relative to the distribution of negative matches can provide a more useful estimate of the overall system performance. For normally distributed data, this is simply the distance between the means of positive and negative populations divided by the standard deviation of the negatives. There are also good practical reasons to adopt this metric, as for a fixed pool of data the population of all false comparisons amongst them is many times larger than the number of positive comparisons. Making the statistics extracted from the negative data more reliable and often more likely to be closely represented by a Gaussian distribution.

12.2 Special Materials

In this section we briefly review the first category of print forensics, that is the use of special materials to aid the forensic security of printed materials. As discussed in the introduction and detailed by Plimmer (2008), special materials are used widely to help forensically identify the origin of documents labels and packaging. Taggants, which are miniature molecular data-carrying platelets, are used to identify product batches and manufacturing sources. They can either be embedded in document substrates for product labels or into ink in the case of currency protection. For example, it is possible to produce microscopic barcodes that consist of minute grains of multilayer coatings that are either added to the paper pulp during the manufacturing process of the substrate or to the ink itself. The multilayer grains are created by spraying multiple coats of coloured inert compounds forming a thin sheet which is crumpled into a fine dust of multi-layered platelets. Commercially available microtags include Secutag® (www.secutag.com) and Microtaggant® (www.microtracesolutions.com).

However this approach can never identify a unique printed item with a high degree of statistical certainty.

An alternative strategy is to generate a unique (non-reproducible) random pattern that can be used to identify a document or product. One example is the Bubble Tag™ (www.prooftag.net) which is formed of a translucent polymer inside which, as a result of a random phenomenon, self-generated bubbles appear. This constellation of bubbles, which are clearly visible to the human observer, forms the basis of the identity given to a product or document. The forensic examiner uses a unique identifier associated with the tag to download an image of it which can then be compared visually against the physical item. Electronic authentication is also available using a variety of desktop appliances to recover an electronic signature from the Bubble Tag itself (though the published details are somewhat sketchy). At a much finer physical scale FiberTag™ (also from Prooftag) uses special paper (Neenah Secure™; www.neenah.com) which incorporates random fibres and prints upon it a QR code along with some fiducial marks to achieve similar results.

12.3   Substrate Forensics

It is desirable to be able to use the randomness of the printed medium itself as forensic signature. Universally applicable, low cost with all expense in the infrastructure (print production and forensic testing) and no change to traditional workflow. In this section we look at some methods for forensically fingerprinting documents by using the random features of the paper substrate.

12.3.1   FiberFingerprint

One of the earliest reported examples of using the random structure of paper to provide a forensic signature for a document is the FiberFingerprint developed at Escher Laboratories (Metois et al. 2002). Their solution used a camera-based device to image a small region of paper (approximately a 5 × 5 mm square) which contained a simple pattern of fiducial marks for automatic registration. The FiberFingerprint in the system they report in their paper is derived from a one-dimensional signal, the Fiber Signal, which is extracted from the two-dimensional capture along a series of linear segments. These linear segments constitute the Signal Path, and they are defined relative to the detected registration marks to provide the relevant translation, rotation and scaling information required. Running along the Signal Path, a raw signal is extracted, filtered, normalized and sampled to generate the final FiberFingerprint (though the details in the paper are somewhat vague).

The authors reported good statistical separation between the distributions of correlation scores for genuine and counterfeit comparisons. They estimated an EER of 10-7 for a 300-byte FiberFingerprint. However, the resolution of the device they present, is only 320 × 240 pixels which by our estimate corresponds to a pixel size on the paper of about 30 microns or an approximate resolution of about 800 ppi (pixels per inch). It is noticeable that the device used to measure the FiberFingerprint used a mechanical alignment scheme that maintained the registration of the sample with respect to the sensor each time it is imaged. It is not known how this affects the statistical performance of the device. It is possible that a more general system, where the sample is free to move and rotate in the field of view of the device, would produce less robust statistical performance. It would seem that this system was purely experimental and has not been incorporated into a product or near product demonstration in the intervening years.

12.3.2   Laser Speckle

In 2005 the company Ingenia Technology (www.ingeniatechnology.com) introduced Laser Surface Authentication (LSA™) which uses a scanning device consisting of one or more laser light sources and a number of photodetectors. Once again the naturally occurring randomness due to the intrinsic roughness present on all non-reflective surfaces is used to provide a unique fingerprint. In this case (see Buchanan *et al.* 2005), the system makes use of the optical phenomenon of laser speckle which scatters the focused laser light in different quantities to the photodetectors that are arranged at different angles with respect to the illumination source. As the unit scans, the fluctuations from mean intensity of each detector are digitized to form the multi-channel signal that forms the fingerprint code of the objects surface.

Buchanan *et al.* (2005) presented results of experiments on 500 sheets of plain paper using such a system. In this case similarity scores are computed by forming the bit pattern of the signature into bytes and counting the number of bytes that are identical between the signatures (the fingerprints are themselves between 200 and 500 bytes long). They reported that the histogram of matching scores for all 124 750 erroneous matches (where the signatures for each of the 500 sheets of paper are compared against each other) is well approximated by a binomial distribution of $n = 280$ samples with a probability of success $p = 0.153$ (which is in turn closely approximated by the continuous Gaussian with the same mean, 0.153, and standard deviation 0.021). They also showed for these examples that the all genuine matches have a probability less than 10^{-72} of occurring by chance from the assumed binomial distribution of erroneous matches (save a single sample that was due to poor alignment of the test sample that was corrected by a second scan). The minimum z-score for the associated Gaussian distribution is approximately 22 which corresponds to an even lower probability of approximately 10^{-127}. While each distribution is an approximation it was clear from the experiment that the probability of a false positive is very low indeed.

Buchanan *et al.*'s experiments did highlight a small but not insignificant issue with the Ingenia system that it relies on the sample alignment being sufficiently accurate so that the extracted fingerprint is representative of the same part of the document or package on each occasion. In the Ingenia system reported by Buchanan *et al.* (2005) the focused laser pattern was 4 mm long and 70 µm wide scanning in the direction orthogonal to the longer axis. In order for signatures to match their cross-section must overlap by 75% (3 mm), making it necessary to align the document to about the nearest millimetre (and according to Buchanan within 2°). While for many documents and regularly shaped parcels this will not be an issue, as mechanical restraints can be used to register documents accurately, it will inevitably limit the utility of the system for irregular or extended printed surfaces.

12.3.3 Substrate Scanning

Clarkson et al. (2009) showed that it is possible to extract a usable forensic signature/fingerprint from a 3×3 inch square of blank paper using 1200 ppi scans from a high-quality commodity document scanner. They actually used four scans to recover the 3D terrain map of the paper using a form of photometric stereo. Between each scan they rotated the document through 90°.

Subtracting each pair of images that are 180° apart allowed the recovery of the component of the surface normal in that direction multiplied by the local surface albedo (i.e. the amount of light the surface material reflects) and a constant fixed scale factor dependent only on the scanner geometry. Thus with four scans it is possible to recover the surface normal's projection into the xy-plane of the scanner up to a scale factor. In their experiments, they down-sampled the 1200 ppi scan by a factor of 8 before recovering the surface normal to reduce the requirement for precise alignment of the scans.

The fingerprint itself is derived from the surface normal by first sampling N 8×8 patches from a set of fixed pseudo-random locations and then projecting each of the vectors (of length 128 as each projected normal contributes two values, one in x and one in y) using a pseudo random orthonormal basis set to get T bits per location (N $\times$ T bits in total). That is, the sign of the dot product of each of the T vectors in the basis set (which are also 128 elements long) provides a single bit of the fingerprint. In their experiments, N and T are chosen to be 100 and 32, respectively, leading to 3200 bit signatures that were compared using their Hamming distance (i.e. count of the number of bits that differ between the two bits strings).

For their experiments (Clarkson et al. 2009) they reported an EER corresponding to a z-score of 26, and a probability of 2.5×10^{-149} for distributions of genuine and erroneous matches that are each well approximated as Gaussian.

This method relies on the fact that the scanner used has a single linear light source parallel to the x-axis of the scanned image. Not all scanners have this asymmetric light bar arrangement. Many have a symmetric illumination sources on both sides of the aperture of the linear sensor for which this method will not work.

12.3.4 PaperSpeckle

The PaperSpeckle method, described in Sharma et al. (2011), uses off-the-shelf commodity USB microscopes (e.g. the DinoLiteTM AM2011 and the Digital Blue QX5) to image small regions of paper (0.5 mm field of view) at a resolution of

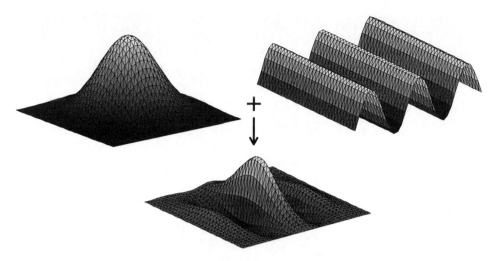

Figure 12.6   A Gaussian envelope and a cosine wave are combined to generate a symmetric Gabor filter that extracts the real part of the local frequency description at an orientation of zero degrees.

512×384 pixels, leading to an impressively small pixel size of just less than a micron on the paper surface. These microscopes have a focus capability that must be used to extract sharp images of the substrate. To aid alignment in the experiments, single ink dots/stains (from a pen) were used for localization purposes (but it is not clear how the registration itself was performed).

To extract the signature/fingerprint from the paper surface, Gabor filters (Gabor 1946) were applied to the microscope image of the paper fibres. Gabor filters are designed to measure the content of an image at a particular spatial frequency, localized in both space (position) and orientation. It is the combination of a sinusoid and Gaussian envelope as shown in Figure 12.6. Wider versions of a Gabor filter measure lower frequencies while rotated versions measure the frequency responses at different orientations. To measure the frequency response at a particular location, a pair of Gabor filters with a $90°$ phase shift of the underlying sinusoid are applied to the image to compute the real and imaginary parts of a complex pair that describe the magnitude and phase of the local image for the given frequency. Figure 12.7 shows a fragment of paper texture along with the magnitude and phase responses for a series of Gabor filters as the wavelength doubles.

In the Paper Speckle method only the sign of the imaginary component is used to construct signature with each location contributing a single bit. This follows on from the method developed by Daugman (1993) for iris recognition which has become the backbone of many government and commercial biometric recognition systems, offering as it does the ability to robustly discriminate many billions of iris patterns. However, in Duagmans method the phase provides 2 bits per location (to what he describes as a phase-code) one each for the signs of the real and imaginary components.

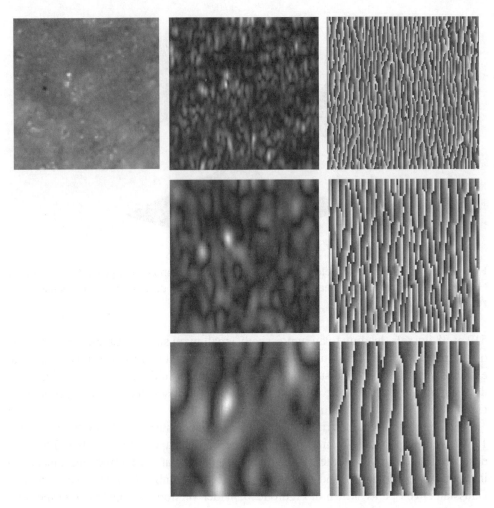

Figure 12.7 A patch of regular 80 g office paper of size 0.3×0.3 mm square captured at high resolution. Top left is a 100×100 image of the paper patch; centre is the magnitude of the Gabor filter; right is the phase of the Gabor filter with discontinuities as the phase wraps around the extreme values; top to bottom wavelengths of two, four and eight pixels respectively.

This quantizes the phase into the four quadrants of the compass. The signature in the PaperSpeckle requires half the computation (as only one Gabor filter is applied at each location) and further quantizes the phase into the northern and southern half circles.

The PaperSpeckle system reduces the length of the forensic signature by performing singular value decomposition (SVD) on the image of the phase bits which are treated as a binary matrix and selecting the 64 or 128 largest singular components as representative of the entire phase code. However, while this approach delivers much shorter code lengths, it also greatly reduces the statistical separation between genuine and erroneous matches. Based on the data presented in Sharma *et al.* (2011), the mean z-score of genuine matches with respect to the pool of erroneous matches reduces from around 55 for the Hamming distance of the raw phase codes to as little as 4.3 for the 64-element singular value codes (the results for 128 element codes were not provided). We are therefore not convinced that the SVD step is a useful aspect of the method. Sharma *et al.* (2011) also reported that similar results had been achieved using an Android 2.1 platform on the Nexus One combined with a consumer grade microscope (Carson MM-200), but this data was not presented in the paper.

12.3.5   Practical Considerations

Each of the substrate forensic methods presented earlier shows that fingerprints can be extracted from naturally occurring substrate materials that by themselves allow forensic levels of discrimination of genuine and erroneous document matches. Furthermore, Buchanan *et al.* (2005), Clarkson *et al.* (2009) and Sharma *et al.* (2011) each reported experiments showing that the extracted signatures were robust to various forms of physical damage. These include scribbling, printing, wetting and drying, ageing (2 years), crumpling tightly, baking at 180°C and rough abrasion. In each case they showed that the separation of the genuine and erroneous similarity scores is maintained, to a high degree of statistical certainty, despite the damage.

One major disadvantage of the substrate only approach is that it is difficult to identify the region of the document associated with the forensic signature. As a result, some external method of registration is generally required to reliably recover the forensic fingerprint. Even in the case of the FibreFingerprint which makes use of fiducial marks, these are not visually prominent and as a result prior knowledge of their approximate location is still necessary. Of course, it is possible to add a visually significant marker to identify the location of the forensic fingerprint. However, it will most likely still be necessary to carefully align the substrate with the imaging device as in each there seems to be a strong dependency of signature extraction on orientation. For example for the scanner, the four images must be acquired along 0°, 90°, 180° and 270°, while for the Laser Speckle the orientation must be within about 2° of true.

12.4   Print Forensics

In this section we turn our attention to forensic signatures recovered from the print itself. At the microscopic ($<10\,\mu m$) level, printing on a substrate results in imperfections that can be used to uniquely identify a printed mark or glyph. All printing technologies exhibit some form of random perturbations that can be used to extract a forensic signature. Inkjet printing, for example, typically shows several classes of imperfections. Inkjet parasitics include extra ink deposited by the nuances of the printer and/or the printing process, which can lead to narrow, curve-like ink imperfections in the boundary of the printed mark. For example, droplet tails can appear outside of the intended boundaries of the glyph–these 'satellite' ink droplets are often detached from the printed character, glyph or other mark with which they are printed. In addition, the interaction of ink with the substrate–card stock, paper, label stock, etc.–for printing, moreover, can lead to preferential wicking along paper fibres.

Cellulose and other organic fibres typically absorb ink in the longitudinal direction (long axis) via capillary movement, leading to pseudopodium-like protuberances from the intended boundary, or periphery, of the printed mark. Differential absorbance of ink along the long axis of the cellulose in paper, for example, leads to 'parasitics' and often creates relatively low-ink containing 'porosities' adjacent to the fibre around the boundary of the intended periphery of the mark.

Other printing processes also exhibit variations. For instance, dry electro-photographic processes (laser) can produce multiple microscopic satellites around the periphery of a dot, and liquid electro-photographic processes (e.g. HP Indigo™) can produce small variations in dot diameter and placement. Idiosyncrasies in printing parasitics are not limited to digital processes, as offset, gravure, flexo-graphic, screen and other traditional printing also generally exhibit substrate-dependent aberrations that can potentially be used for reliable and robust authentication.

The sizes of the parasitics produced by all the printing processes we have examined are in fact much smaller than the physically addressable resolution of the printer. The resolution needed is also higher than the resolutions of all, but the most specialist optical scanners; mainstream scanners are typically restricted to an optical resolution of 1200 ppi (21 × 21 μm pixels).

The first example of using printing paracitics as a forensic mark from which a print signature is extracted was demonstrated by Zhu et al. (2003). They used an IntelPlay QX3 low-cost digital microscope to capture the forensic mark, termed by the authors as a *security pattern*, which is composed of typically between one and four 1/360 of an inch dots but could in theory be any printed glyph. The microscope has a resolution of 320 × 288 pixels at a 200× magnification which the authors estimate to be equivalent to around 21 000 ppi on the paper with the diameter of each dot subtending about 60 pixels. The image of each dot or other glyph is processed to create a binary image and then working from the center of gravity of the thresholded component the shape of the perimeter is characterized according to the average distance to the last thresholded pixel in each radial direction for a set of N radial buckets. This N-dimensional profile vector is then normalized by dividing by the mean of all distances in the profile.

The signature profiles are compared using a sum of squares error (SSE) distance metric between normalized vector elements. While fiducial landmarks are provided to identify the forensic mark, they are not captured by the microscope and so cannot be used to assist alignment of the pattern itself. Instead some (undisclosed) external alignment method was used to physically register the forensic mark within the modest field of view of the microscope and at the correct angle of rotation so that the signature profiles would align properly.

In their experiments Zhu et al. (2003) reported good statistical separation of genuine and erroneous matches. For example, in the case where $N = 32$ for a single printed dot the FPR was about 1 in 10^{14} at the point where the FNR was set to 0.5%. In their analysis they did make certain assumptions about the uniformity of the printing process such that the variance is uniform and equal in all directions, which may not hold in practice, but there final results did show excellent statistical separation of genuine and erroneous SSE scores.

12.4.1 Authenticating Printed Glyphs

Simske and Adams (2010) used the DrCID device, introduced in Section 12.1.1, to explore the use of an individual printable glyph or character as a forensic mark. Images captured by the DrCID hardware were analyzed to generate a set of printed mark features suitable for distinguishing any specific printing of a glyph or character from another. For forensic utility, the DrCID system must be able to determine the difference between two identical glyphs printed twice, and the same glyph imaged by two different DrCID devices, and/or at different times and/or different glyph orientations.

In an approach that builds upon that of Zhu *et al.* (2003), the images captured by the DrCID are analyzed using the following steps: (1) a contrast-insensitive thresholding algorithm to binarize the image; (2) segmentation into connected components, or 'regions'; (3) perimeter determination; and (4) perimeter shape descriptor calculation.

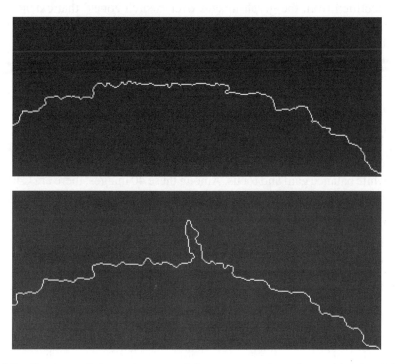

Figure 12.8 Sections of the perimeters of the upper portion of two 14-point Arial letter 'n' characters printed by an HP K5400 thermal inkjet printer. Note the pseudopodium-like 'parasitic' on the lower image.

Examples of a section of the perimeter of 2 printed letter 'n's are shown in Figure 12.8. The shape descriptors for the perimeter are computed as follows. The centroid of the region is computed, and the perimeter divided into sections by angle (e.g. $1°$ increments from $0°$ to $360°$ around the perimeter). After size normalization, for each angular section, the following are recorded:

1. Min-R: minimum radius
2. Max-R: maximum radius
3. Cmplx: complexity (number of changes in the glyph polarity black/white in the radial direction)
4. ShElem: shared elements (number of perimeter points in the section)
5. Uncert: uncertainty (number of perimeter line segments in the section)
6. Neigh-Uncert: neighbourhood uncertainty (moving average of the uncertainty to account for minor, i.e. less than $0.5°$, differences in alignment of the two images with the angular sectioning).

When comparing glyphs the perimeters are aligned by angle (using minimum SSE) and the same set of salient parameters is computed. Significant satellites and porosities are also identified from the glyph images over 'search zones' that extend inward or outward from the periphery where the zone width is determined as a fraction (typically 20%) of the mean radius of the glyph. The Hamming distances between two images are computed based on location differences in satellites and porosities. These are denoted by HD-Sat and HD-Por, respectively.

The results in Table 12.2, which are derived from those reported by Simske and Adams (2010), of an experiment designed to test the relative value of each of the features for identifying authentic glyphs. A DrCID device was used to capture 10 different 6 pt Ariel 'a' characters printed by an HP K5400 thermal inkjet printer. Each unique character was captured four times, with different placement, (small) rotation, white balance gain and focus. Among these 40 images, there are $C40, 2 = 780$ comparisons ($C4, 2 × 10 = 60$ of which compare the exact same printed character and

Table 12.2 Imaging features and their ability to distinguish between genuine and erroneous "a" glyph matches.

Feature name	Genuine matches (60)		Erroneous matches (720)		EER	Mean z-score
	Mean	Std	Mean	Std		
ShElem	0.573	0.020	0.793	0.021	5.4	10.5
Neigh-Uncert	0.971	0.006	0.881	0.016	3.8	5.7
T-Uncert	0.917	0.024	0.809	0.021	2.4	5.1
Max-R	0.073	0.025	0.180	0.020	2.4	5.4
HD-Sat	10.6	5.0	29.1	4.2	2.0	4.4

720 of which compare different printed characters). Assuming Gaussian distributions, the table reports both the EER and the mean z-score of the genuine matches. The number of shared elements (ShElem) which is a measure of the complexity of the perimeter over each radial zone proved to be the best metric in this experiment. These experiments were repeated for other character sets, with similar results. For the widely different 'k', 's' and 'l' characters, for example, the mean z-scores corresponding to ShElem were 9.9, 9.8 and 9.4, respectively, when comparing matched and unmatched characters.

12.4.2   Model-Based Authentication

The method of Zhu *et al.* (2003), and its extension by Simske and Adams (2010) to include additional measures and application to generalized printed glyphs rather than a specific pattern of printed dots, recover signatures from the printed forensic mark that describe both the general shape of the mark and its random variation. In the case of Simske and Adams this information was used to align the signatures, whereas Zhu *et al.* used some external registration method that was not described in the paper. In recent years we have developed a number of techniques that use a model based approach to directly locate the forensic mark in the image under consideration prior to the extraction of the print signature. That is we have prior knowledge of the specific, or class of, printed idem that is being used as a forensic make which is incorporated into a visual model that can be precisely, repeatedly and robustly located in the image of the forensic mark. The print signature is then recovered relative to the registered location of the model with the result that the recovered signature only encodes the residual random information and does not describe the shape of the forensic mark itself. In short the approach that is taken is (1) a modelled item, the forensic mark, is printed; (2) the image of the forensic mark is captured using a suitably high-resolution imaging device; (3) the model is used to accurately locate the forensic mark in the captured image; and (4) a forensic signature is extracted relative the registered location of the printed forensic mark thus identifying what is unique about this specific print.

The use of a generic outline model (e.g. a square or an ellipse) was introduced in Pollard *et al.* (2010a) and Simske *et al.* (2010) and then extended to use specific glyph models in Pollard *et al.* (2010b). In each case the model was used to accurately and reliably locate the outline of the printed forensic mark to measure the precise deviation from the model in order to form the model-based signature profile (MBSP). In Pollard *et al.* (2010a) the model used is an ellipse and the printed forensic mark is an Arial 12 point letter 'o'. In Simske *et al.* (2010) the model used is a square and the forensic mark is a proposed '3D' barcode (the third dimension is colour). In Pollard *et al.* (2010b) the models are Times Roman letter 'a's and 's's, and the forensic marks are 12 point printed versions of the same.

Using a model to extract a signature profile has four major advantages over previous approaches where typically a set of angular regions radiating from the centre of gravity of the forensic mark was used as the basis of its description. First and foremost, the

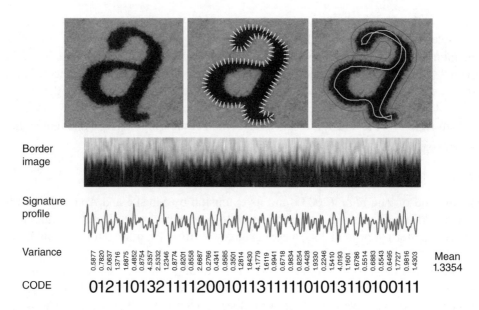

Figure 12.9 The extraction of a model based signature profile for a Times 12 point InkJet letter 'a'. See text for details.

signature profile so extracted comprises only the truly random perturbations introduced by the printing process rather than the general shape conveying properties of the outline. Second, using a model allows forensic comparison between very different images (different resolution, sizes and even distorted images). Thirdly, non-convex shapes, such as the outline of a captured printed letter 'a' have a uniform description free from multiple crossings, critical points and discontinuities. And last but not least, provided the model is free from internal axes of symmetry (not true of the circular 'o' characters used in Pollard *et al.* 2010a) the model-based signature profile recovers a description the order of which is fixed with respect to the model. This makes the matching process simpler and more robust as there is no need to align the signatures in order to compare them, and it also facilitates the extraction of simplified shape warp codes as described in the following text.

Figure 12.9 shows an example of extracting a forensic signature from the outline of a letter 'a' using the approach described in Pollard *et al.* (2010b). Top row left is a 900 × 800 (width × height) region of an image captured by DrCID. Top row middle shows superimposed transformed model data; the model is a set of N uniformly spaced points (*x, y* coordinates) defining the outer edge of a character glyph and associated unit normal vectors (*u, v*). Top row right shows the loci of sampled regions for the extracted normal profile images. Second row shows the profile image extracted between the loci around the outline of the letter 'a', each column of which corresponds to sampling on a vector between the loci along the normal vector (*u, v*) for each individual (x, y) contour point of the model. Third row shows the signature profile extracted from the outline image (in this case 2000 samples long).

Many methods can be used to recover the signature profile from the profile image, including simple thresholding or maximum edge detection. In Pollard *et al.* (2010b) we found the following greyscale edge metric that combines all the data in the profile image to work well. For each column in the profile image the signature profile is defined as follows:

$$p_i = \frac{\sum_j j w_j e_{ij}}{\sum_j w_j |e_{ij}|} \qquad (12.3)$$

where e_{ij} is an edge strength corresponding to the digital derivative (difference of grey level values) of the profile image along column i and w_j is a windowing function (in our case a Gaussian with standard deviation $1/4$ the column height centered on the mid-point of the column). Dividing by a normalizing sum of windowed absolute edge strength results in a measure that achieves robustness to both scene content and illumination variation. The windowing function makes the method less prone to error due to small inaccuracies in the location of the model 'a'.

When comparing forensic marks to test if they are genuine or erroneous we use the following normalized similarity metric:

$$S = 1 - \frac{\text{SAD}}{(\text{SA1} + \text{SA2})/2} \qquad (12.4)$$

where SAD is the sum of absolute differences between the profiles and SA1 and SA2 are the sum of absolute values of the first and second profile measure respectively.

Results are presented in Pollard *et al.* (2010b) for a number of experiments using data collected from an HP Inkjet K5400 office printer. In one experiment, lowercase 12-point Times Roman 'a's and 's's were captured twice using the DrCID device, once approximately vertical and for a second time at a significant angle (about 30° from vertical). This data was used to compare the model-based (MBSP) approach against the Simske and Adams (2010) method described in Section 12.4.1. Specifically, we compared 72 individual 'a' and 's' images to the 71 other images of the same letter (of which just 36 comparisons are valid and 2520 are not). In Figures 12.10 and 12.11, we plot the similarity metric S, for each comparison of each character using each method (the alternative method is the best of the Simske and Adams (2010) metrics; which for this severely rotated data was Max-R metric). As can be seen, the results are far better for MBSP where there is a very clear gap between the distributions of genuine matches and those for the incorrect comparisons. Using the Max-R metric with the Simske and Adams (2010) method is very similar to the earlier Zhu *et al.* (2003) approach, discussed at the start of this section, with $N = 360$ and using the sum of absolute difference rather than sum square difference. Note that the range of similarity is small for the Simske and Adams (2010) method (0.96 to 1) compared to the MBSP

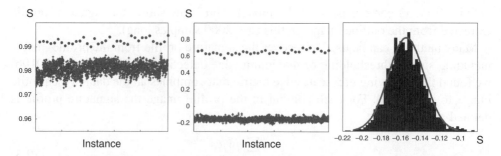

Figure 12.10 Similarity data S for Simske and Adams (2010) method (left) and MBSP (centre) for 12-point Times Roman letter 'a's. Valid matches are solid circles and false matches are stars. The histogram and Gaussian distributions of similarity (S) for all the false matches for the MBSP method are shown on the right (mean = −0.154; standard deviation = 0.018; skew = 0.27; Kurtosis = 3.38).

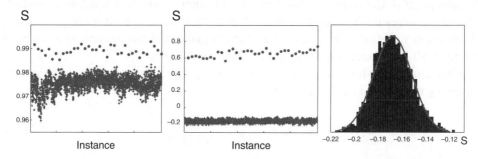

Figure 12.11 The same results as Figure 12.10 but this time for 12-point Times Roman letter 's's (for which the distribution of false matches now has: mean = −0.164; standard deviation = 0.017; skew = 0.17; Kurtosis = 2.93).

(−0.2 to 0.8) as for the former it encodes the shape of the text glyph rather than just the perturbations.

In Figures 12.10 and 12.11, we also show the distributions of similarity scores for the 2520 false matches for the model based approach. Concluding that the distributions are reasonably close to (but not exactly) Gaussian (for which Skew and Kurtosis should be 0.0 and 3.0 respectively) and assuming the same is approximately true for valid matches (where the sample is much smaller), then we can use the z-score to approximate the EER of the MBSP method. These are 18.1 and 13.7, respectively, for the 'a' and 's' data corresponding to very small probabilities of false authentication (see Table 12.1). In comparison, if we compute z-scores for the Simske and Adams (2010) method the values for the respective EERs were only 3.2 and 2.8 for 'a's and 's's. These are much lower than those of the MBSP approach and significantly worse than the results reported in Simske and Adams (2010) where there was no significant rotation of the forensic mark (as is the case for many of the methods reported thus far). The model-based approach is able to better accommodate for the rotation of the forensic mark as the recovered signature profile is referenced with respect to the model itself and only describes the deviations from the model that are largely independent of its orientation.

12.4.2.1 Shape Warp Coding

It is possible to code the signature profile using many fewer bits and still maintain forensic levels of authentication. One such scheme, which we refer to as Shape Warp Coding (SWC), was introduced in Simske *et al.* (2010) for the limited case of micro-colour-tile inspection. It was shown that a shape distortion encoding distance (SDED) based on the SWC allows batch inspection and validation (i.e. the use of a small number of scanned deterrents to determine whether a batch of products is genuine or counterfeit).

Here (and in Figure 12.9) we consider the MBSP as the basis for the SWC for the general case of any irregular text glyph. We first divide the signature profile into N equal segments of length n. Then for each, compute the variance of the signature:

$$V_j = \frac{\sum (p_i - \mu_j)^2}{n - 1} \tag{12.5}$$

where p_i is the signature profile over the segment j and μ_j is its mean value over the that segment. The forth row of Figure 12.9 shows the variance of the signature profile over 40 equally sized (50 samples each) bins.

We then use the mean (or median) value of the variance (or a factor or multiple of it) as an atomic unit of encoding (a 'digit'), to form an N-position string which is the SWC:

$$\text{SWC}(j) = \left\langle \frac{V_j}{V_{mean}} \right\rangle \tag{12.6}$$

where $\langle . \rangle$ is a rounding function. This is shown in the last row of Figure 12.9 for a 40-digit code. The SDED, for comparing the SWCs of any two forensic marks, is defined as follows:

$$\text{SDED} = \sum_j \min \left(|\text{SWC}_1(j) - \text{SWC}_1(j)|, T_{\max} \right) \tag{12.7}$$

where $T_{\max}$ is an optional threshold to improve robustness. The SDED can be considered a form of modified Hamming Distance where the expected value of SWC($*$) is 1 at each digit due to the normalization process described. For example, a pair of SWCs ($N = 40$) extracted from DrCID data for the same printed 'a' and their absolute difference are:

SWC1 = 11011111201101111211211121111111011212112
SWC2 = 11111111210100101211211121111211011111112
DIFF = 00100000011001010000000000000100000101000

for which the SDED is 7 (or 0.175 when normalized by N). In Pollard *et al.* (2010b) we showed that the SDED metric for a SWC of length 200 digits preserved forensic levels of security for a single printed glyph. The SWC has the considerable advantage of being much smaller than full signature profile (which is 2000 floating point values), thus reducing storage and bandwidth requirements for manipulating potentially very large numbers of signature profiles or alternatively making it possible to represent the signature profile in a printed form using an additional 2D barcode.

12.4.2.2   Robustness to Light Fade and Mechanical Wear

One issue with printed material is how robust the print is to ageing. It is known that printed materials can fade over time with prolonged exposure to light. Mechanical wear can also affect the integrity of ink and toner. It should be clear from the foregoing that it is extremely difficult to accidentally or purposefully create a false positive; a false match corresponding to a very high z-score with respect to the underlying distribution of false matches. However, the contrary is not necessarily true. It is possible to either damage a forensic mark or to image it poorly (e.g. poor focus or motion blur during capture) so that the recovered signature is not sufficiently close to the original. Poor imaging is not such a big problem as the capture process can be repeated if a test fails. Damage, on the other hand, is more of an issue as it cannot be repaired. While it is impossible to make a forensic mark totally robust to damage, it is important that some degree of robustness exists so that it will survive every day handling.

To support this, we report in Figure 12.12 the results of an experiment to show the robustness of forensic print signatures to lightness and wear testing for the MBSP method (Pollard *et al.*, 2010b). In the experiment, 40 forensic marks comprising a 4×4 mm square object (DrCID image shown as inset of Figure 12.12) printed on a single HP M4345 Monochrome Laser Printer where the large and small circles are respectively the SDED values before and after an accelerated aging where the print was subject to UV illumination (178 h at 80 lux which is comparable to over 7 years on display in an office environment) and automated mechanical wear (the barcodes were rubbed five times at a pressure of 40 gm/cm^2 using Silbon C abrasive paper, where each rub moved the abrasive paper back and forth across the target). Crosses show SDED values for a random sample of false comparisons of the outline forensic signatures amongst the pool of 40 samples. While this damage reduces the statistical significance of the SDED measurement, as veridical matches move towards the distribution of false matches, the mean z-scores moves from 22.62 to a still highly

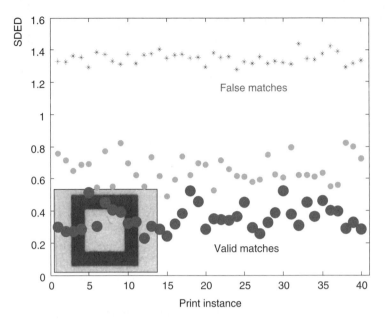

Figure 12.12 Effects of degradation on raw similarity scores for false and valid forensic signature comparisons. Large and small circles show respective SDED signature difference scores before and after damage (simulated 7 years office lighting and mechanical wear), respectively. While stars show signature difference scores for random examples from the pool of erroneous matches.

'statistically forensic' 16.45 and so the experiment showed that the forensic signatures can survive these two types of damage.

12.4.3 Authenticating Printed Halftones

The use of an outline model, presented in Section 12.4.2, is restricted to the case where the forensic mark has a solid identifiable outline against which the signature can be accurately and repeatedly defined. As well as restricting the class of printed object that can be used as a forensic mark, this also limits the amount of image data used to form the forensic print signature to the immediate vicinity of the outline. The work presented in this section, in contrast, utilizes the methodology of iris recognition (Daugman 1993) to derive a more general area-based print biometric that can be applied to halftones images and thus greatly extend the utility and applicability of forensic print authentication. This approach is very similar to the PaperSpeckle method (Sharma *et al.* 2011) described in Section 12.3.4. In that case, however, the analysis was restricted to substrate only materials and required an additional document registration strategy. Our approach was introduced by Pollard *et al.* (2012), where the known forensic mark was a small (4 × 4 mm) printed halftone such as a grey-level logo or more general image fragment. This again provides both a target for aiming the capture device and an implicit frame of reference that allows the recovery of the forensic signature provided the forensic mark lies within the captured image. The approach closely follows the methodology first proposed by Daugman in his 1993 paper on iris recognition and elaborated in subsequent publications (Daugman 2006, 2007). There are three essential elements to this form of iris recognition: (1) registration of the inner and outer iris boundaries; (2) iris encoding by 2-D Gabor filter phase encoding over normalized image coordinates; and (3) testing of statistical independence between encoded feature sequences.

Iris recognition differs from the print authentication of Pollard *et al.* (2012) in three important regards. First, our images were captured using DrCID (Adams 2010) which operates at an almost fixed high resolution (about 7900 ppi), whereas iris images are captured using traditional optics and thus vary in size over a small but significant range. Second, parts of the iris are not properly imaged due to either obscuration (by the eyelids or the eye-lashes) or specular reflections of the near infrared (IR) light sources. Thus encoded features extracted from these damaged regions must be robustly and accurately excluded from the statistical comparison process. Print images, on the other hand, do not generally suffer from such imperfections and the whole encoded sequence can be used. Finally, unique iris features can be encoded across a wide range of spatial frequencies, while the random perturbations associated with printed halftones are more limited. In fact, useful random information can only be recovered over a relatively small range of spatial frequencies. What we found out was that above this frequency range image noise and alignment errors tend to dominate, while below this range the portrayed image content itself is dominant and highly correlated across all signatures (and thus of no value for authentication).

For print signature extraction, it is again important to accurately and repeatedly register the captured halftone image. In Pollard *et al.* (2012), halftone patterns were registered using multi-scale gradient descent (Bouguet 1999) derived from the well-established Lucas and Kanade (1981) method. Figure 12.13 shows three example images: the 'Rainbow Bridge' (top row) and two versions of the HP logo one (called HP Logo 1) with a plain white (no halftone dots) surround and one (called HP Logo 2) that has a grey (halftone dots) background (these images provide a combination of natural and graphic contents and produced results that are representative of a large range of halftone images). In each case, the image was rendered to a size of 4×4 mm^2 as a 600 dpi (dots per inch) halftone (center column) and then printed on an HP LaserJet M4345. The print was captured by DrCID, registered to align with the digital halftone and de-warped to give the images in the right hand column. For the multi-scale representation we normalized band pass filtered images (difference of successive Gaussian filtered images) to have unit standard deviation in order to minimize the difference between the highly stylized 600 dpi half-tone images and the 7900 ppi printed and captured images that are derived from them. Initial approximation is achieved using either image moments (Flusser *et al.* 2009) or by matching image

Figure 12.13   Images on the left are rendered as 4 mm square halftones at 600 dpi (center) and printed, captured with DrCID, registered and dewarped (right). Top row is the Rainbow Bridge, middle row is HP Logo 1 and bottom row is HP Logo 2.

features similar to SIFT (Lowe 2004) recovered from the coarsest scale of the multi-scale representation.

Registration recovers the affine transform that gives the local least sum squared distance between the band-pass filtered images. It is used to dewarp the captured image prior to forensic print signature extraction based on 2-D Gabor filter encoding. Following Daugman's methodology, the random signal is demodulated to extract its phase information using a quadrature pair of 2-D Gabor filters. These were introduced in Figure 12.6 for the symmetrical case (based on the cosine) which provides the real component of the filter pair. The imaginary component is analogous, but is based on the 90° out of phase asymmetric filter (the sine wave) of the same frequency. Thus a pair of filters provides a local estimate of the magnitude (from the sum of squares of the real and imaginary components) and phase of the image signal at a given frequency and orientation. Daugman's phase code only uses the signs of the real and imaginary components to provide a pair of bits h{Re, Im} for each sampled location. The Gabor filters themselves have two major parameters controlling their frequency (or its inverse; wavelength) and orientation.

To construct the complete phase code used to describe the print signature, the h{Re, Im} bit pair for each sample can be combined spatially over an M × M grid defined to cover a pre-described region of the registered image; however, as the sampling frequency (proportional to M) increases beyond a critical value (i.e. dependent on the wavelength of the filter) the neighbouring samples will become increasingly correlated and the effectiveness of the phase code will stop improving as no new information is being added. It is of course also possible to further increase the total number of bits in the phase code by combining the results of Gabor filters with different wavelengths and orientations at each spatial location (provided the different filters are statistically independent). In the case of iris biometrics (Daugman, 1993), the phase code was chosen to have 2K bits (256 bytes) by sampling 1024 filter pairs, though the composition in terms of spatial, frequency and orientation components was not specified.

12.4.3.1   Statistical Independence

In Daugman's method, statistical independence is tested using the norm of the Boolean exclusive OR (XOR) operator applied to the complete code vectors. This quantity is represented as a fractional Hamming distance (HD) by dividing by the code length (following Daugman's lead we label fractional Hamming distance HD and not FHD).

$$\text{HD} = \frac{1}{L}\sum_{i=1}^{L} A_i \oplus B_i \tag{12.8}$$

where A and B are complete phase codes of length L for different captures of the same or different biometric sample (iris or in our case print) and $\oplus$ is the XOR operator. Note that this task is complicated, in the case of iris recognition, by the need to include a mask to represent the valid portions of the code that are free from obscuration and specular reflection. This is not an issue for the forensic print signature since all parts of the code will remain valid.

Provided any given bit in the phase code is equally likely to be 1 or 0 and the phase codes for different samples are uncorrelated, the Hamming distance will correspond to a sequence of Bernoulli trials with equal probability of success and failure of 0.5 and thus be distributed as a Binomial distribution, from which the expected value of the fractional Hamming distance HD is also 0.5. Figure 12.14 shows the results of an experiment on DrCid images for 48 prints of the Rainbow Bridge halftone from the top row of Figure 12.13, with equal numbers of prints from three different HP M4345 laser printers. Performing all 1128 possible false comparisons amongst this corpus we plot, in the figure, the histogram of HD for a single 2-D Gabor filter (of wavelength eight pixels and orientation 0°) sampled over an M = 80 grid (giving a phase code of length 12.8K bits). This distribution has mean value of 0.4916 (which is close to the 0.5 probability we expected) and a standard deviation $\sigma = 0.0058$ for which the corresponding PDF is a binomial having $N = p(1 - p)/\sigma^2 = 7429$ degrees

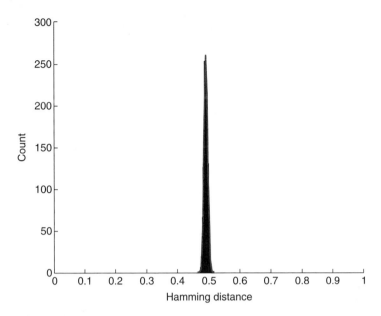

Figure 12.14 Histogram of Hamming distance for erroneous comparisons of phase codes and equivalent overlaid Binomial distribution. See text for details.

of freedom (where p is the mean value of the distribution; 0.4916 in this case). Thus the Hamming distance of the phase code corresponds to an almost perfect sequence of Bernoulli trials of length 7429 with a probability of success (which corresponds to a difference in the phase codes) close to 0.5. This is less than the 12.8K bits in the phase code due to internal correlations amongst the otherwise random trials. This value N represents the effective code length (ECL) of the phase code and is useful to see how much information it encodes. The binomial PDF fits the data in the histogram very well and is shown as an overlay in Figure 12.14. The histogram and its corresponding PDF are very tight, reflecting the fact that for a wavelength of eight pixels the Gabor filter is encoding the purely random and statistically independent features of the printing process. The probability of a significant departure from the expected value of HD close to 0.5 is highly unlikely and hence if such a departure is observed then we can only conclude that the phase codes under comparison are derived from the same printed sample. For example, assuming the measured sample standard deviation to be a good estimate of the population distribution, an HD of 0.3 or less for this data (which corresponds to a Bernoulli trial with just $0.3 \times 7429 = 2229$ successes from a sample of 7429) has a cumulative probability of approximately 4×10^{-247} according to the binomial distribution. So the possibility of it occurring by chance from another print is remote indeed.

Figure 12.15 shows ensemble data as we repeat the experiment of Figure 12.14 varying the wavelength (λ) of the 2-D Gabor filter. One plot shows the mean and

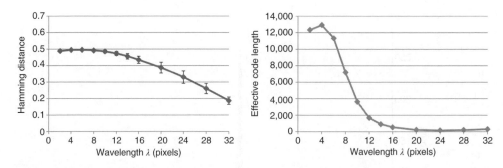

Figure 12.15　Ensemble statistics for erroneous comparisons of phase code. Plot of mean Hamming distance on the left and effective code length on the right (against wavelength of the Gabor filter).

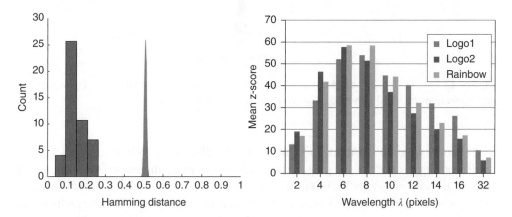

Figure 12.16　HD statistics for valid comparisons with respect to population of erroneous matches. See text for details.

standard deviation (error bars) of HD scores and the other shows the ECL. As the wavelength gets longer the phase coded vector becomes increasingly correlated between the different printed samples, as it conveys the same underlying image content, and the mean HD drops accordingly while the standard deviation increases. The effect on ECL is more dramatic, as the ability to encode large number of bits falls quickly with the standard deviation of the Binomial distribution.

12.4.3.2　Valid Comparisons

Each of the 48 printed halftones of the Rainbow Bridge print was captured for a second time (using a different DrCID imager) in order to compare the HD scores of valid matches with those of the binomially distributed statistically independent false matches. Figure 12.16 left shows a histogram of the HD scores of the valid matches of the 48 printed halftones along with the binomial PDF from Figure 12.14. As can be seen the HD scores are in fact all well separated from the false match population with minimum z-score of 44.8 and a mean z-score of 58.6 (z-score is an appropriate metric as a Binomial distribution with this many degrees of freedom is closely approximated as a Gaussian). While our sample size is small (but significant), these levels of statistical robustness are large even in the context of iris recognition (which is able to discriminate amongst many billions of comparisons; Daugman, 2006).

Figure 12.16 right plots mean z-scores for each of the halftone patterns of Figure 12.13 as we varied just the wavelength of the Gabor filter but kept all other parameters constant. This figure shows that, as discussed previously, there is a relatively small range of wavelength that gives extremely good statistical robustness using a single 2-D Gabor filter. Low frequencies, wavelengths longer than about 12 pixels, correlate between the separate samples and hence do not provide good discrimination as was reflected in their ECL. High frequencies, with wavelengths of less than four pixels, also have poor discrimination despite the fact that they have excellent (but not the best) ECL. The latter is due to a variety of factors including image sampling, noise and stringent alignment requirements required to make use of the fine detail covered at these frequencies.

12.4.4   Authenticating Data-Bearing Halftones

In the previous sub-section we have shown that it is possible to use a digital halftone model to accurately register prints based on it in order to recover repeatable forensic print signatures using phase coding. In Pollard *et al.* (2013), we have also shown that if the printed haftone is modified to carry additional information unknown to the recovery system, it is still possible to recover a reproducible forensic print signature. That is, despite using the original halftone as the model to align the modified print (as that is all that is known to the recovery system) the recognition accuracy is not detrimentally affected. The method of encoding information into the individually printed halftones that we use is called a Stegatone (from Stegenographic Halftone; Ulichney *et al.* 2010). A Stegatone generator takes a data payload and an input image called a 'mule' because it is the vehicle that transports the payload when printed. In Figure 12.17 the reference halftone (a) is the standard clustered-dot halftone generated from the mule image that we also saw in Figure 12.13. All halftone cells are classified in a reference map as either 0-bit, 1-bit, 2-bit, or 3-bit data carriers. These cells are depicted in (b). 0-bit carriers, coloured black, are called 'Reference Cells' because they are unchanged and can be used to aid alignment. The lighter cells represent 1, 2 and 3-bit carriers, (as they can shift in one of 2, 4 and 8 positions). Cells can be reference cells because they are too large to be shifted or too small to be detected. The payload is encoded by means of single pixel shifts of the halftone clusters. The data carrying capacity of this example is 447 raw bits. An example Stegatone is shown in (c) and in (d) is the result after printing, capturing with DrCID and registering using the halftone model from (a). The

(a)   (b)

(c)   (d)

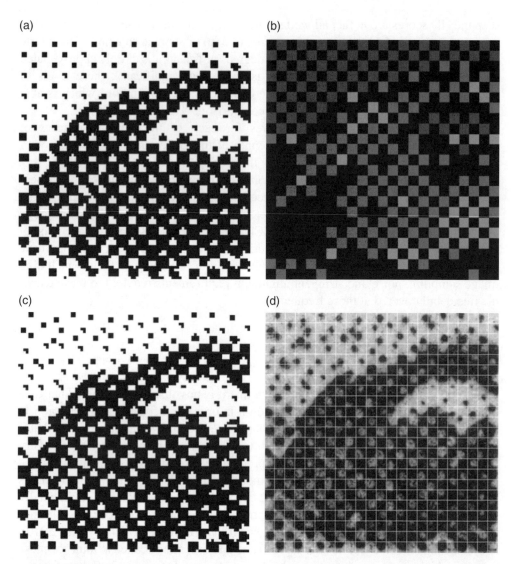

Figure 12.17 (a) classical halftone pattern at 600 dpi; (b) reference map defining payload sizes for each halftone cell; (c) Stegatone derived according to (a), (b) and desired payload; and (d) printed, captured and registered Stegatone with reference cell boundaries highlighted.

Stegatone cell boundaries are also superimposed in (d) to indicate how the cell shift is determined and the error protected payload recovered.

We found that in practice the false match distribution statistics are not significantly altered by the change from halftone to Stegatone printing. This is not very surprising as the local frequency content (amongst all other image properties) of halftones and their related Stegatones is significantly the same and the Gabor filter is essentially

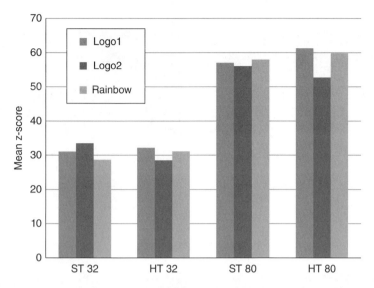

Figure 12.18   Comparison of z-score statistics of Phase Codes extracted from halfones (HT) and Stegatones (STs), where in each case the model used was derived from the halftone.

a measure of the local frequency content of the image. Furthermore, the collection of false match statistics is not dependent on high accuracy registration and so is not likely to be affected by mismatch between the digital halftone used to register the Stegatone image. Of more interest is the effect this mismatch has on the Hamming distance of correct matches. Once again, using the average z-score as a representative shorthand for the statistical robustness of the forensic print signature, Figure 12.18 compares halftone and Stegatone mean z-score values for the two different lengths of phase code. In each case the wavelength was eight pixels and the grid size (M) was either 32 or 80. The value of 32 was chosen as this gives the 1024 samples (or 2048 bits; 256 bytes) used in iris biometrics while 80 is the sampling size beyond which improvements in statistical performance are very limited. It is clear that robustness is maintained for all conditions and image types. This result is a testament to the fact that the repeatability (as opposed to absolute accuracy) of the image registration is maintained despite the fact that the halftone model is being used to register the (locally distorted) Stegatone images.

12.5   Real World Example: Currency Protection

Thus far, the printed glyphs and halftones have been somewhat contrived and printed on Laser based devices. Each has been less than 4 mm on a side to fit within the field of view of the DrCID device and imaged in perfect isolation to aid the image registration process. In this section we consider a more realistic and difficult situation where the chosen forensic mark is an area within a larger printed region. Figure 12.19

(a)   (b)

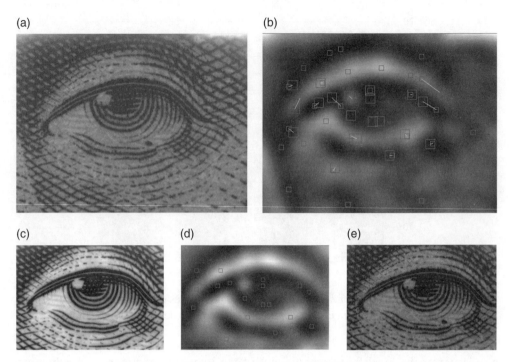

(c)   (d)   (e)

Figure 12.19 Lincoln's left eye captured by DrCID from a US $5 bill. Top left is the raw captured image, bottom left is an image model of the region around the eye for which feature points are located at the coarsest level of the multi scale registrations scheme (bottom center) and matched to similar features in the captured image (top right with transformed features from the model shown large and resident features shown small). After image based registration based on this initial transform, the dewarped registered image (bottom right) is recovered for forensic signature extraction.

shows an interesting example of performance where the left eye of Lincoln on 17 sequentially numbered, pristine US $5 bills were imaged using DrCID with rough manual alignment. A region of interest (1648×1136 pixels which corresponds to approximately 5.3×3.6 mm) centered on the eye is modelled by combining and averaging all 17 images resulting in the image shown to the bottom left in Figure 12.19. The image based alignment scheme used to build the model and to register it against an example image (i.e. the one shown top left in Figure 12.19) starts based on extraction and matching of SIFT (Lowe 2004) like features (shown as small squares) from the band pass images as shown bottom center and top right of Figure 12.19. The larger features shown top right are those corresponding to the best transform found by RANSAC (Random Sample Consensus, Fischler and Bolles 1981) which is in turn used as a starting point for the image-based registration scheme. The resulting final registered and dewarped image region extracted from that in the top left is shown in the bottom right in Figure 12.19.

The 17 extracted and registered regions are then used to form forensic signatures using the 2D Gabor filter-based technique outlined in Section 12.4.3. For consistency, the sample size was 80×80 and the wavelength of the Gabor filter was eight pixels. While the number of printed samples is rather small, the statistics of Hamming Distance are such that the estimate of standard deviation is consistent with the sample size of 136 (that is the total number of false comparisons from within a corpus of 17). In practice, the mean relative Hamming distance for false matches is 0.4642 with a standard deviation of 0.0057 which if we assume is representative of a wider Gaussian population gives a mean z-score for the relative Hamming distance of the correct matches occurring by chance of 58.0743 (which is consistent with the results reported previously for laser printed halftones).

It is our conjecture that all traditionally printed material (independent of perceived quality) is susceptible to forensic signature extraction. Of course, this conclusion would be better supported by this particular experiment if we had only considered $5 bills that were printed by exactly the same (region of the) printing plate so we can separate out those random differences that come from ink and paper variation and those that come from differential wear and manufacturing imperfections of the printing plate itself.

12.6   Summary and Ecosystem Considerations

Finally, we consider how various forms of forensic security printing fit within the wider ecosystem. In forensic authentication, printed information is used to identify with a certain degree of statistical confidence that a document, label, package, or surface is authentic. Forensic materials – such as security substrates, security ink or substrate additives, and secure finishing coatings, laminates or procedures – must be protected from theft, and so constitute controlled substances/procedures. Many different forensic materials exist. However, as we have demonstrated in this chapter, high-resolution imaging obviates the need for 'special' substrates, additives, or finishing by allowing the printed item itself to provide forensic authentication. As we have shown, high-resolution imaging hardware can be used for the highly statistically significant forensic identification of a printed document or label using a print fragment as small as a single printed character. This capability has been extended to other printed content, including small logos containing steganographic data with the same degree of statistical confidence. No special forensics are needed; the printing itself is the forensic.

The level of image analysis is important in security printing as it defines the type of applications that can be initiated by the security printed object. Forensic image analysis, which has been the subject of this chapter, is the most difficult to reproduce but other forms of image authentication are also useful. For example (Simske *et al.* 2009), describe a system that uses a classifier built using scans of large numbers of genuine and counterfeit product packages. The classifier uses many image metrics recovered from the scanned product packages and is able to identify the class to which novel examples correspond with low levels of statistical confidence. What is more as the classifier is allowed to evolve over time it is able to incorporate both new genuine classes of product packaging (due to a process change in the manufacture process) and more importantly identify new sources of counterfeit products.

The role of the person authenticating the security printed object is another means of defining the downstream applications and services that are initiated by decoding the object. Security printing is associated with a supply chain, value chain, or other logistics-managed movement of content between different actors (see Figure 12.20 for a simplified example of a supply chain using forensic print authentication). The seven primary actors, generalized across most domains, are the manufacturer, the warehouser, the distributor, the retailer, the consumer, the inspector, and the forensic analyst. The manufacturer and forensic analyst, one on each end of the overall object lifecycle, are generally interested in the highest level (forensic) of object authentication. It is important, then, that the manufacturer provides forensic-level inspection

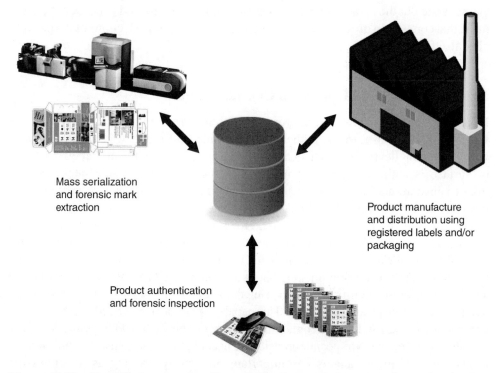

Mass serialization
and forensic mark
extraction

Product manufacture
and distribution using
registered labels and/or
packaging

Product authentication
and forensic inspection

Figure 12.20 Simplified ecosystem for variable data and forensic print authentication. Variable data printing is used to generate unique packaging or labelling that includes serialization data from which an associated forensic mark is extracted and its forensic signature stored on a central database using the serialization data as an index. Throughout the manufacturing, distribution and retailing process the packaging can then be inspected and verified to forensic levels of authenticity.

on the packaging, labels, or other product-associated materials. In between, different motivations – inspection, auditing, track and trace, individual product authentication, supply chain integrity validation – drive the different actors. Hybrid, variable data printing (VDP) allows a single printed region to be used for many purposes simultaneously, allowing each actor to achieve her goals without compromising other actors.

The utility of the encoded information is an important factor when a hybridized security printing design is used. This is preferable when using VDP because the amount of effort to craft multiple variable data regions is only marginally more than the effort required to implement one VDP feature. For this modest upfront effort, a wide variety of downstream advantages are garnered. In order to produce a truly secure printed feature, all three of the following utilities must be provided: (1) unique ID, (2) copy prevention and (3) tamper-evidence.

A unique ID is readily produced using VDP. The simplest unique ID is a serial number. Usually, the order of serial numbers is randomized using a random number generator, encryption, or digital signing. The end result is that each printed item has a unique ID suitable for look up in a (cloud-accessed) database. The unique ID can be used as an entry field for the other printed information–overt, covert, stenographic and/or forensic–associated with the same object. For example, the descriptor for the forensics of a specific printed character can be stored in the database and then compared to the descriptor of the same character on the object tagged with the unique ID. A non-match indicates the object has been copied, reprinted or otherwise counterfeited.

Copy prevention is important for a minimum of one decodable object. Otherwise, a direct copy of the image can be made and falsely 'authenticated.' Clearly, the forensic character descriptors associated with the use of high-resolution images are extremely difficult if not impossible to copy.

The third requirement for a security printing object is tamper-evidence. If an object has a unique ID and cannot be copied, it can still be reused. Tamper-evident objects, however, are compromised when they are authenticated, making them single use. Associating a security feature with the tear strip means that it will be bisected when the package is opened. Scratch-off surfaces that must be rubbed away to access the unique ID underneath are another rational form of tamper-evidence. The extent of print variability is used to craft the analytics for a security printing campaign. The analytics are, of course, the collection and digestion of information associated with the use of the security printing objects.

As argued earlier, VDP opens the door for hybridization, that is using different types of variable regions for different tasks. These include inspection, point of sale, authentication unique ID/mass serialization, forensic authentication, and URL embedding. From a security standpoint, VDP is very powerful: hybridization means that the relationship between multiple VDP objects can be varied from one security printing campaign to the next without requiring a change in the VDP objects used. This is an excellent way to make the counterfeiters spend more in reverse engineering the system. In the context of forensic authentication of printed materials, VDP crucially supports the addition of serialization data that can readily be utilized in the management and communication of forensic signatures associated with the print. It is our belief that this capability will become increasingly important and commercially prevalent over the coming years as the basic infrastructure (VDP printers, databases and mobile devices) becomes the norm. The methods outlined in this chapter could form the basis of this revolution in document and product security.

References

Adams, G., 2010, Hand held Dyson Relay Lens for anti-counterfeiting, Proceedings of the IEEE International Conference on Imaging Systems and Techniques, Thessaloniki, Greece, IEEE, 273–278.

Bouguet, J.-Y., 1999, Pyramid implementation of Lucas Kanade feature tracker: description of the algorithm, OpenCV Documents Intel Corporation, Microprocessor Research Lab.

BS ISO, 2007, Automatic Identification and Data Capture Techniques. QR Code 2005 Bar Code Symbology Specification. London: BSI.

Buchanan, J.D.R., Cowburn, R.P., Jausovec, A.V., Petit, D., Seem, P., Xiong, G., Atkinson, D., Fenton, K., Allwood, D.A., Bryan, M.T., 2005, Forgery: Fingerprinting documents and packaging, Nature, 436–475.

Clarkson, W., Weyrich T., Finkelstein, A., Heninger, N., Halderman J.A., Felten, E.W., 2009, Fingerprinting blank paper using commodity scanners, Proceedings of the 30th IEEE Symposium on Security and Privacy, Berkeley, CA, IEEE, 301–314.

Daugman, J.G., 1993, High confidence visual recognition of persons by a test of visual phase information, IEEE Transactions on Pattern Analysis and Machine Intelligence, 15 (11), 1148–1161.

Daugman, J., 2006, Probing the uniqueness and randomness of IrisCodes: Results from 200 billion comparisons, Proceedings of the IEEE, 94 (11), 1927–1935.

Daugman, J., 2007, New methods in iris recognition, IEEE Transactions on Systems, Man, and Cybernetics, Part B: Cybernetics, 37 (5), 1167–1175

Fischler, M.A., Bolles, R.C., 1981, Random sample consensus: A paradigm for model fitting with applications to image analysis and automated cartography, Communications of the ACM, 24 (6), 381–395.

Flusser, J., Suk, T., Zitová, B., 2009, Moments and Moment Invariants in Pattern Recognition. Chichester: John Wiley & Sons, Ltd.

Gabor, D., 1946, Theory of communication, Journal of Institute of Electrical Engineering, 93, 429–457.

GS1, 2011, GS1 DataMatrix: An introduction and technical overview of the most advanced GS1 Application Identifiers compliant symbology, GS1 AISBL, http://www.gs1.org/docs/barcodes/GS1_DataMatrix_Introduction_and_technical_overview.pdf (Accessed 17 February 2015).

GS1, 2014, EPC Tag Data Standard, version 1.8, GS1 AISBL, http://www.gs1.org/sites/default/files/docs/tds/TDS_1_8_Standard_20140203.pdf (Accessed 17 February 2015).

Lowe, D.G., 2004, Distinctive image features from scale-invariant keypoints, International Journal of Computer Vision, 60 (2), 91–110.

Lucas, B., Kanade, T., 1981, An iterative image registration technique with an application to stereo vision, International Joint Conference on Artificial Intelligence, Vancouver, BC, University of British Columbia, 674–679.

Metois, E., Yarin, P., Salzman, N., Smith, J. R., 2002, Fiber-fingerprint identification, Proceedings of 3rd Workshop on Automatic Identification, Tarrytown, NY, 147–154.

Pizzanelli, D., 2009, The Future of Anti-Counterfeiting, Brand Protection and Security Packaging V. Leatherhead: Pira International.

Plimmer, J., 2008, Choosing correct forensic marker(s) in currency, document and product protection, Proceedings of the SPIE 6075, Optical Security and Counterfeit Deterrence Techniques VI, San Jose, CA, SPIE, 339–347.

Pollard S., Adams G., Simske, S., 2010a, Resolving distortion between linear and area sensors for forensic print inspection, Proceedings of the IEEE International Conference on Image Processing, Hong Kong, IEEE, 1001–1004.

Pollard, S., Simske, S., Adams, G., 2010b, Model based print signature profile extraction for forensic analysis of individual text glyphs, Proceedings of the IEEE nternationa Workshop on Information Forensics and Security, Seattle, WA, IEEE, 1–6.

Pollard, S., Simske, S., Adams, G., 2012, Print biometrics: Recovering forensic signatures from halftone images, Proceedings of the International Conference on Pattern Recognition, Tsukuba, Japan, IEEE, 1651–1654.

Pollard, S., Ulichney, R., Gaubatz, M., Simske, S., 2013, Forensic Authentication of Data Bearing Halftones, VISGRAPP 8, Barcelona, Spain, SciTePress, 109–113.

Sharma, A., Subramanian, L., Brewer, E., 2011, PaperSpeckle: Microscopic fingerprinting of paper, Proceedings of the 18th ACM Conference on Computer and Communications Security, Chicago, IL, ACM, 99–110

Simske, S., Adams, G., 2010, High-resolution glyph-inspection based security system, Proceedings of the IEEE International Conference on Acoustics, Speech, and Signal Processing, Dallas, TX, IEEE, 1794–1797.

Simske, S., Pollard, S., Adams, G., 2010, An imaging system for simultaneous inspection, authentication and forensics, Proceedings of the IEEE International Conference on Imaging Systems and Techniques, Thessaloniki, Greece, IEEE, 266–269.

Simske, S., Sturgill, M., Everest, P., Guillory, G., 2009, A system for forensic analysis of large image sets, Proceedings of the IEEE International Workshop on Information Forensics and Security, London, IEEE, 16–20.

Ulichney, R., Gaubatz, M., Simske, S., 2010, Encoding information in clustered-dot halftones, Proceedings of the IS&T 26th International Conference on Digital Printing Technologies and Digital Fabrication (NIP26), Austin, TX, IS&T, 602–605.

Zhu, B., Wu, J., Kankanhalli, M. S., 2003, Print signatures for document authentication, Proceedings of the 10th ACM Conference on Computer and Communications Security, Washington, DC, ACM, 145–154.

Part Four

Multimedia Content Forensics

13

Digital Image Forensics with Statistical Analysis

Xunyu Pan
*Department of Computer Science and Information Technologies,
Frostburg State University, Frostburg, MD, USA*

13.1 Introduction

13.1.1 Digital Image Forensics

The world is witnessing an unprecedented growth in the popularity of digital imaging devices and their applications in the modern society. The digital technology available on powerful personal computers and high-speed Internet make the production and transmission of digital information possible for a normal person. As one of the major information sources, digital images have evolved to become an essential part of our life in a wide range of fields from entertainment to mass media, from medical diagnosis to criminal justice, and even national security. However, as shown in Figure 13.1, concomitant with the ubiquity of digital images and the increasing sophistication of advanced photo-editing software (e.g. *Adobe Photoshop*), is the rampant problem of digital forgeries, which has seriously debased the credibility of photographic images as definite records of events. Accordingly, digital image forensics has emerged as a new research field that aims to reveal tampering operations in digital images (Farid 2009a).

At the time of the birth of digital image forensics, *active* protection methods such as digital watermarking (Cox *et al.* 2008) and signature (Friedmann 1993) served as major solutions to protect the integrity of digital images. By inserting certain information into images, digital watermarking is considered *invasive* as certain distortion of the

Handbook of Digital Forensics of Multimedia Data and Devices, First Edition.
Edited by Anthony T.S. Ho and Shujun Li.
© 2015 John Wiley & Sons, Ltd. Published 2015 by John Wiley & Sons, Ltd.
Companion Website: www.wiley.com/go/digitalforensics

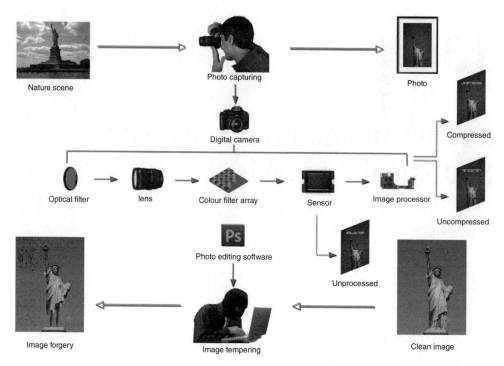

Figure 13.1   Digital camera imaging process and the procedure of making image forgery.

image is unavoidable during the embedding process. Digital signature, on the contrary, is *non-invasive* as only a computed signature is appended to the image as metadata. However, most digital imaging devices currently on the market lack watermarking or signature modules. Meanwhile, the *passive* methods on image tampering detection are more practical than digital watermarking or signature. The purpose of passive methods is to verify the authenticity of the digital images with no prior knowledge. In this chapter, we focus on the digital image forensic techniques based on the passive model.

13.1.2   Background

With the advanced digital image editing tools, various forms of image tampering techniques can be employed to create fake images. Image region duplication and splicing are the two most common manipulations in tampering with digital images, aiming to make highly convincing forgeries. Region duplication copies a continuous portion of pixels and pasted to a different location in the same image. Splicing image is generated by compositing parts from different source images. Meanwhile, image retouching techniques create highly idealized images for magazine covers and advertisements. The computer-generated imagery (CGI) rendering helps to create extremely realistic images which might be impossible to differentiate from photographic images. Finally, the goal of steganography is to embed secret messages within an image or other cover medium so that casual inspection will not reveal the presence of the hidden messages.

A large number of forensic methods have been developed in the past decade to answer a broad range of forensic questions. As a comprehensive survey of this emerging field, Farid (2009b) divided the existing image forensic methods into five categories: (i) pixel-based techniques to find statistical anomalies at the pixels level, (ii) format-based techniques to examine the statistical correlations introduced by compression schemes, (iii) camera-based techniques to exploit artefacts produced by all forms of in-camera and post-camera processing, (iv) physically based techniques to detect inconsistency in lighting and (v) geometry-based techniques to make measurements of objects and their positions relative to the camera. From the perspective of semantic information employed in the detection process Wang *et al.* (2009) summarized the solutions to image tampering problem in a three-level hierarchy: (i) low level where statistical characteristics of image pixels or discrete cosine transform (DCT) coefficients are used as clues for detection, (ii) middle level where simple semantic information is used for detection and (iii) high level where the detection is based on some strong semantic information (e.g. President Barack Obama may not shake hands with Osama bin Laden).

In our opinion, most image forensic tools can be divided into only two simple categories: *semantics-based detection* and *non-semantics-based detection*. A majority of existing detection methods belong to the latter category where the statistical pattern in the image is first modelled and then the inconsistencies in this pattern are inspected across the image to search for clues of tampering. On the other hand, some recent techniques such as those in the last two categories described in (Farid 2009b) can be seen as pilot explorations based on semantic information. As the non-semantics-based methods are becoming more popular and powerful, these semantics-based approaches also deserve further attention from the digital image forensics community. In the following, we briefly review some previous landmark works on digital image forensics based on these two categories.

13.1.2.1 Non-semantics-Based Detection

We first take a close look at the non-semantics-based detection techniques as majority of existing image forensic tools fall into this category. Rather than covering a large number of non-semantics-based forensic methods in an encyclopaedic manner, we have chosen to focus on several representative techniques that are both pervasive and important.

Since 1990s, due to the advantage of excellent image quality under relatively high compression rate, JPEG has become the most popular format for storing and transmitting photographic images over the Internet. As the final processing step, the original images are typically saved in JPEG format only once, while their tampered counterparts require the same step at least twice. With the different artefacts introduced by JPEG compression for the original and manipulated images, a double JPEG quantization is usually a telltale sign of tampering operations (see Chapter 15 in Part IV for more discussion on this topic). In (Popescu and Farid 2004b), the presence of high-frequency peaks in Fourier transforms of DCT coefficients in the tampered images are used as the evidence of image manipulation. A statistical model is created in (Fu *et al.* 2007) to detect double JPEG compression based on the probability distributions of the first digits of the block-DCT and quantized JPEG coefficients.

When capturing scenes from nature with a digital camera, light is focused onto imaging sensors such as CCD or CMOS. In most digital cameras, the colour image is produced by a single sensor together with a colour filter array (CFA). The CFA helps in constructing a periodic pattern such that each sensor element receives a single-colour sample. Due to the periodic pattern among neighbouring pixels in a colour image, these correlations might be altered or completely destroyed during the process of image manipulation (see Chapter 14 in Part IV for more discussion on this topic). Based on this idea, the CFA correlations are computed in (Popescu and Farid 2005b) from the interpolation of the existing colour samples. Since the original images are expected to have a periodic pattern where neighbouring pixels are highly correlated, any deviation from this pattern is an evidence of image tampering. Furthermore, resizing, rotation and stretching of the tampered region, which are necessary operations for producing convincing forgery, can be identified by detecting the specific correlations introduced by these resampling operations (Mahdian and Saic 2008, 2009a; Popescu and Farid 2005a).

The camera-response function is the mapping of irradiance measured by each imaging sensor element and its final pixel value. This function can be estimated (Lin *et al.* 2005) when selecting appropriate patches along edges. Normally, the response function increases monotonically. It should have no more than one inflexion point. The response functions for R, G and B channels should be close to each other, where three features are extracted from the response function to construct a 3D feature vector. By the classification training, significant inconsistencies in the 3D feature space across the image are used for tampering detection.

In (Chen *et al.* 2008a) and (Lukas *et al.* 2006a), the consistency between image and imaging device is verified for camera model identification. The technique detects the presence of camera pattern noise, which is a unique stochastic characteristic of imaging sensors, in each individual region across the image. The forged region is determined as the one that lacks the detected pattern noise. The presence of this type of noise is established using correlation. Another similar idea is proposed in (Swaminathan *et al.* 2008) where distinct intrinsic fingerprint traces introduced by in-camera and post-camera processing operation are used to detect forgery. The method first computes the camera's component parameters and intrinsic fingerprints. The coefficients of manipulation filter for post-camera processing is then estimated. High similarity between the estimated coefficients and the reference pattern, which corresponds to no manipulation, certifies the integrity of the given image.

Natural images are perceptually meaningful to human eyes. Compared to the huge number of randomly generated images, natural images exhibit unique underlying statistical properties. By modelling natural images at the first and higher order statistics upon a multi-scale wavelet decomposition (Farid and Lyu 2003), digital tampering can be detected from natural images based on these statistical features. Forensic tasks ranging from basic image tampering detection (Bayram *et al.* 2006), discriminating natural images against computer generated images (Lyu and Farid 2005), to detecting hidden messages in images (Lyu and Farid 2006) can all be solved using the statistical properties of natural images.

13.1.2.2   Semantics-Based Detection

The non-semantics-based detection tools mostly rely on the modelling of statistical patterns of the image using signal-level information. Based on certain high level cognitive knowledge, some semantic cues introduced by tampering operations can be exposed. Again, we have chosen to only introduce here several representative semantics-based forensic techniques.

In (Johnson and Farid 2005), the light source direction in the scene is estimated using two-dimensional surface normals at a manually selected occluding object boundary. The inconsistency of the light direction across an image is used to make tampering detection decisions. This technique was further extended to estimate the light source direction for 3D surface normals (Johnson and Farid 2007c) and more complex lighting environments (Johnson and Farid 2007b).

As the projection of the camera centre onto the image plane, the principal point should be near the centre of the image. The approach proposed in (Johnson and Farid 2007a) relies on estimating the transformation from world to image coordinates and then factoring this transformation into a product of matrices containing intrinsic and extrinsic camera parameters. With a known focal length, the principal point can be determined from the intrinsic matrix. Given an image of two or more people, the principal points can be estimated from the planar geometric shapes such as the eyes of each person. Inconsistencies in the principal point are then used as evidence of tampering.

In this chapter, we introduce several recently developed techniques to address two critical topics in the field of multimedia security: detecting region duplication and exposing splicing forgery. In the absence of device-dependent protection mechanisms such as digital watermarking or signature, we focus on solving these problems more practically by assuming that most digital manipulations will disturb some statistical property of an image. The computer-generated forgeries can be revealed by detecting these abnormalities with the statistical analysis of data information from images. We introduce in Section 13.2 a method for reliable detection of duplicated image regions with geometric distortions and illumination adjustments. In Section 13.3, we introduce another effective image splicing detection algorithm by exposing inconsistencies in local noise variances. More realistic case studies for these two techniques are demonstrated in Section 13.4. These techniques are further extended to expose forgeries in audio and video signal in Section 13.5. We finally conclude this chapter in Section 13.6 with a summary of these multimedia forensic tools.

13.2 Detecting Region Duplication

13.2.1 Problem Definition

A common manipulation in tampering with digital images is known as *region duplication*, where a continuous portion of pixels are copied and pasted to a different location in the same image. Figure 13.2 exemplifies two main usages of duplicated regions in creating forged images. In the first case (top row), a duplicated region is used

Figure 13.2 (Left) Original untampered images (top: courtesy of A. Popescu and H. Farid, bottom: from (Farbman *et al.* 2009)). (Right) Forgeries created with region duplication (top: manually with PhotoShop using a rotated texture region, bottom: using an algorithm from (Farbman *et al.* 2009), with rotation, scaling, reflection and slight illumination adjustments).

to conceal undesirable contents in the original image. In the second example (bottom row), two duplicated regions are used to create contents that are not in the original image. To make convincing forgeries, the duplicated regions often have geometric or illumination adjustments. For example, in Figure 13.2, the duplicated region in the first forged image is rotated, and those in the second forged image are scaled, rotated and mirrored. As a result, these duplicated regions are well blended into the surroundings at the target locations, and become very difficult to detect visually.[1]

In this section, we introduce an effective method for reliable detection of duplicated regions with geometric distortions and illumination adjustments. Our method starts with the detection of image *keypoints* carrying distinct statistic information of image content. Each keypoint is characterized by a feature vector that consists of a set of image statistics collected at the local neighbourhood of the corresponding keypoint. These feature vectors are efficient to compute and robust to typical geometrical transforms. We formulate detecting region duplication as finding transformed identical regions in an image and use robust estimation to obtain correct keypoints matching and transforms between duplicated regions simultaneously. With the estimated transforms, our method computes the precise location and extent of the duplicated regions to visualize the detection results.

13.2.2   Related Works

Detecting duplicated regions in an image or video corresponds to identifying all pixels in *suspiciously similar* regions. A majority of recently developed methods directly find exact copies of small-pixel blocks in an image. Though the front-end linear image domain is mostly on pixels, some specific image representation schemes such as wavelet (Li *et al.* 2007; Myna *et al.* 2007), bit-plane decomposition (Ardizzone and Mazzola 2009), and multi-scale decomposition (Yang and Huang 2009) are also used. As a brute-force match of all pixel blocks of a given size in an image has a running time quadratic in the size of the image, low dimensional representations of pixel blocks are employed for efficient computation, for example principal component analysis (PCA) (Popescu and Farid 2004a), DCT (Fridrich *et al.* 2003) and singular value decomposition (SVD) (Li *et al.* 2007). A common step used in these methods is to lexicographically sort pixel blocks so that identical blocks end up as adjacent pairs in the sorted list. Recently, kd-tree (Langille and Gong 2006; Mahdian and Saic 2007) and hashing-based Bloom filter (Bayram *et al.* 2009) techniques have been proposed to further speed up the sorting process.

However, in practice, copy–move alone can seldom create plausible forgeries. More likely, as in the example of Figure 13.2, the duplicated regions are subjected to

---

[1] Experiments in psycho-physics have also shown that searching identical regions in an image with a lot of clutter and distractions is a very hard task for human subjects (Bravo and Farid 2004).

geometric and illumination transforms to be better blended into the surroundings at the target location. As such distortions alter the correspondence between pixels in duplicated regions, straightforward matching of blocks of pixels or transform coefficients computed from the pixel values becomes much less effective. For simple transforms such as rotations through fixed angles (e.g. $90°$, $180°$ and $270°$), this problem can be remedied by enumerating these transforms and including the transformed images for detection (Lin *et al.* 2009). Also, using blocks in a log-polar coordinate systems has been proposed to achieve some invariance to rotation and scaling (as both become translations in the log-polar coordinate systems) (Bayram *et al.* 2009; Bravo-Solorio and Nandi 2009; Myna *et al.* 2007). Although they provide more flexibility in detection, the performance of these methods is undercut when detecting duplicated regions with general linear geometric transforms.

As an alternative to the block-based detection method, several recent works propose detection methods that are based on matching image features invariant to general geometric and illumination distortions. In (Huang *et al.* 2008), invariant local image features are used to account for illumination changes to the duplicated regions. However, the invariance of such image features to geometric distortions is not explicitly exploited and the method in this work is for the detection of copy–move with only matched keypoints shown as results. In (Pan and Lyu 2010a), a feature matching based detection method is described to locate duplicated regions distorted with arbitrary rotation or scaling transforms. More recent works (Amerini *et al.* 2010; Pan and Lyu 2009, 2010a,b) take advantage of such image features to recover the parameters of the geometric transform and use such information for more reliable detection of duplicated regions which have undergone flexible geometric transforms. Though these feature matching based region duplication detection methods show promising performance, many of these methods (e.g. Amerini *et al.* 2010; Huang *et al.* 2008) cannot report the location and extent of the detected duplicated regions. Instead, they can only show the matched keypoints, which become much less useful from the practitioners' perspective.

In general, feature-based methods and block-based methods share a common processing framework (Christlein *et al.* 2012). Most of these methods operate on greyscale images by preprocessing the original image with the merging of colour channels. Distinct image features are first collected in subdivided rectangular regions for block-based methods or high entropy regions for feature-based methods. In both cases, a feature vector is then computed for every such image region based on local image statistics. Similar feature vectors within the same image are subsequently matched. A forgery is identified if regions of such matches cluster into larger areas. Further filtering is included in both feature-based and block-based methods for removing spurious matches. In order to group matches that jointly follow a transformation pattern, an optional postprocessing step of the detected regions may also be performed.

13.2.3 Proposed Method

In this section, we describe a generic region duplication detection method (Pan and Lyu 2009, 2010a,b) based on the aforementioned processing framework. Region duplication can be formalized as a 2D linear transform between image regions. Denote pixel locations in the source region and its duplication as Ω_S and Ω_T, respectively. Assuming only gentle changes in the pixel intensities in tampering the image I, region duplication leads to $I(\Omega_T) \approx I(\mathcal{T}(\Omega_S))$, where $\mathcal{T}$ is a linear *manipulation transform* including translation, rotation, scaling, perspective and their combinations as shown in Figure 13.3. Detecting region duplication involves recovering Ω_S and Ω_T, along with the manipulation transform $\mathcal{T}$. This problem bears resemblance to the estimation of camera motion from multiple images (Hartley and Zisserman 2004). The key differences are (1) here we are interested in the transform between different regions in a single image, and (2) we also need to recover the exact locations and extents of the source and duplicated regions. Figure 13.4 illustrates the key steps of our region duplication detection method.

1. We first convert RGB images to greyscale images since the duplicated regions are detected in the illumination domain.
2. Keypoints are detected in the input image using a local image feature description algorithm, where feature vectors are collected. Initial matchings of keypoints are made based on the similarity between the feature vectors.
3. The initial matchings of keypoints are then refined iteratively with the robust RANSAC estimation (Fischler and Bolles 1981), after which only reliable correspondences between keypoints are kept. We further estimate an affine transform between the two corresponding sets of keypoints.
4,5. We then generate two correlation maps, which contain the correlation coefficients of each pixel with its correspondences obtained with the estimated affine transform and its inverse to a pair of duplicated regions.

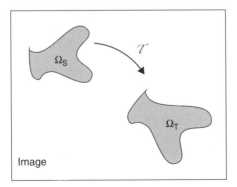

Figure 13.3 A schematic illustration of region duplication operation.

(a)   (b)   (c)

(d)   (e)   (f)

Figure 13.4   Key steps of the proposed method to detect duplicated and distorted regions. (a) An tampered image with region duplication forgery. (b) Detected SIFT keypoints in the image. (c) Matched keypoints after the RANSAC algorithm. (d and e) Region correlation maps generated with the estimated affine transforms. Brighter pixel intensity signifies stronger correlation. (f) Detected duplicated regions.

6. All pixels corresponding to the duplicated regions are found by thresholding the correlation maps. The results are further merged, filtered and smoothed to obtain location and extent of the detected duplicated regions.

In the following, we explain in detail each of these steps.

13.2.3.1   Image Features Collection

The first step in our method is to find image keypoints and collect image features at the detected keypoints. To simplify the procedure of keypoint detection, RGB images are preprocessed by being converted to greyscale images with the merging of colour channels.

Image keypoints and features are essential to many applications in computer vision including image registration, content-based image retrieval, object detection/recognition and visual tracking. Keypoints are locations carrying distinct information of the image content. Each keypoint is characterized by a feature vector that consists of a set of image statistics collected at the local neighbourhood of the corresponding keypoint. Good keypoints and features should represent distinct locations in an image, be efficient to compute and robust to local geometric distortion, illumination variations, noise and other degradations.

One of the most effective keypoint and feature computation algorithms is known as the *scale-invariant feature transform* (SIFT) (Lowe 2004). SIFT keypoints are found by searching for locations that are stable local extrema in the scale space (Lindeberg 1994). At each keypoint, a 128-dimensional feature vector is generated from the histogram of local gradients in its neighbourhood. To ensure that the obtained feature vector is invariant to rotation and scaling, the size of the neighbourhood is determined by the dominant scale of the keypoint, and all gradients within are aligned with the keypoint's dominant orientation. Furthermore, the obtained histograms are normalized to unit length, which renders the feature vector invariant to local illumination changes (Lowe 2004). Figure 13.4b shows the SIFT keypoints detected in an image. The end of each arrow corresponds to the location of one SIFT keypoint. The directions of the arrows show the dominant orientation of each keypoint, and the lengths of the arrows correspond to the dominant scale.

13.2.3.2 Initial Keypoint Matching

The detected SIFT keypoints are then tentatively matched based on their feature vectors using the *best-bin-first* algorithm (Beis and Lowe 1997). For a keypoint at location $\vec{x}$ with feature $\vec{f}$, we match it with keypoint $\tilde{\vec{x}}$, whose corresponding feature vector $\tilde{\vec{f}}$ is the nearest neighbour to $\vec{f}$ corresponding to their l_2 (Euclidean) distance. Due to the smoothness of natural images, the best match of a keypoint usually lies within its close spatial adjacency. To avoid searching for the nearest neighbours of a keypoint from the same region, we perform the search outside a 11×11 pixel window centred at the keypoint. Further, many keypoints can match with each other, but we only keep those with distinct similarities. Specifically, we require that for any feature vector $\vec{f}'$ other than $\vec{f}$ and $\tilde{\vec{f}}$, the distance between $\vec{f}$ and $\tilde{\vec{f}}$ has to be smaller than that between $\vec{f}$ and $\vec{f}'$ by at least a factor of ϵ, as $\|\tilde{\vec{f}} - \vec{f}\|_2 < \epsilon \|\vec{f}' - \vec{f}\|_2$, where $\epsilon \in (0, 1)$ is a preset threshold controlling the distinctiveness of the matching. We use a default $\epsilon = 0.5$ to provide a good trade-off between matching accuracy and ratio of outliers.

13.2.3.3 Estimating Geometric Transform

For realistic region duplication forgeries, the geometric transform between the duplicated regions is typically the combination of several transforms. The duplicated region can be distorted with transforms such as rotation, scaling, shearing, and all combinations that are supported in most photo-editing software. These transforms can be modelled as affine transforms of pixel coordinates. Given two corresponding pixel locations from a region and its duplicate as $\vec{x} = (x, y)^T$ and $\tilde{\vec{x}} = (\tilde{x}, \tilde{y})^T$, respectively, they are related by a 2D affine transform specified by a 2×2 matrix T and a shift vector $\vec{x}_0$ as: $\tilde{\vec{x}} = T\vec{x} + \vec{x}_0$, or more explicitly:

$$\begin{pmatrix} \tilde{x} \\ \tilde{y} \end{pmatrix} = \begin{pmatrix} t_{11} & t_{12} \\ t_{21} & t_{22} \end{pmatrix} \begin{pmatrix} x \\ y \end{pmatrix} + \begin{pmatrix} x_0 \\ y_0 \end{pmatrix}. \tag{13.1}$$

To obtain a unique solution to the unknowns, $(t_{11}, t_{12}, t_{21}, t_{22}, x_0, y_0)$, we need at least three pairs of corresponding keypoints that are not *collinear*. In practice, due to imprecise matching, Equation 13.1 may not be satisfied exactly, and we form the *least squares* objective function using matched keypoints $(\vec{x}_1, \vec{x}_2, \ldots, \vec{x}_N)$ and $(\tilde{\vec{x}}_1, \tilde{\vec{x}}_2, \ldots, \tilde{\vec{x}}_N)$, as follows:

$$L(T, \vec{x}_0) = \sum_{i=1}^{N} \|\tilde{\vec{x}}_i - T\vec{x}_i - \vec{x}_0\|_2^2, \tag{13.2}$$

and searching for solutions to T and $\vec{x}_0$ that minimize $L(T, \vec{x}_0)$.

13.2.3.4 Robust Transform Estimation

We can use the initial matchings of SIFT keypoints to estimate the affine transform parameters, but the obtained results are inaccurate due to the large number of mismatched keypoints. To prune out unreliable keypoint correspondences and obtain accurate transform parameters simultaneously, we employ a widely used robust estimation method known as the *random sample consensus* (RANSAC) algorithm (Fischler and Bolles 1981). The main advantage of RANSAC matching algorithm is that it can estimate the model parameters with a high degree of accuracy even when a significant number of mismatched pairs are present. Using the initial matching of SIFT keypoints, we run the following two steps N times:

1. Randomly select three or more pairs of matched keypoints that are not collinear. Using the chosen pairs of keypoints, estimate T and shift vector $\vec{x}_0$ by minimizing the objective function given in Equation 13.2.
2. Using the estimated T and $\vec{x}_0$, classify all pairs of matched SIFT keypoints into *inliers* or *outliers*. Specifically, a pair of matched keypoints $(\vec{x}, \tilde{\vec{x}})$ is an inlier if $\|\tilde{\vec{x}} - T\vec{x} - \vec{x}_0\|_2 \leq \beta$; otherwise, it is an outlier. The threshold $\beta > 0$ controls the maximum distance of inliers.

The RANSAC algorithm returns the estimated transform parameters that lead to the largest number of inliers. In our experiments, we choose default values for $N = 100$ and $\beta = 3$ as they lead to better empirical performance. Figure 13.4c shows the SIFT keypoint correspondences after the RANSAC estimation.

13.2.3.5 Locating Duplicated Regions

With the estimated affine transform, we compare each pixel to its transformation to find identical regions. In practice, because the estimated affine transform can be the inverse of the actual transform (from pixel level, we cannot differentiate which region is the source and which one is the duplicate), we check the correspondence of $\vec{x}$ using both the estimated affine transform, $\vec{x}_f = T\vec{x} + \vec{x}_0$ and its inverse, $\vec{x}_b = T^{-1}(\vec{x} - \vec{x}_0)$.

Taking the forward transform as example, the similarity between $\vec{x}$ and $\vec{x}_f$ is evaluated with the *correlation coefficients* between the pixel intensities within small neighbouring areas of each location. Denoting the pixel intensity at location $\vec{x}$ as $I(\vec{x})$, and $\Omega(\vec{x})$ as the 5×5 pixels neighbouring area centred at $\vec{x}$, the correlation coefficient between the two pixel locations is computed as Equation 13.3:

$$c_f(\vec{x}) = \frac{\sum_{\vec{s} \in \Omega(\vec{x}), \vec{t} \in \Omega(\vec{x}_f)} I(\vec{s})I(\vec{t})}{\sqrt{\left[\sum_{\vec{s} \in \Omega(\vec{x})} I(\vec{s})^2\right]\left[\sum_{\vec{t} \in \Omega(\vec{x}_f)} I(\vec{t})^2\right]}}. \tag{13.3}$$

The correlation coefficient for the inverse transformed $\vec{x}_b$ is computed in a similar manner. The correlation coefficient is in the range of $[0, 1]$, with larger value indicating higher level of similarity. Further, it is invariant to local illumination distortions – any illumination changes consistent within the local neighbourhood will cancel out each other. The computed $c_f(\vec{x})$ and $c_b(\vec{x})$ are placed into correlation maps, shown in Figures 13.4d and e.

The correlation maps are then smoothed and merged to obtain the duplicated regions. First, we apply a Gaussian filter of size 7×7 pixels to reduce the noise in the correlation maps. Next, with a threshold $c \in [0, 1]$, the correlation maps are discretized to binary images. Experimental results suggest that a value of 0.3 can provide a good trade-off between detection accuracy and false detection rate. The obtained binary maps for c_f and c_b are then combined into a single map by a union of the binary values. Next, we use an area threshold, $A = 0.1\%$ of the total area of the image to remove small isolated regions. As a final post-processing step, we use mathematical morphological operations to smooth and connect the boundaries of the detected duplicated regions. Shown in Figure 13.4f are the final detected regions.

13.2.4 Performance Analysis

In this section, we describe an experimental evaluation of the performance of the proposed region duplication detection method. As there are several adjustable parameters in our method, we list their default values in Table 13.1. Unless specified otherwise, the results reported in the following sections are based on this set of parameters.

Table 13.1 Default values for parameters used in the implementation.

$\epsilon = 0.5$	Threshold in keypoint matching
$\beta = 3$	Maximum distance of inliers in RANSAC
$N = 100$	Number of RANSAC iterations
$c = 0.3$	Threshold of correlation map
$A = 0.1\%$	Area threshold (as a fraction of total image area) for duplicated regions

To evaluate the performance of the proposed region duplication detection method, we created a set of automatically generated forged images with duplicated and distorted regions. Forged images were generated based on 25 uncompressed PNG true colour images of size 768×512 pixels^2 (Franzen 1999). The main reason for choosing this set of images is that they have not been subjected to any known digital tampering operations. Using these untampered images, we created several sets of forged images with duplicated regions. In our experiments, to test the effect of the sizes of the duplicated regions on the final detection, we used three different block sizes (32×32, 64×64 and 96×96 pixels) corresponding to 0.26%, 1.04% and 2.34% of total image area, respectively.

We then distort the duplicated regions using several types of transforms, with the shift vector chosen randomly in the same image. These affine transforms are representatives of the most frequent manipulations in creating region duplication forgeries provided in most photo-editing tools. Affine transforms between regions are defined with regard to the local coordinate system originating at the geometric center of the corresponding source regions, and implemented with bi-cubic interpolations using MATLAB.

1. *Copy-move*: The duplicated region is translated to the target location without geometric or illumination distortion, corresponding to $T = \begin{pmatrix} 1 & 0 \\ 0 & 1 \end{pmatrix}$.
2. *Rotation*: The duplicated region is rotated with a random angle $\theta \in [0°, 360°)$, corresponding to $T = \begin{pmatrix} \cos\theta & \sin\theta \\ -\sin\theta & \cos\theta \end{pmatrix}$.
3. *Scaling*: The duplicated region is scaled up or down with a random but reasonable scaling factor $s \in [0.8, 2.0]$, corresponding to $T = \begin{pmatrix} s & 0 \\ 0 & s \end{pmatrix}$.
4. *Free-form*: The duplicated region is distorted with a linear transform of a random matrix $T = \begin{pmatrix} t_{11} & t_{12} \\ t_{21} & t_{22} \end{pmatrix}$.
5. *Illumination adjustment*: The affine transform is the same as in the case of copy–move, but all pixels in the duplicated region have intensities modulated to 80% of their original values.

For each of the five types of region duplication and size of duplicated regions, we generated 4 tampered images using each of the 25 images in the Kodak database, resulting in a total 1500 forged images.

To evaluate the performance of the detected duplicated regions in these forged images, we used two quantitative measures to evaluate the performance of our method. Denote Ω as pixels in the true duplicated regions ($\Omega = \Omega_S + \Omega_T$, as shown in Figure 13.3), and $\tilde{\Omega}$ as pixels in the detected duplicated regions, we define:

---

^{2} These images were released by the Kodak Corporation for unrestricted research usage.

1. *Pixel detection accuracy* (PDA) rate: The fraction of pixels in duplicated regions that are correctly identified, as $\text{PDA} = \frac{|\tilde{\Omega} \cap \Omega|}{|\Omega|}$,

2. *Pixel false positive* (PFP) rate: The fraction of pixels in untampered regions that are detected as from duplicated regions, as $\text{PFP} = \frac{|\tilde{\Omega} - \Omega|}{|\tilde{\Omega}|}$, where $|\cdot|$ denotes the cardinality.

A comprehensive performance evaluation of any region duplication detection method must consider both the PDA and the PFP rates. On the one hand, the algorithm should detect as many as possible pixels in the duplicated regions. On the other hand, it should reduce the number of pixels in untampered regions that are detected as from duplications. In our method, PDA/PFP rates can be manipulated by adjusting the correlation threshold c.

The trade-offs between the PDA and PFP rates are completely described with the *receiver-operator characteristics* (ROC) curve. To generate the ROC curves in our experiment, we produced different PDA/PFP rates by changing the threshold in the correlation map c in the range of $0.00 - 0.95$ with step size 0.05. To reduce the effect of random samples, each pair of PDA/PFP rates was computed as the averages over all 100 forged images of each distortion and block size. Figure 13.5 shows the ROC curves

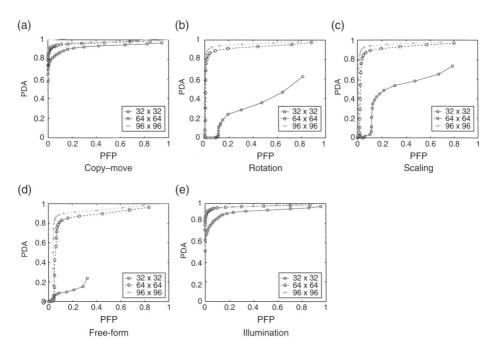

Figure 13.5   ROC curves of different tampering operations and different block sizes. Results are averaged over 100 randomly synthesized forgeries with region duplications.

of PDA/PFP rates corresponding to five distortions and three sizes of the duplicated regions.

These ROC curves demonstrate that for block sizes of 64×64 and 96×96, with a PFP rate around 5%, our method achieves a PDA rate greater than 85%. Such a level of PDA/PFP rates is usually sufficient to identify the duplicated regions visually. On the other hand, there is a clear effect of the size of the duplicated blocks on the detection performance, the larger the block size, the better the overall performance (as the area under the ROC curve is larger). This is to be expected, as larger duplicated regions include more SIFT keypoints, which makes the matching and transform estimations more reliable. Also, there is a difference in performance for different types of region duplication manipulation. In particular, the simple cases of copy–move and illumination distortion are the easiest to detect, while the most difficult case is when the duplicated regions are subject to free-form affine transforms.

Our next experiment addresses the robustness and sensitivity of our method to different levels of lossy (JPEG) compression, and noise levels. Shown in Figure 13.6 are the ROC curves of PDA/PFP for the detection of duplicated regions with rotation, free-form affine transform and illumination distortion under different JPEG QF (quality factor) values (top row) and signal-to-noise-ratios (SNRs) of additive white Gaussian noises (AWGN) (bottom row). The forged images were created with duplicated regions

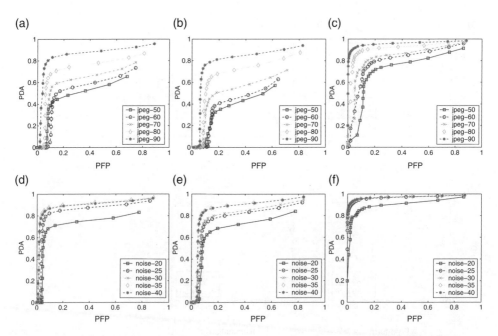

Figure 13.6 ROC curves of PDA and PFP rates for different JPEG QF values (**Top**) and SNRs (**Bottom**) of region duplications with (a,d) rotation, (b,e) free-form transform and (c,f) illumination distortion.

of size 64 × 64 pixels, and the distortions of these duplicated regions were applied as in the previous experiment. The forged images were then converted to JPEG images (Quality Factor = 50, 60, 70, 80, 90) or contaminated with AWGN (SNR = 20, 25, 30, 35, 40 dB).

As shown in the ROC curves in Figure 13.6, the overall detection performance of our method is relatively robust to these degradations. Even with low image qualities, more than 70% of the pixels in duplicated regions can still be detected with less than 20% of PFP rates. In general, the performance tends to decrease for lower image quality. The main reason is that artefacts such as the "blockiness" in low-quality JPEG compression or high-level noise interfere with the SIFT algorithm in detecting keypoints. As less reliable keypoints are available in such cases, the detection performance is strongly affected.

Area-based performance measures such as PDA/PFP rates are useful when we know that the tested image is a forgery, that is $\Omega_S + \Omega_T \neq \emptyset$. Yet, in practice, this is usually not known a priori. In the next set of experiments, we tested the overall image-level detection performance of our method. Specifically, for a forged image, a successful detection occurs when our method detects a duplicated region larger than the area threshold A (see Table 13.1). For an untampered image, a true negative occurs when our

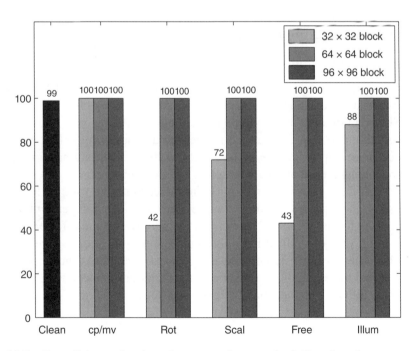

Figure 13.7 Overall image-level performance of our method. For clean images, we report the true negative rates. For forged images of different duplicated region sizes and operations, we report the detection accuracies. Both rates are given in percentages.

method does not detect any duplicated region. Figure 13.7 shows the overall detection performance of our method on the 1500 synthesized images, as measured by the rate of successful detections. Also shown is the true negative rate for the 325 untampered (clean) images.

As these results show, our method correctly identified all forged images with duplicated blocks of size 64×64 or 96×96 pixels, and the average PDA/PFP rates for these detections are 83.5% and 8.8%, respectively. The detection performance for duplicated blocks of size 32×32 pixels are significantly inferior, which corroborates the ROC curves in Figure 13.6. For untampered images, our method achieved high true negative rates (on average 99.08%), as for most untampered images, intrinsically similar regions (e.g. textures and other repetitive structures) usually lack the high similarity resulting from identical duplicates.

Finally we created a set of convincing forgeries based on the untampered image shown in Figure 13.2. Irregular regions are chosen, distorted with rotation, scaling, free-form affine transform, perspective projection, reflection or illumination adjustments and then pasted to target locations. The created forgeries, along with the detection results of our method, are shown in Figure 13.8. As the visual results and the accompanying PDA/PFP rates show, our method can reliably detect these duplicated regions.

13.3   Exposing Splicing Forgery

13.3.1   Problem Definition

In this section, our target is another more popular image tampering technique known as *splicing*, where a forged image is generated by compositing parts from different source images. Shown in Figure 13.9 is a popular example of splicing, which was created by pasting a section (the shark) from one image into another using Adobe Photoshop.[3] With good choices of the source images and careful manual editing, forged images made with splicing are usually hard to detect by untrained eyes.

Most digital images have inherent noise introduced either during their acquisition or by the subsequent processing (e.g. compression). For an untampered photographic image, the noise levels (measured by the noise standard deviation) of different regions across the whole image usually differ only slightly. But with regions spliced from other images with different intrinsic noise levels, or small amounts of noise intentionally added to conceal traces of editing operations, the variations in local image noise levels could become telltale evidence that the image has been tampered.

---

[3] This image was circulated on the Internet in 2002 and rumoured to be a candidate of *Photo of the Year* for the *National Geographic* magazine (National Geographic News 2002). It was later identified as forged as both the original images were found.

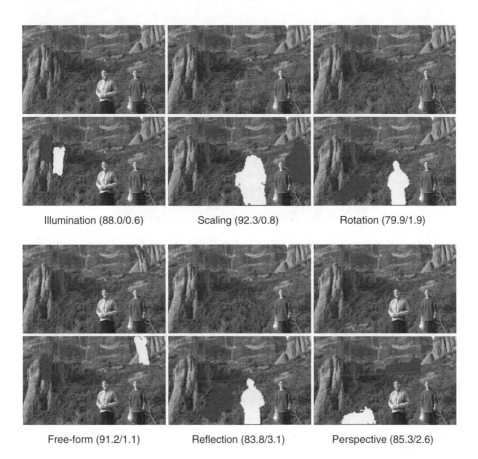

Illumination (88.0/0.6)   Scaling (92.3/0.8)   Rotation (79.9/1.9)

Free-form (91.2/1.1)   Reflection (83.8/3.1)   Perspective (85.3/2.6)

Figure 13.8 Detection results (Bottom rows) of our method for a set of forgeries (Top rows) manually made with Photoshop and different types of distortions based on the untampered image shown in the left panel of the top row of Figure 13.2. In the parentheses are the PDA/PFP rates of the detected duplicated region in percentage. Colours are added to differentiate the regions detected as duplicates of each other.

13.3.2   Related Works

In digital image forensics, image noise has been widely used for source camera identification and image manipulation detection. Related prior work falls roughly into three major categories. We have chosen to focus on reviewing several representative methods for each category as discussed in the following text.

In the first category, noise is used as a distinct feature for camera model identification. In (Lukás *et al.* 2006b), the photo-response nonuniformity (PRNU), which is a unique stochastic characteristic of imaging sensors, is employed as an intrinsic fingerprint to identify the source camera for a given image by pattern correlation. The method is improved in (Chen *et al.* 2008b) for PRNU estimation with fewer training

Figure 13.9 The composite image, which was rumoured to be a candidate of National Geographic's "Photo of the Year" was spliced together from a U.S. Air Force photo taken near San Francisco's Golden Gate Bridge and a photo of a shark from South Africa.

images and further used for image tampering detection. The major limitation of this type of methods is that they depend on the knowledge of specific camera models (see Chapter 9 in Part III for more discussion on this topic).

Recently, another category of methods has been developed, which uses extraction of additional noise features with the aid of supervised learning algorithm. In (Sutcu *et al.* 2007), demosaicing characteristics are combined with PRNU in a two-round learning process to identify camera model. The method is further integrated into a single classification model (Filler *et al.* 2008) using machine learning techniques. This class of techniques was extended to digital scanner identification in (Gou *et al.* 2009; Khanna *et al.* 2007). By extracting the statistical noise features from image denoising operations, wavelet analysis, and neighbourhood predication, another feature-based approach (Gou *et al.* 2007) is proposed to detect image tampering. However, the supervised learning method does not provide the exact extent and location of the tampered regions. Another limitation is that only several specific camera models were examined by the authors of the learning algorithm, while the detection performance on other camera models is unknown.

In the last category, image noise variance is estimated at local image blocks to locate suspicious regions. In (Popescu and Farid 2004b), noise variance is estimated by computing the second and fourth moments at each local image block. But the method assumes that the kurtosis values of the original signal and of the noise are known. Another recently developed method (Mahdian and Saic 2009b) uses a median-based estimator to compute the variance of noise at each image block in a high frequency sub-band of the wavelet transformed image. These image blocks are then merged by examining noise differences between neighbouring blocks to form various homogenous regions. Although this method can locate tampered regions, the detection accuracy and the false positive rate are not extensively evaluated in (Mahdian and Saic 2009b). Meanwhile, the computation time of the proposed algorithm can be high due to the inefficient region merging algorithm.

13.3.3 Proposed Method

In this section, we introduce a blind but effective method (Pan *et al.* 2011, 2012b) to detect image splicing forgery by detecting inconsistencies in local noise variances. Our method estimates local noise variances based on an observation that kurtosis values of natural images in band-pass filtered domains tend to concentrate around a constant value, and is accelerated by the use of dynamic programming.

13.3.3.1 Kurtosis Concentration

For a random variable x, we define its kurtosis as $\kappa = \frac{\tilde{\mu}_4}{(\sigma^2)^2} - 3$, where $\sigma^2 = E[(x - \mu)^2]$ and $\tilde{\mu}_4 = E[(x - \mu)^4]$ are the variance and fourth-order central moment of x, respectively. By this definition, a Gaussian variable has kurtosis value zero.

It has been widely observed that the second-order statistics of natural images are invariant with regard to scale (Burt and Adelson 1981). Figure 13.10 illustrates the kurtosis values (related to the fourth-order cumulant) of the discrete cosine transform (DCT) responses of a natural image selected from the Kodak dataset (Franzen 1999). More specifically, the image is convolved with 63 DCT alternating current (AC) filters to produce the responses, where the kurtosis values of these response signals are computed and sorted. The results suggest that the kurtosis of natural images exhibits near constancy property across different scales.

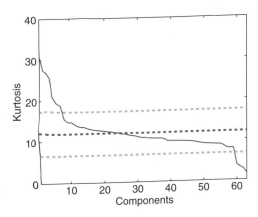

Figure 13.10   The near constancy property of kurtosis of nature image in band-pass filtered domain.

13.3.3.2 Global Noise-Level Estimation

By definition, kurtosis and variance are closely related. Denote $\vec{y} = \vec{x} + \vec{z}$ as the result of contaminating a natural image $\vec{x}$ with white Gaussian noise $\vec{z}$ of zero mean and unknown variance σ^2. Our goal is to estimate σ^2 from the noise corrupted image $\vec{y}$ alone with some general statistical knowledge of $\vec{x}$. We would like to point out that the Gaussian assumption is not as restrictive as it looks, in particular when the image is transformed from pixel domains to band-pass filtered domains such as DCT and wavelet. This is because the non-Gaussian-independent noise in the pixel domain approaches Gaussian after being linearly mixed by the band-pass filters, a direct result of the central limit theorem.

We first transform the image into a band-pass filtered domain. We find that the particular form of the band-pass filter transforms does not affect the algorithm and performance, and use the AC filters from a fixed-point DCT decomposition for its efficiency. Further, if the band-pass filters are normalized to have unit one, the variance of the noise is same as that in the original pixel domain.

For a band-pass filtered domain of K channels (i.e. the responses of image convolved with K different band-pass filters), we denote the kurtosis of the original and the noisy image in the kth channel as κ_k and $\tilde{\kappa}_k$, respectively. We use $\tilde{\sigma}_k^2$ for the variance of the kth channel of the noisy image $\vec{y}$. These statistics are related (Benedict and Soong 1967), as follows:

$$\tilde{\kappa}_k = \kappa_k \left(\frac{\tilde{\sigma}_k^2 - \sigma^2}{\tilde{\sigma}_k^2} \right)^2 . \tag{13.4}$$

Equation 13.4 can be further simplified if we consider the statistical regularity of natural images in the band-pass filtered domains – they tend to have super-Gaussian marginal distributions (Burt and Adelson 1981), or equivalently, have positive kurtosis values ($\kappa_k > 0$). Also, $\tilde{\sigma}_k^2 - \sigma^2$ is positive as it is the variance of the noise free image in the kth channel. Therefore, we can take square root on both sides of Equation 13.4 to yield

$$\sqrt{\tilde{\kappa}_k} = \sqrt{\kappa_k} \left(\frac{\tilde{\sigma}_k^2 - \sigma^2}{\tilde{\sigma}_k^2} \right) . \tag{13.5}$$

Now we take advantage of the kurtosis concentration behaviour of natural images in the band-pass filtered domains (Section 13.3.3.1). In particular, this suggests that the kurtosis of the noise-free natural image $\vec{x}$ across the K band-pass filtered channels can be approximated by a constant, or $\kappa_k \approx \kappa$ ($k = 1, \ldots, K$). We then form an objective function of minimizing the difference of the two sides of Equation 13.5, as follows:

$$\min_{\sqrt{\kappa}, \sigma^2} \sum_{k=1}^{K} \left[\sqrt{\tilde{\kappa}_k} - \sqrt{\kappa} \left(\frac{\tilde{\sigma}_k^2 - \sigma^2}{\tilde{\sigma}_k^2} \right) \right]^2 , \tag{13.6}$$

whose optimal solution provides an estimation to the noise variance. It turns out that this optimal solution is in closed-form, as follows:

$$\sqrt{\kappa} = \frac{\left\langle \sqrt{\tilde{\kappa}_k} \right\rangle_k \left\langle \frac{1}{(\tilde{\sigma}_k^2)^2} \right\rangle_k - \left\langle \frac{\sqrt{\tilde{\kappa}_k}}{\tilde{\sigma}_k^2} \right\rangle_k \left\langle \frac{1}{\tilde{\sigma}_k^2} \right\rangle_k}{\left\langle \frac{1}{(\tilde{\sigma}_k^2)^2} \right\rangle_k - \left\langle \frac{1}{\tilde{\sigma}_k^2} \right\rangle_k^2}$$

(13.7)

$$\sigma^2 = \frac{1}{\left\langle \frac{1}{\tilde{\sigma}_k^2} \right\rangle_k} - \frac{1}{\sqrt{\kappa}} \frac{\left\langle \sqrt{\tilde{\kappa}_k} \right\rangle_k}{\left\langle \frac{1}{\tilde{\sigma}_k^2} \right\rangle_k},$$

where $\langle \cdot \rangle_k$ is a shorthand notation for averaging over the K band-pass filtered channels (Pan *et al.* 2012b). We should point out that Equation 13.7 is key to the extension of the global noise variance estimation to efficient local noise variance estimation.

The global noise estimation algorithm over the image y, as shown in Figure 13.11, can be summarized as the following major steps:

1. *Conversion to DCT domain*: Produce the response image y_i by the convolution of y with each filter i from the 8×8 DCT basis.
2. *Computation on response images*: Compute variance $\sigma_{y_i}^2$ and kurtosis κ_{y_i} for each response image y_i.

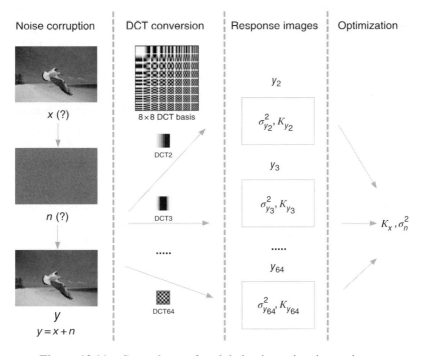

Figure 13.11 General steps for global noise estimation on image.

3. *Noise estimation*: The variance of the added noise and the kurtosis of the original clean image can be estimated by minimizing Equation 13.6 using the closed-form solution Equation 13.7.

13.3.3.3   Local Noise Variance Estimation

The global noise variance estimation method assumes that the noise variance σ^2 is a constant across the image. However, our goal is to expose image splicing by detecting the inconsistencies in local noise variances. In this section, we extend the global noise variance estimator given in Equation 13.7 to an efficient local noise variance estimation method. To this end, we estimate noise variance $\sigma^2(i,j)$ at each pixel location (i,j) using Equation 13.7, with statistics collected from all surrounding rectangular windows of (i,j), $\Omega^k_{(i,j)}, k = 1, \ldots, K$, in all band-pass filtered channels. In doing so, we first express the variance and kurtosis using the raw (un-centred) moments, $\mu_m = \mathrm{E}[x^m]$, as:

$$\sigma^2 = \mu_2 - \mu_1^2$$

$$\kappa = \frac{\mu_4 - 4\mu_3\mu_1 + 6\mu_2\mu_1^2 - 3\mu_1^4}{\mu_2^2 - 2\mu_2\mu_1^2 + \mu_1^4} - 3. \tag{13.8}$$

A naive implementation, applying Equation 13.8 and then Equation 13.7 to each local window $\Omega^k_{(i,j)}$, will lead to an overall running time of $\mathcal{O}(MNK)$, where M and N are the sizes of the image and local window in pixel. This will become inefficient when M is relatively large. On the other hand, we can use *integral image* (Viola and Jones 2002) to accelerate the overall algorithm to a running time of $\mathcal{O}(NK)$. Integral image is a dynamic programming technique for efficient computation of sum values in rectangular regions in an image (or one channel in a band-pass filtered domain). Using integral image, we can efficiently estimate local noise variance in each overlapping local window of an image (Pan *et al.* 2012b).

13.3.4   Performance Analysis

In this section, we evaluate the local noise-level estimation method with a set of experiments and demonstrate the efficacy of the method on detecting image splicing forgery. Our first experiment is based on a set of images with artificially added spatially varying additive white Gaussian noise (AWGN). The top row of Figure 13.12 shows three images corrupted with spatially varying noise. Specifically, the leftmost image is generated with horizontal stripes of AWGN with uniformly increasing peak signal-to-noise ratios (PSNRs) (from 13 dB at the bottom to 20 dB at the top with 1 dB step). The images in the middle and right are generated with AWGN of 20 dB PSNR of annular and checkerboard structures, respectively. The bottom row of Figure 13.12 shows the local noise level estimation of these images, with intensities proportional to the estimated noise level measured as the standard deviations of the AWGN. In the

Figure 13.12   Local noise-level estimation for three different images with different AWGN noise patterns.

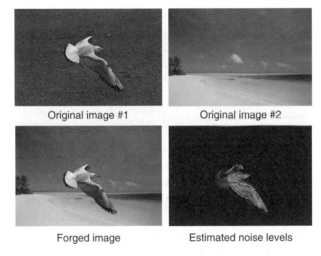

Figure 13.13   Detecting image forgeries with the local noise-level analysis. The two original images were taken with a Canon 400D digital camera with an ISO speed 1600 (top left) and an unknown camera (top right), respectively. All original images are downloaded from Flickr.com.

analysis, we used 63 8×8 random filters, with overlapping sliding windows of 5×5 pixels in each channel for local estimation. As these results show, our method is effective in revealing spatially varying noise levels in these data. Furthermore, the dynamic programming implementation achieved significant speedup – analyzing a 768×512 pixel image on a machine with an Intel CPU of 2.4 GHz and 2 GB memory took about 20 s, while evaluating each sliding window individually took more than 2.8 h.

The local noise-level estimation method can be employed directly for exposing image splicing. Figure 13.13 shows experimental results of using local noise-level estimation to detect image forgeries. We created these forged images by splicing regions from two original images using Photoshop. The two original images were captured with different camera models and ISO speeds, and therefore are expected to have different noise characteristics, which are usually non-Gaussian and not simply additive. Finally, we show the results of running our local noise-level estimation method on these images. Note that the difference in the noise characteristics of the original images and spliced regions are clearly revealed as a result of our analysis.

13.4 Case Studies

In this section, we apply the techniques introduced early in this chapter to solve some real-world problems. Though we have already shown some simple examples created by our own image tampering tools to demonstrate the effectiveness of the introduced methods, it will be more convincing if we can use them to detect image tampering with practical examples. We employ the developed image forgery detection methods to expose several realistic image forgeries which have raised the public's attention. Widely collected from academic articles, news media and public websites, most forged images used in this section were created with known or unknown state-of-the-art tools.

13.4.1 Region Duplication Forgeries

In the actual practice of creating convincing digital forgeries, it is common to pick irregular source regions, apply affine transform and/or illumination distortions to the duplicated region, and process the boundaries of the duplicated regions to better blend it with the surroundings at the target location. With the aid of sophisticated photo-editing software, such as the *healing brush* in Photoshop (Georgiev 2004) and the *smart fill* tool in Image Doctor 2 (Alien Skin Software LLC 2007), forged images can be created with convincing visual appearance using duplicated and distorted regions. And seamless region splicing is an active area of development in Computer Graphics, (Agarwala 2007; Farbman *et al.* 2009; Jia *et al.* 2006; Pérez *et al.* 2003). In this section, we test the introduced region duplication detection method with several convincing digital forgeries made with state-of-the-art image retouching algorithms and tools.

Shown in Figure 13.14 are the results of the region duplication detection method described in Section 13.2.3 on some realistic forgeries. Forged images in the first two rows are generated with the splicing algorithm developed by Farbman *et al.* (2009), which creates a natural transition between duplicated region and the surroundings at the target location. In the "deer" image, the duplicated region is rotated, scaled and mirrored; in the "cherry" image, the duplicated region is rotated and overlapped with the source region. The third row of Figure 13.14 shows a forgery created with the *Smart Fill* tool in the Image Doctor 2 software (Alien Skin Software LLC 2007). The unpublished algorithm used to create this forgery is more sophisticated: instead of using a continuous duplicated region of relatively large size, smaller regions containing mostly textures (sometimes the selected region has a size of less than 20 pixels) are

Figure 13.14 Detection results of the introduced method for a set of more challenging forgeries. The top two rows are based on the technique in (Farbman *et al.* 2009), the bottom row uses images made with the *Smart Fill* tool of Alien Skin Software LLC (2007). (Left): Original image. (Middle) Tampered image. (Right) Detection using the introduced method. Colours are added to differentiate the regions detected by the introduced method as duplicates of each other.

combined and arranged to cover larger region at the target location. This makes the visual detection of the duplicated regions significantly more difficult. It also poses as a challenge to the introduced method, especially due to that the smaller identical regions provide less reliable keypoint matchings. However, the detection result of the introduced method as shown in the right column can still provide considerable clues to draw an inspector's attention for scrutiny.

In Figure 13.15, we further show the detection result of the same method on an alleged forged image that has recently raised the public's attention. The image shown in the left panel appeared on the front pages of several internationally recognized newspapers including *The Los Angeles Times*, *The Financial Times*, and *The Chicago Tribune* as well as several major news websites in 2008 (Times 2008). Shortly after

Figure 13.15 (Left) An image doubted for digital tampering. (Middle) A photograph that is believed to be taken at about the same time. (Right) Detected duplicated regions using the introduced method. Colours are added to differentiate the regions detected by the introduced method as duplicates of each other.

this image was published, doubts were raised regarding its authenticity. It was later confirmed as a tampered image by inspection of photography experts and the appearance of another photograph that was believed to have been taken at about the same time (middle column). Consistent with the analysis of photography experts, the introduced method is able to recover the two major regions that are believed to have been duplicated from other parts of the image.

Finally, we demonstrate the detection capability of the proposed method based on a very interesting case. In November 2011, a photograph titled *Rhein II* by German artist Andreas Gursky was auctioned for $4.3 million at Christie's New York, making it the most expensive photograph ever sold in the world (BBC News 2011). Due to the stunning price, this photo has attracted a lot of attention internationally. Though most related news only focus on the price, very few of them doubt the integrity of the photo, where the artist actually removed some intrusive features such as dog walkers, cyclists and factory building from the original photo. Furthermore, the exact location and extent of the forgery, to the best of our knowledge, have not been revealed yet. Fortunately, the introduced method can successfully identify three pairs of duplicated regions in *Rhein II*.[4]

13.4.2 Splicing Forgeries

Splicing is another more common form of image tampering operation. We test the splicing detection method described in Section 13.3.3 on some images collected from Worth1000.com, a popular image manipulation contest website. Figure 13.16 shows the detection results of several sample images downloaded from this website, where all forged images are spliced by regions from two original images.[5] For each forged image shown in Figure 13.16, the relatively small splicing region has higher local noise level. Note that these forgeries are carefully manipulated and processed and have

---

[4] Due to copyright restrictions, the original photo is only available at Sprüth Magers Berlin London. You can find the detection results in our recent work (Pan 2011).

[5] Due to copyright restrictions, the original forged images are only available at Worth1000.com. You can also find these images in our recent work (Pan *et al.* 2012b).

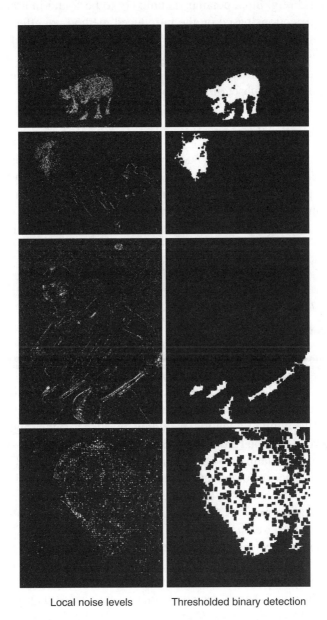

Local noise levels   Thresholded binary detection

Figure 13.16 Detection results of the introduced method on a set of realistic forgeries. Each forged image is spliced by regions from two original images, where the relatively small image region has higher local noise level.

realistic appearance, many can only be exposed based on conceptual knowledge of the physical world (e.g. hippopotamus is unlikely to be found in arctic regions). On the other hand, detections based on the introduced method are effective in revealing spliced regions using only low-level image statistics, albeit occasional false detections or mis-detections also occur. These signal statistics based detection results can be used to automatic examinations over a large set of images, therefore human inspection can be focused on a reduced subset of suspicious images.

13.5 Other Applications

We present in this section some additional important applications based on the techniques introduced early in this chapter. In Section 13.5.1, We first describe a method to detect the splicing forgery in digital audios by the conversion of the local noise-level estimation algorithm to 1D form, which is an extension of the local noise level estimation algorithm proposed in Section 13.3.3. we also show in Section 13.5.2 that the region duplication detection algorithm described in Section 13.2.3 can be employed to expose region duplication forgery in videos.

13.5.1 Detecting Audio Splicing

13.5.1.1 Introduction

Digital audios have become ubiquitous with the popularity of the Internet and portable digital devices such as personal music players and smartphones. Further, rapid developments of low-cost and sophisticated editing software make the modification of audio file much easier even for untrained users. There have been several recent cases of audio forgery drawing the public's attention, including the alleged tampering of the recorded audio of actor Mel Gibbson (ABC News 2010), and the controversy over the authenticity of the audio files claimed to be the voice of Osama Bin Laden (BBC News 2002). The increasing number of forged audios calls for more effective tools for the authentication and forgery detection for digital audios.

In recent years, several active forensic detection methods for audio signals have been developed (Krätzer *et al.* 2007, 2011; Malik and Farid 2010; Yang *et al.* 2008). For instance, acoustic devices, for example microphones, are identified (Krätzer *et al.* 2007, 2011) by extracting background features of audio streams. Similar forensic tools based on the amount of sound reverberation, which uniquely decides the shape and composition of a room where the audio signal was recorded, is proposed in (Malik and Farid 2010). In another work (Yang *et al.* 2008), the digital tampering in MP3 audio data is identified by checking the inconsistency of frame offsets. However, most of these methods assume some knowledge of the recording device or the specific file format. On the other hand, we may obtain more general forgery detection methods using common statistical properties of digital audios, independent of specific recording devices or file formats.

13.5.1.2 Methodology

We describe a new method (Pan *et al.* 2012a) that can be applied to detect splicing forgery in digital audio signals, where sections from one audio are inserted into another audio. This is achieved by detecting abnormal differences in the local noise-levels in an audio signal. The estimation of local noise levels is based on an observed property of audio signals – they tend to have kurtosis close to a constant in the band-pass filtered domain. The variance of noise in the audio signal is estimated by minimizing an objective function that has a closed-form optimal solution. We examine the noise-level inconsistency within the audio file, which can be used to detect the location and length of suspicious audio clips.

Similar to the image noise-level estimation algorithm proposed in Section 13.3.3, let us denote a clean audio signal as $\vec{x}$, and let $\vec{y} = \vec{x} + \vec{z}$ be the result of $\vec{x}$ contaminated by an additive white Gaussian noise (AWGN) $\vec{z}$ of unknown variance σ^2. Our goal is to estimate σ^2 from $\vec{y}$. To this end, we produce the response signal y_k by the convolution of y with the kth order filter from the $1 \times N$ DCT basis. We further denote κ_k, $\tilde{\kappa}_k$ and $\tilde{\sigma}_k^2$ as the kurtosis of x_k and y_k, and the variance of y_k, respectively. The kurtosis of x_k and y_k, and the variance of y_k and σ^2 are related as $\tilde{\kappa}_k = \kappa_k \left(\frac{\tilde{\sigma}_k^2 - \sigma^2}{\tilde{\sigma}_k^2} \right)^2$. Assuming κ_k values are approximately constant across different DCT bands, we can estimate the kurtosis of the audio signal κ and its variance σ^2 by minimizing their squared difference:

$$L \left(\sqrt{\kappa}, \sigma^2 \right) = \sum_{k=1}^{N^2} \left(\sqrt{\tilde{\kappa}_k} - \sqrt{\kappa} + \frac{\sqrt{\kappa}\sigma^2}{\tilde{\sigma}_k^2} \right)^2. \tag{13.9}$$

The global noise-level estimation method can be further extended for the estimation of locally varying noise levels in audio signals. Figure 13.17 demonstrates the proposed local noise-level estimation method, which can be employed for audio forgery detection.

As a demonstration of the effectiveness of the proposed method, we created realistic audio forgeries using audio editing software GoldWave from GoldWave Inc. (2010). More specifically, we selected two source audio clips in WAV format with sampling rate 16 KHz and data rate 256 Kbps from the TIMIT dataset (Garofolo *et al.* 1993). We also selected an original sound track in MP3 format with sampling rate 22 KHz and data rate 24 Kbps for the popular episode *The Marine Biologist* of American television sitcom *Seinfeld*. For fair comparison, we first compressed the TIMIT WAV file to MP3 file with the same sampling and data rates as those of the *Seinfeld* MP3 sound track. We next performed two experiments with insertion or substitution of chosen word segments from the *Seinfeld* episode into the two compressed TIMIT audio clips. In the experiment as shown in Figure 13.18, we substituted two words in another TIMIT audio signal for two word segments cropped from the episode of *Seinfeld*. During the manipulation process, we carefully chose the tampering section so that the resulting sentence is still meaningful. We also tuned both the volume and speed of the splicing

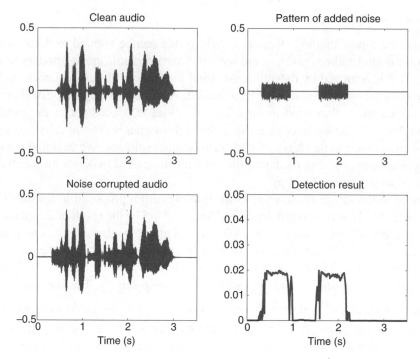

Figure 13.17 Illustration of local noise-level estimation on a noise corrupted audio signal.

audio components using GoldWave Inc. (2010) to make the forged audio signal that one hears more realistic. The detection results demonstrated that the individual splicing segments in the forged audio signal exhibit significant noise-level differences, which provides strong evidence of audio tampering.

13.5.2 Exposing Video Forgery

Video forensics is a relatively new research topic in the area of multimedia forensics. As one very common and simple trick in video tampering, region duplication is the manipulation that deliberately removing undesired objects or events from a video sequence. The forensic technique proposed in Section 13.2.3 can also be extended to detect duplicated region in videos.

Shown in the top row of Figure 13.19 are three frames of a video that has regions duplicated from the same frame. The duplicated regions in each frame were generated with the algorithm proposed in (Farbman *et al.* 2009), with illumination and slight geometric adjustments. Shown in the bottom row of Figure 13.19 are the detected duplicated regions highlighted. Note that the first frame is untampered, and running the introduced method on it shows no false detection. For the other frames, the described method correctly identified duplicated regions in each frame as Figure 13.19 shows.

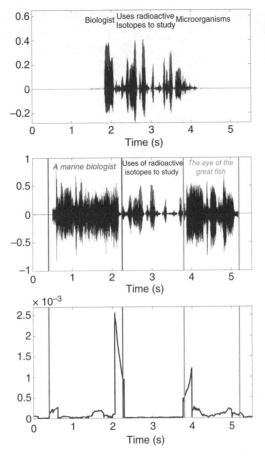

Figure 13.18   Example of substitution tampering, where the words "*biologist*" and "*microorganisms*" in a source TIMIT audio signal (top) are replaced by "*A marine biologist*" and "*the eye of the great fish*" cropped respectively from an episode of *Seinfeld*. In forged audio signal (middle) and detection result (bottom), the locations of the two substituted audio segments are marked between the two pairs of vertical lines, respectively.

13.6   Summary

The past decade has witnessed remarkable growth in our ability to capture, manipulate and distribute digital audios, images and videos. The average user today has access to powerful computers, high-quality digital audio recording devices and digital photo/video cameras, fast broadband network and sophisticated media-editing software. While developments in digital technology have in many ways made our lives convenient, they have also introduced many new challenges. In particular, multimedia products in digital formats are much more vulnerable to malicious tampering compared to their non-digital counterparts. These digital forgeries, by misleading our perception, are having an increasingly negative social impact.

Figure 13.19 Detection results (bottom row) of the introduced method for region duplication forgery in a video sequence (top row). The original and the tampered videos are courtesy of the authors of (Farbman *et al.* 2009).

We have presented, in this chapter, a set of effective and efficient methods to address these new challenges. In the absence of device-dependent protection mechanisms such as digital watermarking or signature, we make the assumption that most digital manipulations will disturb some statistical property of a natural image. The computer-generated forgeries can be revealed by detecting these abnormalities with the statistical analysis of data information from images. More specifically, we focus on solving two common digital image forensic problems: (i) detecting region duplication and (ii) exposing splicing forgery. First of all, a recent forensic method is introduced to locate the duplicated image regions that have undergone geometric and illumination adjustments. The duplicated regions are matched using image features carrying distinct statistic information of image content. Second, we describe a closed-form solution to estimate local noise levels in images based on a novel statistical theorem on kurtosis concentration. Exposing image splicing forgery is achieved through the identification of image regions with abnormal noise levels. Finally, the proposed image forgery detection algorithms are naturally extended to two other important digital forensic areas: (i) detecting audio splicing and (ii) exposing video forgery.

We summarize here the major contributions of each introduced method.

- *Detecting region duplication*: Region duplication is a simple and effective operation to create digital image forgeries, where a continuous portion of pixels in an image, after possible geometric and illumination adjustments, is copied and pasted to a different location in the same image. Most existing region duplication detection methods are based on directly matching blocks of image pixels or transform coefficients, and are not effective when the duplicated regions have geometric or illumination distortions. We describe an efficient region duplication detection method that is robust to distortions of the duplicated regions. This method starts by estimating the transform between matched SIFT keypoints, which are insensitive to geometric and illumination distortions, and then finds all pixels within the duplicated regions after discounting the estimated transforms. The introduced method shows effective detection results on an automatically synthesized forgery image database with duplicated and distorted regions. We further demonstrate its practical performance with several challenging forged images created with state-of-the-art image editing tools.

- *Exposing splicing forgery*: Another more common image tampering operation is known as splicing, where a region of one image is pasted into another image, with the aim of changing the content of original image. Digital images typically have intrinsic noise introduced during the capturing or post-processing. Spliced images from different origins tend to have different intrinsic noise levels. Based on the kurtosis concentration property of natural images in band-pass filtered domains, we describe an effective algorithm to estimate image noise variance. The proposed method can be further extended to efficient estimation of local noise variances, which is applied to the detection and the locating of spliced regions in digital images.

- *Other applications*: As a nature extension of the image splicing detection technique, the local noise levels in audios are estimated with a 1D algorithm derived from its counterpart in the 2D image domain. One advantage of the proposed method is that it requires no specific knowledge of the recording devices or the format of audios. Second, it is sufficiently fast for forgery detection on a large repository of audio files. In addition, we also show that the region duplication detection method for images can be applied to the detection of forgeries in videos.

In conclusion, forensic tools aiming to verify the integrity of the digital images are in high demand and have hence drawn significant attention from academia, government and industry. Though some available forensic tools can achieve high performance on exposing one or more digital tampering attacks, we believe that an important direction of digital image forensics is to improve the detection performance and to increase the range of the forensic applications by combining all the current available forensic techniques.

We would like to point out that as most techniques in digital image forensics, there exists counter-measures to the current state-of-the-art forgery detection methods (Gloe *et al.* 2007) (see Chapter 16 in Part IV for more discussion on this topic). Tomorrow's technology will definitely allow more complicated digital manipulation operations and challenge the capabilities of current forensic tools. As the race between the forger and the forensic expert continues to escalate, our goal is not to completely eradicate or prevent digital tampering, but to make it more time-consuming and painful!

References

Agarwala A 2007 Efficient gradient-domain compositing using quadtrees. *ACM SIGGRAPH*, San Diego, CA.

Alien Skin Software LLC 2007 *Image Doctor 2: Restore, Retouch, Remove and Repair*. www.alienskin.com (Accessed 23 February 2015).

Amerini I, Ballan L, Caldelli R, Del Bimbo A and Serra G 2010 Geometric tampering estimation by means of a sift-based forensic analysis. *Proceedings of IEEE International Conference on Acoustics, Speech, and Signal Processing (ICASSP)*. IEEE Computer Society, Dallas, TX.

Ardizzone E and Mazzola G 2009 Detection of duplicated regions in tampered digital images by bit-plane analysis. *ICIAP '09: Proceedings of the 15th International Conference on Image Analysis and Processing* Springer-Verlag, Berlin, Germany, pp. 893–901.

Bayram S, Avcibas I, Sankur B and Memon ND 2006 Image manipulation detection. *Journal Electronic Imaging* **15**(4), 041102.

Bayram S, Taha Sencar H and Memon N 2009 An efficient and robust method for detecting copy-move forgery. *IEEE International Conference on Acoustics, Speech and Signal Processing*, Washington, DC.

Beis J and Lowe D 1997 Shape indexing using approximate nearest-neighbour search in high-dimensional spaces. *CVPR* San Juan, Puerto Rico, pp. 1000–1006.

Benedict TR and Soong TT 1967 The joint estimation of signal and noise from the sum envelope. *IEEE Transactions on Information Theory* **13**, 447–454.

Bravo M and Farid H 2004 Recognizing and segmenting objects in clutter. *Vision Research* **44**(4), 385–396.

Bravo-Solorio S and Nandi AK 2009 Passive forensic method for detecting duplicated regions affected by reflection, rotation and scaling. *17th European Signal Processing Conference*. Glasgow, Scotland, pp. 824–828.

Burt P and Adelson E 1981 The Laplacian pyramid as a compact image code. *IEEE Transactions on Communication* **31**(4), 532–540.

Chen M, Fridrich J, Goljan M and Lukas J 2008a Determining image origin and integrity using sensor noise. *IEEE Transactions on Information Forensics and Security* **3**(1), 74–90.

Chen M, Fridrich JJ, Goljan M and Lukás J 2008b Determining image origin and integrity using sensor noise. *IEEE Transactions on Information Forensics and Security* **3**(1), 74–90.

Christlein V, Riess C, Jordan J, Riess C and Angelopoulou E 2012 An evaluation of popular copy-move forgery detection approaches. *IEEE Transactions on Information Forensics and Security* **7**(6), 1841–1854.

Cox IJ, Miller ML, Bloom JA, Fridrich J and Kalker T 2008 *Digital Watermarking and Steganography*. Elsevier, Amsterdam, the Netherlands.

Farbman Z, Hoffer G, Lipman Y, Cohen-Or D and Lischinski D 2009 Coordinates for instant image cloning. *ACM Transaction on graphics* **28**(3), 1–9.

Farid H 2009a Photo fakery and forensics. *Advances in Computers*. **77**, 1–55.

Farid H 2009b A survey of image forgery detection. *IEEE Signal Processing Magazine* **2**(26), 16–25.

Farid H and Lyu S 2003 Higher-order wavelet statistics and their application to digital forensics. *IEEE Workshop on Statistical Analysis in Computer Vision (in conjunction with CVPR)*, Madison, WI.

Filler T, Fridrich JJ and Goljan M 2008 Using sensor pattern noise for camera model identification. *IEEE International Conference on Image Processing*, San Diego, CA.

Fischler MA and Bolles RC 1981 Random sample consensus: A paradigm for model fitting with applications to image analysis and automated cartography. *Communications on ACM* **24**(6), 381–395.

Franzen R 1999 Kodak lossless true colour image suite source. http://r0k.us/graphics/kodak. Accessed on 31 January 2015.

Fridrich J, Soukal D and Lukas J 2003 Detection of copy-move forgery in digital images *Digital Forensic Research Workshop*, Cleveland, OH.

Friedmann GL 1993 The trustworthy digital camera: Restoring credibility to the photographic image. *IEEE Transactions on Consumer Electronics* **39**(4), 905–910.

Fu D, Shi YQ and Su W 2007 A generalized benford law for JPEG coefficients and its applications in image forensics *Proceedings of SPIE Security, Steganography, and Watermarking of Multimedia Contents IX*, San Jose, CA.

Garofolo JS, Lamel LF, Fisher WM, Fiscus JG, Pallett DS and Dahlgren NL 1993 DARPA TIMIT acoustic-phonetic continuous speech corpus CD-ROM. National Institute of Standards and Technology, Gaithersburg, MD. NISTIR 4930.

Georgiev T 2004 Photoshop healing brush: A tool for seam-less cloning. *Workshop on Applications of Computer Vission (ECCV)*, Praque, Czech Republic.

Gloe T, Kirchner M, Winkler A and Böhme R 2007 Can we trust digital image forensics? *Proceedings of the 15th international conference on Multimedia (MULTIMEDIA '07)*. ACM, New York, pp. 78–86.

GoldWave Inc. 2010 *GoldWave v5.58*. www.goldwave.com (Accessed 23 February 2015).

Gou H, Swaminathan A and Wu M 2007 Noise features for image tampering detection and steganalysis. *IEEE International Conference on Image Processing*, San Antonio, TX.

Gou H, Swaminathan A and Wu M 2009 Intrinsic sensor noise features for forensic analysis on scanners and scanned images. *IEEE Transactions on Information Forensics and Security* **4**(3), 476–491.

Hartley RI and Zisserman A 2004 *Multiple View Geometry in Computer Vision*, 2nd edn. Cambridge University Press, Cambridge, UK.

Huang H, Guo W and Zhang Y 2008 Detection of copy-move forgery in digital images using SIFT algorithm. *IEEE Pacific-Asia Workshop on Computational Intelligence and Industrial Application*, Wuhan, Chaina.

Jia J, Sun J, Tang CK and Shum HY 2006 Drag- and-drop pasting *ACM SIGGRAPH*, Boston, MA.

Johnson M and Farid H 2005 Exposing digital forgeries by detecting inconsistencies in lighting. *ACM Multimedia and Security Workshop*, New York.

Johnson M and Farid H 2007a Detecting photographic composites of people. *6th International Workshop on Digital Watermarking*, Guangzhou, China.

Johnson M and Farid H 2007b Exposing digital forgeries in complex lighting environments. *IEEE Transactions on Information Forensics and Security* **3**(2), 450–461.

Johnson M and Farid H 2007c Exposing digital forgeries through specular highlights on the eye. *9th International Workshop on Information Hiding*, Saint Malo, France.

Khanna N, Mikkilineni AK, Chiu GTC, Allebach JP and Delp EJ 2007 Scanner identification using sensor pattern noise. *Proceedings of the SPIE International Conference on Security, Steganography, and Watermarking of Multimedia Contents IX*, San Jose, CA.

Krätzer C, Oermann A, Dittmann J and Lang A 2007 Digital audio forensics: a first practical evaluation on microphone and environment classification. *ACM MM&Sec*, Dallas, TX.

Krätzer C, Qian K, Schott M and Dittmann J 2011 A context model for microphone forensics and its application in evaluations. *Proceedings of SPIE Media Watermarking, Security, and Forensics III*, San Francisco, CA.

Langille A and Gong M 2006 An efficient match-based duplication detection algorithm. *CRV '06*. IEEE Computer Society, Washington, DC, p. 64.

Li G, Wu Q, Tu D and Sun S 2007 A sorted neighborhood approach for detecting duplicated regions in image forgeries based on DWT and SVD. *ICME*, Beying, China, pp. 1750–1753.

Lin Z, Wang R, Tang X and Shum H 2005 Detecting doctored images using camera response normality and consistency. *CVPR*, San Diego, CA.

Lin HJ, Wang CW and Kao YT 2009 Fast copy-move forgery detection. *WSEAS Transactions on Signal Processing* **5**(5), 188–197.

Lindeberg T 1994 *Scale-Space Theory in Computer Vision*. Kluwer Academic Publishers, Dordrecht, the Netherlands.

Lowe D 2004 Distinctive image features from scale-invariant keypoints. *IJCV* **60**(2), 91–110.

Lukas J, Fridrich J. and Goljan M 2006a Detecting digital image forgeries using sensor pattern noise. *Proceedings of SPIE Security, Steganography, and Watermarking of Multimedia Contents VIII*, San Jose, CA.

Lukás J, Fridrich JJ and Goljan M 2006b Digital camera identification from sensor pattern noise. *IEEE Transactions on Information Forensics and Security* **1**(2), 205–214.

Lyu S and Farid H 2005 How realistic is photorealistic? *IEEE Transactions on Signal Processing* **53**(2), 845–850.

Lyu S and Farid H 2006 Steganalysis using higher-order image statistics. *IEEE Transactions on Information Forensics and Security* **1**(1), 111–119.

Mahdian B and Saic S 2007 Detection of copy-move forgery using a method based on blur moment invariants *Forensic Science International* **17**(2–3) 180–189.

Mahdian B and Saic S 2008 Blind authentication using periodic properties of interpolation. *IEEE Transactions on Information Forensics and Security* **3**(3), 529–538.

Mahdian B and Saic S 2009a Detection and description of geometrically transformed digital images. *Proceedings of SPIE Media Forensics and Security*, San Jose, Ca.

Mahdian B and Saic S 2009b Using noise inconsistencies for blind image forensics. *Image and Vision Computing* **27**(10), 1497–1503.

Malik H and Farid H 2010 Audio forensics from acoustic reverberation. *ICASSP*, Dallas, TX.

Myna AN, Venkateshmurthy MG and Patil CG 2007 Detection of region duplication forgery in digital images using wavelets and log-polar mapping. *ICCIMA '07: Proceedings of the International Conference on Computational Intelligence and Multimedia Applications*. IEEE Computer Society, Washington, DC, pp. 371–377.

BBC News 2002 Bin Laden Tape "Not Genuine". http://news.bbc.co.uk/2/hi/middle_east/2526309.stm (Accessed 23 February 2015).

National Geographic News 2002 Shark "Photo of the Year" is E-Mail Hoax. http://news.nationalgeographic.com/news/2002/08/0815_020815_photooftheyear.html (Accessed 23 February 2015).

ABC News 2010 Did Someone Mess with Mel Gibson's Audio Recordings? http://abcnews.go.com/Entertainment/mel-gibsons-rants-messed/story?id=11169736 (Accessed 23 February 2015).

BBC News 2011 Andreas Gursky's Rhein II Sets Photo Record http://www.bbc.co.uk/news/entertainment-arts-15689652 (Accessed 23 February 2015).

Pan X 2011 *Digital Forensics Using Local Signal Statistics*. PhD thesis, State University of New York at Albany, Albany, NY.

Pan X and Lyu S 2009 Exposing doctored images by detecting duplicated and distorted regions. Technical Report SUNYA-CS-09-02. Department of Computer Science, SUNY Albany, Albany, NY.

Pan X and Lyu S 2010a Detecting image region duplication using SIFT features. The 35th *IEEE International Conference on Acoustics, Speech, and Signal Processing (ICASSP)*, Dallas, TX.

Pan X and Lyu S 2010b Region duplication detection using image feature matching. *IEEE Transactions on Information Forensics and Security* **5**(4), 857–867.

Pan X, Zhang X and Lyu S 2011 Exposing image forgery with blind noise estimation. *The 13th ACM Workshop on Multimedia and Security*, Buffalo, NY.

Pan X, Zhang X and Lyu S 2012a Detecting splicing in digital audios using local noise level estimation. *The 37th IEEE International Conference on Acoustics, Speech and Signal Processing (ICASSP)*, Kyoto, Japan.

Pan X, Zhang X and Lyu S 2012b Exposing image splicing with inconsistent local noise variances. *The 4th IEEE International Conference on Computational Photography (ICCP)*, Seattle, WA.

Pérez P, Gangnet M and Blake A 2003 Poisson image editing. *ACM SIGGRAPH*, San Diego, CA.

Popescu A and Farid H 2004a Exposing digital forgeries by detecting duplicated image regions. Technical Report TR2004-515. Department of Computer Science, Dartmouth College, Hanover, NH.

Popescu A and Farid H 2004b Statistical tools for digital forensics. *6th International Workshop on Information Hiding*, Toronto, Ontorio, Canada.

Popescu A and Farid H 2005a Exposing digital forgeries by detecting traces of re-sampling. *IEEE Transactions on Signal Processing* **53**(2), 758–767.

Popescu A and Farid H 2005b Exposing digital forgeries in colour filter array interpolated images. *IEEE Transactions on Signal Processing* **53**(10), 3948–3959.

Sutcu Y, Bayram S, Sencar HT and Memon ND 2007 Improvements on sensor noise based source camera identification. *IEEE International Conference on Multimedia and Expo*, Beijing, China.

Swaminathan A, Wu M and Liu K 2008 Digital image forensics via intrinsic fingerprints. *IEEE Transactions on Information Forensics and Security* **3**(1), 101–117.

Times NY 2008 In an Iranian Image, a Missile Too Many. http://thelede.blogs.nytimes.com/2008/07/10/in-an-iranian-image-a-missile-too-many (Accessed 23 February 2015).

Viola P and Jones M 2002 Robust real-time object detection. *International Journal of Computer Vision* **57**(2), 137–154.

Wang W, Dong J and Tan T 2009 A survey of passive image tampering detection *IWDW '09: Proceedings of the 8th International Workshop on Digital Watermarking*. Springer-Verlag, Berlin, Germany, pp. 308–322.

Yang QC and Huang CL 2009 Copy-move forgery detection in digital image. *PCM '09: Proceedings of the 10th Pacific Rim Conference on Multimedia*. Springer-Verlag, Berlin, Germany, pp. 816–825.

Yang R, Qu Z and Huang J 2008 Detecting digital audio forgeries by checking frame offsets. *ACM MM&Sec*, Oxford, UK.

14

Camera-Based Image Forgery Detection

Hagit Hel-Or and Ido Yerushalmy
Department of Computer Science, University of Haifa, Haifa, Israel

14.1 Introduction

Given the ubiquitous and influential nature of digital media and, specifically, digital images, it is natural that the notion of image forgery and its detection becomes of significant importance. Image authentication and detection of forgery typically aim to extract unique identifiers of the image and/or acquisition devices or find irregular characteristics of an image in order to define a reliable measure of suspected forgery. In many cases, these algorithms are even able to point out the suspected region and forgery type. To place this in context, forgery detection and image authentication can be considered as based on three levels of assumptions:

1. Rules and models of the physics of the scene
2. The inherent characteristics of the acquisition system itself, namely, the camera components and imaging pipeline used to acquire the images
3. Statistics of natural images

The first level of assumptions relies on laws of physics and nature being preserved under their projection into the image. Inconsistencies and breaking of these rules form a basis for forgery detection. Inconsistencies in size of objects in the image form a basis for forgery detection (Johnson and Farid 2007a; Wu *et al.* 2010; Zhang *et al.* 2010). Additional forgery detection methods have been developed based on inconsistency of

---

Handbook of Digital Forensics of Multimedia Data and Devices, First Edition.
Edited by Anthony T.S. Ho and Shujun Li.

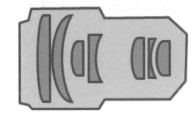

Figure 14.1 The camera lens system captures and redirects light into the camera. The system is typically composed of a set of interacting optical lenses.

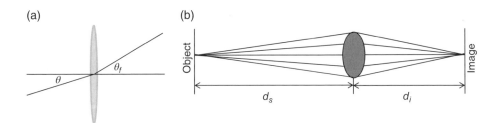

Figure 14.2 Lens equations. (a) Snell's law relates the angle of incidence with the angle of refraction. (b) The Lensmaker equation relates the lens strength, the distance of object to lens and the distance of lens to image plane.

Light falling on a lens refracts to a degree dependent on lens power (which in turn is dependent on lens material and surface curvature). Snell's law explains the refraction of light on the boundary between two media (Figure 14.2a):

$$n * \sin(\theta) = n_f * \sin(\theta_f), \tag{14.1}$$

where θ is the angle of incidence, θ_f the angle of refraction and n and n_f the refractive indices of the two media, in our case air and the lens material.

The camera and lens system are designed so that objects at specific distances (determined by the focusing system) will converge into a sharp image representation on the sensor plane. Lens positioning relative to the sensor plane and lens power determine the distance of objects that will be in focus. This is given by the Lensmaker equation (see Figure 14.2b):

$$\frac{1}{d_s} + \frac{1}{d_i} = \frac{1}{f}$$

where d_s and d_i are the distance from the lens to the object and from the lens to the image plane, respectively, and f is the focal length representing the lens power. The camera motor determines the position of the lens (or lenses) relative to the sensor plane. Improper adjustment of lens position relative to the sensor plane results in a defocused image. For camera optic systems with more than one lens, these equations are more complex.

The quality of the end image strongly depends on the quality of the camera optics and specifically the quality of the lenses themselves. Imperfection in the surface curvature and in the lens material may strongly affect the resulting end image. These affects include geometric aberrations (also called monochromatic aberrations), colour aberrations and lighting flow aberrations.

14.2.1.1 Geometric Aberrations

When lens surface curvature is imperfect, either by design or due to poor production, geometric aberrations occur (Bilissi and Langford 2007); these include (Figures 14.3 and 14.4):

Spherical aberration – Arises due to a spherical lens surface rather than the optimal aspherical surface (which is more difficult and expensive to produce) that provides

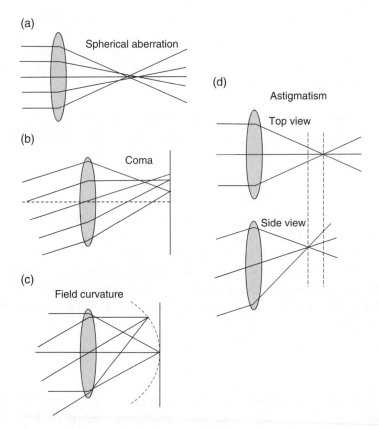

Figure 14.3 Lens geometric distortions. (a) Spherical Aberration. (b) Coma. (c) Field curvature (d) Astigmatism.

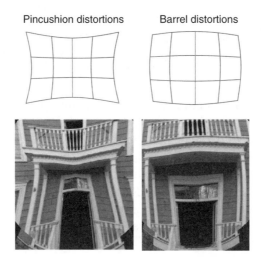

Figure 14.4   Radial distortions: (Left) pincushion distortion (Right) barrel distortion.

for good focusing. Under spherical aberrations, rays of light originating from a scene point on or near the lens axis do not converge into a sharp point resulting in a blurred end image.

Coma – Light rays originating from off-axis objects are refracted inconsistently at the lens centre and lens periphery causing a non-symmetric comet-shaped blur.

Astigmatism – Light rays from off-centred objects are refracted inconsistently for different angles and inclinations from the lens optical axis. The resulting blur is asymmetric and scales with distance from the optical axis.

Field curvature – Due to the lens spherical structure and due to the Lensmaker equation (Eq. 14.2), planar objects positioned normal to the optical axis have increased object distance at points of the object distal from the optic axis and thus have their image focused on a plane closer to the lens than points of the object closer to the optic axis. Thus, the planar object cannot be brought into focus on the camera's typically flat image plane.

Radial distortion (pincushion and barrel distortions) – Due to similar reasons as above for field curvature, radial distortion is due to change of magnification along the radial direction from the lens axis. Rather than blur, as a function of off-axis distance as in field curvature, radial distortion is characterized by off-axis straight lines being projected into curved lines in the end image. Increased magnification with radial distance produces the pincushion distortion, while decreased magnification produces the barrel distortion.

14.2.1.2   Chromatic Aberrations

Chromatic aberrations are due to wavelength-dependent effects of the lens on the end image and are characterized by chromatic (colour) artefacts. The materials used in lens production are characterized in that the lens has a different refractive index for different spectral wavelength. Thus, following Snell's law (Eq. 14.1), a single polychromatic ray of light that enters a camera lens is refracted differently per spectral wavelength, causing focusing at different planes or on different points on the same plane. Shorter wavelength light (blue and violet) has a larger refractive index and so bends more and thus converges on a plane closer to the lens than the middle wavelength light (yellow and green), whereas long wavelength light has a smaller refractive index and so focuses on a plane more distal to the lens than the middle and short wavelength lights (see Figure 14.5a). This causes chromatic aberration in the end image which is visible as colour fringes especially at high contrast edges in the image (see Figure 14.6). Although achromatic lenses (achromatic doublet) and other complex lens systems can compensate for some of the inconsistencies in refraction across different wavelengths, they do not overcome the chromatic aberration completely.

There are two basic types of chromatic aberrations (see Figure 14.5):

Longitudinal (axial) – Polychromatic light originating from objects on or near the lens optic axis focuses at different planes along the optic axis according to wavelength. This results in different levels of blur for the colour channels of the end image. The effect is mainly seen in the centre regions of the image.

Lateral (transverse) – Polychromatic light originating from off-centered objects has different wavelengths of light focused in the same focal plane but at different positions on the focal plane. The effect can be viewed as a magnification of the red

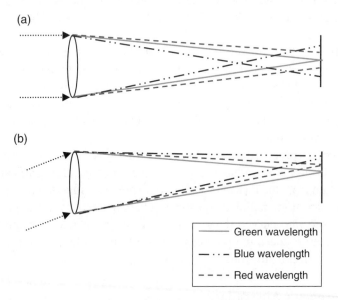

Figure 14.5 Chromatic aberrations: (a) Axial aberration occurs when different wavelengths converge at different depths from the lens. (b) Lateral aberration occurs when different wavelengths converge at different points on the image plane.

Figure 14.6   Chromatic aberration can be seen as the bright blue stripes along the edges of the building (Image from http://en.wikipedia.org/wiki/File:Chromatic_aberration_ (comparison).jpg.)

(a)   (b)

Figure 14.7   Lighting flow aberrations. (a) Vignetting. (b) Light flare.

and blue image channels compared to the green channel of the end image resulting in colour fringes at high contrast edges in the image. The effect is weak in the centre regions of the image and increases with distance from image centre.

14.2.1.3   Lighting Flow Aberrations

Lighting flow aberrations arise when complex systems of lenses are used. Light rays entering the camera optic system interact with the different lenses and may affect the amount of light arriving at different locations on the imaging plane and thus in the end image. We note two types of lighting flow aberrations (see Figure 14.7):

Vignetting – This is the effect in which less light reaches the periphery of the image plane compared to its centre, resulting in a darkening of the edges of the end image. Light rays entering a complex camera lens system may be blocked by filters and different parts of each lens in the optic system. Additionally, rays arriving at oblique angles have a natural light drop-off which increases with angle. Wide-angle lenses are specifically prone to vignetting.

Light flare (lens flare) – Light from a very bright light source, either visible in the scene or immediately outside the camera view, often causes light flare which is characterized by light speckles in the end image or by haze covering regions of the end image. This effect is caused by light entering the complex camera optics and reflecting and scattering internally between the lenses and filters of the system before exiting and impinging on the image plane.

The effects and aberrations described here appear to lesser extent in high-end cameras with high-quality optics. Although described here as due to optics of the camera, many of these affects interact with effects due to the sensors of the camera (see following text).

For additional information on camera optics and optics in general, see Kingslake (1992), Hecht (2002) and Ray (2002).

14.2.2   Sensors

The sensor system of the camera converts physical entities, namely, photons, into a corresponding electrical signal. An image sensor is a grid of silicon-based photodiodes, each of which has the ability to capture photons, convert and store their equivalent in electric charge Figure 14.8). Within the camera, the sensor is positioned at the focal plane of the camera optics, thus capturing the light intensity of the image scene. The electric charge produced by the sensor is directly proportional to the light intensity focused at each sensor unit (pixel) and is ultimately represented as a single image pixel value.

Two main types of image sensors are used in cameras: the CCD and the CMOS. The CCD sensors are characterized by the method of electric charge readout, in which electric charge is shifted along the columns of sensor units to be outputted into an amplifier and then an analogue-to-digital converter that maps the charge of each sensor unit into a digital value.

CCD   CMOS sensor

Figure 14.8   Camera sensor system converts photons into a corresponding electrical signal. Image sensors are in the form of a grid of silicon-based photodiodes.

The CMOS sensor is an active pixel sensor in which every sensor unit (pixel) is part of an electric circuit. This allows amplification and analogue-to-digital conversion to be performed at each sensor unit individually.

Due to the manufacturing process, CMOS sensors are less expensive than CCD sensors, and they consume less power due to the efficiency of their circuitry. However, CMOS sensors have circuitry adjacent to each sensor unit, obscuring and hindering light path to the photodiode. Thus, CMOS sensors traditionally tend to have lower light sensitivity than CCD sensors and higher noise levels as well. Though CMOS sensor quality has greatly improved over the years and is now on par with the more mature CCD sensors that have been around longer, the CCD sensors still tend to have higher quality and more pixels. CCD sensors are often associated with high-end cameras, whereas CMOS sensors with the less expensive low-end cameras.

The photodiodes of the sensor unit respond to photons independent of the spectral wavelength. The end image should be trichromatic, and thus, sensor absorption must be directed to produce values for the three colour channels of the end image. Colour filters are placed over the sensor units to filter the incoming light and produce sensor responses to certain spectral wavelengths; a single filter is used per sensor unit. Typically, three or four different filters are used in a fixed pattern across the sensor array producing a **colour filter array (CFA)**. Camera systems differ in the CFA pattern used and in the spectral sensitivities of the filters. Figure 14.9 shows a few CFA patterns commonly used in cameras. The output value per sensor unit is per a single spectral band or channel, defined by the filter covering the unit. To produce the end image, the two missing spectral bands per unit must be estimated. This process is called demosaicing and is part of the image pipeline of the camera as discussed in Section 14.2.3.

A characterization of the camera sensors is in the noise and the aberrations they introduce into the final end image. Noise is dependent on sensor size, well capacity, fill factor, quantum efficiency and other factors. Noise may be temporal, varying with each image capture and per pixel, or may be of a fixed pattern type which remains consistent across image captures. Some of the major noise types include:

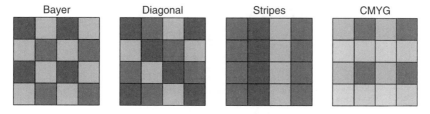

Figure 14.9   Colour filter array (CFA). Colour filters are placed over the sensor units to filter the incoming light and produce sensor responses to specific spectral wavelengths. Typically, three or four different filters are used in a fixed pattern across the sensor array producing a CFA.

Dark noise – Sensor units are sensitive to thermal conditions. In addition to the response to photons, sensor units may also register a charge from thermal electrons known as 'dark current', producing additive charge and consequently noise in the pixel values of the end image. Dark noise is a fixed pattern noise, consistent over image captures. Subtracting a dark image (image captured without shutter opening) from the captured image can reduce the dark noise; however, since dark noise increases with exposure time and with temperature increase, removal is never complete.

Readout noise – Inconsistencies in the readout circuitry of each sensor unit (specifically in a CMOS sensor) introduces variability in the readout value per unit. This noise varies per readout and per pixel and so cannot be removed. Readout noise can be reduced with better performing circuitry.

Reset noise – Following readout, sensors must be restored to initial state (of zero charge). Similar to readout noise, the reset process and circuitry may not be consistent and may produce stray charge after reset, thus introducing noise in the end image. This noise varies per reset and per pixel and so cannot be removed. Reset noise can be reduced with better performing circuitry.

Shot noise (photon noise) – Due to the stochastic nature of photons, the number of photons arriving at a sensor unit varies between image acquisitions even on the exact same scene. The shot noise of a pixel is the difference between the actual photon count and the expected mean at a given intensity. The variable number of photons reaching the sensor is described by the Poisson distribution in which the mean equals the variance of the distribution. This implies an increase in noise with higher intensities; however, the signal-to-noise ratio (SNR) (= ratio of mean to standard deviation) is actually higher at high intensities. Thus, shot noise is more dominant (and visible) in low-intensity images than at high intensities (see Figure 14.10).

Hot pixels – The photodiodes in each sensor unit may respond slightly differently to light. This may be due to errors in the production process, charge leakage between photodiodes and more. This results in individual pixels that appear darker or brighter within an otherwise uniform intensity region of the image. This effect is enhanced

Figure 14.10 Shot noise is visible in images of low scene illumination. Noise appears as speckled bright and dark pixels around the local mean intensity.

when image is captured under long exposure. Since the pixel irregularities are consistent across image captures, the effect can be reduced using image processing techniques (e.g. within the camera processing unit).

Fixed pattern noise – Mismatch and inconsistencies in circuit component size and performance may cause constant drift/bias in senor unit output. The noise may appear as variations of pixel value at a single pixel, along a column of pixels or in blocks of pixels. This noise is constant over image acquisitions and so can be reduced using subtraction of noise image.

Blooming – When sensor units are oversaturated, charge may 'spill over' to neighbouring units. This produces incorrect increased intensities at neighbouring pixels. This type of noise is expressed in the end image as a region effect similar to the effect of glare and light flare (see preceding text) or can be expressed locally as a colour bleeding effect at high contrast edges. The latter is due to neighbouring pixel values representing different colour channels (due to the CFA; see preceding text). Advanced circuitry and isolation mechanisms may control or limit blooming effects.

For additional information on camera sensors, see Holst (1998), Janesick (2001) and Nakamura (2005).

14.2.3   Image Processing Pipeline

The image processing unit within every camera transforms the sensor outputs into the final digital image. The image processing pipeline largely varies between cameras as well as within cameras for different camera settings. There are numerous modules involved in the pipeline. In the context of this paper, we mention the most common modules.

Image demosaicing (CFA interpolation) – The output from sensors which are overlaid with a CFA (Figure 14.9) is a sampled 3-channel colour image with a single value (red, green or blue intensity) at each pixel. Demosaicing is the process of interpolating the missing colour values to produce a full 3-channel colour image. Demosaicing methods exploit spatial and spectral correlations between neighbouring pixels and across colour channels. Image gradients and pixel similarities are used to enhance the process and reduce artefacts. Poor demosaicing results in colour artefacts known as colour aliasing, which are visible in the form of sporadic colours appearing at high frequency and high contrast edges. Demosaicing methods vary widely and are often unique to camera models. For further details and a survey on demosaicing methods, see Menon and Calvagno (2011) and Li *et al.* (2008a).

Gamma correction (gamma encoding) – Sensor output is linear with intensity of the scene and, accordingly, with the number of photons impinging on the sensor.

However, it is well known that human brightness sensitivity is non-linear with intensity (Hunt 2005); thus, for efficient encoding of scene intensity, the linear output of the sensors is non-linearly transformed to produce the end image. The transformation is typically a gamma function or power function of the form *out = in$^\gamma$* with the factor γ being the gamma of the transformation. Gamma less than 1 gives the desired non-linear transformation with typical values around $1/2.2$ (e.g. standard for sRGB (1999)).

White balancing (colour correction) – Due to scene illumination and due to camera's colour filters, the image obtained from the sensor outputs is often colour biased with an overall off-white tone colouring. White balancing (colour correction) is the process of transforming the image colour values so that colours in the end image appear as if acquired under neutral (white) illumination. Specifically, achromatic colours (whites and greys) should appear neutral in the end image. User involvement can greatly assist this process: users may indicate a camera setting that reflects the scene illumination (tungsten, sunlight, flash) and may also provide the camera with calibration information by pointing out (or capturing an image of) a white object in the scene. Most often, however, white balancing must be performed in the image pipeline based on the sensor output alone. Various methods of white balancing have been suggested, each relying on assumptions on the scene (e.g. grey world assumption (Land 1977; Buchsbaum 1980)), on the illumination (linear models (Cohen 1964)) or on specific objects in the scene (skin-based illumination detection (Hel-Or and Wandell 2002)) and more. Pixel colour value manipulation is typically performed via LUT or matrix multiplications. For a review on white balancing and colour correction, see Lamand and Fung (2008).

Image enhancement – As part of the imaging pipeline, every camera incorporates various image enhancement modules which may include contrast enhancement, sharpening, noise reduction and more.

Correction for optics and sensor aberrations (see preceding text) may also be incorporated in the pipeline. See Ramanath *et al.* (2005) for a survey.

Image compression – One of the last stages of the imaging pipeline is the preparation of the end image for storage and/or transmission. This typically involves compression of the image. The most common compression standard used in cameras is the Joint Photographic Experts Group image compression format, commonly known as JPEG (Wallace 1991; Pennebaker and Mitchell 1992, 1993). JPEG is a lossy compression technique that can achieve a compression factor of 10–20 with very little effect on image quality. The JPEG compression scheme flow is shown in Figure 14.11. Colour conversion and down-sampling are first applied to the input image to exploit the compression capabilities of the colour data. Image is then split into 8×8 non-overlapping blocks to each of which the discrete cosine transform (DCT) (Ahmed *et al.* 1974) is applied which represents the image in terms of frequency content. The 64 DCT coefficients undergo quantization according to 64 quantization factors which are stored in a quantization table (QT). Each DCT

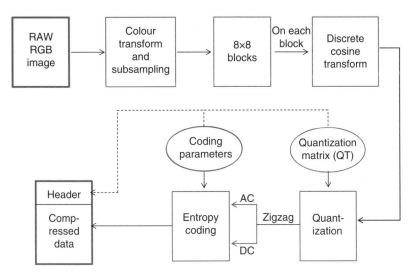

Figure 14.11   JPEG compression algorithm flow.

coefficient is divided by its corresponding quantization factor and then rounded to closest integer. Higher-frequency coefficients are quantized more strongly than lower-frequency coefficients in consistency with human sensitivity to errors in images. In fact, many of the high-frequency coefficients are quantized to zero.

Following quantization, lossless entropy encoding is applied to produce the compressed data. This data is typically stored in a JPEG formatted image file together with a header which stores the JPEG parameters used for that image, including the QT as well as other parameters.

Various modes and variations of JPEG compression have been developed including lossless, progressive and hierarchical as well as 12-bit JPEG, JPEG 2000 and motion JPEG (M-JPEG). Due to its lossy characteristic, JPEG compressed images often show compression artefacts, especially blockiness (correlating with the 8×8 boundaries of the blocks).

For additional information on JPEG compression see Wallace (1991) and Pennebaker and Mitchell (1992, 1993).

14.3   Camera-Based Forgery Detection Methods

As mentioned earlier, each of the three basic units of the camera system, namely, optics, sensors and image pipeline systems, introduces unique effects which may leave telltale artefacts in the final output image. These artefacts may then be exploited in attempts to validate image authenticity and to detect forgery. These artefacts may also assist in determining the source camera of an image which in turn may also serve as a technique for forgery detection (see Chapter 9).

14.3.1 Optics-Based Forgery Detection

The optic system of a camera introduces geometric distortions, chromatic aberrations and lighting flow artefacts in an image. These are exploited in the following studies on forgery detection.

Radial distortion produced by the lens system causes straight lines to appear as curved lines in the output image. In Choi *et al.* (2006a, 2006b), geometric distortions due to the camera optics, specifically radial distortion in images, are used to determine the source camera of an image. The authors argue that different manufacturers employ different lens system design to compensate for radial distortion and that the lens focal length affects the degree of radial distortion. Thus, each camera model will express a unique radial distortion pattern that can be used to identify it from its image. In the study, radial distortion parameters are evaluated per image, and these parameters (together with many other features) are used to train a classifier to determine the source camera of the image.

In Chennamma and Rangarajan (2010), radial distortion is used to detect splicing forgery. Image splicing is defined as a joining of two or more different images or image segments. Assuming that every image is distinguished by its radial distortion parameters, determining inconsistencies in the radial distortion parameters across an image can indicate forgery. Copy–paste forgery (where image segments are pasted into another location in the image or into a different image) can similarly be detected.

Chromatic aberrations are used to determine copy–paste and splicing forgeries. In Johnson and Farid (2006), chromatic aberration is modelled as a spatial expansion of one colour band relative to the other (e.g. B vs. G and G vs. R). Expansion parameters are considered as the chromatic aberration parameters of a given image. Inconsistencies in these parameters across an image may indicate suspected regions of forgery.

In Lanh *et al.* (2007b), this approach is extended to determine source camera of an image. The study focuses on cell phone cameras for which chromatic aberrations are more severe due to poorer quality of the lens system. The authors assume that the different lens structure of every camera induces unique and identifiable chromatic aberrations. The expansion parameters, used earlier, are used to train a classifier that will then detect the source camera of a given image from the image's expansion parameters.

In Yerushalmy and Hel-Or (2011), the directionality of the chromatic aberration artefacts (specifically the purple fringing aberration (PFA)) is exploited to detect forgery. PFA is expressed as a blue–purple halo on the distal side of bright objects (or on the proximal side of dark objects) relative to the image centre, the effect increasing in strength with distance from image centre. These characteristics are used to determine image centre. Inconsistent estimates of the image centre may indicate forgery. This approach is further detailed as a case study in Section 14.4.

Finally, in Lyu (2010), lighting flow aberrations, specifically vignetting, are used to determine the source camera of an image. Different lens combinations create unique vignette patterns in the image; thus, vignette pattern parameters may serve as camera/lens signatures. In the study, vignette pattern parameters are estimated based on statistics of natural non-vignetted images. The method can deal with vignetting functions that are not centred at image centre as well as not necessarily circular, thus allowing robustness in cases when image has been cropped and/or scaled.

14.3.2   Sensors-Based Forgery Detection

Image artefacts due to the sensor system of the camera can be exploited to determine source camera of an image as well as detect forgeries in image. Two main directions have been taken. One class of methods relies on patterns of sensor noise, specifically fixed pattern noise which appears consistently in all images of a given camera yet differs among cameras. Another class of methods relies on chromatic artefacts and chromatic pattern uniqueness due to sensitivities of the sensor filters in the CFA and due to the demosaicing techniques in the camera.

Hot pixels (bad pixels) are pixels with abnormally increased or decreased values. They may appear as individuals, columns or blocks of pixels. In an early study (Geradts *et al.* 2001), the authors assume that the spatial pattern of hot pixels is unique to each camera. The spatial pattern is used as a signature to determine the source camera of an image. In this study, the hot pixel patterns were manually extracted, and statistical methods were used to classify between two possible camera sources.

Another similar method is used in Dirik *et al.* (2007b, 2008) to detect source camera from an image based on sensor dust. In DSLR cameras with interchangeable lens, dust often seeps into the system. Dust particles settle on the sensors (or between the sensors and the optic system of the camera) and produce minute shadows in the end image. Suspected dust spot locations are detected in images of a given camera, from which a sensor dust pattern is calculated which is unique to the camera. Camera source can then be determined from the dust pattern in a given image based on statistical correlation with the camera dust pattern.

The innovative idea proposed in Lukas *et al.* (2006b) for identifying source camera of an image relies on the fact that some of the noise due to the sensor system is fixed and recurring such as dark noise, hot pixels and other fixed pattern noise (see preceding text). The basic assumption is, however, that the noise pattern is unique per camera and thus can be used as a signature or fingerprint of the camera. This unique pattern per camera is termed photo response non uniformity (PRNU) pattern. In Lukas *et al.* (2006b), the noise pattern of an image is obtained by taking the residual of the image after subtraction of a denoised version of the image. The PRNU of a specific camera is obtained by averaging the noise patterns of a large set of images acquired by that camera. Given a new image, its source camera is determined by correlating the image noise pattern with the PRNUs of the possible cameras. The camera with highest correlation is considered the source camera of the image. The authors showed that the method reliably predicts source camera and performs well even when images had undergone JPEG compression and gamma correction.

However, it was found that the correlation can easily be affected by the image content. Thus, the method of Lukas *et al.* (2006b) was extended in Chen *et al.* (2007a) and Li (2009) so that the PRNU is weighted in different regions of the image, according to the reliability of the region in predicting correlation which in turn is dependent on local image intensity and texture.

In Lukas *et al.* (2006a), the method of Lukas *et al.* (2006b) was exploited to detect forgeries in images rather than determining the source camera. Image regions suspected of forgery are those in which the noise pattern is inconsistent with the camera noise pattern. In Goljan *et al.* (2007), the method was exploited to determine whether two images are from the same source camera without access to the camera nor to other images acquired by the camera. This is performed by detecting common regions in both images' noise patterns.

Both forgery detection and device identification have been unified in a common framework in Chen *et al.* (2008).

The PRNU-based source camera identification was improved using machine learning techniques. In Chan *et al.* (2012), weighted correlations as in Chen *et al.* (2007a) as well as the actual correlation between PRNU of camera and the estimated noise of an image block were used as features to train a 2D classifier for camera identification. Similarly, in Gou *et al.* (2007) and Filler *et al.* (2008), these and other image features (such as image statistics, wavelet coefficients, neighbourhood predictions, etc.) were used to train classifiers to determine the source camera of an image.

Additional forgery detection studies based on PRNU include using different denoising techniques to extract the PRNU (Khanna *et al.* 2009; Chierchia *et al.* 2010), combining the method with CFA-based methods (Sutcu *et al.* 2007) (see following text), extensions to scanners (Khanna *et al.* 2007b, 2009) and videos (Kurosawa *et al.* 1999; Chen *et al.* 2007b) and distinguishing between sensor types (scanners vs. cameras) (Khanna *et al.* 2007a).

14.3.3   Image Processing Pipeline-Based Forgery Detection

The image processing module of the camera is highly complex, containing many sub-modules and algorithms, and varies greatly between cameras. Thus, it is not surprising that many studies in forgery detection have been based on effects of this module on the end image. The studies in this context aim to determine the source camera of a given image or to determine spliced images.

The methods reviewed in this section rely on determining the unique signature of a camera based on the camera response function (CRF), the demosaicing technique, the white balancing method and JPEG compression artefacts.

14.3.3.1   CRF Methods

A basic approach to evaluating a camera signature is to determine the CRF which is the function that maps the light intensity of the scene impinging on the sensor to the pixel intensity value in the end image. The CRF is largely dependent on the image processing module with its various algorithms that aim to produce a high-quality end image. Thus, the CRF is often unique for a specific camera model and can be used as a signature. Without a specific formulation of the CRF, the methods for camera source determination reduce to the class of methods based on statistics of natural images (see Chapter 13). However, considering the camera goals, the CRF is typically non-linear so as to efficiently compress the dynamic range of intensities and to produce pleasing and high-quality images when later transmitted, displayed and printed. Denote by $f(r)$ the CRF that maps the scene light intensity r to pixel intensity R.

The mapping f can be a simple gamma transform:

$$R = f(r) = r^{\gamma}$$

where parameter gamma is a constant value per colour channel or a more complex mapping of the form:

$$R = f(r) = r^{(\alpha + \beta * \gamma)}$$

with α, β constant per channel.

Assuming these models of CRF, several approaches to determining forgery and camera source have been suggested.

It has been shown in Ng *et al.* (2007) that when the CRF is modelled as a gamma function, then geometric invariants can be obtained in image regions which are linear in pixel intensity. From the geometric invariants and using machine learning-based methods, the CRF parameters can be estimated. In Lin *et al.* (2004) and Lin and Zhang (2005), it was shown that this approach can be practised at edges in the image. The study assumes that pixel values at an edge are equal to a linear blend of pixel values on either side of the edge and that co-linearity exists between colour channels at these edge pixels (i.e. the blending parameters are the same across colour channels). It can be shown that a non-linear mapping of the image values (i.e. CRF) will distort this assumption. Thus, CRF parameters can be estimated by correcting for the CRF until the co-linear relationship holds. Similarly, in Ng (2009) and Ng and Tsui (2009), the geometric invariants of Ng *et al.* (2007) were computed at edges in the image that were modelled as a sigmoidal function with a clear linear profile in the centre. The approach in Ng *et al.* (2007) was used in Hsu and Chang (2006, 2007) to detect splicing forgery by comparing the CRF parameters found at either side of the suspected splicing location. Cross-fitting techniques are used to compare CRF parameters and together with machine learning tools are used to classify authentic versus spliced images.

Similarly, the approach in Lin and Zhang (2005) and Ng *et al.* (2007) was used in Lin *et al.* (2005) to detect suspected spliced regions based on inconsistencies in the CRF parameters.

A different approach to forgery detection based on CRF was suggested in Popescu and Farid (2004) where bicoherence (a statistical relationship between frequency bands of the image) is used to measure the amount of quadratic phase coupling (QPC) effect (a special relation between triplets of frequency bands). QPC is directly related to CRF in that it increases with increased non-linearity of CRF. Thus, the CRF parameter γ can be estimated by searching for a gamma transform which minimizes the bicoherence magnitude.

14.3.3.2 Demosaicing and CFA-Based Methods

The sensors, each overlaid by a spectral colour filter, produce a single value per pixel array, and demosaicing must be used to interpolate the missing colour values to produce a full 3-channel colour image. Demosaicing methods used by cameras vary widely; thus, the demosaicing parameters can be exploited to uniquely determine source camera of an image as well as determine forgery in images. Interpolation algorithms in general and in demosaicing specifically introduce correlations between neighbouring pixels. Furthermore, due to the repetitive pattern of the CFA, the correlation patterns between pixels will be periodic as well.

The interpolation kernel (CFA pattern), interpolation coefficients as well as periodicity pattern may serve as indicators of the demosaicing algorithm used and, consequently, the source camera.

Numerous approaches based on demosaicing have been proposed in the context of image forgery. These include:

1. Determining whether CFA interpolation was performed at all and thus distinguishing between synthetic (computer graphics-based) images and real camera acquired photos
2. Determining the CFA pattern (e.g. Bayer), allowing classification of camera types and determining the source camera of an image
3. Determining the interpolation coefficients and thus the specific demosaicing algorithm used which can serve as a camera signature and enabling determining the source camera of an image as well as detecting forged regions in an image, when these coefficients differ

A straight forward approach proposed in Choi *et al.* (2011) is based on the assumption that interpolated pixel values are between the maximum and minimum values of their neighbours. Non-interpolated pixels are not necessarily so. Given a set of possible CFA patterns, the number (probability) of cases where intermediate values appear at interpolated pixels is counted per CFA. The highest number indicates the correct

CFA pattern. This study has been extended in Choi *et al.* (2013) to detect colour manipulations in images.

In its simplest form, the demosaicing operation is modelled as a linear interpolation that is a linear combination of the neighbouring pixels:

$$I(x, y) = \sum_{u,v \in \Omega} \alpha_{u,v} * I(x + u, y + v) + N(x, y)$$

where I is the image, Ω defines the neighbourhood of a pixel, $\alpha_{u,v}$ are the linear combination coefficients and N is the additive noise. The interpolation is valid at pixel locations x, y where specific colour values are missing; these as well as the neighbourhood Ω on which interpolation is performed are dependent on the CFA pattern. Both the CFA and the coefficients alpha are unknown; however, the interpolated pixels are known to be correlated with their neighbouring pixels. Furthermore, due to the periodicity of the CFA, the pattern of correlated pixels is periodic. In Popescu and Farid (2005b) based on Popescu and Farid (2005a), expectation maximization (EM) technique is used to evaluate both the CFA pattern and the interpolation coefficients by evaluating correlations between pixels and their neighbourhood. The periodicity of the correlation pattern indicates authenticity of the image. Disrupted periodicity indicates suspected forgery such as sliced image or copy–pasted image region. The detected CFA and coefficients can serve to uniquely define a camera and thus determine the source camera of an image. Similarly, in Swaminathan *et al.* (2007), the demosaicing interpolation is assumed to be linear but represented globally using matrix equation:

$$Ax = b$$

where b are the interpolated pixels, x the neighbouring pixels and A the linear coefficients. Assuming a specific CFA, coefficients can be determined using a matrix decomposition technique (SVD). Various possible CFA patterns are tested and that with the minimum interpolation error are determined to be the correct CFA. To improve evaluation, pixels are divided into three sets according to local dominant gradient direction (horizontal, vertical and no-dominant direction), and the method is applied to each set independently. The CFA pattern and the coefficients were used to distinguish between different demosaicing algorithms. In Swaminathan *et al.* (2006, 2008), it was shown that correct source camera can be detected from tampered images and even under slicing and copy-paste forgery. The method in Popescu and Farid (2005a) and (2005b) was extended in Bayram *et al.* (2005) by testing a number of different kernel sizes as well as dividing the image into smooth and textured regions and applying the estimation methods independently on each. In Long and Huang (2006) and Huang and Long (2008), interpolation is assumed to be quadratic rather than linear, and a coefficient matrix is obtained. A neural network is used to train on the coefficient matrix and then used to determine the source camera for a given image.

Another approach relies on the idea that variance of the second-order derivatives at interpolated pixels is lower than the variance at non-interpolated pixels. The pattern of variance change exhibits periodicity. The peak frequency of the variance pattern, indicates the CFA pattern, and the magnitude of the peak frequency indicates the method of interpolation (Gallagher 2005). Based on this idea, Gallagher and Chen (2008) distinguish between real and synthetically generated images. High pass filtering is applied, and variance of the resulting values is computed along the image diagonals. Frequency analysis is then performed on the sequence of diagonal variances. A peak frequency implies periodicity in the variances which, in turn, indicates that demosaicing has been performed on the image. The method is used to detect forgery by computing periodic frequency locally.

A forged region of an image will have a different periodicity than the rest of the image.

A variant of this method in Takamatsu *et al.* (2010) extracts noise variance from neighbouring pixels rather than from the filtered image as well as using the ratio of the mean noise variances of interpolated pixels to observed pixels.

In Cao and Kot (2009), a second-order partial derivative is applied to the data (to remove the DC component that represents image content) so that correlation within colour channel as well as across colour channels is captured. EM is used to perform classification of samples that have been demosaiced using the same interpolation. For each class, interpolation coefficients are calculated from which features are selected including the coefficients themselves as well as statistic measures of the interpolation error. These features were used to classify different demosaicing algorithms using a trained classifier. The approach was also used to detect tampered regions in images (Cao and Kot 2010, 2012).

In Bayram *et al.* (2006) and in Bayram *et al.* (2008), CFA parameter estimation using EM as used in Popescu and Farid (2005b) and in Bayram *et al.* (2005) and the inter-pixel analysis approach used in Gallagher (2005) were combined. The former was used in textured regions and the latter in smooth regions. A single combined classifier was used to determine the source camera of an image. Detection accuracy was further improved by incorporating sensor noise-based source camera detection (Lukas *et al.* 2006b) to first classify to camera and then verify using the CFA approach.

14.3.3.3 Additional approaches

The approach in Ho *et al.* (2010) relies on inter-channel correlation and is based on the assumption that the variance of the Fourier coefficients is the same over the different colour channels. Fourier coefficients of the green–red and the green–blue difference images are computed and the variances calculated from which different demosaicing interpolation algorithms can be classified. In Hu *et al.* (2012), the method was extended to use shape and texture features of the variance maps in order to classify the demosaicing algorithms.

In Dirik *et al.* (2007a) and Dirik and Memon (2009), discrimination between real and synthetic images is performed as well as determining whether an image has been tampered. Two methods are used:

1. Pattern estimation – Assumes a set of probable interpolation kernels. For each kernel, it re-interpolates the image using the kernel and measures the MSE between the re-interpolated image and the original. The assumption is that correct interpolation kernel will give significantly lower MSE than the other kernels. Large variability of MSE values across the different kernels indicates greater likelihood that the image was not tampered.
2. CFA-based noise analysis – CFA noise is estimated by taking the difference between original and denoised images. Similar to Gallagher and Chen (2008), it is assumed that variance of the noise is lower at interpolated pixels than at non-interpolated pixels. Large differences in variances indicate that no tampering has been performed on the image.

In Kirchner (2010), the raw sensor data is estimated from the full colour image which, in turn, determines the CFA and interpolation parameters that were used by the demosaicing algorithm. Straight forward computation is highly complex (requiring inversion of a large matrix), so efficient methods are implemented based on the fact that the interpolation matrix is sparse (small kernel support per pixel) and that the CFA pattern is repetitive and thus periodic.

Finally, in Celiktutan *et al.* (2006, 2008), source cell phone cameras are determined based on measuring the similarity between different bit planes of an image. Numerous features including binary similarity measures, image quality measures and higher-order wavelet statistics are extracted and used with a classifier to identify the source camera.

14.3.3.4   White Balance methods

Due to imbalance of the red green and blue colour filters in the sensor unit as well as due to the scene illumination spectra, the raw image acquired by the sensors may have a colour bias (e.g. a blue or green tint). Thus, the image processing unit performs white balancing or colour correction to correct the colour tones of the image such that achromatic (grey and white) tones will indeed appear achromatic. Numerous white balancing algorithms exist and cameras adopt such unique algorithms. Thus, the algorithm or its parameters may serve as camera signature and can be used for source camera identification from an image. In Deng *et al.* (2011), source camera identification was performed based on white balancing. It is assumed that re-applying the same white balance algorithm on an image which has already been white balanced will result in little change. Thus, an image is white balanced using several different methods, and image quality measures are extracted from the difference between original and re-white balanced images. These measures are used as features to train a classifier and then determine the source camera of novel images.

14.3.3.5   JPEG Compression-Based Methods

In the last stages of the image processing pipeline, the end image is compressed and transmitted or stored in memory typically using the JPEG compression scheme (Wallace 1991; Pennebaker and Mitchell 1992). The fact that the compression is lossy and is composed of known algorithmic steps and controlled by flexible, specifiable parameters allows a wide range of analysis in the field of image forgery. The flexibility of the algorithm and its parameters allows individual cameras to use their own inbred JPEG compression scheme which can then serve as a camera signature and allow camera source identification. The fact that the compression is lossy implies that compressed images display compression artefacts which can also serve to detect the source camera as well as detect image forgery and tampering.

The studies mentioned in this section deal with determining the source camera of a given compressed image, detecting compression history of an image (whether image was compressed, double compressed or more than twice compressed) and detecting tampered image, namely, whether portions of an image have been copy–pasted or sliced into the image (either from the same image or from a different image). The reader is referred to the Chapter 15 in this book for additional information.

In the context of image forgery detection, three major strategies are employed:

1. Based on artefacts expressed in the pixels of the compressed image, basically exploiting blockiness artefacts in the image
2. Based on artefacts found in the frequency (DCT) coefficients of the compressed image
3. Based on artefacts found when image is doubly compressed

Of course, the most straight forward and naïve approach is to consider the header of the JPEG image file. In Kee *et al.* (2011), JPEG compression parameters in the header, including the QT, Huffman codes, thumbnails, etc., are used as signatures that uniquely define each camera. Source camera can then be identified. However, since JPEG headers are easy to manipulate, they do not form a reliable basis for forgery detection, and more advanced techniques based on image content must be used.

The first class of methods is based on the visual artefacts in the image due to JPEG compression, namely, the blockiness artefacts in the image. JPEG's division into 8×8 blocks and the per-block quantization error produce discontinuities along the block boundaries (and thus produce a distinctive grid pattern of discontinuities). The blockiness in an image is analyzed in Fan and Queiroz (2003) by considering the statistics of the intensity differences between neighbouring pixels. If image is compressed, then these differences should be greater at pixels located along the border of the 8×8 blocks defined by the JPEG compression than at pixels within the blocks.

Using this approach, an image can be determined as compressed or not, the block grid location can be determined, and regions inconsistent with the grid can be determined as suspected forgery. In Luo *et al.* (2007), this approach is extended by defining the blocking artefact characteristics matrix which encodes the pixel differences under numerous location shifts. This matrix shows a symmetric pattern in compressed images and loses its symmetry when image is tampered or forged. In Li *et al.* (2009), the block grid is determined based on the fact that high-frequency DCT coefficients within a block are close to zero, whereas across block boundary, these coefficients will be larger due to JPEG artefact at the block boundary. Inconsistencies in the block grid are used to determine copy–paste forgery locations in images.

A second class of image forgery detection studies is based on characteristics and artefacts that appear in the DCT coefficients of the compressed image. Since these quantized coefficients can be reconstructed from the compressed image (via inverse DCT), these characteristics can be exploited to determine the QT used in the compression which, in turn, can be used to determine the source camera, to detect forged image regions (based on inconsistencies of the QT) and to determine the compression history of the image. The variation in QTs has been statistically analyzed and shown to be very useful in the context of image forgery (Farid 2006, 2008; Kornblum 2008; Mahdian *et al.* 2010).

The quantization process induces special characteristics in the quantized DCT coefficients. Due to quantization, the DCT coefficients of a compressed image will be clustered around multiples of the quantization factors used per each frequency.

To estimate the QT from a given image, a straight-forward method involves guessing different quantization factors and estimating how close the DCT coefficients of the image blocks are to a multiple of this quantization factor. The value producing minimal distance over a majority of the image blocks is determined as the correct quantization factor. This is performed independently for each DCT frequency (Fridrich *et al.* 2001; Fan and Queiroz 2003). Given the QT, every image block can be tested for consistency to detect suspected forged regions in the image. Considering a histogram of all DCT coefficients of a specific frequency over all image blocks, the non-zero bins of the histogram are located at (or around) multiples of the quantization factor associated with that frequency. Thus, the histogram will display periodic peaks and zeros, with the period dependent on the quantization factor. This is exploited in Luo *et al.* (2010) to detect whether an image has been compressed. The Fourier spectrum of the histogram will display a strong peak indicating the period. This is exploited in Ye *et al.* (2007) to estimate the quantization factor for each DCT frequency independently and construct the QT. Given the QT, quantization error can be evaluated in each block (as in Fridrich *et al.* (2001) and Fan and Queiroz (2003)), and blocks with large errors are deemed forged.

A third class of approaches exploits the effects that occur when an image is compressed twice or more. This is important in order to determine image compression history in general but also allows determining the QT used in the compressions.

Furthermore, considering a compressed tampered image with copy–pasted regions, it will contain regions that are doubly compressed (non-tampered regions) and regions that are singly compressed (tampered regions originating in an non-compressed image) or doubly compressed with different parameters (tampered regions originating in a compressed image). These differences will allow detections of tampered regions in images. The basic approach in this class of methods involves recompressing the original compressed image and evaluating its effect. The basic notion is that if recompression was performed with the same QT as the initial compression, then the difference between original and recompressed images will be minimal. In Luo *et al.* (2010), the QT was determined by recompressing the given image using a set of possible QTs. In each case, the number of pixels whose values remain the same under recompression is counted. The QT that results in largest number of similar pixels is considered the correct QT. In Farid (2009a) and Zhao *et al.* (2010), the sum of differences is computed rather than the number of similar pixels. This sum is expected to be small when recompression is with the same QT as the original. Furthermore, computation of sum of differences in local regions will reveal regions that differ from the rest of the image and thus indicate tampered regions. This approach was extended in Zach *et al.* (2012) where features were extracted from the difference image and used to learn and later classify between single and doubly compressed image regions (tampered vs. non-tampered regions). In Wang *et al.* (2010), PCA is used on the differences between original and recompressed images to extract the high-frequency component. Image regions that show large values of high-frequency noise are deemed tampered.

Another approach in this class relies on the DCT coefficients having specific statistical characteristics. The most significant digit of the DCT coefficients of a natural image has been shown to satisfy Benford's law (Benford 1938) which states that the distribution of values is logarithmic. However, coefficients of doubly compressed images do not satisfy Benford's law. In Fu *et al.* (2007) the correct QT was determined by recompressing using various QTs and determining the QT for which Benford's law remains satisfied. In a more complex situation where the original image has been doubly compressed, the first QT used in double compression is determined in Lukas and Fridrich (2003) by reproducing the double compression and evaluating its effect using coefficient histogram matching. Another characteristic of the DCT coefficients is known as the double quantization (DQ) effect. The analysis in Lukas and Fridrich 2003, Popescu and Farid (2004), Popescu (2005) and He *et al.* (2006) shows that the DQ effect is expressed as periodic peaks and valleys in the histogram of the DCT coefficients (per DCT frequency across all image blocks) that are dependent on the quantization factors of the first and second compressions. Frequency analysis (DFT) of the absolute values of the DCT coefficients will show distinct peaks indicating the periodicity. This was exploited in Popescu (2005) and Mahdian and Saic (2009) by estimating a measure of periodicity of these DFT coefficients. A measure greater than a set threshold implies the image has been doubly compressed. The study in Lin *et al.* (2009) and Mahdian and Saic (2009); exploits DQ to determine locations of tampered regions in images. When an image is spliced or copy–pasted and then JPEG compressed, the tampered region will be compressed once while the original non-tampered regions will be doubly compressed and the DQ effect will appear. For these images, the histogram of DCT coefficients can be viewed as a sum of two histograms: a periodic histogram with high peaks and valleys (due to the double compression) and a random histogram with low non-periodic data (due to singly compressed blocks). Using Bayesian analysis, a probability map for the image blocks is produced from which features are extracted and used to train a classifier for determining tampered images as well as determining the specific tampered blocks.

Finally, we mention two adjoining sets of studies. The first deals with the basic question of whether an image has been at all compressed or doubly compressed. One approach relies on feature extraction and machine learning tools. In Pevn'y and Fridrich (2008), the DCT histograms for low frequencies are used as features to train a classifier and then classify images as singly or doubly compressed. An additional set of multi-classifiers determines the pair of primary and secondary QTs used in the double compression. In Fu *et al.* (2007) and Li *et al.* (2008b), features are based on the distribution of the most significant digit of the DCT coefficients of the image and rely on Benford's law Benford (1938) to determine whether an image has been singly or doubly compressed. In Li *et al.* (2010), this feature set was extended to include moments of the DCT distribution itself. In Chen and Hsu (2011), features were extracted to represent both the periodicity of an image in the spatial domain (periodicity of compression artefacts in the image) and the periodicity of the spectrum of DCT coefficients. The study in Chunhua *et al.* (2008) extracts features based on differences between neighbouring DCT coefficients of an image. The authors stated that these differences are correlated in compressed images, yet correlation is affected in doubly compressed images.

In Huang *et al.* (2010) and Manimurugan and Jose (2012), multiple compressions using the same QT are determined. They assume that when image is multiply compressed with the same QT, then DCT coefficients become increasingly stable. They show that the number of DCT coefficients that change between sequentially compressing an image can be used to determine the number of compressions applied on an image. A different approach in Yanga *et al.* (2011) uses statistics per DCT frequency that determines the probability of different quantization factors of the first QT given quantization factors of the second QT as well as the DCT coefficients themselves. These statistics are used to determine whether an image is doubly compressed or not.

Finally, we mention studies that deal with double compression when image has been tampered between compressions. In Qu *et al.* (2008), images are determined to be cropped as well as doubly compressed by exploiting the fact that in this case, the DCT coefficients can be expressed as a linear combination of DCT coefficients in four underlying blocks in the original uncropped image. By testing over all possible shifts, the mixing matrix for the linear combination can be found, thus determining that the image has been cropped as well as determining the block grid shift. In Bianchi and Piva (2012), images are determined to be resized as well as doubly compressed based on the assumption that without resizing, JPEG compressed images display a near lattice distribution property (NLDP), namely, periodic in a grid structure. The image is inversely resized at various scales and the NLDP tested. The scale producing the largest above threshold value is determined as the scaling parameter used. In Ferrara *et al.* (2013), images are determined to be enhanced as well as doubly compressed. Both the periodic pattern of the histogram of doubly compressed DCT coefficients as in Lin *et al.* (2009) and the generalized Benford's law (Fu *et al.* 2007) are used to detect whether image is compressed or doubly compressed with enhancement.

14.4   Forgery Detection Based on PFA: A Case Study

As a case study, we present here a novel forgery detection scheme first presented in Yerushalmy and Hel-Or (2011), which exploits the camera's effect on the end image; specifically, chromatic lens aberration is exploited to detect image cropping and copy–paste forgeries.

   As described in Section 14.2.1.1, lens aberration is due to the fact that the refractive index which determines the bending angle of incoming light rays is dependent on the spectral wavelength of the light. Shorter wavelengths (blue and violet) has a larger refractive index than middle (yellow and green) and long wavelengths (red) and so bend more than other wavelengths. This implies that light of different wavelengths passing through a lens will focus at different points or on different imaging planes relative to the lens (see Figure 14.5). The effects of chromatic aberration are expressed in the end image as colour fringes at high contrast edges (see Figure 14.6). However, in practice, chromatic aberration effects are typically confounded with numerous aberrations including lateral chromatic aberration (LCA), axial chromatic aberration, purple blooming aberration and others (Ray 2002; Smith 2007; Pedrotti and Pedrotti 2008; Van Walree 2014) that affect the image chromatically. Furthermore, it has been observed (Smith 2007) that spatial and geometric aberrations such as spherical aberration, coma aberration and astigmatism aberration have a secondary effect on chromatic aberrations. Both lens and camera sensors are considered as sources of these aberrations.

   In the study presented here, we focus on the general effect in an image termed *PFA* which is expressed as a blue–purple halo near the edges of objects in the image (Figure 14.12).

   PFA encompasses both chromatic aberration effects due to the lens and effects produced by the interaction of the camera sensor system with the lens (see Yerushalmy and Hel-Or (2011) for more details).

   PFA expresses several unique characteristics that cannot be explained by any single aberration; specifically, the PFA encompasses LCA yet differs from it. The most significant difference is that while both types of aberrations frequently appear near

Figure 14.12 Enlarged part of two images displaying purple blooming at the edges marked by the arrows.

Figure 14.13 Schematic diagram that depicts the difference between LCA and PFA (image centre is depicted by '+'). While LCA is a pure expansion–contraction effect of the blue versus green channels, PFA is characterized by a blue–purple halo on the distal side of bright objects (sometimes also accompanied by a minor yellow tint on the opposite side, due to the LCA effect).

edges in the image, PFA does not exhibit an expansion–contraction transformation as the LCA does. In the latter, the blue colour channel of the image is magnified about the centre, with respect to the green channel. In PFA, the blue–purple halo appears on the distal side of bright objects (or on the proximal side of dark objects) relative to the image centre, while the opposite side remains almost unchanged (see Figure 14.12). It should be noted that the purple–blue halo is more prominent than the blue halo of the LCA due to the additional factors described earlier, which intensify the aberration. Figure 14.13 shows a schematic example of LCA versus PFA.

14.4.1 Forgery Detection Based on PFA

The purple–blue halo of the PFA is directional, namely, it appears on the distal side of bright objects (or on the proximal side of dark objects) relative to the optical image

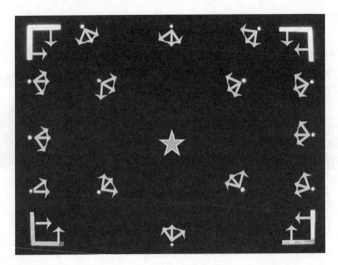

Figure 14.14 Image with arrows indicating the 'direction' of the PFA. The optical image centre can then be found (depicted by the star).

centre (see Figure 14.12). Note that the optical image centre does not necessarily align with the geometric centre of the image (e.g. in cropped images). The directional characteristic of PFA is exploited to locate the optical image centre. Regions of the image in which PFA does not point to the common optical image centre can be marked as suspected forged areas. See Figure 14.14 where arrows indicate the 'direction' of the purple halo. The centre star depicts the optical image centre.

In addition to directionality, two additional characteristics of the PFA aberration are exploited in the proposed approach:

1. PFA increases in strength with distance from the optical image centre (Figure 14.15a).
2. PFA increases in strength with increased intensity contrast at edges (Figure 14.15b).

Taking these characteristics into account, a reliability measure for every PFA event can be introduced. The presented approach attempts to detect all PFA events in the image and determine their directions together with such a measure of reliability. The combined information from all detected PFA directions is used to determine the optical image centre. If a region contains inconsistent PFA directions, it can be concluded that either local image noise has affected the results or that the region is forged. The measure of reliability assists in overcoming such noise, allowing the algorithm to perform well even on JPEG compressed images. Figure 14.19 shows an example of an authentic digitally acquired image, with image centre correctly detected according to the PFA.

(a)

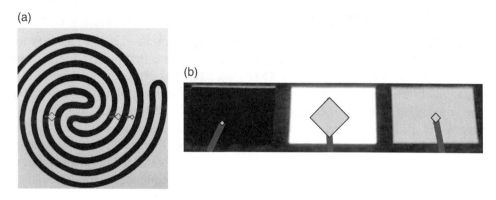

(b)

Figure 14.15 (a) Aberration strength increases with distance from image centre. An enlarged region of an image is shown (image centre is located to the right of the region). PFA at three edges with equal colour contrast is marked with a diamond-headed arrow whose size corresponds to the strength of aberration. The arrows point in the direction of the image centre. (b) Aberration strength increases with contrast. An enlarged region of an image is shown (image centre is above the region) with three patches equally distant from the image centre with increasing contrast. The diamond-headed arrows point in the direction of the image centre. The size of the diamond corresponds to the strength of aberration.

14.4.2 Algorithm

Using the PFA properties, the optical image centre can be determined. Regions that do not point to the centre suggest that they are not original and have been tampered or completely implanted there after the actual scene was acquired by the camera. Furthermore, if centre is found to be misaligned with the geometric image centre, suspected cropping may be assumed. The following steps describe the proposed method:

1. Identify PFA events along edges.
2. Determine PFA direction for each detected PFA event.
3. Initialize the optical image centre (x_0, y_0) to be the geometric image centre.
4. Repeat the following until convergence or stopping criteria reached.
5. Assign a reliability measure to each PFA event according to aberration strength, edge contrast and distance from the current optical image centre (x_0, y_0).
6. From the collection of PFA directions and their associated reliability measure, robustly determine an updated optical image centre (x_0, y_0).
7. Repeat steps 5–6 until convergence. Following convergence.
8. Analyze the PFA directions to determine regions in which these directions are inconsistent with the evaluated image centre. These regions are marked as suspected forgery.

The algorithm steps are detailed in the following.

14.4.2.1   Identifying PFA Events

PFA can be found along edges, especially where high contrast exists. To identify the aberration indicators, all significant edges in the image are first detected (e.g. using Canny edge detector (Canny 1986)). Every edge pixel is evaluated to determine if it is a viable PFA event. For each edge pixel, the transition of colour across the edge is analyzed in the xyY colour space (Wyszecki and Stiles 1967) which allows independent analysis of the luminance Y and the chromatic content x, y of a signal which is typically displayed using the chromaticity plane (Figure 14.16). It is assumed that in natural images, the change in colour across the boundary between different colour regions is linear in chromaticity. Thus, a pixel sequence across a boundary should contain a linearly varying mixture of the two bordering colours, (perhaps with a change of luminosity). In the xyY colour space, this assumption translates to a linear transition between the two colour points in the chromaticity $(x - y)$ plane (see Figure 14.17a), and perhaps a change in the luminance (Y) parameter. In the case of PFA, this assumption is violated, since a blue–purple hue affects the edge. This will cause the transition in the chromaticity plane to behave non-linearly with a bias towards the blue–purple region of the plane. Figure 14.17 shows an example of the transition

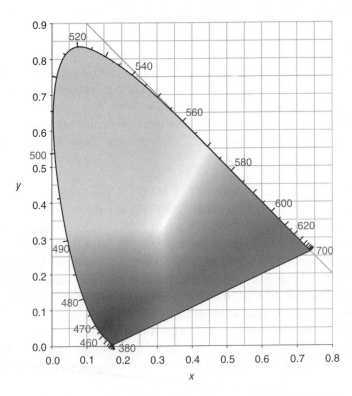

Figure 14.16 Chromaticity plane of the xyY colour space. The purple–blue locus is at 450 nm.

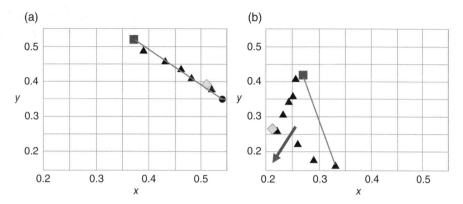

Figure 14.17 PFA analysis in the *xy* chromaticity plane. The chromaticities of image pixel colours across an edge boundary are shown for a non-aberrated edge (a) and for an edge with PFA (b). The start and end points of the pixel sequence are marked with a square and a dot respectively. The grey line is the expected linear transition between the two end points. The left plot shows an almost linear transition between colours on both sides of the edge. The right plot shows a deviation towards the blue–purple values (indicated by arrow) of the chromaticity plane. The diamond depicts the furthest point from the expected segment.

in chromaticity plane across a non-aberrated edge (Figure 14.17a) and across an edge with PFA (Figure 14.17b). Pixel values along a segment perpendicular to the edge were uniformly sampled (their chromaticity marked as Δ in the plots). The chromaticity of the two bordering colour regions is marked with a square and a dot in the chromaticity plane, corresponding to the start and end points of the sampled pixel sequence, respectively. The non-aberrated edge (Figure 14.17a) displays a linear chromaticity transition, while the PFA edge (Figure 14.17b) displays a deviation towards the blue–purple region.

A PFA event is determined at an edge pixel if the chromaticity values across the edge deviate significantly towards the blue–purple chroma. This is evaluated by estimating the PFA strength as described in the following text. It should be noted that due to the presence of LCA together with PFA (see preceding text), some images also display the complement of a blue–purple halo, namely, yellow-tinted edges. These effects will appear at objects' edges on the side opposing the purple halo (i.e. on the proximal side of bright objects). For these edges, a similar non-linear transition will be observed in the chromaticity plane with deviation towards the yellow region. However, this deviation is usually much weaker than the blue–purple deviation due to PFA, which is intensified by CCD associated aberrations (see preceding text).

14.4.2.2   Determining PFA Direction

The PFA effect is displayed as a blue–purple halo on the distal side of bright objects and on the proximal side of dark objects. Thus, the intensity gradient across the edge where

PFA was detected indicates the direction towards the image centre. We define the PFA direction as a unit vector $\overline{N}$ perpendicular to the edge in the direction corresponding to increasing intensity across the edge. In case of aberrations that produce yellow artefacts at the edge (corresponding to deviation towards yellow tones in the chromaticity plane), the PFA direction will be reversed.

14.4.2.3 Determining PFA Strength

The strength of a PFA event is dependent on the magnitude and direction of the chromatic deviation in the chromaticity plane (Figure 14.17b). Let c_a and c_b be the chromaticities on either side of the edge. They define an expected linear transition segment in the chromaticity plane (grey segments in Figure 14.17). To determine PFA strength at an edge, a point-to-segment distance is calculated for each pixel chromaticity value across the edge, and the point with maximum distance is determined (diamond in Figure 14.17). The vector in the chromaticity plane between the expected transition segment and the most distant point of the sequence is projected onto the blue–purple direction, determined as the unit vector from mid-segment to the blue–purple wavelength of 450 nm ($x = 0.2\, y = 0.1$ in the chromaticity plane; see Figure 14.16). The magnitudes of the projections along the blue–purple direction are normalized:

$$\delta = \left(\frac{s - s_{\min}}{s_{\max} - s_{\min}} \right), \tag{14.2}$$

where s is the projected value and $s_{\max}(s_{\min})$ is the maximum (minimum) value found in the image. The normalized projection value, δ, is defined as the PFA strength. If aberration is present at the edge, then the PFA strength is expected to be large.

14.4.2.4 PFA Reliability Measure

The reliability measure quantifies the consistency of the calculated PFA strength with the expected values. As described earlier, the PFA is expected to increase in strength with distance from image centre as well as with increase in edge contrast. These characteristics are used to define a measure of reliability for each detected PFA region. The edge contrast t associated with a PFA event is defined as the absolute difference in intensity across the edge, and the distance d of the PFA event is the linear distance to the optical image centre. The reliability of a PFA region is then defined as:

$$\rho = \left(\frac{t - t_{\min}}{t_{\max} - t_{\min}} \right) * \left(\frac{d - d_{\min}}{d_{\max} - d_{\min}} \right), \tag{14.3}$$

where $t_{\max}(t_{\min})$ is the maximum (minimum) edge contrast found in the image and $d_{\max}(d_{\min})$ is the maximal (minimal) distance to the optical image centre.

14.4.2.5 Calculating Location of Image Centre

Using the collection of PFA events, the location of the image centre can be estimated from the event **location** ($\mathbf{X}$), the **direction** ($\vec{N}$), the **strength** (δ) and the **reliability** (ρ) of each PFA event (Eqs. 14.2–14.3).

The detected PFA direction is always perpendicular to the analyzed edge. Thus, it does not necessarily indicating the actual direction towards the image centre. This problem, called the 'aperture problem', is well known in computer vision, where local movement of objects can be obtained only in the direction perpendicular to their boundary edge, creating a 'normal flow' map (Jahne 1995). To obtain the actual motion per pixel in a scene, 'optical flow' algorithms have been proposed (Horn and Schunck 1981); these however are computationally expensive and provide more than the image centre. A simpler approach is adopted in this study based on focus of expansion (FOE) detection (Negahdaripour and Horn 1989). This method assumes that the underlying flow field displays a centralized flow. The method attempts to locate the FOE point based on the PFA directions associated with the PFA events, thus determining a central image point which is consistent with the centralized flow.

Determining the FOE based on the PFA 'flow field' takes into account both the strength and the reliability of the detected PFA events. Let $X_i = (x_i, y_i)$ be the edge pixel coordinates of the ith detected PFA event and let $X_0 = (x_0, y_0)$ be the image centre to be determined. The point X_i is associated with a line passing through the point in the direction $\vec{N}$. The line is of the form $a_i x + b_i y + c_i = 0$ with (a_i, b_i) normalized to unit vector. The process finds the point X_0 which minimizes the squared perpendicular distance to all the lines, normalized by the PFA strength and the reliability factor:

$$X_0 = X_0(x_0, y_0) = \arg \min_{x,y} \sum_i \frac{(a_i x + b_i y + c_i)^2}{(1 - \delta_i \rho_i + \varepsilon)^2} + R_i(x, y) \qquad (14.4)$$

where δ_i is the PFA strength, ρ_i is its reliability as described earlier and ε (small value) is used to avoid singularity. $R_i(x, y)$ is a penalty factor associated with the ith PFA event that is assigned to location (x, y) if it is inconsistent with the ith PFA direction (i.e. (x, y) is 'behind' the ith edge relative to the PFA direction):

$$R_i(x, y) = \begin{cases} \alpha K_i & \vec{N}_i \bullet (X - X_0) < 0 \\ 0 & \text{otherwise} \end{cases} \qquad (14.5)$$

where $\bullet$ denotes inner product, K_i is the normalized point-to-segment distance between the ith edge and the optical image centre X_0 and α is a scale factor. The penalty is added, and not multiplied by the remaining parameters, to avoid a scenario in which the algorithm converges to an image centre located 'behind' any PFA edge and thus is inconsistent with the PFA effect. In all experiments, α was set to 5×10^5. A multi-resolution descent search is used to minimize the function and determine the coordinates of the image centre.

The above method produces good results, based on the normal flow map. However, there are two concerns that must be taken into account. First, the map may contain outliers due to noise in the image (e.g. due to JPEG compression). To overcome this, a robust-regression method, the median outlier filter (Rousseeuw and Leroy 2003), is used. The data points proximal to the median (70% of the points) are considered valid, while the rest are marked as outliers. In the process of centre detection, the weighted distances of the normal-flow vectors relative to the calculated centre (based on aberration strength, reliability and penalty) are the input to the filter. The process is run iteratively and outliers are removed at each iteration until the calculated centre location stabilizes.

The second concern is that the reliability measure assigned to every PFA event (Eq. 14.3) depends on the distance to the optical image centre. However, this centre is unknown and is computed iteratively using Equation 14.4 and using the outlier removal process mentioned above. This issue is specifically significant in cases of cropped images when the optical image centre may differ significantly from the geometric image centre (see following text). To overcome this concern, at each iteration, the reliability measure is updated based on the image centre found at the previous iteration. The process iterates until the calculated optical image centre location stabilizes.

14.4.2.6   Detecting Traces of Forgery

The analysis of the PFA direction map as well as the calculated centre, allows detection of image regions that have been tampered since their acquisition by the camera.

PFA regions are analyzed to determine inconsistencies between their direction and the calculated centre. These regions, when removed, will maintain a consistent flow. To identify these regions, the weighted distance from the calculated centre of the image (Eq. 14.4) is calculated independently for each PFA event. The penalty factor (Eq. 14.5) assures that PFA events pointing to an illegal direction have a much higher weighted distance than others. This implies that such PFA events can be distinguished from others (using a threshold), even if the latter do not point directly to the calculated centre due to the aperture problem. A map of these values is formed, from which regions highly suspicious of forgery are detected. An example of such a map is shown in Figure 14.18. Note that in the case of forgery involving duplicated image regions, only the duplicated region will be marked as suspicious, since it alone will be inconsistent with the geometric centre, allowing the system to distinguish the source region from its copies (see Figure 14.21). Additionally, it is able to detect copy-move forgeries even when the copy is magnified or reduced in size (Figure 14.21).

Another scenario is where the optical image centre is detected at a distance from the geometric image centre and yet is accompanied by a consistent PFA direction map. This is a unique constellation that may indicate that the image has been cropped from its original size. According to the evaluated optical image centre, an estimate of the original image size can be deduced by the algorithm (see Figure 14.22). This

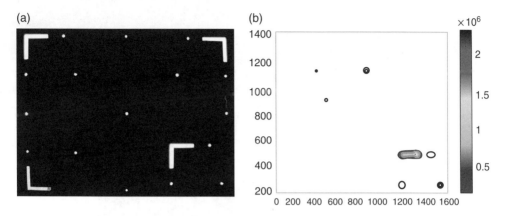

Figure 14.18 (a) A forged version of Figure 14.14. (b) The map of weighted distances of PFA events. Note that the map clearly indicates the forged region.

capability, although unable to restore the missing data, allows the user to obtain a sense of the original size of the image and what portion was cropped.

Finally, if the calculated optical image centre was found to be located near the geometric centre with strong supporting PFA events and there is no significant region in which PFA events are inconsistent with the globally evaluated optical image centre, then it may be assumed that the image is authentic.

14.4.3 Test Results

In this section, the performance of the presented algorithm is demonstrated on various types of images, all compressed using standard JPEG format. No camera calibration is assumed nor is any knowledge on the camera parameters used to acquire the images. Images included indoor and outdoor images, portraits, scenic views, urban scenery and more. Images are of size ranging between 1536×2048 and 640×480. Primary camera models are Samsung S630, Canon SD200 and Canon Power-Shot A520.

14.4.3.1 Examples

Specific examples are shown to demonstrate the capabilities of the proposed algorithm. Figure 14.19 shows the results of applying the proposed algorithm on an authentic image. PFA events are shown with diamond-headed arrows indicating strength and direction. Detected image centre is shown as green star; its deviation from the true image centre is 5%. Figure 14.20 shows additional examples where image centre was calculated with angular errors $5.71°$ (left) and $8.9°$ (right). Figure 14.21-left shows an example of an authentic image which was edited to include a copy of the original person (Figure 14.21, right). The algorithm detected the suspicious section while correctly discriminating between the original and its copy. Additional examples can be found in the supplementary material.

Figure 14.19 An authentic image. Diamond-headed arrows indicate a selection of PFA events detected by the proposed method. Diamond size corresponds to the strength of aberration. The star depicts the detected image centre.

Figure 14.20 Authentic images. Star depicts the calculated centre.

Figure 14.21 Example of copy-paste forgery detection using the proposed algorithm. The man in the original image (left) was duplicated and scaled to create a forged image (right). Shaded box marks the suspicious region detected by the proposed algorithm.

The capabilities of the approach on cropped images are shown in Figure 14.22 where two images have been cropped by 33% on the right side (left image) and by 71% of the top and right sides (right image). The location of the optical image centre (star) was determined using the suggested algorithm. The fact that most PFA vectors are consistent with the centre, which is not aligned with the geometric centre, suggests that the images were cropped. Dashed lines mark the full image frame as estimated using the calculated centre. The proposed algorithm was able to restore both size and shape of the un-cropped image with a success rate of 93% (left image) and 85% (right image). Note that image centre of right image was detected beyond the bound of the image. Additional results can be found in the supplementary material.

14.4.3.2 Rigorous Testing

The image forgery detection method described earlier is shown to perform well in rigorous testing. In the first test, a set of 45 JPEG compressed images were used to test the basic capability of the algorithm, namely, to correctly identify that the optical and geometric image centres are aligned as expected in authentic, un-edited images. An *angular-error* measure as defined in Johnson and Farid (2006), was used to quantify the level of accuracy of centre detection. This measure evaluates the angle between

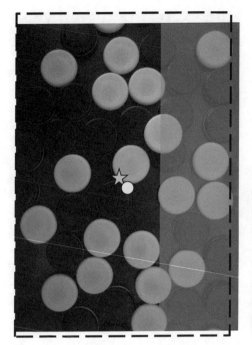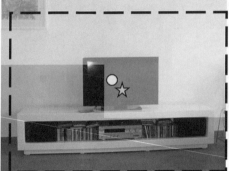

Figure 14.22 Detecting cropped images. Left: test image with 33% removed from the right side (shown as desaturated). The proposed algorithm was able to restore the original size and shape with a 93% success rate, marked with a dashed line. The dot marks the genuine image centre and the star the calculated centre using the proposed algorithm. Right: 71% of the image was cropped, including the centre. The algorithm was able to restore original size and shape with an accuracy of 85%.

the lines connecting a PFA point to the true centre and to the calculated optical centre averaged over all PFA points. Figure 14.23 presents the results showing that 51% of the images were analyzed with an average angular-error of up to 15°. When considering a larger, but still acceptable, value of 25°, 87% of the images reached this accuracy. For comparison, results are shown for image forgery detection based on the approach in Johnson and Farid (2006), where only 9% of the images reached the accuracy of 15° and 33% reached 25°.

In another experiment, crop forgery detection was tested. A set of 120 cropped images were analyzed, in order to test the ability to detect cropping and restore of the original image size. The images were produced automatically using a set of 30 authentic JPEG compressed images from various cameras. Each authentic image had 33% eliminated (from the top, bottom, right or left portions of the image), producing four different cropped images. For each image, the algorithms calculated the optical image centre and determined the size and shape of the un-cropped image. The size-restoration success rate was evaluated using the *average-precision* measure defined in

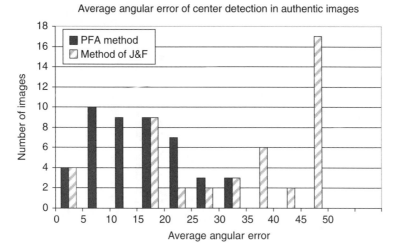

Average angular error of center detection in authentic images

Figure 14.23 Performance results over a set of authentic images. Histogram of the average angular error between optical and geometric image centres is shown. Solid bars represent the results of the method presented earlier, while striped bars represent results for the method in Johnson and Farid (2006).

Muller *et al.* (2001) which evaluates the ratio between the intersection and the union of the true un-cropped image area and the estimated image area. Figure 14.24 presents the results showing that size-restoration success rate is between 70 and 95% for over 89% of the images, with a total average of 80% success rate. For comparison, results are shown for the approach in Johnson and Farid (2006), where the average success rate is 53%.

In the third experiment, copy–paste forgery was tested. A set of 45 forged images was created by replacing four patches in each image with patches extracted from a different image. Each patch size was 4% of the total image size. This test sets a major challenge to the algorithm due to the small size of the forged region. Forgery detection was applied on the images and regions suspected as forged were marked. For the proposed algorithm, forged regions were marked using the following settings:

1. PFA events for which the PFA direction formed more than 90° angle with the calculated centre were marked as suspected forgery.
2. Smoothing was applied to the map of weighted distances from the PFA events to the calculated centre of the image to remove remaining noise.
3. Filtered PFA events with a weighted distance greater than a predefined threshold were marked as suspected forged regions. This threshold was set to be 0.0014% of the maximum penalty factor.

The image was segmented into regions the same size as the forged patches. An authentic region marked as forged is considered a false positive, while a forged patch

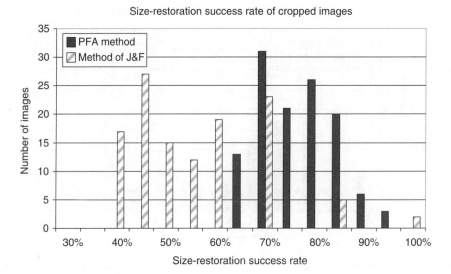

Figure 14.24 Performance results over a set of cropped images. Histogram of success rate in size restoration of the cropped images is shown.

marked as such was considered a true positive. An example is shown in Figure 14.25. An original image with four pasted regions is shown, with the regions detected as suspected forgery marked. Note that a false-positive window was detected in the bottom-right corner due to image noise. In our experiment, a region was considered forged if the calculated centre was not within 4% of the image centre. The results for the proposed method showed a very low false-positive rate (6%), accompanied by close to 70% true-positive rate; in comparison, the method in Johnson and Farid (2006) suffers from over-detection of image regions as suspected forgery. Consequently, it has a false-positive rate of 89%, with true-positive rate of 96%.

14.4.4   Discussion

Some of the results of the previous tests share a common characteristic that indicates the general strengths and weaknesses of the algorithm. Across all tests, it can be seen that images with stronger contrast generate better results (see Figure 14.26). This can be seen in Figure 14.27 (left) which shows the distribution of angular error as a function of image contrast measured as sum of squared gradients in the image. This is not surprising, as PFA appears more strongly over edges with greater contrast as described in Section 14.4.1 and supported in Figure 14.27 (right). In addition, as the sharpness of the image decreases, there are fewer distinct boundaries between objects in the image. Since this is the grounds on which the PFA is sought (see Section 14.4.2), the possible number of PFA events decreases as well as their reliability, and the analysis may become degraded. Furthermore, the number of PFA events detected in an image

Figure 14.25 An example of a synthetically forged image, with four patches inserted from a different image. Results of image forgery detection are shown: three true positives are marked in solid green; a false positive is marked in dashed red outlines.

Figure 14.26 Contrast effects. Left: image with high contrast and clear boundaries generated a good result (average angular error of 8.47°). Right: blurred image with low contrast and a relatively small number of objects (average angular error of 20°).

and their distribution within the image affect the results of the algorithm. The location of PFA events is highly dependent on image content. In our tests, PFA events ranged in number between 500 and 12 000. The distribution of PFA events was found to be uneven.

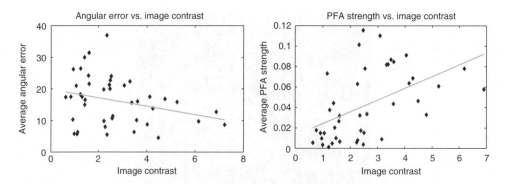

Figure 14.27 Left: distribution of angular error as a function of image contrast. As contrast increases, the angular error decreases since additional PFA events are found. Right: average PFA strength of all events in an image, as a function of image contrast. As described in Section 14.4.1, PFA is stronger across edges with greater contrast. In both graphs, a linear trend line was added for emphasis purposes.

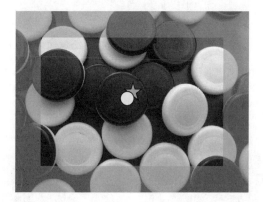

Figure 14.28 Symmetric cropping about the image centre is difficult to detect.

The weaknesses of the algorithm beyond the statistics of PFA events in the image involve the specifics of the forgery. First, the method is not sensitive to symmetric cropping about the image centre as shown in Figure 14.28. Additionally, insertion of a region extracted from a source image into the same relative location in another image results in a forged image that is difficult to detect by the proposed method. These drawbacks arise from the fact that PFA is based on local indicators, each of which is evaluated relative to the image centre.

14.5 Conclusion

In this chapter, we reviewed the topic of image forgery based on inherent characteristics of the acquisition system. The camera and most acquisition devices are comprised of three basic units: the optics, the sensors and the image processing unit.

Each unit forms a stage in the acquisition process; however, each unit also introduces unique effects (whether positive or negative) which may leave a telltale signature in the final output image. These affects may then be exploited in attempts to determine the source camera of an image as well as to detect forgery in images. In this chapter, we reviewed studies dealing with forgery based on each of the three camera units. We also presented a case study which described a specific forgery detection method based on chromatic aberrations.

References

N. Ahmed, T. Natarajan, and K.R. Rao, 'Discrete cosine transform', IEEE Transactions on Computers, vol. C-23, no. 1, pp. 90–93, 1974.

S. Bayram, H.T. Sencar, N. Memon, and I. Avciba, 'Source camera identification based on CFA interpolation', in IEEE International Conference on Image Processing. Genoa, Italy: IEEE, 2005.

S. Bayram, H.T. Sencar, and N. Memon, 'Improvements on source camera model identification based on CFA interpolation', in International Conference on Digital Forensics, 2006.

S. Bayram, H.T. Sencar, and N. Memon, 'Classification of digital camera models based on demosaicing artifacts', Digital Investigation, vol. 5, pp. 46–59, 2008.

F. Benford, 'The law of anomalous numbers', Proceedings of the American Philosophical Society, vol. 78, no. 4, pp. 551–572, 1938.

T. Bianchi and A. Piva, 'Reverse engineering of double JPEG compression in the presence of image resizing.' in Workshop on Information Forensics and Security, pp. 127–132, 2012.

E. Bilissi and M. Langford, "Advanced Photography", London/New York: Focal Press, 2007.

G. Buchsbaum, 'A spatial processor model for object colour perception', Journal of the Franklin Institute, vol. 310, no. 1, pp. 1–26, 1980.

J. Canny, 'A computational approach to edge detection', IEEE Transactions on Pattern Analysis and Machine Intelligence, vol. 8, pp. 679–698, June 1986.

H. Cao and A.C. Kot, 'Accurate detection of demosaicing regularity for digital image forensics', IEEE Transactions on Information Forensics and Security, vol. 4, no. 4, pp. 899–910, 2009.

H. Cao and A.C. Kot, 'Detection of tampering inconsistencies on mobile photos', in International Workshop on Digital Watermarking, pp. 105–119, 2010.

H. Cao and A.C. Kot, 'Measuring the statistical correlation inconsistencies in mobile images for tamper detection', Transactions on Dating Hiding and Multimedia Security, vol. 7, pp. 63–81, 2012.

O. Celiktutan, I. Avciba, B. Sankur, and N. Memon, 'Source cell-phone identification', IEEE Signal Processing and Communications Applications, pp. 1–3, 2006.

O. Celiktutan, B. Sankur, and I. Avciba, 'Blind identification of source cell-phone model', IEEE Transactions on Information Forensics and Security, vol. 3, no. 3, pp. 553–566, 2008.

L.H. Chan, N.F. Law, and W.C. Siu, "A two dimensional camera identification method based on image sensor noise," in IEEE International Conference on Acoustics, Speech and Signal Processing, pp. 1741–1744, 2012.

Y.L. Chen and C.T. Hsu, 'Detecting recompression of JPEG images via periodicity analysis of compression artifacts for tampering detection', IEEE Transactions on Information Forensics and Security, vol. 6, no. 2, pp. 396–406, 2011.

M. Chen, J. Fridrich, and M. Goljan, 'Digital imaging sensor identification (further study),' in SPIE Conference on Security, Steganography, and Watermarking of Multimedia Contents (E.J. Delp and P.W. Wong, eds.), vol. 6505, San Jose, CA. Bellingham, WA: SPIE, 2007a.

M. Chen, J. Fridrich, M. Goljan, and J. Lukas, 'Source digital camcorder identirfication using sensor photo-response non-uniformity,' in SPIE Conference on Security, Steganography, and Watermarking of Multimedia Contents (E.J. Delp and P.W. Wong, eds.), vol. 6505, 2007b.

M. Chen, J. Fridrich, M. Goljan, and J. Lukas, 'Determining image origin and integrity using sensor noise', IEEE Transactions on Information Forensics and Security, vol. 3, no. 1, pp. 74–90, 2008.

H.R. Chennamma and L. Rangarajan, 'Image splicing detection using inherent lens radial distortion,' International Journal of Computer Science Issues, vol. 7, no. 6, pp. 149–158, 2010.

G. Chierchia, S. Parrilli, G. Poggi, C. Sansone, and L. Verdoliva, 'On the infiuence of denoising in PRNU based forgery detection,' in ACM Workshop on Multimedia in Forensics, Security and Intelligence, Firenze, Italy, pp. 117–122, New York: ACM Press, 2010.

K.S. Choi, E.Y. Lam, and K.K.Y. Wong, "Automatic source camera identification using intrinsic lens radial distortion, " Optics Express, vol. 14, no. 24, pp. 11551–11565, 2006a.

K.S. Choi, E.Y. Lam, and K.K.Y. Wong, 'Source camera identification using footprints from lens aberration', in SPIE Conference on Digital Photography (N. Sampat, J.M. DiCarlo, and R.A. Martin, eds.), vol. 6069, San Jose, CA. Bellingham, WA: SPIE, 2006b.

C.H. Choi, J.H. Choi, and H.K. Lee, 'CFA pattern identification of digital cameras using intermediate value counting', in Proceedings of the Thirteenth ACM Multimedia Workshop on Multimedia and Security, Buffalo, NY, pp. 21–26, New York: ACM Press, 2011.

C.H. Choi, H.Y. Lee, and H.K. Lee, 'Estimation of colour modification in digital images by CFA pattern change', Forensic Science International, vol. 226, no. 1–3, pp. 94–105, 2013.

C. Chunhua, Y.Q. Shi, and S. Wei, 'A machine learning based scheme for double JPEG compression detection', in International Conference on Pattern Recognition, pp. 1–4, 2008.

J. Cohen, 'Dependency of the spectral reflectance curves of the munsell color chips', Psychonomic Science, vol. 1, pp. 369–370, 1964.

Z. Deng, A. Gijsenij, and J. Zhang, 'Source camera identification using auto-white balance approximation', in IEEE International Conference on Computer Vision, pp. 57–64, 2011.

A.E. Dirik and N. Memon, 'Image tamper detection based on demosaicing artifacts', in IEEE International Conference on Image Processing, pp. 1509–1512, 2009.

A.E. Dirik, S. Bayram, H.T. Sencar, and N. Memon, 'New features to identify computer generated images', in IEEE International Conference on Image Processing, vol. 4, pp. IV–433–IV–436, 2007a.

A.E. Dirik, H.T. Sencar, and N. Memon, 'Source camera identification based on sensor dust characteristics', in IEEE Workshop on Signal Processing Applications for Public Security and Forensics, pp. 1–6, 2007b.

A.E. Dirik, H.T. Sencar, and N. Memon, 'Digital single lens reflex camera identification from traces of sensor dust', IEEE Transactions on Information Forensics and Security, vol. 3, no. 3, pp. 539–552, 2008.

Z. Fan and R. de Queiroz, 'Identification of bitmap compression history: JPEG detection and quantizer estimation', IEEE Transactions on Image Processing, vol. 12, no. 2, pp. 230–235, 2003.

W. Fan, K. Wang, F. Cayre and Z. Xiong, '3D lighting-based image forgery detection using shape-from-shading,' in 20th European Signal Processing Conference, Bucharest, Romania, pp. 1777–1781, 2012.

H. Farid, 'Digital image ballistics from JPEG quantization', Tech. Rep. TR2006-583, Department of Computer Science, Dartmouth College, 2006.

H. Farid, 'Digital image ballistics from JPEG quantization: a follow up study', Tech. Rep. TR2008-638, Department of Computer Science, Dartmouth College, 2008.

H. Farid, 'Exposing digital forgeries from JPEG ghosts', IEEE Transactions on Information Forensics and Security, vol. 1, no. 4, pp. 154–160, 2009a.

H. Farid, 'A survey of image forgery detection', IEEE Signal Processing Magazine, vol. 2, no. 26, pp. 16–25, 2009b.

H. Farid and M. Bravo, 'Image forensic analyses that elude the human visual system,' in SPIE Symposium on Electronic Imaging, San Jose, CA, 2010.

P. Ferrara, T. Bianchi, A. DeRosaz, and A. Piva, 'Detection and estimation of contrast enhancement in double compressed images', in IEEE International Workshop on Multimedia Signal Processing (MMSP2013), 2013.

T. Filler, J. Fridrich, and M. Goljan, 'Using sensor pattern noise for camera model identification', in IEEE International Conference on Image Processing, pp. 1296–1299, 2008.

J. Fridrich, M. Goljanb, and R. Du, 'Steganalysis based on JPEG compatibility', in SPIE Multimedia Systems and Applications IV, pp. 275–280, 2001.

D. Fu, Y.Q. Shi, and W. Su, 'A generalized Benford's law for JPEG coefficients and its applications in image forensics', in SPIE Conference on Security, Steganography, and Watermarking of Multimedia Contents (E.J. Delp and P.W. Wong, eds.), vol. 6505, 2007.

A.C. Gallagher, 'Detection of linear and cubic interpolation in jpeg compressed images', in Second Canadian Conference on Computer and Robot Vision, pp. 65–72, 2005.

A.C. Gallagher and T.H. Chen, 'Image authentication by detecting traces of demosaicing', in IEEE Workshop on Vision of the Unseen (in conjunction with CVPR), pp. 1–8, 2008.

Z.J. Geradts, J. Bijhold, M. Kieft, K. Kurosawa, K. Kuroki, and N. Saitoh, 'Methods for identification of images acquired with digital cameras', in SPIE Conference on Enabling Technologies for Law Enforcement and Security (S.K. Bramble, E.M. Carapezza, and L.I. Rudin, eds.), vol. 4232, Boston, MA, pp. 505–512, Bellingham, WA: SPIE, 2001.

M. Goljan, M. Chen, and J. Fridrich, 'Identifying common source digital camera from image pairs', in IEEE International Conference on Image Processing, 2007.

H. Gou, A. Swaminathan, and M. Wu, 'Robust scanner identification based on noise features', in SPIE Conference on Security, Steganography, and Watermarking of Multimedia Contents, (E.J. Delp and P.W. Wong, eds.), February 2007.

J. He, Z. Lin, L. Wang, and X. Tang, 'Detecting doctored JPEG images via DCT coefficient analysis', in European Conference on Computer Vision, 2006.

E. Hecht, Optics. San Francisco, CA: Pearson Education/Addison-Wesley, 2002.

H. Hel-Or and B.A. Wandell, 'Object-based illumination classification', Pattern Recognition, vol. 35, no. 8, pp. 1723–1732, 2002.

J. Ho, O.C. Au, and J. Zhou, 'Inter-channel demosaicking traces for digital image forensics', in IEEE International Conference on Multimedia and Expo, 2010.

G.C. Holst, CCD Arrays, Cameras, and Displays. Winter Park, FL: JCD Publishing, 1998.

B.K.P. Horn and B.G. Schunck, 'Determining optical fiow', Artificial Intelligence, vol. 17, pp. 185–203, 1981.

Y.F. Hsu and S.F. Chang, 'Detecting image splicing using geometry invariants and camera characteristics consistency', in IEEE International Conference on Multimedia and Expo, July 2006.

Y.F. Hsu and S.F. Chang, 'Image splicing detection using camera response function consistency and automatic segmentation', in IEEE International Conference on Multimedia and Expo, July 2007.

Y. Hu, C.T. Li, X. Lin, and B.B. Liu, 'An improved algorithm for camera model identification using inter-channel demosaicking traces', in Eighth International Conference on Intelligent Information Hiding and Multimedia Signal Processing (IIH-MSP) (G.A. Tsihrintzis, J.S. Pan, H.-C. Huang, M. Virvou, and L.C. Jain, eds.), Piraeus-Athens, Greece, pp. 325–330, Los Alamitos, CA: IEEE, 2012.

Y. Huang and Y. Long, 'Demosaicking recognition with applications in digital photo authentication based on a quadratic pixel correlation model', in Computer Vision and Pattern Recognition, pp. 1–8, 2008.

F. Huang, J. Huang, and Y.Q. Shi, 'Detecting double JPEG compression with the same quantization matrix', IEEE Transactions on Information Forensics and Security, vol. 5, no. 4, pp. 848–856, 2010.

R.W. Hunt, The Reproduction of Colour. Hoboken, NJ: John Wiley & Sons, Inc., 2005.

B. Jahne, Digital Image Processing. Berlin/New York: Springer, 1995.

J.R. Janesick, Scientific Charge-Coupled Devices. Bellingham, WA: SPIE Press Monographs, Society of Photo Optical, 2001.

M. K. Johnson and H. Farid, 'Exposing digital forgeries by detecting inconsistencies in lighting', in ACM Multimedia and Security Workshop, pp. 1–10, 2005.

M.K. Johnson and H. Farid, 'Exposing digital forgeries through chromatic aberration', in ACM Multimedia and Security Workshop, pp. 48–55, 2006.

M.K. Johnson and H. Farid, "Detecting photographic composites of people', in Proceedings of the International. Workshop on Digital Watermarking, 2007a.

M. K. Johnson and H. Farid, 'Exposing digital forgeries in complex lighting environments', IEEE Transactions on Information Forensics and Security, vol. 3, no. 2, pp. 450–461, 2007b.

E. Kee and H. Farid, 'Exposing digital forgeries from 3-D lighting environments', in Workshop on Information Forensics and Security, 2010.

E. Kee, M.K. Johnson, and H. Farid, 'Digital image authentication from JPEG headers', IEEE Transactions on Information Forensics and Security, vol. 6, no. 3, pp. 1066–1075, 2011.

E. Kee, J. O. 'Brien, and H. Farid, 'Exposing photo manipulation with inconsistent shadows', ACM Transaction on Graphic, vol. 32, no. 3, pp. 1–12, 2013.

N. Khanna, A.K. Mikkilineni, G.T.C. Chiu, J.P. Allebach, and E.J. Delp, 'Forensic classification of imaging sensor types', in SPIE Conference on Security, Steganography, and Watermarking of Multimedia Contents (E.J. Delp and P.W. Wong, eds.), vol. 6505, 2007a.

N. Khanna, A.K. Mikkilineni, G.T.C. Chiu, J.P. Allebach, and E.J. Delp, 'Scanner identification using sensor pattern noise', in SPIE Conference on Security, Steganography, and Watermarking of Multimedia Contents (E.J. Delp and P.W. Wong, eds.), vol. 6505, 2007b.

N. Khanna, A.K. Mikkilineni, and E.J. Delp, 'Scanner identification using feature based processing and analysis', IEEE Transactions on Information Forensics and Security, vol. 4, no. 1, pp. 123–139, 2009.

R. Kingslake, Optics in Photography. Bellingham, WA: SPIE Press Monographs, SPIE Optical Engineering Press, 1992.

M. Kirchner, 'Efficient estimation of CFA pattern configuration in digital camera images', in SPIE Conference on Media Forensics and Security, vol. 7541, pp. 11–22, 2010.

J.D. Kornblum, 'Using JPEG quantization tables to identify imagery processed by software', Digital Investigation, vol. 5, Supplement 1, pp. S21–S25, 2008.

K. Kurosawa, K. Kuroki, and N. Saitoh, 'CCD fingerprint method identification of a video camera from videotaped images', in IEEE International Conference on Image Processing, vol. 3, pp. 537–540, 1999.

E.Y. Lamand, G.S. Fung, 'Automatic white balancing in digital photography', in Single-Sensor Imaging: Methods and Applications for Digital Cameras (R. Lukac, ed.), pp. 267–294, Boca Raton, FL: CRC Press, 2008.

E.H. Land, 'The retinex theory of color vision', Scientific American, vol. 237, no. 6, pp. 108–128, 1977.

T.V. Lanh, K.S. Chong, S. Emmanuel, and M.S. Kankanhalli, 'A survey on digital camera image forensic methods', in IEEE International Conference on Multimedia and Expo, pp. 16–19, 2007a.

T.V. Lanh, S. Emmanuel, and M.S. Kankanhalli, 'Identifying source cell phone using chromatic aberration', in IEEE International Conference on Multimedia and Expo, July 2007b.

C.T. Li, 'Source camera linking using enhanced sensor pattern noise extracted from images', in International Conference on Imaging for Crime Detection and Prevention, 2009.

X. Li, B. Gunturk, and L. Zhang, 'Image demosaicing: a systematic survey', Electronic Imaging, vol. 6822, pp. 1–15, 2008a.

B. Li, Y. Q. Shi, and J. Huang, 'Detecting doubly compressed JPEG images by using mode based first digit features', in IEEE Workshop on Multimedia Signal Processing, pp. 730–735, 2008b.

W. Li, Y. Yuan, and N. Yu, 'Passive detection of doctored JPEG image via block artifact grid extraction', IEEE Transactions on Signal Processing, vol. 89, no. 9, pp. 1821–1829, 2009.

B. Li, Y.Q. Shi, and J. Huang, 'Detecting double compressed JPEG images by using moment features of mode based DCT histograms', in IEEE Intlernational Conference on Multimedia Technology, pp. 23–26, 2010.

S. Lin and L. Zhang, 'Determining the radiometric response function from a single grayscale image', in Proceedings of the 2005 IEEE Computer Society Conference on Computer Vision and Pattern Recognition (CVPR'05)-Volume2-Volume02, CVPR'05, Washington, DC, pp. 66–73, IEEE Computer Society, 2005.

S. Lin, J. Gu, S. Yamazaki, and H.Y. Shum, 'Radiometriccalibration from a single image', in Proceedings of the 2004 IEEE Computer Society Conference on Computer Vision and Pattern Recognition, CVPR'04, Washington, DC, pp. 938–945, IEEE Computer Society, 2004.

Z. Lin, R. Wang, X. Tang, and H.Y. Shum, 'Detecting doctored images using camera response normality and consistency', in IEEE Conference on Computer Vision and Pattern Recognition, 2005.

Z. Lin, J. He, X. Tang, and C.K. Tang, 'Fast, automatic and fine-grained tampered JPEG image detection via DCT coefficient analysis', Pattern Recognition, vol. 42, no. 11, pp. 2492–2501, 2009.

Q. Liu, X. Cao, C. Deng, and X. Guo, 'Identifying image composites through shadow matte consistency', IEEE Transactions on Information Forensics and Security, vol. 6, no. 3, pp. 1111–1122, 2011.

Y. Long and Y. Huang, 'Image based source camera identification using demosaicking', in IEEE Workshop on Multimedia Signal Processing, pp. 419–424, 2006.

J. Lukas and J. Fridrich, 'Estimation of primary quantization matrix in double compressed JPEG images', in Digital Forensic Research Workshop, Auguat 2003.

J. Lukas, J. Fridrich, and M. Goljan, 'Detecting digital image forgeries using sensor pattern noise', in SPIE Conference on Security, Steganography, and Watermarking of Multimedia Contents (E.J. Delp and P.W. Wong, eds.), vol. 6072, 2006a.

J. Lukas, J. Fridrich, and M. Goljan, 'Digital camera identification from sensor noise', IEEE Transactions on Information Forensics and Security, vol. 1, no. 2, pp. 205–214, 2006b.

W. Luo, Z. Qu, J. Huang, and G. Qiu, 'A novel method for detecting cropped and recompressed image block', in IEEE Conference on Acoustics, Speech and Signal Processing, pp. 217–220, 2007.

W. Luo, J. Huang, and G. Qiu, 'JPEG error analysis and its applications to digital image forensics', IEEE Transactions on Information Forensics and Security, vol. 5, no. 3, pp. 480–491, 2010.

S. Lyu, 'Estimating vignetting function from a single image for image authentication', in ACM Workshop on Multimedia and Security, pp. 3–12, 2010.

B. Mahdian and S. Saic, 'Detecting double compressed JPEG images', in International Conference on Imaging for Crime Detection and Prevention, 2009.

B. Mahdian and S. Saic, 'A bibliography on blind methods for identifying image forgery', Image Communication, vol. 25, no. 6, pp. 389–399, 2010.

B. Mahdian, S. Saic, and R. Nedbal, 'JPEG quantization tables forensics: a statistical approach', in International Workshop on Computational Forensics, pp. 150–159, 2010.

S. Manimurugan and B. Jose, 'A novel method for detecting triple JPEG compression with the same quantization matrix', International Journal of Engineering Trends and Technology, vol. 3, no. 2, pp. 94–97, 2012.

D. Menon and G. Calvagno, 'Color image demosaicking: an overview', Journal of Image Communication, vol. 26, no. 8–9, pp. 518–533, 2011.

J. Nakamura, Image Sensors and Signal Processing for Digital Still Cameras. Optical Science and Engineering. Boca Raton, FL: Taylor & Francis, 2005.

S. Negahdaripour and B.K.P. Horn, 'A direct method for locating the focus of expansion', Computer Vision, Graphics, and Image Processing, vol. 46, no. 3, pp. 303–326, 1989.

T.T. Ng, 'Camera response function signature for digital forensics part II: signature extraction', in IEEE Workshop on Information Forensics and Security, pp. 161–165, December 2009.

T.T. Ng and M.P. Tsui, 'Camera response function signature for digital forensics-part I: theory and data selection', in IEEE Workshop on Information Forensics and Security, pp. 156–160, December 2009.

T.T. Ng, S.F. Chang, and M.P. Tsui, 'Using geometry invariants for camera response function estimation', in IEEE Conference on Computer Vision and Pattern Recognition, June 2007.

J. O'Brien and H. Farid, 'Exposing photo manipulation with inconsistent reflections', ACM Transactions on Graphics, vol. 31, no. 1 pp. 4:1–4:11, 2012.

F. Pedrotti and L. Pedrotti, Introduction to Optics. Upper Saddle River, NJ: Pearson Education, 2008.

W.B. Pennebaker and J.L. Mitchell, JPEG Still Image Data Compression Standard, 1st ed. Norwell, MA: Kluwer Academic Publishers, 1992.

W.B. Pennebaker and J.L. Mitchell, JPEG Still Image Data Compression Standard. New York: Van Nostrand Reinhold, 1993.

T. Pevn'y and J. Fridrich, 'Detection of double-compression in JPEG images for applications in steganography,' IEEE Transactions on Information Forensics and Security, vol. 3, pp. 247–258, June 2008.

A. Piva, 'An overview on image forensics', ISRN Signal Processing, vol. 2013, p. 22, 2013.

A.C. Popescu, Statistical Tools for Digital Image Forensics. PhD thesis, Department of Computer Science, Dartmouth College, Hanover, NH, 2005.

A.C. Popescu and H. Farid, 'Statistical tools for digital forensics,' in International Workshop on Information Hiding, pp. 128–147, 2004.

A.C. Popescu and H. Farid, 'Exposing digital forgeries by detecting traces of resampling', IEEE Transactions on Signal Processing, vol. 53, no. 2, pp. 758–767, 2005a.

A.C. Popescu and H. Farid, "Exposing digital forgeries in colour filter array interpolated images," IEEE Transactions on Signal Processing, vol. 53, no. 10, pp. 3948–3959, 2005b.

Z. Qu, W. Luo, and J. Huang, 'A convolutive mixing model for shifted double JPEG compression with application to passive image authentication', in IEEE International Conference on Acoustics, Speech, and Signal Processing, pp. 1661–1664, 2008.

R. Ramanath, W.E. Snyder, Y. Yoo, and M.S. Drew, 'Color image processing pipeline in digital still cameras', Signal Processing Magazine Special Issue on Color Image Processing, vol. 22, no. 1, pp. 34–43, 2005.

S.F. Ray, Applied Photographic Optics: Lenses and Optical Systems for Photography, Film, Video, Electronic and Digital Imaging. Oxford: Focal Press, 2002.

P.J. Rousseeuw and A.M. Leroy, Robust Regression and Outlier Detection. Hoboken, NJ: John Wiley & Sons, Inc., 2003.

L. Seidel, 'Ueber die theorie der fehler, mit welchen die durch optische instrumente gesehenen bilder, behaftet sind, und uber diemathematischen bedingungen ihrer aufhebung', Abhandlungen der naturwissenschaftlich-technischen Commission der Bayerischen Akademieder Wissenschaften, vol. 1, pp. 227–267, 1857. Translation: Essays in the scientific-technical Commission of the Bavarian Academy of Sciences.

H.T. Sencar and N. Memon, "Overview of state-of-the-art in digital image forensics, "in Algorithms, Architecture and Information Systems Security (B.B. Bhattacharya, S. Sur-Kolay, S.C. Nandy, and A. Bagchi, eds.), pp. 325–348, Hackensack, NJ: WorldScientific, 2007.

W. Smith, Modern Optical Engineering, 4th ed. New York: McGraw Hill Professional, McGraw Hill, 2007.

sRGB, Multimedia systems and equipment – colour measurement and management – Part 2-1: colour management – default RGB colour space – sRGB. IEC 61966-2-1 (1999–10), 1999.

Y. Sutcu, S. Bayram, H.T. Sencar, and N. Memon, 'Improvements on sensor noise based source camera identification', in IEEE International Conference on Multimedia and Expo, pp. 24–27, 2007.

A. Swaminathan, M. Wu, and K.J.R. Liu, 'Component forensics of digital cameras: a non-intrusive approach', in Information Sciences and Systems, pp. 1194–1199, March 2006.

A. Swaminathan, M. Wu, and K.J.R. Liu, 'Non intrusive component forensics of visual sensors using output images', IEEE Transactions on Information Forensics and Security, vol. 2, pp. 91–106, 2007.

A. Swaminathan, M. Wu, and K.J.R. Liu, 'Digital image forensics via intrinsic fingerprints', IEEE Transactions on Information Forensics and Security, vol. 3, no. 1, pp. 101–117, 2008.

J. Takamatsu, Y. Matsushita, T. Ogasawara, and K. Ikeuchi, 'Estimating demosaicing algorithms using image noise variance', in IEEE Conference on Computer Vision and Pattern Recognition, 2010.

P. Van Walree, 'Photographic Optics', 2014. toothwalker.org/optics.html (Accessed 4 February 2015).

G.K. Wallace, 'The JPEG still picture compression standard', Communications of the ACM, vol. 34, pp. 30–44, 1991.

W. Wang, J. Dong, and T. Tan, 'A survey of passive image tampering detection', in International Workshop on Digital Watermarking (A.T.S. Ho, Y.Q. Shi, H.J. Kim, and M. Barni, eds.), Guildford, UK, pp. 308–322, Berlin: Springer-Verlag, 2009.

W. Wang, J. Dong, and T. Tan, 'Tampered region localization of digital colour images based on JPEG compression noise', in International Workshop on Digital Watermarking, pp. 120–133, 2010.

Worth1000, image manipulation and contest website. Opened January 2002. http://.www.worth1000.com (Accessed 4 February 2015).

L. Wu, X. Cao, W. Zhang, and Y. Wang 'Detecting image forgeries using metrology', Machine Vision and Applications, vol. 23, no. 2, pp. 363–373, 2010.

G. Wyszecki and W.S. Stiles, Color Science: Concepts and Methods, Quantitative Data and Formulae, 2nd ed. New York: John Wiley & Sons, Inc., 1967.

J. Yanga, G. Zhua, and J. Huang, 'Detecting doubly compressed JPEG images by factor histogram', in Annual Conference of the Asia Pacific Signal and Information Processing Association, October 2011.

S. Ye, Q. Sun, and E.C. Chang, 'Detecting digital image forgeries by measuring inconsistencies of blocking artifact', in IEEE International Conference on Multimedia and Expo, pp. 12–15, 2007.

I. Yerushalmy and H. Hel-Or, 'Digital image forgery detection based on lens and sensor aberration', International Journal of Computer Vision, vol. 92, no. 1, pp. 71–91, 2011.

F. Zach, C. Riess, and E. Angelopoulou, 'Automated image forgery detection through classification of JPEG ghosts', in Pattern Recognition, Joint 34th DAGM and 36th OAGM Symposium, pp. 185–194, 2012.

W. Zhang, X. Cao, J. Zhang, J. Zhu, and P. Wang, 'Detecting photographic composites using shadows', in Proceedings of the IEEE International Conference on Multimedia and Expo (ICME'09), pp. 1042–1045, July 2009.

W. Zhang, X. Cao, Y. Qu , Y. Hou, H. Zhao and C. Zhang 'Detecting and extracting the photo composites using planar homography and graph cut', IEEE Transactions on Information Forensics and Security, vol. 5, pp. 544–555, 2010.

Y.Q. Zhao, F.Y. Shih, and Y.Q. Shi, 'Passive detection of paint-doctored JPEG images', in International Work shop on Digital Watermarking, pp. 1–11, 2010.

15

Image and Video Processing History Recovery

Tiziano Bianchi[1] and Alessandro Piva[2]
[1]*Department of Electronics and Telecommunications, Politecnico di Torino, Italy*
[2]*Department of Information Engineering, University of Florence, Italy*

15.1 Introduction

When observing an image or a video on a web site, often people do not realize that such media have undergone a long series of transformations before appearing in the current form. Many of such processing steps are not necessarily malicious. Images and videos are compressed to save bandwidth and storage space, are often resized and colour corrected to improve their quality, may be re-encoded to meet the requirements of different platforms.

Recovering the correct sequence of processing steps or, in short, the processing history of an image or a video, is an important task in multimedia forensics (Piva 2013; Stamm *et al.* 2013). Even the recovery of a subset of the parameters characterizing the processing history can provide important information on the image or video life cycle. Furthermore, a natural application is to verify whether the recovered history is consistent in different spatial or temporal portions of the same image or video, which can be a direct proof of manipulation of the original medium.

The main challenge in image and video processing history recovery is that one usually does not have access to the original media. Nevertheless, each of the processing operations usually leaves a trace on the final signal. By tracking those traces, it is often possible to recover a sequence of processing steps, that, even if approximated, is still useful to reconstruct the actual processing history.

Handbook of Digital Forensics of Multimedia Data and Devices, First Edition.
Edited by Anthony T.S. Ho and Shujun Li.
© 2015 John Wiley & Sons, Ltd. Published 2015 by John Wiley & Sons, Ltd.
Companion Website: www.wiley.com/go/digitalforensics

In this chapter, we will describe the most useful traces that can be used for image and video processing history recovery, and the main forensic techniques that are based on them. Due to the ubiquitous diffusion of compressed images and videos, the most important traces are those relying on specific properties of the encoding process. However, coding artefacts are rarely the only trace of processing. Editing operations, like resizing and colour enhancement, are commonly applied to both images and video.

In Sections 15.2 and 15.3, we will introduce the main existing approaches used in the forensic literature to reveal the presence of coding and editing artefacts. Then, in Section 15.4 we will show how to employ such tools to recover several parameters characterizing the processing history. Our approach is to give more importance to techniques that are most suitable in practical scenarios. Therefore, we will consider only techniques dealing with compressed images or videos, since, especially in the case of videos, it is highly unrealistic to have access to the original raw signal. Interestingly enough, in some cases the presence of coding artefacts seems to ease the estimation of editing parameters, so that this is a scenario where interesting results can be observed.

Section 15.5 is devoted to two interesting case studies, showing possible applications of processing history recovery to tamper detection. Namely, we will show how the illustrated forensic techniques can be applied in realistic settings in order to partly recover the processing history, and how this information can provide useful hints about the authenticity of the analyzed media.

15.2   Coding Artefacts

Lossy compression is one of the most common operations when dealing with digital images and videos, due to the convenience of handling smaller amounts of data, for both storage and transmission. As a matter of fact, with the exception of high-end models, in most digital cameras each picture is compressed directly after taking a shot, whereas video content is almost uniquely available in lossy compression format, since motion pictures would require a huge bit rate to be represented either in an uncompressed or a lossless format. Due to its lossy nature, image and video coding leaves detectable artefacts on the signals reconstructed after compression: revealing such coding-based artefacts in digital images and videos is a relevant task in history recovery, as well as a powerful tool for detecting traces of forgery.

In the following, we will discuss the main coding artefacts due to JPEG compression and generic video compression. Sections 15.2.1 and 15.2.2 provide a description of the main artefacts due a single JPEG compression and a double JPEG compression, respectively. Sections 15.2.3 and 15.2.4 are devoted to generic video encoding and re-encoding, respectively. The reader should be aware that the following sections do not cover all possible coding artefacts: less common coding artefacts, such as those due to an hybrid image compression (e.g. JPEG/JPEG2000) and multiple (i.e. more than two) encodings are only briefly referenced.

15.2.1 JPEG Compression

Despite its being introduced more than 20 years ago by the Joint Photographic Experts Group, JPEG is still today's the most common and widespread image compression standard (ISO/IEC 1991). In a nutshell, JPEG compression in its lossy version is performed following three basic steps:

1. *Discrete Cosine Transform (DCT)*: The image is divided into 8×8 non-overlapping blocks. For each block, pixel values are first shifted from unsigned integers in the range $[0, 2^b - 1]$ to signed integers in the range $[-2^{b-1}, 2^{b-1} - 1]$, where b is the number of bits per pixel (typically $b = 8$). Each block is then DCT transformed in order to obtain the coefficients $D(i,j)$, where i and j ($1 \le i, j \le 8$) are the row and column indexes within the block. In the case of colour images, red, green and blue channels are first converted into luminance and blue/red chrominance channels, chrominance channels are optionally downsampled, and the aforementioned process is applied independently on each channel.

2. *Quantization*: The DCT coefficients obtained in the previous step are quantized according to a quantization table which must be specified as an input to the encoder. Quantization is implemented as division of each DCT coefficient $D(i,j)$ by the corresponding quantizer step size $\Delta(i,j)$, followed by rounding to the nearest integer. That is,

$$C(i,j) = \text{sign}(D(i,j))\text{round}\left(\frac{|D(i,j)|}{\Delta(i,j)}\right). \tag{15.1}$$

Thus, the reconstructed value at the decoder is

$$D_Q(i,j) = \Delta(i,j) \cdot C(i,j). \tag{15.2}$$

The quantization table is *not* specified by the standard. In many JPEG implementations, it is usual to define a scalar quality factor Q corresponding to a set of predefined tables. This is the case, for instance, of the quantization tables adopted by the Independent JPEG Group,[1] which are obtained by properly scaling the image-independent quantization table suggested in Annex K of the JPEG standard (ISO/IEC 1991) according to a quality factor $Q \in [1, 100]$.

Since quantization is not invertible, this operation introduces characteristic errors on the reconstructed image. As we will see, most forensic methods rely on those errors to recover image processing history.

3. *Entropy Coding*: DCT quantized coefficients are losslessly coded and written to a bitstream, using either variable length coding by means of properly designed Huffman tables (the most common option) or arithmetic coding.

---

[1] www.ijg.org.

Probably the most simple scenario in which JPEG coding artefacts come into play regards digital images available in the pixel-domain as bitmaps, without any knowledge about prior processing. In these cases, it can be interesting to detect whether a particular image has been previously compressed and which were the compression parameters being used. The underlying idea of forensic methods coping with this problem is that JPEG leaves characteristic compression traces either in the pixel domain or in the transform domain.

15.2.1.1   Pixel Domain-Based Features

Block-based image coding schemes like JPEG introduce visible discontinuities at block boundaries, especially when the image is compressed at low bit rates. Several methods proposed in the literature leverage on this footprint in order to understand whether an image has been block-processed. The common idea of these methods is that discontinuities due to block processing can be revealed by computing local differences, or a derivative operator, in the pixel domain and evaluating the periodicity of the output. It is also interesting to note that similar features can be used to detect other periodic artefacts, like those caused by resampling, as will be discussed in Section 15.3.

Fan and de Queiroz (2000, 2003) described a method explicitly tailored to 8×8 block processing and capable of revealing artefacts even when very light JPEG compression is applied. The proposed algorithm is based on the idea that in a compressed image, the pixel differences across 8×8 block boundaries should be larger than those within a block. Such a behaviour can be measured by computing two signals, Z_1 and Z_2, taking into account inter and intra-block pixel differences. The energy of the difference between the histograms of Z_1 and Z_2 is compared to a threshold, and if it is higher that this threshold, the presence of prior compression is deduced.

Other methods consider the problem of estimating the block size and location. In Wang *et al.* (2000) the authors modelled discontinuities at block boundaries by introducing an interfering blocky signal. The absolute value of the gradient between each column or row of the image is used to estimate the power of the blocky signal and to reveal its presence. In a similar way, in Liu and Heynderickx (2008) the vertical and horizontal gradients are computed and their periodicity due to gradient peaks at block boundaries is estimated in the frequency domain using the discrete Fourier transform (DFT).

The robustness of such techniques can be improved with several expedients. For example, Tjoa *et al.* (2007) proposed to subtract a median filtered version to the gradient, in order to enhance the peaks, and then apply a threshold based on the sum of the gradients, aimed at avoiding spurious peaks caused by edges from objects in the image. The period of the resulting function is computed using a maximum likelihood estimation scheme adopted for pitch detection.

A different approach was developed in Chen and Hsu (2008), where a linear dependency model of pixel differences for within-block and across-block pixels was introduced. The probability of each pixel following this model is estimated using an expectation-maximization (EM) algorithm. Finally, the periodicity of the blocking artefacts is estimated by computing the spectrum of the probability map obtained in the previous step.

15.2.1.2 Transform Domain-Based Features

In the transform domain, lossy image coding schemes modify the histogram of transformed coefficients; more specifically, due to the scalar quantization operation, the histogram of each coefficient $D_Q(i,j)$ shows a typical comb-like distribution, in which the peaks are spaced apart by $\Delta(i,j)$, instead of a continuous distribution. Ideally, the distribution can be expressed as follows:

$$p(D_Q; \Delta) = \sum_k w_k \delta(D_Q - k\Delta), \tag{15.3}$$

where δ is the Dirac delta function and w_k are weights that depend on the original distribution (indexes (i,j) are omitted for the sake of clarity). The aforementioned feature is a powerful means for identifying lossy image compression, like JPEG. Moreover, the quantization step $\Delta(i,j)$ can be recovered by estimating the distance between peaks in these histograms.

This characteristic is partially attenuated when a JPEG image is reconstructed in the pixel domain, since pixel values are rounded to integers. As a consequence, the histograms of DCT coefficients ($\hat{D}_Q(i,j)$) computed from decoded pixel values are not exactly comb-shaped, but are blurred with respect to those obtained directly after quantization ($D_Q(i,j)$). An example of this behaviour is shown in Figure 15.1.

Several methods analyzing some characteristics of the shape of DCT histograms have been proposed in the literature that allow a previous JPEG compression to be detected; a simple example is given by the following method.

Luo et al. (2010) observed that in a JPEG-compressed image, the number of DCT coefficients with values in the range $(-1, +1)$ is greater than the number of DCT coefficients with values in the range $(-2, -1] \cup [+1, +2)$, when using quantization steps that are equal to or larger than 2. The authors defined the ratio between these two numbers as a feature, and verified that its value, in case of JPEG-compressed images, is usually close to zero and in general much smaller than that obtained on corresponding uncompressed images. JPEG compression can be then detected when the ratio is smaller than a given threshold.

A more powerful feature can be obtained by jointly considering several DCT coefficients quantized with different quantization steps. For example, in the case of JPEG compression, each 8×8 block in the reconstructed image is obtained by multiplying the quantized DCT coefficients by step factor and applying an inverse DCT. As a result, such blocks can be modelled as a point in a 64-dimension lattice, generated by the composition of the JPEG quantization matrix and the DCT matrix, perturbed by a noise due to rounding and truncation errors.

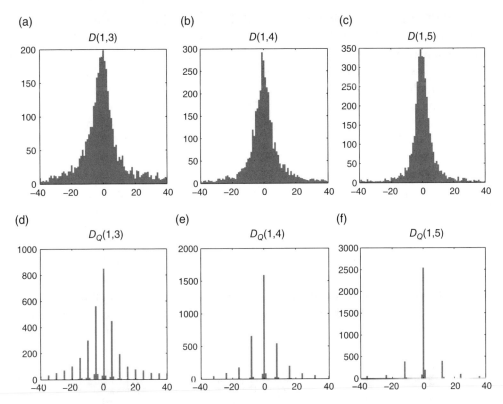

Figure 15.1 Histograms of three DCT coefficients of the popular Lena image ($D(1,3)$, $D(1,4)$, $D(1,5)$) before (a–c) and after (d–f) quantization. The quantization step $\Delta(i,j)$ can be estimated by the gaps between consecutive peaks. In the example $\Delta(1,3) = 5$, $\Delta(1,4) = 8$ and $\Delta(1,5) = 12$.

The distribution of such a signal can be modelled as a mixture given by several components, each one having its mean equal to a lattice point. Such a behaviour can be considered as a very powerful feature for detecting JPEG compression, since it is very unlikely that an uncompressed signal will exhibit such a distribution. Moreover, this near-lattice distribution property (NLDP) of JPEG images is expected to hold even if the decompressed image undergoes subsequent processing steps, which will be investigated in the following sections.

15.2.2 Double JPEG Compression

Since the JPEG format is adopted in most of the digital cameras and image processing tools, we can expect that some analyzed images may actually be recompressed JPEG images. Detecting the presence of artefacts introduced by JPEG recompression is a powerful tool for analyzing the processing history of a digital image. In this section we will mainly consider a single recompression step, meaning that we only have a first JPEG compression and a second JPEG compression. However, some of the techniques described hereafter can also be used to detect whether a JPEG image has been compressed more than twice (Milani *et al.* 2012c).

Recompression artefacts can be categorized into two classes, according to whether the second JPEG compression adopts a DCT grid aligned with the one used by the first compression (as shown in Figure 15.2) or not (as shown in Figure 15.3). The first case will be referred to as aligned double JPEG (A-DJPG) compression, whereas the second case will be referred to as non-aligned double JPEG (NA-DJPG) compression. NA-DJPG compression artefacts will normally occur when the image is cropped or padded before recompression.

JPEG recompression artefacts can be detected by applying specific models, which in general are valid only for one of the two possible classes outlined earlier. In the following, such models will be reviewed, and the main algorithms in the literature that are based on those model will be briefly described.

When the statistical modelling of coding artefacts becomes complicated, as in the case of double JPEG compression, a convenient approach is also to resort to machine learning techniques. In this framework, a classifier is trained on a suitable dataset providing examples of the different classes (e.g. singly JPEG compressed and doubly JPEG compressed images), so that it is able to learn the statistical model directly

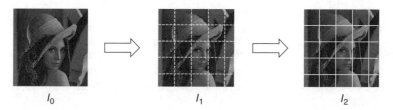

Figure 15.2 An example of aligned double JPEG (A-DJPG) compression: the uncompressed image I_0 is first compressed, with a block grid shown in dashed lines, obtaining a single compressed image I_1; this image is again compressed, with a block grid shown in solid lines, aligned with the previous one, obtaining the final image I_2.

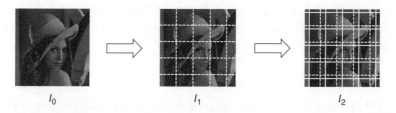

Figure 15.3 An example of non-aligned double JPEG (NA-DJPG) compression: the uncompressed image I_0 is first compressed, with a block grid shown in dashed lines, obtaining a single compressed image I_1; this image is again compressed, with a block grid shown in solid lines, misaligned with the previous one, obtaining the final image I_2.

from the data. Support vector machines (SVMs) are a powerful tool for implementing classifiers based on machine learning (Cortes and Vapnik 1995) and are widely used in multimedia forensics.

15.2.2.1 Detection of A-DJPG Compression

Most of the algorithms dealing with A-DJPG compression rely on the shape of the histogram of DCT coefficients. As noted by Lukáš and Fridrich (2003), aligned double JPEG compression can be approximated by double quantization (DQ) of transform coefficients $D(i,j)$, such that

$$D_{Q_1,Q_2} = \Delta_2 \cdot \text{sign}(D) \cdot \text{round} \left(\frac{\Delta_1}{\Delta_2} \text{round} \left(\frac{|D|}{\Delta_1} \right) \right), \qquad (15.4)$$

where indexes (i,j) have been omitted for the sake of clarity.

Re-quantizing already quantized coefficients with different quantization steps affects the histogram of DCT coefficients, introducing the so-called DQ effect. Characteristic periodic peaks in the histogram are introduced, which modify the original statistics according to different configurations, which depend on the relationship between the quantization step sizes of consecutive compression operations, that is respectively, Δ_1 and Δ_2, as shown in Figure 15.4. For this reason, most solutions are based on statistical footprints extracted from such histograms.

Popescu and Farid (2005c) reported the presence of periodic artefacts in the histogram of DCT coefficients as strong peaks in medium and high frequencies of the Fourier transformed histogram and proposed a statistical model to characterize such artefacts. This method was further explored by He *et al.* (2006), providing a way

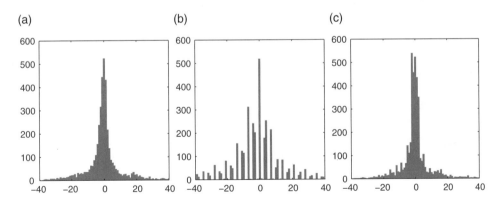

Figure 15.4 Histogram of quantized DCT coefficients $C = D/\Delta_1$ in the case of single compression (a) compared to the histograms of quantized DCT coefficients $C = D_{Q_1,Q_2}/\Delta_2$ in the case of A-DJPG compression: $\Delta_2 < \Delta_1$ (b) and $\Delta_2 > \Delta_1$ (c). It can be noticed that double quantization effects are much more evident when $\Delta_2 < \Delta_1$. In the example $\Delta_1 = 7$, whereas $\Delta_2 = 2$ (b) or $\Delta_2 = 9$ (c).

to automatically detect and locate regions that have have been doubly compressed. Interesting results were also obtained by computing a set of features measuring the periodicity of DCT coefficients, which is perturbed when an A-DJPG compression is applied, and training a classifier on those features, as shown in Chen and Hsu (2011).

Double quantization can be observed also in the pixel domain. Feng and Doërr (2010), starting from the observation that recompression induces periodic artefacts and discontinuities in the image histogram, derived a set of features from the pixel histogram and showed that an SVM trained on such features can detect A-DJPG compression.

Several improvements and modifications of the aforementioned techniques have been proposed in the literature. A very interesting extension, as will be specified in Section 15.4.3, is that fine-grained maps indicating the tampering probabilities for each 8×8 image block can be derived from the statistical analysis of these artefacts, as done by Lin *et al.* (2009b), Bianchi *et al.* (2011) and Bianchi and Piva (2012b).

Another widely-adopted strategy for the detection of A-DJPG compression relies on the so-called Benford's law or first digit law (Benford 1938): the distribution of the most significant decimal digit m (also called 'first digit') of the absolute value of quantized transformed coefficients, in the case of an original uncompressed image, is closely related to the Benford's equation or its generalized version, that is

$$p(m) = N \log_{10} \left(1 + \frac{1}{m} \right) \text{ or } p(m) = N \log_{10} \left(1 + \frac{1}{\alpha + m^{\beta}} \right), \tag{15.5}$$

respectively (where N is a normalizing constant). Whenever the empirical distribution deviates significantly from that predicted by the aforementioned equations, one can reasonably doubt that the image is compressed twice, and it is also possible to estimate the compression parameters of the first coding stage.

Most of the methods based on Benford's law actually compute the distribution of first digits either globally for all DCT coefficients, as in Fu *et al.* (2007), or separately for each DCT coefficient, as in Li *et al.* (2008), and then feed the resulting feature vector to a classifier, following a machine learning approach. The detection approaches based on such features prove to be very reliable, giving detection accuracy higher than 90%. More recently, this approach has also been extended to the case of multiple JPEG compression steps since in many practical cases images and videos are compressed more than twice (Milani *et al.* 2012c).

A different strategy that works well in the case of A-DJPG compression exploits the property of idempotency that characterizes the operators involved in the coding process. If an image has been encoded using a certain configuration, re-applying the same coding operations will lead to a new image that is highly correlated with the image under examination. In Farid (2009), the authors propose to recompress the image under analysis at several quantization factors, and then compare these differently compressed versions of the image with the input image; if the same quality factor of the one used for the double compressed area is adopted, a spatial local minima, the so-called JPEG ghosts, will appear in correspondence of that area. When used to detect possibly doubly compressed regions, this method works only if the region under analysis has a lower quality factor than the rest of the image, and it requires the suspect region to be known in advance.

15.2.2.2   Detection of NA-DJPG Compression

When the second JPEG compression uses a DCT grid that is not aligned to that of the first JPEG compression, DCT coefficients will not exhibit the typical DQ artefacts that are proper of A-DJPG compression. Therefore, the analysis of NA-DJPG compression requires specific tools based on different coding artefacts.

A possible way to detect the presence of NA-DJPG is to exploit blocking artefacts. The idea is that a singly compressed JPEG image exhibits regular blocking artefacts, whereas NA-DJPG compression usually disrupts this regularity, since the blocking artefacts due to the second JPEG compression interferes with the blocking artefact left by the first compression. Several methods are based on this kind of model.

Starting from the idea proposed by Fan and de Queiroz (2003), Luo *et al.* (2007) computed an 8×8 blocking artefact characteristics matrix (BACM) in the pixel domain to measure the symmetrical property of the blocking artefacts in a JPEG image; NA-DJPG compression is revealed by an asymmetric BACM. Some features, cumulated over the whole image, are extracted from the BACM and fed to a classifier in order to distinguish whether blocking artefacts are present or not.

Results demonstrated that the aforementioned method has good performance only when the quality of the last compression is higher than the quality of the first one. Furthermore, the method is reliable only when the analyzed region is above 500×500 pixels. In order to localize possibly tampered regions, the previous algorithm can be independently applied to different parts of an image obtained after segmentation, as shown in Barni *et al.* (2010).

Other methods based on blocking artefacts in the pixel domain include Chen and Hsu (2008) and Chen and Hsu (2011). Such methods are based on more sophisticated models than the BACM; however, the basic principle remains the same: a signal is extracted from the image, highlighting different behaviours in the case of single JPEG compression and NA-DJPG compression. Then, some statistical features are extracted from such a signal and machine learning techniques are used to classify them.

A slightly different approach is proposed in Qu *et al.* (2008), where the authors observed that each 8×8 image block, after NA-DJPG compression, is the result of the superposition of four different 8×8 blocks from the original JPEG image. The different contributions are then separated into independent signals by using independent component analysis (ICA)[2] and NA-DJPG compression is identified by means of a classifier applied to the result of the decomposition.

---

[2] ICA refers to a set of techniques used to separate independent and non-Gaussian components which are linearly mixed (Comon 1994).

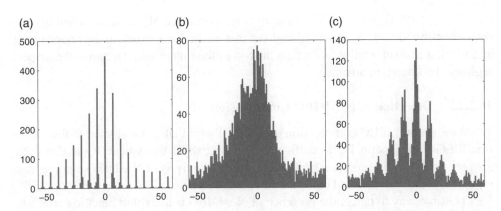

Figure 15.5 Histograms of DCT coefficients D computed with different 8×8 DCT grids in the presence of NA-DJPG compression: aligned with the last compression (a), random grid (b), aligned with the previous compression (c). It is evident that DCT coefficients computed according to the grid of the previous compression tend to cluster around the point of a lattice defined by Δ_1.

Another effective footprint for revealing NA-DJPG compression is the NLDP property discussed in Section 15.2.1. The main idea behind this method is that of detecting NA-DJPG compression by measuring how DCT coefficients cluster around a given lattice, defined by the JPEG quantization table, for any possible grid shift. As illustrated in Figure 15.5, when the DCT coefficients are computed according to the grid of the previous compression NA-DJPG can be detected, and the parameters of the lattice also give the values of the primary quantization table.

Bianchi and Piva (2011, 2012a) proposed a simple scalar measure for NLDP in the case of NA-DJPG compression. A simple threshold detector was employed, which does not rely on any classifier. Results obtained in this work showed an improvement with respect to previous approaches: an analyzed region of 256×256 pixels is sufficient to match the best results presented in previous works, and over 90% detection performance is obtained even in the presence of similar quantization factors in the first and second compressions. As a consequence, this method retains good performance even in the case of low-quality JPEG images. Moreover, in some cases the statistical properties of DCT coefficients can be used to detect NA-DJPG compression artefacts even on single 8×8 blocks (Bianchi and Piva 2012b), which permits a fine-grained localization of doubly compressed areas in an image.

Finally, idempotency can be successfully applied also in the case of NA-DJPG compression. Meng *et al.* (2010) presented a method for identifying tampering and recompression in a JPEG image based on the requantization of transform coefficients. Similarly to Farid (2009), the main idea relies on the fact that, when the image has been compressed twice, an image recompressed using the same quantization steps and the same grid of the first JPEG compression, will be highly correlated with the analyzed image. However, copied parts of the image might exhibit poor correlation due to the desynchronization of DCT blocks.

15.2.3 Video Compression

Video coding architectures are in general more complex than those adopted for still images. During the past two decades, a wide set of video coding algorithms have been standardized, introducing several degrees of freedom in the processing steps of a standard video encoder (Sullivan and Wiegand 2005). Although most of the widely used coding standards, like those of MPEG-x or H.26x families, inherit the use of block-wise transform coding from the JPEG standard, the architecture is complicated by several additional coding tools, since video compression aims at reducing both spatial and temporal redundancy in the captured video sequence. Moreover, often the transform adopted within the same coding standard is not unique.

The main modules in a conventional video coding architecture are illustrated in a simplified block diagram shown in Figure 15.6. First, the encoder splits the video sequence into frames, and each frame is divided into blocks of pixels. Each block is then predicted by a prediction module, exploiting spatial and/or temporal correlation. Finally, the prediction residual is encoded following a sequence of steps similar to those adopted by the JPEG standard.

Within video coding standards, MPEG-2 (ISO/IEC 2008) is one of the more widely employed, and so we will take it as a reference to illustrate more in detail the different coding steps. The standard defines different types of pictures: intra-coded pictures, referred to as I-frames; predictive-coded pictures, commonly named P-frame, when the prediction is based only on previous frames; and B-frames, when the prediction is based on both previous and future frames. Following the block-based coding principle, each frame of a video sequence is divided into macroblocks (MBs) of 16×16 samples, which are encoded according to different coding modes depending on the selected type of frame.

MBs in I-frames are encoded without considering other frames, in a similar way as it happens with JPEG images: each MB of the luminance component (a similar processing is applied to chrominance components) is divided into blocks of 8×8 pixels, DCT is applied to each block, and the resulting coefficients are quantized and then entropy encoded. For example, the so-called Motion-JPEG encoding (Sullivan

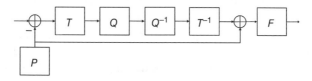

Figure 15.6 Simplified block diagram of a conventional video codec. P models motion compensated prediction, T is the orthonormal transform, Q is the quantizer, and F models rounding and in-loop filtering. Entropy encoding is omitted for brevity.

and Wiegand 2005) is obtained by compressing the whole video using only intra-coded pictures, without exploiting temporal redundancy.

Although being very similar to the JPEG compression scheme, the aforementioned procedure uses a slightly different quantization function. In MPEG-2, the coarseness of the quantization is selected by the encoder through the quantizer scale factor, denoted as S, that ranges from 0 to 31 and maps the values of the multiplier that is applied to the quantization matrix. The factor S controls the trade-off between quality and bitrate: if the value of S is constant, then a fixed quantizer will be used, leading to a Variable BitRate (VBR); if S is adapted on a frame by frame (or even on a MB by MB) basis, then a Constant BitRate (CBR) is achieved.

Temporal redundancy, that is the strong correlation between adjacent frames, is exploited through motion compensation. When encoding a picture as a P-frame, each MB is compared with the MBs lying in a corresponding area within the previously encoded and reconstructed frame (reference frame), in order to find the region that better resembles the MB to encode. If a good match is found, then the MB is predictively coded: the displacement between the current position and the position in the reference frame is encoded as a motion vector, and the difference between the actual MB and the reference MB is 8×8-DCT transformed and quantized. Predictively coded MB are denoted as P-MB. When a good match is not available, the MB is intra-coded like in an I-frame and we refer to this type of macroblock as I-MB. Finally, if the resulting motion vector is null and the residual difference after quantization is also negligible, then the MB is skipped and is denoted as S-MB. Temporal redundancy is further exploited by considering B-frames: the only difference between P-frames and B-frames is that the MBs on B-frames are bidirectionally predictively coded, that is, motion compensation can be carried out from both past and/or future reference frames. Encoded I-frame can only contain I-MBs, while encoded P-frames or B-frames may contain any of the aforementioned types of macroblocks.

In MPEG-2, each video sequence is divided into several groups of pictures (GOPs), that is, sequences of I-frames, P-frames and/or B-frames; each GOP starts with an I-frame, whereas the remaining frames are either P-frames or B-frames. The I-frames provide random access points in the compressed video data, since they can be decoded without referencing to other pictures. Moreover, any error propagation due to transmission errors in previous frames will be terminated by an I-frame. The length and structure of the GOP in a video sequence is usually fixed.

Different features of the encoding process can be exploited for forensic analysis of video sequences. When each frame is considered as a single image, it is possible to apply image-based forensic techniques, like block processing artefacts detection and DCT coefficient analysis. Due to the high specialization of video encoders, some of these features may even enable a further forensic analysis. For example, in many video encoders the block partitioning strategy is not fixed, as it depends on the specifications of coding standard and on the adopted rate-distortion optimization policy. Therefore, block processing artefacts can be used to identify the adopted codec.

However, the major difference between JPEG forensics and video forensics is that coding operations along the temporal dimension enable a more thorough analysis for the latter. For instance, different codec implementations may adopt diverse spatial or temporal prediction strategies, according to rate-distortion requirements and computational constraints. The identification of the adopted motion vectors and coding modes provides relevant information to the forensic analyst, that can be used to trace the processing history of the analyzed video.

In the following, we provide a survey of forensic tools aimed at reconstructing the coding history of video content. Since uncompressed video is extremely difficult to handle, when dealing with a video sequence we can safely assume that it is a compressed signal. Hence, most of existing video forensic approaches deal with video re-encoding, since the problem of identifying a video codec from an uncompressed video sequence has little practical value and is usually neglected by the forensic literature.

15.2.4   Video Re-encoding

Every time a video sequence is edited, it is usually safe to assume that we have an already compressed signal that is re-encoded. The detection of double encoding or, more generally, transcoding, is thus one of the most studied problems in video forensics. Due to the complexity and diversity of video coding architectures, there are far few results in the literature regarding the detection of double video compression with respect to the detection of double compression in images. On the one hand, a video sequence usually has a large number of frames and since some of these frames are encoded in a JPEG-like fashion, JPEG forensic techniques can be successfully used also in this case. On the other hand, the detection of double video compression becomes significantly more difficult whenever two different codecs are involved, since different codecs may involve very different parameter settings. It is worth noting the the problem of hybrid double compression, which exists in both image and video compression, is more important in video forensics, where different codecs are routinely employed, than in image forensic, where JPEG is by far the most common codec.

A large part of the solutions proposed so far in the literature are mainly focused on MPEG video, and they exploit the same ideas originally used for double JPEG compression. Luo *et al.* (2008) proposed a method for detecting double MPEG compression based on blocking artefacts. Inspired by the method in Fan and de Queiroz (2003), the authors defined a metric for computing the block artefact strength (BAS) for each frame. The mean BAS is computed on different sequences obtained by removing from the original sequence a number of frames ranging from one to the size of the GOP minus 1, obtaining a feature vector of BAS values. Whenever the sequence has been previously modified by frame removal and re-compression, the feature vector presents a characteristic behaviour.

Wang and Farid (2006) considered two scenarios, depending on whether the GOP structure used in the first compression is preserved or not. In the former situation, every frame is re-encoded in a frame of the same kind. Since encoding I-frames is not dissimilar from JPEG compression, when an I-frame is re-encoded as an I-frame at a different bitrate, DCT coefficients are subject to double quantization and the histograms of DCT coefficients assume, as already described, a characteristic shape. In the latter situation, which typically arises in the case of frame removal or insertion, I-frames can be re-encoded into another kind of frame. However, the authors observed that this usually produces larger prediction errors after motion compensation. When observing the Fourier transform of this prediction error, the presence of spikes reveals a change in the GOP structure, which is a clue of video re-encoding.

As already described, Benford's law can be used for double compression detection. An approach based on Benford's law is presented in Sun et al. (2012), where the first-digit distribution of DCT coefficients of I-frames is considered and a 12-dimensional feature is extracted to be classified using SVMs. Besides detecting double encoding, the method also classifies the second encoding as being at a higher or lower bitrate with respect to the first one. On the other hand, this method may not work when the two encodings are performed using different implementations of the MPEG-2 standard. Recently, Benford's law together with SVMs has been employed by Milani et al. (2012a) for detecting more than two encodings of the same video.

With the aim of generalizing the double encoding detection to a scenario with several codecs, different GOP sizes and distinct target bitrates, Vázquez-Padín et al. (2012) proposed to use a robust and very distinctive footprint based on the variation of the macroblock prediction types in the re-encoded P-frames. The authors considered the following scenario. During the capture of a scene, a first compression is performed with an arbitrary GOP size, denoted by G_1, and a fixed constant bitrate B_1. Then, the video sequence is re-encoded with a different GOP size G_2 and a fixed constant bitrate B_2, that can be equal to or different from B_1. By assuming that a baseline encoding profile is adopted, supporting only I-frames and P-frames, a specific variation of the number of macroblocks coded as I-MB and S-MB is observed in P-frames previously encoded as I-frames. This feature is an indication of a double compression; furthermore, since this feature is visible on P-frames that were intra-coded during the previous encoding, the size of the GOP used in the previous encoding can be estimated as well.

15.3 Editing Artefacts

Most of the images that can be found on public distribution channels, like the Internet, very rarely appear in the exact format in which they were acquired. It is a common practice to edit such images, for example to reduce the image size or to correct some image features. Several forensic tools have been developed aiming at revealing traces of editing in digital images. In the following, we will discuss two important classes of forensic tools, dealing with the detection of resampling and image enhancement operations. This section mainly covers image specific editing, since very few papers considered the forensic analysis of video specific editing. However, a few examples exist, as in the case of the forensic analysis of frame rate conversion (Bian *et al.* 2013).

15.3.1 Resampling

Geometric transformations, like resizing and/or rotation, are commonly applied to images. These operators modify the position of samples in the pixel domain, requiring that the original image be resampled according to a new sampling lattice. Resampling introduces specific correlations in the image samples, which can be used as evidence of both benign editing, like scaling or rotation of the whole image, as well as malicious editing, by checking whether only a certain region of the image has been resampled.

In mathematical terms, the process of resampling can be modelled as the convolution of the image pixels by an interpolation kernel $\phi(\mathbf{s})$, followed by resampling on the desired grid. More formally, the resampled image is given by

$$\mathbf{I}_2(\mathbf{m}) = \sum_{\mathbf{n}} \mathbf{I}_1(\mathbf{n})\phi(\mathbf{Am} + \mathbf{b} - \mathbf{n}) \qquad (15.6)$$

where $\mathbf{n} = (n_1, n_2)$ and $\mathbf{m} = (m_1, m_2)$, $n_1, n_2, m_1, m_2 \in \mathbb{N}$ are the pixel coordinates of the original and resized image, respectively, whereas $\mathbf{A}$ is a 2×2 matrix and $\mathbf{b}$ is a 2×1 shift vector defining the resampling grid. For example, in the case of simple image resizing we have $\mathbf{A} = \gamma^{-1}\mathbf{I}$, where γ is the resize factor and $\mathbf{I}$ is the identity matrix and we can assume that $\mathbf{b} = 0$ whenever image resizing preserves the position of the upper left pixel. The above formalism can be used to model other generic affine transformations of the image plane, like rotation and shearing.

The key idea of many resampling detection methods is that resampling leaves periodic traces in a high pass filtered version of the image, that is, it can be detected by analyzing the periodicity of the residual image

$$\mathbf{R} = \mathcal{F}(\mathbf{I}) \qquad (15.7)$$

where $\mathcal{F}$ can be implemented either as the difference between the image and its local prediction or as a derivative operator. An example is given in Figure 15.7, where $\mathbf{R}$ is computed for both a non-resampled and a resampled image, using a Laplacian operator, and the corresponding spectra are shown. It can be noticed how the residual computed on the resampled image shows a periodic behaviour, clearly visible in the peaks appearing in the magnitude of its spectrum.

Popescu and Farid (2005b) proposed a method to detect periodic correlations introduced in resampled images by common resampling kernels, similar to the one introduced by the same authors in Popescu and Farid (2005a). They estimated the interpolation kernel parameters through the expectation–maximization (EM) algorithm and obtained a probability map that for each pixel indicates its probability of

(a)   (b)   (c)   (d)

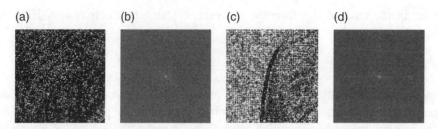

Figure 15.7   Residual images **R** computed on a non resampled image (a) and a resampled image (c), together with the corresponding spectra (b) and (d), respectively. For better showing the periodic behaviour, the residual images have been enhanced as $e^{-\mathbf{R}^2}$. It can be noticed the periodic behaviour of the residual of a resampled image, clearly visible in the four peaks of the rightmost spectrum.

being correlated to neighbouring pixels. The presence of interpolated pixels results in a periodic map, clearly visible in the frequency domain. The method has an accuracy close to 100%, but only on uncompressed images.

Alternatively, Gallagher (2005) observed that the variance of the second derivative of a resampled signal is periodic. Hence, resampling can be revealed by analyzing the Fourier transform spectrum of the second derivative of the overall image. Based on the aforementioned ideas, the periodicity of the second, or other order, derivative was also studied by other researchers in Dalgaard *et al.* (2010), Mahdian and Saic (2007), Prasad and Ramakrishnan (2006), Song *et al.* (2011), Wei *et al.* (2008).

The methods based on linear prediction and those based on image derivatives are closely related, as demonstrated in Kirchner (2008a) and Kirchner and Gloe (2009). As a matter of fact, the second derivative can be approximated by the difference between the image and a fixed predictor. The use of a fixed predictor, instead of the optimal predictor depending on the actual interpolation kernel parameters, permits to achieve a simplified detector, much faster than the one in Popescu and Farid (2005b), while achieving similar performance, as reported in Kirchner (2008a) and Kirchner and Gloe (2009), and in further studies by the same authors (Kirchner 2008b, 2010).

Among methods that deviate from the prediction paradigm, an interesting approach to resampling detection has been proposed by Mahdian and Saic (2008). They studied the periodic properties of the covariance structure of interpolated signals and their derivatives and applied a Radon transform to the derivative of the analyzed signal to reveal a periodic behaviour. In Mahdian and Saic (2009), the same authors also proposed to detect periodic patterns introduced in images by interpolation by using cyclostationarity analysis, detecting specific correlations between its spectral components. Further studies on the application of cyclostationarity analysis to the resampling detection problem can be found in Vázquez-Padín and Pérez-González (2011) and Vázquez-Padín *et al.* (2010).

15.3.2   Image Enhancement

Today, it is difficult to find published images that are not processed by at least some enhancement operators, like smoothing, contrast enhancement, histogram equalization and median filtering.

An interesting approach to the detection of median filtering has been proposed in Kirchner and Fridrich (2010). The basic idea is that pixels in adjacent rows or columns of median filtered images often share the same value, the so called 'streaking artefacts'. These artefacts can be revealed by considering first-order differences for groups of two pixels and studying the corresponding histograms. This simple approach obtains a perfect detection for a false positive rate <1.8%, provided that images are not compressed. In order to cope with JPEG compression, the same authors proposed an alternative first-order difference-based detector using the subtractive pixel adjacency matrix (SPAM) features defined in Pevný *et al.* (2010).

The above method is outperformed by the algorithm proposed in Yuan (2011), which works even in the case of low resolution and JPEG-compression. For JPEG-compressed images at a quality factor equal to 70, the aforementioned approach obtains a perfect detection for a false positive rate of 10%. The key observation is that the two-dimensional median filter significantly affects either the order or the quantity of the grey levels contained in an image area with the same size as the filter window. Median values originating from overlapping windows are thus dependent, and this local dependence is identified by introducing a new feature set and applying machine learning techniques.

As to the detection and estimation of contrast enhancement and histogram equalization in digital images, several works have been proposed by Stamm and Liu (2008, 2010b). The first of these works targets the detection of the contrast enhancement operation. The key idea is to reveal footprints left in the image by the operator, which consist in the formation of sudden peaks and zeros in the histogram of pixel values. An example of the aforementioned footprint is given in Figure 15.8, where we show the histogram of the grey levels of the popular image Lena, together with the histogram of the same image after contrast enhancement.

In order to detect the presence of contrast enhancement in an image, Stamm and Liu (2008) proposed to analyze the energy of the DFT of the histogram. It was argued that natural images tend to have a smooth histogram: even in the presence of scenes with sharp transitions, the effects of the illumination, lens resolution and sensor sampling will produce slight and continuous variations on the pixel values, yielding a smoothing of the histogram values. As a result, the histogram of a natural image can be considered as a typically low pass signal. On the contrary, after contrast enhancement the formation of sudden peaks and valleys in the histogram values will increase the energy of the high frequency components of the histogram. Hence, the energy of these high-frequency components can be measured and compared with a threshold in order to detect the presence of contrast enhancement.

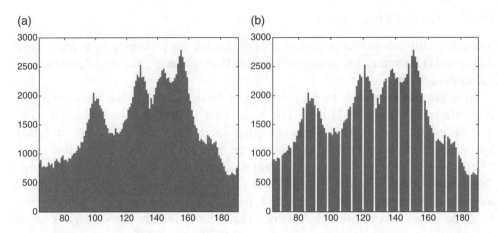

Figure 15.8 A particular of the histogram of the popular greyscale Lena image before (a) and after contrast enhancement (b).

Even if these techniques were originally thought for contrast enhancement detection, Stamm and Liu (2010b) also showed how the techniques can be successfully applied to other kinds of histogram processing, as well as to splicing localization.

15.4 Estimation of Processing Parameters

A classical application scenario for the tools described in the previous sections is image tampering detection, by verifying whether a given image presents either coding or editing artefacts, and whether those artefacts exhibit spatial inconsistencies. However, a much more ambitious goal is that of moving a step further, in order to collect information about the processing chain which led to a specific observed image or video. In this respect, a first example can be provided by forensic tools that not only detect double JPEG compression but also estimate previous compression parameters, like (Farid 2009) or (Bianchi and Piva 2012a).

Unfortunately, the aforementioned tools suffer from severe limitations in real-life scenarios. For example, if the image is resized between successive JPEG compressions, which is often the case when digital images are posted on photo sharing applications, the models the above methods rely on are no longer valid. In the following sections, we will present some interesting approaches for the estimation of either coding parameters or editing parameters that can be successfully used in realistic scenarios. We will also show how some forensic tools can be extended to provide information about local anomalies in images or video frames.

15.4.1 Estimation of Coding Parameters

In image and video coding architectures, the choice of the coding parameters is not dictated by the standard, but depends on the specific implementation of the codec and on the characteristics of the coded signal. As to JPEG compression, user-defined coding parameters are limited to the selection of the quantization matrices, which are adopted to improve the coding efficiency based on the psycho-visual analysis of human perception. Conversely, in the case of video compression there is a significantly wider number of coding parameters that can be adequately tuned. As a consequence, forensic analysis of video codecs needs to take into account a larger number of degrees of freedom. The methods aiming at estimating different coding parameters and syntax elements characterizing the adopted encoders can be grouped into several categories, depending on the type of media analyzed and on the specific parameters they look for. A brief survey is given in the following text.

15.4.1.1 Encoder Identification

The identification of a specific, vendor-dependent, implementation of an image or video encoder is an important piece of information in recovering the coding history of digital media, since it is often the first step to be taken before estimating further parameters.

Lin *et al.* (2009a) proposed a method that aims at identifying the history of source coding operations applied to digital images. Three different image source encoders were considered: transform-based coding, subband coding and differential image coding. The designed algorithm first looks for the presence of footprints left by a general block-based encoder, by using the techniques described in Section 15.2. If evidence of block-based coding is found, a similarity measure for each of the previous coding schemes is computed in order to detect the one being used: transform coding is characterized by comb-shaped histograms of the coefficients in the transform domain; subband coding is characterized by the presence of ringing artefacts near image edges; differential image coding is characterized by the whiteness of the residual obtained from the difference between the encoded image and its denoised version. The candidate encoder is the method giving the highest similarity measure.

The approach in Bestagini *et al.* (2012) presents an effective codec identification strategy that allows determining the codec used in the first compression stage in the case of double video compression, noting that the codec used in the second compression stage is known since the bitstream is usually available. The proposed algorithm relies on the assumption that quantization is an idempotent operator, that is, whenever a quantizer is applied to a value that has already been previously quantized and reconstructed by the same quantizer, the output value is highly correlated with the input value. As a matter of fact, it is possible to identify the adopted codec and its configuration by re-encoding the analyzed sequence a third time, with different codecs and parameter settings. Whenever the output sequence presents the highest correlation with the input video, one can infer that the adopted coding set-up corresponds to that of the first compression.

15.4.1.2   Block Size Estimation

The estimation of the block size can be particularly important in the case of video coding, since most video coding architectures encode frames using different block-by-block decompositions. For this reason, artefacts at block boundaries can be exploited to reveal traces of a specific previous compression step.

In Li and Forchhammer (2009), the block size in a compressed video sequence is estimated by analyzing the reconstructed picture in the frequency domain and separating the peaks that are related to discontinuities at block boundaries from those related to intrinsic features of the underlying scene.

However, in some modern video coding architectures, including, for instance, H.264/AVC (ITU-T/ISO/IEC 2003) as well as the recent high efficiency video coding (HEVC) standard under development (ITU-T/ISO/IEC 2012), a deblocking filter is used to smooth artefacts at block boundaries, and variable block sizes, with possibly non-square blocks, can be adopted. For such codecs, traditional block detection methods fail, leaving this as an open research issue.

15.4.1.3   Quantization Step Estimation

Since the quantization step is a user-defined parameter both in JPEG compression and in video compression, many forensics methods have been proposed in the literature to estimate such parameter from either an uncompressed or a re-encoded signal. Most of the methods are based on the observation that the histogram of DCT coefficients has a characteristic comb-like shape, where the spacing between successive peaks is related to the adopted quantization step.

The scheme proposed in Fan and de Queiroz (2000, 2003) approximates the histogram of DCT coefficients before quantization by means of a Gaussian distribution for the DCT coefficient $D(1, 1)$ (DC coefficient), and a Laplacian distribution for the other 63 DCT coefficients (AC coefficients). According to this model, the maximum likelihood (ML) estimator of the quantization is defined.

Lukáš and Fridrich (2003) used instead a non-parametric method. As a first step, the histograms of absolute values of all analyzed DCT coefficients are computed from the image under investigation. The image is then cropped, so as to disrupt the structure of JPEG blocks, and compressed with a set of candidate quantization tables. The cropped and compressed images are then recompressed using the same quantization matrix as the input image and the histograms of absolute values of DCT coefficients are computed again from this double compressed and cropped images. The estimator chooses the quantization table corresponding to the recompressed image whose histogram is most similar to the histogram computed on the analyzed image. A similar solution was proposed by Pevný and Fridrich (2008), where only the histograms related to the nine most significant DCT frequencies were considered, since those frequencies are expected to have a high number of non-zero coefficients.

The corresponding quantization steps employed in the first compression are estimated via an SVM classifier, whereas the remaining quantization steps are estimated via an ML estimator.

In Ye *et al.* (2007) the authors proposed a method for estimating the whole quantization table. Separate histograms are computed for each DCT frequency, and the corresponding quantization steps $\Delta(i,j)$ are estimated by analyzing the periodicity of the power spectrum of each histogram. Periodicity is detected with a method based on the second-order derivative applied to the histograms. Luo *et al.* (2010) improved the method by taking into account rounding of the pixel values, which happens when a JPEG image is reconstructed in the pixel domain. The consequence is that the histograms of DCT coefficients $(\hat{D}_Q(i,j))$ computed from decoded pixel values are blurred with respect to those obtained directly after quantization $(D_Q(i,j))$. Hence, it is possible to consider such rounded-coefficients histograms when estimating the quantization step for each DCT frequency.

In the case of colour image compression, distinct quantization tables can be used for each colour component. The problem of estimating these quantization tables was addressed by Neelamani *et al.* (2006). First, the authors introduced a maximum a posteriori (MAP) estimation method for extracting the quantization step size in greyscale images, by refining the algorithm already proposed in Fan and de Queiroz (2003). Then, they extended the solution to colour images, by observing that the periodicity of the histogram is revealed only when the image is transformed to the correct colour space and interpolation artefacts are removed.

In Lin *et al.* (2009a), the authors proposed to estimate the quantization table as a linear combination of existing quantization tables. A first estimate of the quantization step size for each DCT band is obtained by using the methods proposed above. Then, the authors observed that in most cases high-frequency coefficients do not contain enough information for a correct estimation, and proposed to estimate the missing quantization steps as a linear combination of other existing quantization tables collected into a database.

Similar arguments can be used to estimate the quantization parameters in video coding, when the same quantization matrix is used for all blocks in a frame. In Chen *et al.* (1998) and Tagliasacchi and Tubaro (2010), the authors considered the case of MPEG-2 and H.264/AVC coded video, respectively. The major difference with respect to JPEG-based methods is that the histograms are computed from DCT coefficients of prediction residuals. To this end, motion estimation is performed at the decoder side to recover an approximation of the motion compensated prediction residuals available at the encoder.

An interesting application of quantization step estimation in video history recovery is to track how quantization parameters vary over time, in order to get information about the rate-control algorithm applied at the encoder side, which could be used to identify vendor-specific codec implementations (Milani *et al.* 2012b).

15.4.1.4 Identification of Motion Vectors

As explained in Section 15.2, a significant difference between image and video coding is the use of predictors exploiting temporal correlation between consecutive frames. Similarities among neighbouring video frames are exploited by constructing a predictor of the current video frame using motion estimation and compensation. In most video coding architectures, a block-based motion model is adopted, where a motion vector (MV) is estimated for each block, so as to implement a motion-compensated predictor.

Valenzise *et al.* (2010) showed how to estimate the motion vectors originally adopted by the encoder, when the bits storing the information are missing at the decoder. The key technique is to perform motion estimation by maximizing, for each block, an objective function that measures the comb-like shape of the histogram of the resulting prediction residuals in the DCT domain.

15.4.1.5 GOP Size Estimation

In the case of re-encoded video, an important parameter is the size of the group of pictures used by the first coding algorithm. Such information can be used to select particular frames undergoing specific double encoding patterns, for example frames that are doubly encoded as I-frames, which enables further forensic analyses. The method proposed by Bestagini *et al.* (2012) can be used in principle to obtain such information: by re-encoding the analyzed video using different GOP sizes, the output sequence obtained using the same GOP size as the first encoding will show a higher correlation with the input sequence than other output sequences, and will allow the parameter configuration of the first encoding to be estimated. Nevertheless, the aforementioned technique requires re-encoding the analyzed sequence using several different codecs and several different GOP sizes, which may become a computationally intensive task.

A simpler technique is to use the variation of prediction footprint (VPF) proposed by Vázquez-Padín *et al.* (2012), based on the observation that P-frames that were previously encoded as I-frames will show a higher number of I-MB and a lower number of S-MB with respect to regular P-frames. The technique consists in counting the number of I-MB and S-MB in consecutive frames and computing a signal that is proportional to the difference of those quantities. In the presence of re-encoding, the aforementioned signal will present periodic peaks corresponding to the re-encoded I-frames of the previous video. The aforementioned signal will usually have some missing peaks due to the fact that not all the I-frames are re-encoded as P-frames and that the variability of the video content may also alter the number of I-MB and S-MB. Hence, the size of the GOP is estimated by computing a suitable energy function for a number of candidate periods. Since the candidate periods are integer values, and some candidates can be immediately discarded according to the peak configuration, the search of the previous GOP size is very efficient. Another advantage of using the

VPF is that the above footprint holds for different video coding standards and can be used even in the case of re-encoding using different codecs, which makes it very suitable for practical forensic scenarios.

15.4.2   Estimation of Editing Parameters

In principle, several of the methods used to detect the presence of image editing can also be used to estimate the parameters of the editing operation. Nevertheless, most of the approaches described in Section 15.3 fail when applied in realistic scenarios, since they usually have very low performance in the presence of compressed images. For example, in Stamm and Liu (2010a), the authors extended the methodology discussed in Section 15.3.2 to provide an estimate of the actual mapping induced by the contrast enhancement operator. Although their method appears successful on uncompressed images, usually even a small amount of compression tends to destroy the peak-to-valley behaviour of the image histogram, so that the method is not suitable in realistic settings.

In the following, we will discuss a couple of techniques that can be used to estimate the parameters of possible editing operations when such processings have been applied on compressed images. Namely, the presence of a previous JPEG compression seems to ease the detection of editing operations, even in the presence of successive recompressions, which is a very interesting scenario for practical applications.

15.4.2.1   Image Enhancement

In Ferrara *et al.* (2013), the authors considered a chain composed by double JPEG compression interleaved by a linear contrast enhancement. The following processing chain was considered: the luminance of a JPEG colour image with quality Q_1 is linearly stretched and then re-saved in another JPEG colour image with quality Q_2. Two approaches, borrowed from double JPEG compression detection, were extended for the identification of the considered chain; furthermore, assuming Q_2 to be known, the methods provide the joint estimation of the chain operator parameters, that is the first quality Q_1 and the amount of contrast enhancement.

The first approach exploits the periodic pattern of the histogram of doubly compressed DCT coefficients, as shown in Section 15.2.1. It was demonstrated that in the presence of a linear contrast enhancement aforementioned the periodic pattern is essentially maintained, although with a different period depending on the amount of contrast enhancement. By estimating such a period, it is usually possible to derive the parameters of the contrast enhancement operation.

The second approach is based on Benford's law, namely on the mode-based first digit features (MBFDFs) proposed in Li *et al.* (2008), which are based on the distribution of the first digit of DCT coefficients for each separate DCT frequency, or mode. Even if contrast enhancement is expected to modify the distribution of the first digit, the resulting distribution will still violate the generalized Benford's law, so that MBFDFs can be used to distinguish singly and doubly compressed images. Moreover, different parameters of the contrast enhancement operator will produce different patterns on the distribution of the first digit of DCT coefficients. Hence, MBFDFs can also be used to discriminate different parameters of the processing chain.

In order to distinguish enhanced and recompressed images from singly compressed images, MBFDFs are classified according to Fisher's linear discriminant analysis (LDA). The parameters of the processing chain, that is, the quantization step of the previous compression and the amount of contrast enhancement, can be estimated by using a 'one-against-one' multi-classification strategy, where each possible combination of values is considered as a different class. Given N_C possible classes, $N_C(N_C - 1)/2$ two-class LDA classifiers are constructed, considering every possible combination of two classes. Each classifier 'votes' for its winning class, and the class obtaining more votes corresponds to the estimated values for the parameters of the processing chain.

The aforementioned approach works well in presence of a finite set of possible parameters, like in the case of the quantization steps. However, for continuous valued parameters, like the amount of contrast enhancement, it requires a binning of the parameter space, with a proper choice of the bin size. It is worth noting that a fine search of parameter values may be impractical, due to the fact that the number of required classifiers grows quadratically with the number of bins.

15.4.2.2   Image Resizing

One of the first forensic techniques aiming at the detection of resampling in JPEG images was presented in Kirchner and Gloe (2009). The key observation is that a previous JPEG compression may help in detecting a successive image resizing, even after a second compression, since the artefacts of the first compression act as a sort of pilot signal carrying the resampling artefacts. In Bianchi and Piva (2012c), the authors moved a step further, introducing a forensic technique that also provides an estimation of both the resize factor and the compression parameters of the previous JPEG compression. Such additional information is important, since it can be used to reconstruct the history of an image and perform a more detailed forensic analysis.

The approach of Bianchi and Piva (2012c) exploits the NLDP introduced in Section 15.2.1 and is based on the extension of the technique proposed in Bianchi and Piva (2012a) for non-aligned double JPEG compression. Let us consider the case of image resizing followed by JPEG recompression. Observing the NLDP directly on the resampled pixels is not an easy task. However, if we are able to reverse the resizing operation, which corresponds to estimating an approximation of the intermediate continuous image surface $\xi(\mathbf{s}) = \sum_{\mathbf{n}} \mathbf{I}_1(\mathbf{n})\phi(\mathbf{s} - \mathbf{n})$ and resampling it according to the original pixel grid, we will again observe the NLDP of the resulting pixels. An example is given in Figure 15.9, where the distributions of DCT coefficients obtained

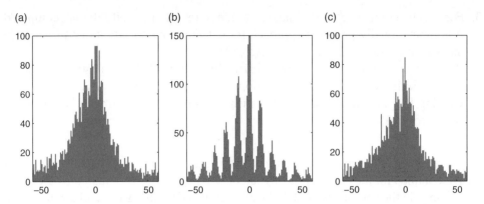

Figure 15.9 Histograms of DCT coefficients D computed after reversing the resizing operation: $\gamma = 1.15$ (a), $\gamma = 1.2$ (b), $\gamma = 1.25$ (c). The NLDP is clearly visible when $\gamma = 1.2$, which is the actual resizing factor used in the processing chain.

after reversing the resizing step, using different candidate γ-s, are shown: when γ matches resizing factor used in the processing chain, the NDLP is evident.

Although the NLDP is a very discriminative feature, detecting it in practice can be a difficult task. First, the lattice defined by the first JPEG compression depends on hidden parameters. In some cases, these parameters can be estimated by exploiting the fact that they can assume only a limited number of possible values. For example, in the case of non-aligned double JPEG compression there are only 64 possible grid shifts between the first and second compressions (Bianchi and Piva 2012a). Nevertheless, estimating the quantization matrix of the first JPEG compression may still be a problem.

Secondly, when the image has been resized after the first compression, the parameter γ may have a very large set of possible values, in which case the aforementioned approach may become unpractical. Nevertheless, in some relevant scenarios we can assume to have some prior knowledge about the possible resizing factors. For example, most photo sharing applications use a limited number of possible image sizes when resizing the original uploaded pictures. Since also commercial cameras are based on image sensors providing a set of standard image sizes, the respective image sizes can be paired in order to enumerate a set of possible resizing factors. Hence, we can output a set of candidate resizing factors by choosing all the image size pairs which result in the size of the observed image.

In the following, we illustrate a simple algorithm exploiting the NLDP for the reverse engineering of a processing chain composed by a first JPEG compression, an image resizing, and a final JPEG compression. The algorithm can be summarized by the following steps:

1. A number of candidates for the resizing factor are estimated.
2. For each candidate, the resizing step is reversed.

3. For each counter-resized image, a measure of the NLDP is computed. A computationally practical approach for detecting the NLDP is that of using the measure proposed in Bianchi and Piva (2012a).
4. If the measure of one of the counter-resized images is greater than a given threshold, the image is labelled as doubly compressed with resizing factor equal to that yielding the maximum value of the measure, otherwise the image is labelled as singly compressed.

The rationale of the above algorithm is that, in the presence of NLDP, the measure of NLDP obtained for the correct resizing factor will be much higher than the other ones. Conversely, in the absence of NLDP the probability of finding a particular γ for which the measure is higher than the threshold is negligible, since the probability of verifying the NLDP for a generic signal is very low. The results in Bianchi and Piva (2012c) demonstrated that in the presence of prior knowledge regarding the possible resizing factors, the algorithm is able to detect a resized and recompressed image with an accuracy between 80% and 95%, provided that the quality of the second compression is not lower than the quality of the first compression.

15.4.3 Artefact Localization

Several existing forensic techniques need to manually select a suspect region in order to test the presence or the absence of a particular artefact. A natural extension is to provide a localization algorithm that, unlike previous approaches, automatically derives the probability of each pixel to exhibit a given feature. An interesting approach, proposed for the first time in Lin *et al.* (2009b), can be applied to JPEG encoded images in order to obtain the probability to be forged of each 8×8 block.

The approach is based on the following forgery scenario: an original JPEG image is decompressed, a part of it is tampered with, and the modified image is again saved in JPEG format. According to the aforementioned forgery scenario, a tampered image will consist of two groups of pixels: the one that has not been modified, thus undergoing a double quantization, and the one that has been introduced between the encodings, which with high probability will not show traces of double quantization after the second encoding, thus making localization possible. A thorough explanation of this model is given in Lin *et al.* (2009b).

If we consider the histogram of a specific DCT coefficient (e.g. the one in position (0,1) in all 8×8 blocks), we should see a mixture of two components: a comb-shaped component due to double compressed regions, and a smooth component due to regions that have been compressed only once. An example is given in Figure 15.10.

In Bianchi *et al.* (2011) and Bianchi and Piva (2012b) a Bayesian inference method for images was proposed that first assigns to each DCT coefficient of a block its probability of belonging to each one of these components, and then accumulates these probabilities for all the coefficients within a block, producing an aggregated probability

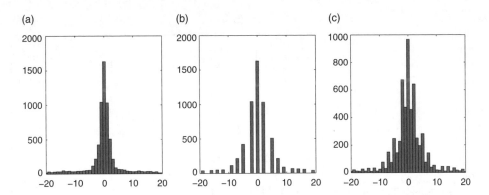

Figure 15.10   Histogram of a DCT coefficient from a single compressed image (a), double compressed image (b) and tampered image (c). Notice that the tampered histogram can be seen as a mixture of the previous two histograms.

for the block of being/not being doubly compressed. The output is a map associating to each 8×8 block of pixels its probability of being modified.

In order to compute such a map, the mentioned algorithm basically performs the following steps for each group of DCT coefficients sharing the same position within a block:

1. From the observed DCT coefficients, estimate the histogram $\tilde{h}$ that would result after a single encoding with the quantization step used in the second compression. This can be obtained by using the calibration technique in Lukáš and Fridrich (2003): the image is cropped by one row and one column, and the resulting image is quantized with the second quantization matrix;
2. Estimate the quantization step that was used during the first compression;
3. Knowing both quantization steps (the quantization step of the second compression is obtained from the header of the JPEG file), compute a function $n(x)$ that gives the number of bins of the original histogram that are mapped in the bin corresponding to the value x in the doubly quantized histogram.

By denoting with $\mathcal{H}_0$ and $\mathcal{H}_1$ the hypothesis of being tampered and original, respectively, for each coefficient D Bianchi *et al.* (2011) obtained:

$$p(D|\mathcal{H}_0) = \tilde{h}(D) \tag{15.8}$$

and

$$p(D|\mathcal{H}_1) \approx n(D) \cdot \tilde{h}(D), \quad D \neq 0. \tag{15.9}$$

As an illustrative example, in Figure 15.11 the models proposed in Bianchi and Piva (2012b) are compared with the histograms of quantized DCT coefficients of

(a)   (b)

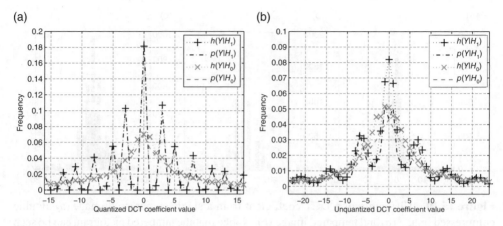

Figure 15.11 Example of double JPEG compression models: $h(Y|\mathcal{H}_1)$ and $h(Y|\mathcal{H}_0)$ denote the histograms of either quantized (A-DJPG) or unquantized (NA-DJPG) DCT coefficients of a doubly compressed and a singly compressed image, respectively, for a single DCT frequency, whereas $p(Y|\mathcal{H}_1)$ and $p(Y|\mathcal{H}_0)$ are the corresponding models derived in Bianchi and Piva (2012b): A-DJPG (a), NA-DJPG (b).

doubly compressed and singly compressed images, for both aligned and non-aligned double JPEG compression, considering a single DCT frequency in both cases. The sub-figures show that there is a good agreement between the proposed models and the real distributions and that the distributions of singly compressed and doubly compressed DCT coefficients can be effectively separated.

Once $p(D|\mathcal{H}_1)$ and $p(D|\mathcal{H}_0)$ have been estimated, the coefficient D can be classified as belonging to one of the two models according to the value of the likelihood ratio

$$\mathcal{L}(D) = \frac{p(D|\mathcal{H}_1)}{p(D|\mathcal{H}_0)}. \tag{15.10}$$

It is worth noting that the aforementioned approach can be easily extended to any forensic trace, even to traces that are not left by JPEG recompression, provided that the appropriate probability models for both $p(D|\mathcal{H}_0)$ and $p(D|\mathcal{H}_1)$ are defined.

These steps are carried out separately for each group of DCT coefficients. If multiple DCT coefficients within the same block are considered, by assuming that they are independently distributed we can express the likelihood ratio corresponding to the kth block as

$$\mathcal{L}(k) = \prod_{(i,j)} \mathcal{L}(D_k(i,j)) \tag{15.11}$$

where $D_k(i,j)$ denotes the DCT coefficient at position (i,j) within the kth block. Such values form a likelihood map of the JPEG image with resolution 8×8 pixels, which can be used to localize possibly forged regions within the image. Since the areas of interest are usually connected regions with area greater than 8×8 pixels, the likelihood map can be further processed by cumulating the likelihoods on a local window, by assuming that if neighbouring blocks are doubly compressed, the likelihood that the reference block is also doubly compressed will increase.

In Bianchi and Piva (2012b), the validity of the aforementioned method was assessed by evaluating the performance of a detector based on thresholding the likelihood map: the results showed that, defined as Q_1 and Q_2 the quality factors of the first and second compression, the method is able to correctly identify traces of A-DJPG compression unless $Q_2 = Q_1$ or $Q_2 \ll Q_1$, whereas it is able to correctly identify traces of NA-DJPG compression whenever $Q_2 > Q_1$ and there is a sufficient percentage of doubly compressed blocks.

Automatic localization of double compression artefacts can be extended to video sequences as well. In this sense, intra-frame artefact localization is probably the less studied field in video forensics today, and most of the existing approaches work only under strict assumptions. The most recent method is the one proposed by Wang and Farid (2009), where a DQ analysis is applied on a macroblock-by-macroblock basis; since this analysis makes sense only on frames that have been encoded twice as intra, they worked around this problem by assuming that Motion-JPEG encoding has been performed (i.e. only intra-coded pictures are used), thus heavily restricting the applicability of the method. Furthermore, the double quantization analysis is performed separately on each MB, leading to a computationally intensive analysis.

15.5   Case Studies

This section is devoted to describe two case studies where some of the forensic tools previously described have been applied to detect and localize the presence of manipulations in properly tampered JPEG images and MPEG-2 video sequences.

15.5.1   Localization of Forgeries in JPEG Images

We describe here a simulation of legal case organized on 2011 by the Laboratory of Forensic Science (FORLAB) of the University of Florence[3]: the legal case has been created by a group of students along with a set of digital images forming a key part of evidence, for the resolution of the case itself. In this scenario, it was assumed that data were acquired, according to the best forensic practice, from a computer and two cameras found in the victim's home by the technical advisor of the public prosecutor, who then delegated to experts in multimedia forensics their content analysis. This analysis was requested to discriminate between intact and possibly altered images for future use as evidence or not, and to extract some details on possibly manipulated images useful for the investigation. The experts had no knowledge about the legal

---

[3] FORLAB, led by Dr. A. Piva, studies and exploits methodologies and techniques for the analysis and processing of audio-visual data for forensic purposes (www.forlab.org).

case, nor about the source devices and the possible manipulations undergone by the images to be studied. The available dataset was composed by 46 JPEG images, having resolution 3264×2448 or 2816×2112. An analysis of the header files, which is out of the scope of this chapter, allowed the experts to verify that the images were acquired from two different cameras (a Nikon Coolpix L1 and a Nikon Coolpix L19), and to select a set of suspicious images, candidates of possible manipulations. To these images, the image tampering localization algorithm proposed by Bianchi *et al.* (2011) was applied to generate a tampering probability map. The results of the analysis showed that some images are probably manipulated. In the following, the most interesting results are shown.

In Figure 15.12b the probability map clearly reveals an area where a third person may have been removed. A visual inspection seems to confirm the suspects raised by the image forensic tool: in Figure 15.12c a detail of the previous image shows a blending operation on the lower left part of the scene.

In Figure 15.13b the probability map reveals a rectangular region that could have been pasted from another image; again, a visual inspection seems to confirm the suspects: in Figure 15.13c it has been highlighted an area where a straight line along which the leaves positioned on the ground do not match and there is a rush of hues. In addition, an irregularity in the car parked in the background and on the bag is visible, so identifying a rectangular area, which may be the result of a cut and paste operation.

It is interesting to note that at the end of the challenge, the original images were released, confirming the results obtained by the forensic analysis and by the visual inspection; the original versions of the two previous images are shown in Figure 15.14.

(a)   (b)   (c)

Figure 15.12 Image under analysis (a), probability map of Bianchi *et al.* (2011) (b), and a detail representing the lower left part of the scene (c). Lighter areas correspond to higher probability of being tampered. The map clearly shows an elongated area having a high probability of being tampered, which is confirmed by a visual analysis of the detail.

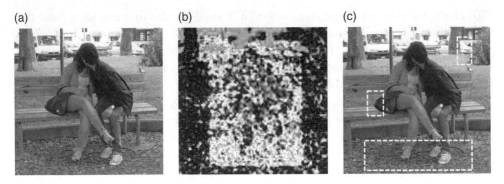

Figure 15.13 Image under analysis (a), probability map of Bianchi *et al.* (2011) (b), and image with highlighted suspect areas of the scene (c). Lighter areas correspond to higher probability of being tampered. The map clearly shows a rectangular region having a high probability of being tampered.

Figure 15.14 Original versions of the images shown in Figure 15.12 (a) and Figure 15.13 (b).

15.5.2 Localization of Forgeries in MPEG-2 Videos

In this section, we present a case study in which the aim is the localization of forgeries in the frames of an MPEG-2 video. It is assumed that, starting from an MPEG-2 video sequence, an attacker decodes the video, alters the visual content of a group of frames (e.g. by adding an object not present in the acquired scene), and finally encodes the resulting sequence again with MPEG-2, using a different GOP size; moreover, it is assumed that a fixed quantizer with the default quantization matrix is employed, leading to a VBR coding.

A possible solution to this problem basically can be split into two phases: given a video file under analysis, first, frames that have been intra-coded twice are individuated; next, on these frames traces of double quantization at a spatial level are looked for, building a fine-grained probability map of tampering for each analyzed frame. The two steps can be solved thanks to the application of two forensic methods described in the previous sections: the method presented by Vázquez-Padín *et al.*

(2012) allows to estimate the size of the GOP used for the first encoding, and thus to find those frames that have been encoded as intra both in the first and second encoding; on these frames, the localization of regions of the frame that exhibit DQ effect, by adapting and extending the method proposed by Bianchi *et al.* (2011) for the localization of forgeries in JPEG images to the MPEG-2 encoding scheme.

Let us now make some assumptions on the case study: we suppose that the video under analysis, composed of N frames, has been encoded twice using G_1 and G_2 as the GOP size for the first and second encoding, respectively, where $G_1 \neq m \cdot G_2, \forall m \in \mathbb{N}$, and that the GOP structure is fixed. Thus, the frames that have been intra-coded twice will be the ones with indices

$$\mathcal{C}_{G_1,G_2} = \{n \in \mathbb{N} : n = m \cdot \mathrm{lcm}(G_1, G_2) \wedge n \leq N, \forall m \in \mathbb{N}\},$$

(where $\mathrm{lcm}(a, b)$ is the least common multiple between a and b), and the number of frames in this set is:

$$|\mathcal{C}_{G_1,G_2}| = 1 + \left\lfloor \frac{N}{\mathrm{lcm}(G_1, G_2)} \right\rfloor,$$

where $\lfloor \cdot \rfloor$ stands for the flooring function. In other words, forgery localization can be performed every $\mathrm{lcm}(G_1, G_2)$ frames. Therefore, for relatively prime values of G_1 and G_2 the analysis can be carried out only once every $G_1 \cdot G_2$ frames; at a frame rate of 25 fps, combinations of usual values for G_1 and G_2 (like 12 for PAL videos, 15 for NTSC videos) result in a satisfactory time resolution for the analysis.

Since DCT coefficients quantization is a key step both in JPEG and MPEG-2 coding, methods relying on DQ effect can been borrowed from image to video forensics. However, some significant differences must be considered to devise a correct model for MPEG-2:

1. The dequantization formula in JPEG differs from that of MPEG-2 (ISO/IEC 2008);
2. In JPEG, the 8×8 quantization matrix is declared in the header and is fixed; in MPEG-2, instead, the matrix is parameterized by the scale factor S, to adjust the quantization strength;
3. In JPEG, the quantization matrix is the same for all the image; this also holds for MPEG-2 when a fixed quantizer is used, while the quantization matrix may change from frame to frame, or from MB to MB, in CBR coding.

Each of these facts has a direct implication on the model described in Bianchi *et al.* (2011). Since a different quantization formula is used, the function $n(x)$ in Equation 15.9 will change; its value will depend on each element in the 8×8 quantization matrix, and on the scale factors S_1 and S_2 that parameterized the quantization matrix in the first compression and in the second compression, respectively, as demonstrated in Labartino *et al.* (2013). Given that S_2 and the quantization matrix are available from the bitstream, S_1 is the only parameter that has to be estimated.

An effective way to get an estimate of S_1 is to iteratively search the value $\hat{S}_1$ that minimizes the difference between the observed histogram $h(D)$ and the theoretical probability distribution $p(D; \hat{S}_1)$. The details of the above procedure can be found in Bianchi *et al.* (2011) and Labartino *et al.* (2013). Using $\hat{k}_1$ and the $n(x)$ function derived in Labartino *et al.* (2013) for the MPEG-2 case, we can compute the probability in Equation (15.9). Finally, the probability for each 8×8 block of being tampered is obtained through Equation (15.11).

A complete experimental validation has been carried out and reported in Labartino *et al.* (2013) to demonstrate the performance of the proposed solution. Here, we are more interested in analyzing the practical efficiency of the two-phases scheme for a real tampered video.

A video sequence representing a landscape was acquired, having a spatial resolution of 720×576 pixels, and a GOP size $G_1 = 3$. The video was then tampered by splicing on it an unidentified flying object (UFO) that seems flying on the sky, and by recompressing it with a new GOP size $G_2 = 4$. By applying the method presented by Vázquez-Padín *et al.* (2012), we were able to detect a GOP size $\hat{G}_1 = 3$, thus a correct detection was obtained. Then, we were able to deduce that a double intra coding happened from the first frame every 12 frames, and thus to them the forgery localization algorithm was applied, by considering the first 5 DCT frequencies, taken in zig-zag order.

The proposed method allowed us to detect the presence of the tampering in the double intra-coded frames, as demonstrated in Figure 15.15, where a forged frame along with the probability map generated by the proposed method are shown: the map, even if postprocessed with just a median filtering, highlights the presence of a region composed by blocks with high probability of tampering in correspondence of the UFO.

15.6 Conclusions

This chapter is devoted to the analysis of the most useful traces left into a digital image and video clip that are used by forensic techniques for processing history recovery. Since most of the images and video clips that we can find in real applications are in compressed format, a particular attention has been given to traces introduced during the encoding process, although footprints left by editing operations have also been studied.

It should be clear that image and video forensics is still a young research area where several issues are still open. In particular, two problems are of paramount importance.

It has been highlighted in the current literature that in most of the presented works each operation in the image life cycle has been considered in isolation, in such a way that each digital footprint has been analyzed regardless of the remaining processing stages. Only few papers until now tried to reach a much more ambitious goal represented by the extraction of information about the whole processing chain

(a)   (b)

(c)   (d)

Figure 15.15 Two examples of intra-frame tampering (a–c) and the produced probability maps (b–d). Lighter areas represent higher probabilities of presence of tampering. For showing purposes, a 3 × 3 median filter has been applied to the maps.

which led to a specific observed image or video, but the analyzed chains are usually rather simple. This leaves scope for a more complicated analysis of operator chains.

Another open problem is how to exploit the output of a set of forensic tools to make a decision about the authenticity of an image or video clip: indeed, in real cases, a manipulated content is obtained by applying a small set of processing tools, hence only a part of the available trace detectors will reveal the presence of tampering, whereas some others could give dubious answers due to their low reliability. It may also happen that the positive answer of one algorithm inherently implies the negative answer of another because they search for mutually excluding traces. In the literature a few researchers (Barni and Costanzo 2012; Fontani *et al.* 2011, 2013) have proposed a decision fusion framework to properly exploit the output of several forensic tools in the case of digital images. These are the first proposals in this area, but much work is still to be carried out to obtain an effective and performing forensic tool working in real applications, in particular for what concerns the localization of a tampering in the content.

References

Barni M and Costanzo A 2012 A fuzzy approach to deal with uncertainty in image forensics. *Signal Processing: Image Communication* **27**(9), 998–1010.

Barni M, Costanzo A and Sabatini L 2010 Identification of cut & paste tampering by means of double-JPEG detection and image segmentation *Proceedings of 2010 IEEE International Symposium on Circuits and Systems (ISCAS)*, Paris, France, pp. 1687–1690.

Benford F 1938 The law of anomalous numbers. *Proceedings of the American Philosophical Society* **78**(4), 551–572.

Bestagini P, Allam A, Milani S, Tagliasacchi M and Tubaro S 2012 Video codec identification. *Proceedings of 2012 International Conference on Acoustics, Speech, and Signal Processing (ICASSP)*, Kyoto, Japan, pp. 2257–2260.

Bian S, Luo W and Huang J 2013 Detecting video frame-rate up-conversion based on periodic properties of inter-frame similarity. *Multimedia Tools and Applications* **72**(1), 437–451.

Bianchi T and Piva A 2011 Detection of non-aligned double JPEG compression with estimation of primary compression parameters. *Proceedings of 2011 IEEE International Conference on Image Processing (ICIP)*, Brussels, Belgium, pp. 1929–1932.

Bianchi T and Piva A 2012a Detection of nonaligned double JPEG compression based on integer periodicity maps. *IEEE Transactions on Information Forensics and Security* **7**(2), 842–848.

Bianchi T and Piva A 2012b Image forgery localization via block-grained analysis of JPEG artifacts. *IEEE Transactions on Information Forensics and Security* **7**(3), 1003–1017.

Bianchi T and Piva A 2012c Reverse engineering of double JPEG compression in the presence of image resizing *Proceedings of 2012 IEEE International Workshop on Information Forensics and Security (WIFS)*, Tenerife, Spain.

Bianchi T, De Rosa A and Piva A 2011 Improved DCT coefficient analysis for forgery localization in JPEG images. *Proceedings of 2011 International Conference on Acoustics, Speech, and Signal Processing (ICASSP)*, Prague, Czech Republic.

Chen YL and Hsu CT 2008 Image tampering detection by blocking periodicity analysis in JPEG compressed images. *Proceedings of 2008 IEEE Workshop on Multimedia Signal Processing (MMSP)*, Cairns, Queensland, Australia, pp. 803–808.

Chen YL and Hsu CT 2011 Detecting recompression of JPEG images via periodicity analysis of compression artifacts for tampering detection. *IEEE Transactions on Information Forensics and Security* **6**(2), 396–406.

Chen Y, Challapali KS and Balakrishnan M 1998 Extracting coding parameters from pre-coded MPEG-2 video. *Proceedings of 1998 International Conference on Image Processing (ICIP)*, Chicago, IL, vol. 2, pp. 360–364.

Comon P 1994 Independent component analysis, a new concept? *Signal Processing* **36**(3), 287–314.

Cortes C and Vapnik V 1995 Support-vector networks. *Machine Learning* **20**(3), 273–297.

Dalgaard N, Mosquera C and Pérez-González F 2010 On the role of differentiation for resampling detection. *Proceedings of 2010 International Conference on Image Processing (ICIP)*, Hong Kong, China, pp. 1753–1756.

Fan Z and de Queiroz RL 2000 Maximum likelihood estimation of JPEG quantization table in the identification of bitmap compression history. *Proceedings of 2000 International Conference on Image Processing (ICIP)*, Vancouver, British Columbia, Canada, vol. 1.

Fan Z and de Queiroz RL 2003 Identification of bitmap compression history: JPEG detection and quantizer estimation. *IEEE Transactions on Image Processing* **12**(2), 230–235.

Farid H 2009 Exposing digital forgeries from JPEG ghosts. *IEEE Transactions on Information Forensics and Security* **4**(1), 154–160.

Feng X and Doërr G 2010 JPEG re-compression detection. *Proceedings SPIE 7541, Media Forensics and Security II* (eds. Memon ND, Dittmann J, Alattar AM and Delp EJ), San Jose, CA, vol. 7541, p. 75410J.

Ferrara P, Bianchi T, De Rosa A and Piva A 2013 Reverse engineering of double compressed images in the presence of contrast enhancement. *Proceedings of 2013 IEEE International Workshop on Multimedia Signal Processing (MMSP)*, Pula, Italy.

Fontani M, Bianchi T, De Rosa A, Piva A and Barni M 2011 A Dempster-Shafer framework for decision fusion in image forensics. *Workshop on Information Forensics and Security*, Foz do Iguaçu, Brazil.

Fontani M, Bianchi T, De Rosa A, Piva A and Barni M 2013 A framework for decision fusion in image forensics based on Dempster-Shafer theory of evidence. *IEEE Transactions on Information Forensics and Security* **8**(4), 593–607.

Fu D, Shi YQ and Su W 2007 A generalized Benford's law for JPEG coefficients and its applications in image forensics. In *Proceedings SPIE 6505, Security, Steganography, and Watermarking of Multimedia Contents IX* (eds. Delp EJ and Wong PW), San Jose, CA, vol. 6505, p. 65051L.

Gallagher A 2005 Detection of linear and cubic interpolation in JPEG compressed images. *Proceedings of the 2nd Canadian Conference on Computer and Robot Vision, 2005*, Washington, DC, pp. 65–72.

He J, Lin Z, Wang L and Tang X 2006 Detecting doctored JPEG images via DCT coefficient analysis. *Computer Vision ECCV 2006, Lecture Notes in Computer Science Volume 3953*, pp. 423–435. Springer, Berlin, Germany.

ISO/IEC 1991 Digital Compression and Coding of Continuous tone Still Images, Part 1, Requirements and Guidelines. ISO/IEC IS 10918-1, International Organization for Standardization, Geneva, Switzerland.

ISO/IEC 2008 Information technology – generic coding of moving pictures and associated audio information – Part 2: Video. ISO/IEC IS 13818-2, International Organization for Standardization, Geneva, Switzerland.

ITU-T/ISO/IEC 2003 Advanced Video Coding for Generic Audio-Visual Services. Technical Report ITU-T Rec. H.264 and ISO/IEC 14496-10 (AVC), International Telecommunication Union and International Organization for Standardization, Geneva, Switzerland.

ITU-T/ISO/IEC 2012 High Efficiency Video Coding (HEVC) Text Specification Draft 9. ISO/IEC JCTVC-K1003, International Telecommunication Union and International Organization for Standardization, ITU-T/ISO/IEC Joint Collaborative Team on Video Coding (JCT-VC), Geneva, Switzerland.

Kirchner M 2008a Fast and reliable resampling detection by spectral analysis of fixed linear predictor residue *Proceedings of the 10th ACM Workshop on Multimedia and Security (MM&Sec)*, Oxford, UK, pp. 11–20.

Kirchner M 2008b On the detectability of local resampling in digital images. *Proceedings of SPIE 6819, Security, Forensics, Steganography, and Watermarking of Multimedia Contents X* (eds. Delp EJ, Wong PW, Dittmann J and Memon N), San Jose, CA, vol. 6819, p. 68190F.

Kirchner M 2010 Linear row and column predictors for the analysis of resized images. *Proceedings of the 12th ACM Workshop on Multimedia and Security (MM&Sec)*, Rome, Italy, pp. 13–18.

Kirchner M and Fridrich JJ 2010 On detection of median filtering in digital images *Proceedings SPIE 7541, Media Forensics and Security II*, San Jose, CA, vol. 7541, p. 754110.

Kirchner M and Gloe T 2009 On resampling detection in re-compressed images *Proceedings of 2009 IEEE International Workshop on Information Forensics and Security (WIFS)*, London, UK, pp. 21–25.

Labartino D, Bianchi T, De Rosa A, Fontani M, Vázquez-Padín D, Piva A and Barni M 2013 Localization of forgeries in MPEG-2 video through GOP size and DQ analysis. *Proceedings of IEEE 2013 International Workshop on Multimedia Signal Processing (MMSP)*, Pula, Italy.

Li H and Forchhammer S 2009 MPEG2 video parameter and no reference PSNR estimation. *Proceedings of Picture Coding Symposium, 2009 (PCS)*, Chicago, IL, pp. 1 –4.

Li B, Shi YQ and Huang J 2008 Detecting doubly compressed JPEG images by using mode based first digit features. *Proceedings of 2008 International Workshop on Multimedia Signal Processing (MMSP)*, Cairns, Queensland, Australia, pp. 730–735.

Lin WS, Tjoa SK, Zhao HV and Liu KJR 2009a Digital image source coder forensics via intrinsic fingerprints. *IEEE Transactions on Information Forensics and Security* **4**(3), 460–475.

Lin ZC, He JF, Tang X and Tang CK 2009b Fast, automatic and fine-grained tampered JPEG image detection via DCT coefficient analysis. *Pattern Recognition* **42**(11), 2492–2501.

Liu H and Heynderickx I 2008 A no-reference perceptual blockiness metric *Proceedings of 2008 International Conference on Acoustics, Speech, and Signal Processing (ICASSP)*, Las Vegas, NV, pp. 865–868.

Lukáš J and Fridrich J 2003 Estimation of primary quantization matrix in double compressed JPEG images. *Proceedings of Digital Forensic Research Workshop*, Cleveland, OH.

Luo W, Qu Z, Huang J and Qiu G 2007 A novel method for detecting cropped and recompressed image block. *Proceedings of 2007 International Conference on Acoustics, Speech, and Signal Processing (ICASSP)*, Honolulu, HI, pp. 217–220.

Luo W, Wu M and Huang J 2008 MPEG recompression detection based on block artifacts. *Proceedings SPIE 6819, Security, Forensics, Steganography, and Watermarking of Multimedia Contents X*, San Jose, CA, vol. 6819, p. 68190X.

Luo W, Huang J and Qiu G 2010 JPEG error analysis and its applications to digital image forensics. *IEEE Transactions on Information Forensics and Security* **5**(3), 480–491.

Mahdian B and Saic S 2007 On periodic properties of interpolation and their application to image authentication. *Proceedings of the Third International Symposium on Information Assurance and Security, 2007 (IAS)*, Manchester, UK, pp. 439–446.

Mahdian B and Saic S 2008 Blind authentication using periodic properties of interpolation. *IEEE Transactions on Information Forensics and Security* **3**(3), 529–538.

Mahdian B and Saic S 2009 A cyclostationarity analysis applied to image forensics. *Proceedings of 2009 IEEE Workshop on Applications of Computer Vision (WACV)*, Snowbird, UT, pp. 389–399.

Meng XZ, Niu SZ and Zou JC 2010 Tamper detection for shifted double JPEG compression. *Proceedings of 2010 Sixth International Conference on Intelligent Information Hiding and Multimedia Signal Processing (IIH-MSP)*, Darmstadt, Germany, pp. 434–437.

Milani S, Bestagini P, Tagliasacchi M and Tubaro S 2012a Multiple compression detection for video sequences *Proceedings of 2012 IEEE International Workshop on Multimedia Signal Processing (MMSP)*, Banff, Alberta, Canada, pp. 112–117.

Milani S, Fontani M, Bestagini P, Barni M, Piva A, Tagliasacchi M and Tubaro S 2012b An overview on video forensics. *APSIPA Transactions on Signal and Information Processing* **1**, 22.

Milani S, Tagliasacchi M and Tubaro M 2012c Discriminating multiple JPEG compression using first digit features. *Proceedings of 2012 International Conference on Acoustics, Speech, and Signal Processing (ICASSP)*, Kyoto, Japan, pp. 2253–2256.

Neelamani R, de Queiroz RL, Fan Z, Dash S and Baraniuk RG 2006 JPEG compression history estimation for colour images. *IEEE Transactions on Image Processing* **15**(6), 1365–1378.

Pevný T and Fridrich J 2008 Estimation of primary quantization matrix for steganalysis of double-compressed JPEG images. *Proceedings SPIE 6819, Security, Forensics, Steganography, and Watermarking of Multimedia Contents X*, San Jose, CA, vol. 6819, p. 681911.

Pevný T, Bas P and Fridrich J 2010 Steganalysis by subtractive pixel adjacency matrix. *IEEE Transactions on Information Forensics and Security* **5**(2), 215–224.

Piva A 2013 An overview on image forensics. *ISRN Signal Processing* **2013**, 1–22.

Popescu A and Farid H 2005a Exposing digital forgeries in colour filter array interpolated images. *IEEE Transactions on Signal Processing* **53**(10), 3948–3959.

Popescu AC and Farid H 2005b Exposing digital forgeries by detecting traces of resampling. *IEEE Transactions on Signal Processing* **53**(2–2), 758–767.

Popescu AC and Farid H 2005c Statistical tools for digital forensics. *Information Hiding, Lecture Notes in Computer Science*, Volume 3200, Springer, Berlin, pp. 128–147.

Prasad S and Ramakrishnan KR 2006 On resampling detection and its application to detect image tampering. *Proceedings of 2006 IEEE International Conference on Multimedia and Expo*, Toronto, ON, Canada, pp. 1325–1328.

Qu Z, Luo W and Huang J 2008 A convolutive mixing model for shifted double JPEG compression with application to passive image authentication. *Proceedings of 2008 International Conference on Acoustics, Speech, and Signal Processing (ICASSP)*, Las Vegas, NV, pp. 1661–1664.

Song GS, Yun YI and Lee WH 2011 A new estimation approach of resampling factors using threshold-based peak detection. *Proceedings of 2011 IEEE International Conference on Consumer Electronics (ICCE)*, Las Vegas, NV, USA, pp. 731–732.

Stamm MC and Liu KJR 2008 Blind forensics of contrast enhancement in digital images. *Proceedings of 2008 International Conference on Image Processing (ICIP)*, San Diego, CA, pp. 3112–3115.

Stamm M and Liu KJR 2010a Forensic estimation and reconstruction of a contrast enhancement mapping. *Proceedings of 2010 International Conference on Acoustics, Speech, and Signal Processing (ICASSP)*, Dallas, TX, pp. 1698–1701.

Stamm MC and Liu KJR 2010b Forensic detection of image manipulation using statistical intrinsic fingerprints. *IEEE Transactions on Information Forensics and Security* **5**(3), 492–506.

Stamm M, Wu M and Liu K 2013 Information forensics: An overview of the first decade. *IEEE Access* **1**, 167–200.

Sullivan G and Wiegand T 2005 Video compression – from concepts to the H.264/AVC Standard. *Proceedings of the IEEE* **93**(1), 18–31.

Sun T, Wang W and Jiang X 2012 Exposing video forgeries by detecting MPEG double compression. *Proceedings of 2012 International Conference on Acoustics, Speech, and Signal Processing (ICASSP)*, Kyoto, Japan, pp. 1389–1392.

Tagliasacchi M and Tubaro S 2010 Blind estimation of the QP parameter in H.264/AVC decoded video. *Proceedings of 2010 11th International Workshop on Image Analysis for Multimedia Interactive Services (WIAMIS)*, Desenzano del Garda, Italy, pp. 1–4.

Tjoa S, Lin WS, Zhao HV and Liu KJR 2007 Block size forensic analysis in digital images. *Proceedings of 2007 International Conference on Acoustics, Speech, and Signal Processing (ICASSP)*, Honolulu, HI.

Valenzise G, Tagliasacchi M and Tubaro S 2010 Estimating QP and motion vectors in H.264/AVC video from decoded pixels. *Proceedings of the 2nd ACM Workshop on Multimedia in Forensics, Security and Intelligence MiFor '10. ACM*, New York, pp. 89–92.

Vázquez-Padín D and Pérez-González F 2011 Prefilter design for forensic resampling estimation. *Proceedings of 2011 IEEE International Workshop on Information Forensics and Security (WIFS)*, Foz do Iguaçu, Brazil.

Vázquez-Padín D, Mosquera C and Pérez-González F 2010 Two-dimensional statistical test for the presence of almost cyclostationarity on images. *Proceedings of 2010 International Conference on Image Processing (ICIP)*, Hong Kong, China.

Vázquez-Padín D, Fontani M, Bianchi T, Comesaña P, Piva A and Barni M 2012 Detection of video double encoding with GOP size estimation. *Proceedings of 2012 IEEE International Workshop on Information Forensics and Security (WIFS)*, Tenerife, Spain, pp. 151–156.

Wang W and Farid H 2006 Exposing digital forgeries in video by detecting double MPEG compression. *Proceedings of the 8th workshop on Multimedia and Security (MM&Sec)*, Geneva, Switzerland, pp. 37–47.

Wang W and Farid H 2009 Exposing digital forgeries in video by detecting double quantization. *Proceedings of the 11th ACM workshop on Multimedia and Security (MM&Sec'09)*. Princeton, NJ, USA, pp. 39–48.

Wang Z, Bovik AC and Evans BL 2000 Blind measurement of blocking artifacts in images. *Proceedings of 2000 International Conference on Image Processing (ICIP)*, Vancouver, BC, Canada.

Wei W, Wang S and Tang Z 2008 Estimation of rescaling factor and detection of image splicing. *Proceedings of 11th IEEE International Conference on Communication Technology, 2008 (ICCT)*, Hangzhou, China, pp. 676–679.

Ye S, Sun Q and Chang EC 2007 Detecting digital image forgeries by measuring inconsistencies of blocking artifact. *Proceedings of 2007 IEEE International Conference on Multimedia and Expo (ICME)*, Beijing, China, pp. 12–15.

Yuan HD 2011 Blind forensics of median filtering in digital images. *IEEE Transactions on Information Forensics and Security* **6**(4), 1335–1345.

16

Anti-Forensics of Multimedia Data and Countermeasures

Giuseppe Valenzise[1], Stefano Tubaro[2] and Marco Tagliasacchi[2]
[1]*CNRS LTCI, Télécom ParisTech, Paris, France*
[2]*DEIB, Politecnico di Milano, Milano, Italy*

16.1   Introduction

In order to assess the authenticity of a digital image, forensic analysts rely on a key tenet of forensics: an (illegitimate) alteration of multimedia data is bound to leave some characteristic and detectable traces in the tampered content (Milani *et al.* 2012; Stamm and Liu 2010a). These fingerprints are distinctive of the specific acquisition, coding and editing operations carried out on the multimedia content, and have been described in detail in other chapters of this book (especially those in Parts III and IV). In the case of audio, for instance, the reverberation time or the microphone used to grab the signal leave predictable and detectable fingerprints (Kraetzer *et al.* 2007; Malik and Farid 2010). For images and video, it is possible, for example to identify the camera that shot a picture through the non-uniformity of its sensor's noise (Chen *et al.* 2008b); or to detect traces of coding (Fan and de Queiroz 2003; Ferrara *et al.* 2012; Lin *et al.* 2009; Valenzise *et al.* 2010), re-sampling (Popescu and Farid 2005a; Vazquez-Padin and Comesana 2012), cropping (Bianchi and Piva 2012; Luo *et al.* 2007) and point-wise processing such as contrast enhancement and histogram equalization (Fontani and Barni 2012; Stamm and Liu 2010a).

While the forensic analysis tools progress with newer and more powerful fingerprint detectors, a knowledgeable adversary might attempt to remove the traces of his forgeries, or increase the number of false alarms detected by the forensic analyst, in

Handbook of Digital Forensics of Multimedia Data and Devices, First Edition.
Edited by Anthony T.S. Ho and Shujun Li.
© 2015 John Wiley & Sons, Ltd. Published 2015 by John Wiley & Sons, Ltd.
Companion Website: www.wiley.com/go/digitalforensics

such a way to make the existing forensic tools less accurate and practically useless. The techniques aimed at concealing the fingerprints left by multimedia processing operators are known as *anti-forensic* or counter-forensic techniques. Studying anti-forensics is of paramount importance in the field of forensics for a number of reasons. First of all, it enables to understand the limitations of current forensic approaches, for example by clarifying up to which extent one can trust upon the evidence provided by certain multimedia fingerprints. Second, anti-forensic algorithms leave as well some characteristics fingerprints which can be detected in order to show that the content has been tampered with. Finally, being aware of the weaknesses of current analysis tools enables to improve them and create new methods which are more reliable and less prone to be fooled.

In this chapter we focus on the specific case of anti-forensics of image and video content, although the basic principles outlined earlier apply to audio as well. We start by reviewing in Section 16.2 the anti-forensic techniques proposed in the literature so far. Next, we analyze in depth in Section 16.3 one particular example, namely, the anti-forensics of JPEG compression fingerprints. We show as well how this anti-forensic technique can be detected with anti-forensic countermeasures. Finally, we discuss in Section 16.4 how anti-forensic techniques can be evaluated with respect to forensic detectors, and illustrate how the interplay between the forger and the forensic analyst can be cast in a game-theoretical framework.

16.2 Anti-forensic Approaches Proposed in the Literature

Despite the important development of forensics in the past decade, it has been only recently that forensic researchers have started investigating the *reliability* of forensic techniques as tools to generate evidence about the authenticity and integrity of digital images. Thus, while there are a number of forensic detectors for a fairly variegate set of possible signal manipulations, examples of anti-forensic techniques (and their possible countermeasures) are quite limited. In the following we try to provide a structured review of these methods, leaving for the next sections a more in-depth discussion on the design of anti-forensic attacks and countermeasures with an example on JPEG compression.

In their seminal work on tamper hiding, Kirchner and Böhme (2007) proposed to classify anti-forensic approaches according to two dimensions: the generality of the anti-forensic technique and the position of the attack in the processing chain. In the first case, one can distinguish between *targeted* and *universal* attacks. Targeted techniques try to hide those fingerprints detectable by a specific forensic tool, normally known by the attacker. While these techniques are normally quite effective with respect to the targeted forensic tool, another analysis algorithm might still detect the manipulation. On the other hand, universal attacks aim at restoring some natural and plausible image statistics, which are modified with respect to the original content in the forgery process. The rationale is that, since forensic fingerprints are designed to capture these anomalies, one could fool a large number of forensic tools by manipulating as many image statistics as possible in such a way to sufficiently lower the detection rate of the analyst. An example of universal attack is low-quality compression of doctored images, which enables to remove possible traces of tampering in the frequency domain. In general, designing universal attacks could be challenging since maintaining a large number of image statistics might produce a sensible loss of quality in the tampered image (as in the case of low-quality image compression attack).

As a second dimension of classification of anti-forensic techniques, Kirchner and Böhme (2007) divide *post-processing* and *integrated* methods, according to their position in the processing chain. Attacks based on post-processing try to remove the traces of the forgery *after* it has been applied in order to modify the content. A typical example of post-processing methods is removing the traces of JPEG compression in a given compressed picture. Conversely, integrated attacks are designed to interact directly with the manipulation operation, in such a way to modify the content without leaving any fingerprints on it. A tool which enables undetectable copy-move forgeries is an example of integrated attack. Integrated attacks could be more effective than post-processing methods, but developing them could be much harder in many cases.

While the taxonomy proposed earlier describes well the properties of anti-forensics, we propose a simpler yet functional classification based on which kind of fingerprints the anti-forensic attack is aimed at concealing. Thus, we consider anti-forensic methods that remove fingerprints left during either acquisition, coding or editing of the content.

16.2.1 Anti-forensics of Acquisition Fingerprints

The acquisition process of a digital picture/video is a rather complex process which includes several steps: first, light is converted to current through the image sensors in conjunction with a colour filter array (CFA). Then, the analogue signal is digitized and the missing colours are interpolated. Also, other forms of post-processing or denoising can be operated by the camera/scanner software. From the point of view of anti-forensics, the two most relevant acquisition fingerprints are the image sensor noise and the CFA interpolation.

16.2.1.1 Image Sensor Noise

The image sensor, which is a matrix of sensor elements (pixels) that capture photons and convert them to voltages, introduces some characteristic noise patterns which are independent from the particular captured scene and depend on, for example imperfections in the manufacturing process of the CCD. The noise pattern is the sum of two components, the fixed pattern noise (FPN) and the photo-response non-uniformity (PRNU). FPN is caused by small electric currents that flow into the CCD sensors even in the absence of photons and depends on the temperature and exposure time. The FPN can be easily measured by capturing a 'dark frame', i.e., a photo taken in the dark (e.g.

with the shutter closed). On the other hand, the dominant part of pattern noise is given by PRNU, which is partly due to the different sensitivity of each pixel to light (pixel non-uniformity or PNU), and partly to low-frequency defects given by dust particles on the lenses or optical effects. Of these two, only the PNU is characteristic of the sensors and acts as a multiplicative noise pattern with respect to the light received on the sensor. In order to estimate the PNU, one can apply a flat-fielding algorithm (Janesick 2001) on the raw pixel values output from the sensor.

If the raw pixels are not available, for example, because the camera is not accessible or is unable to store raw values, Lukáš *et al.* (2006b) have proposed to recover the PNU by averaging multiple images taken from the same sensor in order to obtain the camera reference pattern. In order to enhance noise estimation and reduce the dependence on scene content, a smoothed version of the image (obtained with a wavelet denoising algorithm) is subtracted from the original picture to obtain a noise residual. To decide whether a given test image has been shot with the camera, Lukáš *et al.* (2006b) proposed to compute the correlation coefficient between the noise residual and the camera reference pattern, and if this correlation is higher than a certain threshold the image is deemed to have been taken with that specific camera. Given the PNU signal has a very low signal-to-noise ratio, small values of correlations (and thus of the threshold) are expected, and thus a correct choice of the decision threshold is of extreme importance. Lukáš *et al.* (2006b) consider the threshold selection problem in the framework of statistical hypothesis testing, by modelling the distribution of the correlation coefficients as a generalized Gaussian distribution. Then, the optimal threshold can be found by maximizing the probability of detection given an upper bound on the false acceptance rate. As an alternative to simple correlation, statistics proposed in the literature include the generalized matched filter (Goljan *et al.* 2007) and the peak to correlation energy (Chen *et al.* 2008c).

The forensic detector described earlier achieves very good detection performance (e.g. Lukáš *et al.* (2006b) reported false rejection rates lower than 10^{-3} for a false acceptance rate fixed to 10^{-3}), and is rather robust to gamma correction and JPEG compression. Furthermore, based on the PRNU a number of digital image forgeries can be accurately discovered (Lukáš *et al.* 2006a). Therefore, it is potentially a very powerful forensic tool to be used in courtrooms. However, a malicious attacker could operate in such a way to partially remove the PRNU fingerprint and lower the correlation with the camera reference pattern under the detection threshold. To this end, one could apply a denoising algorithm to the image in order to remove traces of the PRNU, or even a simple desynchronization operation such as rotation or scaling. Gloe *et al.* (2007) have shown that it is actually possible to hinder the detection of the PRNU, if the camera is available to the attacker, using the following procedure: first, the dark and flat fields of the camera are estimated (in some restricted conditions, for example known exposure time and shutter speed); afterwards, the dark field is subtracted from the image, and the result is divided element-wise by the field frame (given that the PRNU noise is multiplicative).

As a consequence, a main security flaw of PRNU-based forensic tools is that it is possible to use inverse flat-fielding to plant a fingerprint of a camera C' into an image taken with camera C – a technique known as *fingerprint-copy attack* (Goljan *et al.* 2011). In this way, one could frame an innocent victim by passing her as the author of a picture. More in detail, the attacker estimates the pattern noise of the victim's camera C', for example the dark and flat-field frame as suggested in (Gloe *et al.* 2007). This can be done, for example if the attacker has access to a set of pictures taken from the victim (e.g. posted on the Internet). Then, he multiplies (element-wise) the image taken by C by the flat-field frame, and adds the dark frame. There is evidence that, using one of the threshold-based PRNU detectors described earlier, a fingerprint-copy attack may be effective in making a fake fingerprint appear as a genuine one (Gloe *et al.* 2007; Steinebach *et al.* 2010).

Goljan *et al.* (2011) proposed as a countermeasure to the fingerprint-copy attack a procedure known as *triangle test*. The basic intuition behind the triangle test is that, in order to frame the victim by copying the PRNU fingerprint of her camera C, the attacker has to use some of her pictures. The victim has access to a set of images which include the ones used by the attacker, although she does not know which ones they are. The forged image $\mathbf{J}$ will contain traces of the PRNU $\mathbf{P}$ of C (the camera of the victim), *plus* other traces of noise which are specific to the ensemble of pictures used by the attacker for the forgery. In other words, the correlation between the forged image and those victim's pictures used by the attacker will be higher than the correlation with images shot by C not used in the forgery (which only share the PRNU noise). In order to defend herself, the victim starts by estimating $\mathbf{P}$ using pictures which are sure to not have been used by the adversary, for example she can take new photos with her camera. Then, for each possible image $\mathbf{I}$ that could have been used by the attacker to produce the forged picture $\mathbf{J}$, she performs the triangle test as follows. She computes the correlation $c_{\mathbf{IP}}$ between the noise residual of candidate image and the camera reference PRNU, as well as the correlation $c_{\mathbf{JP}}$ between the noise residual of the forged image and the reference pattern. Through these two values, it is possible to estimate the correlation $\hat{c}_{\mathbf{IJ}}$ between the noise residual of the candidate image and the one of the forged image, under the hypothesis that $\mathbf{I}$ was *not* used in the attack. This correlation is then compared with the experimental correlation $c_{\mathbf{IJ}}$ through a composite hypothesis test. Briefly, when $\mathbf{I}$ has been used by the attacker, the correlation $c_{\mathbf{IJ}}$ is expected to be significantly larger than $\hat{c}_{\mathbf{IJ}}$, due to the presence of an additional noise component which is correlated with $\mathbf{I}$ and which would not be justified by the PRNU noise alone. This test can be performed on single images, although its accuracy may be rather low; or it could be done testing a number of candidate images all at once, in order to increase the reliability of the test (pooled test). Furthermore, the triangle test can be extended to the case when multiple images have been forged by the adversary. The triangle test can be applied when the adversary uses a high-quality fingerprint estimated from 300 images, and shows that planting a fake sensor fingerprint into an image without leaving traces is not such an easy task as it could be erroneously believed.

16.2.1.2   CFA Interpolation

Due to physical constraints and in order to reduce sensor costs, most digital cameras employ a single sensor in conjunction with a CFA to capture colour images. The layout of the colour filter with respect to the pixel grid can greatly vary across the camera brand or model, with the most popular scheme being the Bayer pattern. In all the cases, in order to reconstruct a high-resolution colour image, the three-colour channels have to be interpolated using a CFA interpolation algorithm. This interpolation introduces some characteristic correlations between the pixels of the image. When an image is tampered with, these correlation can be locally destroyed, and this fact can be used as a tell-tale of forgery. Popescu and Farid (2005b) proposed an expectation–maximization procedure to detect the correlation produced by CFA interpolation. The core observation of their method is that, since the colour filters in a CFA are typically arranged in a periodic pattern, also the resulting correlations will be periodic. The detection procedure is thus similar in principle to that of detecting resampling (and in general linear space-invariant filtering) described in Section 16.2.3. The absence of CFA correlation may imply the presence of local tampering.

In order to hinder the detection of forgeries through the CFA fingerprint described earlier, one could synthesize locally the missing correlation pattern by restoring the specific linear dependencies between pixels which are characteristic of a given CFA. Kirchner and Böhme (2009) solved the optimal CFA synthesis problem in two steps. First, the sample signal for one colour, $\mathbf{y}$, is assumed to be obtained through a linear filtering of the original sensor signal, that is $\mathbf{y} = \mathbf{Hx} + \varepsilon$, where noise ε models the part of the image which is not in the span of $\mathbf{H}$. The mixing matrix $\mathbf{H}$ corresponds to the weights of the target CFA pattern. Then, an approximation of the original sensor signal $\hat{\mathbf{x}}$ is found through least-squares as $\hat{\mathbf{x}} = (\mathbf{H}^T\mathbf{H})^{-1}\mathbf{H}^T\mathbf{y}$. Finally, the estimated $\hat{\mathbf{x}}$ is mapped to a CFA compatible signal $\hat{\mathbf{y}} = \mathbf{H}\hat{\mathbf{x}}$. Kirchner and Böhme (2009) showed also how to solve the least-squares problem efficiently using the structure of $\mathbf{H}$ and proposed an approximate solution with complexity which grows linearly with the number of pixels.

16.2.2   Anti-forensic of Compression Fingerprints

During the lifetime of a digital multimedia object, lossy compression is an almost ubiquitously present processing block, due to storage and bandwidth constraints in typical application scenarios. Thus, a number of forensic methods to detect image (Chen *et al.* 2008a; Fan and de Queiroz 2003; Farid 2009; Huang *et al.* 2010; Lukáš and Fridrich 2003) and video (Bestagini *et al.* 2012; Luo *et al.* 2008; Wang and Farid 2006) compression have been proposed in the literature.

Image compression typically leaves both *statistical* fingerprints and *compression artefact* traces. Statistical fingerprints are linked to the fact that quantization – an operation present in any lossy compression scheme – changes the intrinsic statistics of the image (Stamm and Liu 2010a). Specifically, in a general picture coding scheme, an image is first transformed through some decorrelating transform (e.g. DCT or Wavelet), and the transform coefficients are quantized (using a scalar quantizer or a perceptually motivated quantization matrix). After quantization, the decoded transform coefficients can only take values that are integer multiples of the quantization step size. This introduces a characteristic comb-shaped pattern in their histogram, which is detectable by a forensic analyst.

Methods aimed at removing this pattern (Stamm and Liu 2010c; Stamm *et al.* 2010a) restore the original distribution by first estimating the distribution of the original coefficients using, and then by adding a properly designed dithering noise to the quantized transform coefficients, similarly to what is done to remove traces of histogram editing (Section 16.2.3). The added dither is a noise which may deteriorate the quality of the image and alter other natural statistics in the pixel domain. Thus, countermeasures have been proposed to detect compression anti-forensics (Lai and Böhme 2011; Valenzise *et al.* 2013). Based on these studies, recently Fan *et al.* (2013b) have improved the JPEG anti-forensic attack by using a non-parametric DCT model of transform coefficients, and have proposed a general variational framework (Fan *et al.* 2013a) that enables to fool most of the proposed JPEG antiforensic countermeasures.

In the case of video, a typical tampering operation that can be applied to a video sequence consists of adding or deleting frames, for example to hide the presence of some object or person. Wang and Farid (2006) proposed a forensic tool to detect double MPEG compression and frame insertion/deletion using both spatial fingerprints (similarly to image double-compression traces) and temporal clues. The temporal fingerprint left by re-compression after frame insertion/deletion is a periodic pattern in the inter-frame prediction error energy sequence $e(t)$. Indeed, when the structure of the group of pictures (GOPs) is altered, frames previously belonging to the same GOP will be spread on two different GOPs. It turns out that P frames whose anchor has been reallocated to another GOP will exhibit a larger prediction error with respect to other frames, thus leaving a signature which enables to identify the forgery. Stamm *et al.* (2012) reformulated the problem in a more general statistical hypothesis testic framework, and extended the detector to the case of adaptive GOP size. In addition, they proposed an anti-forensic technique which consists in adding noise to the prediction residuals in such a way to raise the noise energy level in $e(t)$ and thus eliminating the peaks due to re-compression. To this end, they alter motion vectors of specific P frames, by setting some motion vectors to zero in order to intentionally increase the prediction residual energy. While this does not leave traces into the quality of the reconstructed video (due to the closed-loop prediction scheme of any hybrid video codec), it is true that the altered motion vectors can be detected, e.g. by performing motion estimation at the decoder. Since there will be a large mismatch between true motion and the tampered one, a forensic analyst could be able to detect the traces of an anti-forensic manipulation.

16.2.3   Anti-forensic of Editing Fingerprints

Thanks to the availability of easy-to-use photo-editing software, producing photore-alistic fakes has become a rather simple task, even for non-professional users. There is a number of forensic tools designed to detect fingerprints left by common editing operations, including histogram processing, filtering, copy-paste forgeries, etc. In the following we describe the basic anti-forensic techniques developed to counter these detectors, in the case of images. As for video, so far no anti-forensics of editing fin-gerprints has been developed, also because performing complex and accurate video editing (taking into account motion) is still more complicated than forging still images.

16.2.3.1   Histogram-Based Tools

Many editing operations have a direct effect on the first-order statistics of the modified images (in particular, the histogram). Examples of this kind of processing include point-wise transformations (gamma correction, contrast stretching) or histogram equalization and matching (Stamm and Liu 2010a). As seen in the previous section, coding also affects in some way the histogram in the transform domain. Therefore, several forensic techniques use first-order statistics to find traces of forgeries. For instance, Stamm and Liu (2008) proposed to detect the application of a global contrast enhancing (GCE) operator based on the characteristic artefacts it produces on the image histogram. A GCE operator is a nonlinear point-wise mapping followed by rounding to integer pixel values. When applied to the pixels of an image, these will be remapped in two possible ways: when the GCE is contractive, multiple pixels in the original image are mapped to the same pixel values in the transformed one, causing peaks in the final histogram; conversely, when the mapping is expansive, some pixel values will be skipped over, leading to gaps in the resulting histogram. These artefacts can be effectively detected by a frequency analysis of the histogram of the forged image. In their follow-up work, Stamm and Liu (2010b) improved their detector in such a way that it can estimate as well the original GCE function used for the forgery.

Cao *et al.* (2010a) described both an integrated and a post-processing attack to remove the peaks and gaps in the histogram left by a GCE operator. The basic idea is to add local random dithering to the modified pixel values in such a way to redistribute pixels and smooth out the contrast-enhanced histogram. The variance of the dithering noise is larger for the values which are mapped to the neighboiurhood of potential peaks and gaps. With this technique the authors showed that it is possible to reduce the detection rate of the forensic detectors described above to that of a random detector.

While the previous technique is targeted to fool the detector in Stamm and Liu (2008), Barni *et al.* (2012) proposed a universal method to conceal traces of histogram processing. The rationale of this approach is to transform any tampered image in such a way that its histogram looks similar to that of a natural (authentic) image. This is performed in three steps. In the first stage (histogram retrieval), the attacker retrieves from a sufficiently large database of images a target image $\mathbf{I}_t$ whose histogram is the most similar to the histogram-processed image $\mathbf{I}_p$ to be attacked. The χ^2 distance is used as a measure of similarity between histograms. Once a template histogram has been found, the second phase of the algorithm (histogram mapping) consists in finding a displacement matrix such that applying it to $\mathbf{I}_p$ will produce an image with histogram as similar as possible to that of $\mathbf{I}_t$. This is obtained by solving an optimization problem in which the cost function is the Kullback–Leibler divergence between the target histogram and the current solution, with a constraint on the maximum per pixel distortion that can be introduced in the process. Finally, in the third step (pixel remapping), the pixel mapping defined by the optimal displacement matrix is implemented in a perceptually convenient way, by displacing pixels in those areas of the pictures which they are less likely to be noticed. This is achieved by computing a variance map and computing the structural similarity index (SSIM) at each iteration for controlling the areas of the image which are less sensible to noise. Experimental results using the detector in Stamm and Liu (2008) showed that it is possible to considerably reduce the detection performance of histogram processing while maintaining the visual quality above a minimum desired distortion. The advantages of this approach include the fact that the attacker does not need to know in advance how the forensic detector works; and that this method can be applied even if the adversary does not have the control over the tampering operation applied. On the other hand, the possibility of finding a good untampered histogram to match with strongly depends on the variety and size of the image database where histogram retrieval is performed.

16.2.3.2   Filtering-Based Tools

Image filtering can be used to enhance image quality, or to remove undesired noise or traces of tampering, or as a processing step in task such as decimation/interpolation. One of the most remarkable forensic tools based on filtering fingerprints is the one proposed by Popescu and Farid (2005a) to detect traces of image resampling. When an image is resampled, for example in order to scale or rotate the picture, the grid of pixel is changed in order to align it to the one of the new image. In doing so, a low-pass filter is applied in order to avoid aliasing and visually unpleasant artefacts in the result. Assuming a linear, space-invariant low-pass filtering (i.e. linear interpolation), it is apparent that the output pixels will be linear combinations of neighbouring pixels. As a result, low-pass filtering introduces specific correlations which were not present in the original signal, and that are periodic across the image (since the filter coefficients are constant and space invariant). In a practical forensic scenario, however, the low-pass filter used for interpolation – and thus the nature and form of correlation between pixels – is not known. Therefore, in order to determine whether a signal has been resampled, Popescu and Farid (2005a) proposed an expectation–maximization procedure which simultaneously estimate a set of periodic samples that are correlated to their neighbours, and the specific form of these correlations.

In order to remove these correlations, Gloe *et al.* (2007) and Kirchner and Böhme (2008) proposed an integrated attack by applying a geometric distortion to pixel position in such a way to perturb the regularity of the pixel grid spacing and break the periodic correlations which constitute the resampling fingerprint. Since simple geometric distortion may produce visible artefacts in the tampered image, the strength of the distortion is modulated through a Sobel edge detector in order to avoid jitter across edges. The distortion introduced by the tampering operation can be further controlled using a *dual-path* approach, in which a different processing is performed on low- and high-frequency components of the image by following two different paths. In the low-frequency path, the resampled image is smoothed through a median filter. Then, the high-frequency component is found by subtracting the original image from the median-filtered edition. In the high-frequency path, this component is resampled with geometric distortion and edge modulation, where the edge information is obtained from the resampled image prior to the median filter. Finally, the two paths are summed up in order to compute the final image.

A very common tool used in image editing and forgery (and a good example of anti-forensics) is the *median filtering*. While there have been several forensic techniques proposed to detect median filtering (see, e.g. Cao *et al.* (2010b) or Kirchner and Fridrich (2010)), a more recent approach is the work by Yuan (2011), which is able to detect locally the traces of median filtering with high accuracy. The core of this method is to extract from a test image a set of features, designed to target specific traces linked to median filtering (distribution of block median, occurrence of block center grey level, quantity of grey levels in a block, etc.). It is possible to combine these features into an aggregate measure f which proves to be an excellent discriminant for separating the median-filtered from unmodified images. Fontani and Barni (2012) proposed a post-processing attack to deceive this detector. The problem is set up as an optimization problem, where one minimizes a cost function which is the sum of two terms: a cost related to the presence of median filtering, measured through the aggregate feature f described earlier, and a fidelity term, that the PSNR between the filtered image and the candidate image. The result of the optimization is an optimal sharpening filter which enables to remove the statistical traces left by the median filter without introducing undesirable artefacts into the tampered image.

16.2.3.3 Copy-Move Forgeries

A typical editing operation used to forge the content of an image consists in duplicating portions of an image in order to move, duplicate or conceal an object from the scene. This forgery is known as *copy-move*, if both the origin and the target patch belong to the same image, or *splicing* in case the origin and the destination are in two different pictures. In the literature many forensic techniques have been presented to detect copy-move forgeries (Bayram *et al.* 2008). Among them, a recent and interesting approach is based on the use of local features such as the scale-invariant feature transform

(SIFT) (Amerini *et al.* 2011). When a portion of an image is cloned to make a forgery, often the author of the editing is forced to apply some geometric transformation to the moved object so that it can fit appropriately into the modified picture. Methods based on SIFT features are especially useful for matching pixels regions which have been geometrically transformed (e.g. rotated or scaled) and outperform simpler matching-based methods. There are two main elements that enable to specify local features: the key-point detector and the local descriptor. Key-points are detected as local extrema in the scale-space, and generally correspond to corners or high-contrast points of the picture. Once the key-points have been selected, a local descriptor is computed for each of them over a window centered in the key-point. An example of descriptor is the histogram of gradients used in the SIFT features (Lowe 2004). As SIFT-based copy-move detectors use the matching of SIFT features as a fingerprint of an attack, an adversary may try to remove as many as possible key-points from the cloned object, or change their descriptors so that the matching fail.

The problem of SIFT *security* has been discussed by Hsu *et al.* (2009), who have been amongst the first to show that removing SIFT features from a picture while preserving a good image quality – a task considered very challenging before – is actually possible. They achieved this result with two kinds of attacks. The first anti-forensic technique is the block-based *collage attack*, which consists in finding, from a large database of images, all the image blocks where no key-points are detected. Given this set $\mathcal{K}$ of key-point-free blocks, one can detect SIFTs in the image to be attacked, and for each block which contains a key-point, substitute it with the most similar block in $\mathcal{K}$ as far as this does not incur in a too high quality cost. The second anti-forensic attack to SIFT detection, called *extrema duplication*, aims at deceiving the core part of the key-point detector, which find local maxima/minima in the scale space, by duplicating the value of each local extremum. In this way, the extremum is no longer unique and no key-point is selected.

Do *et al.* (2010) showed how it is possible to fool a content-based image retrieval (CBIR) system by removing or modifying the key-points in a query image. In order to remove key-points, one can locally tamper with the image, for example by smoothing out the area around a key-point in such a way that it will not satisfy the conditions to be selected by the key-point detector. Other approaches to eliminate key-points consist in applying a geometric distortion to the attacked image as described by Caldelli *et al.* (2012).

One important observation is that simply removing key-points from an image is in general neither sufficient nor desirable for hiding a forgery. In fact, when local features are compared with a database of images to find the closest match, true positives tend to have a much higher score compared to other images in the list of retrieved pictures. Since it is sufficient that a few key-points are matched in order to have a fairly high score, all the key-points should be removed to guarantee the undetectability of the forgery. This is in general not feasible for two reasons: (i) key-point removal degrades image quality and (ii) an image with very few or no key-points would easily increase suspicion of forgery. Therefore, not only should key-points be removed, but new ones have to be inserted in such a way to move the true match down in the ranking of nearest neighbours of the query image. Amerini *et al.* (2013) described how to inject SIFT key-points into an image by means of local contrast enhancement without deteriorating the resulting image quality too much.

16.3   Case Study: JPEG Image Forensics

In this section we present a rather simple, though effective, anti-forensic technique to eliminate *statistical fingerprints* of JPEG compression, which is by far the most popular coding tool for images.[1] We do not consider other compression artefact traces, such as blocking artefacts, for which the interested reader can refer to, for example the anti-forensic deblocking filter described by Stamm *et al.* (2010b). The example of JPEG anti-forensics and its countermeasures is instructive for a number of reasons. First of all, several forensic analysis tasks leverage JPEG compression fingerprints, including the identification of which camera took a picture (Farid 2006), or the detection of double JPEG compression (Chen *et al.* 2008a; Huang *et al.* 2010; Lukáš and Fridrich 2003); also, localized evidence of double compression can reveal copy-move forgeries (Bayram *et al.* 2008; Bianchi *et al.* 2011; Fridrich *et al.* 2003; Luo *et al.* 2007). Therefore, studying in detail JPEG anti-forensics is helpful to show how a targeted anti-forensic attack aimed to conceal a given fingerprint can be conceived, and how the analysis of anti-forensic attacks can help discover their limitations and produce more robust forensic tools. Finally, as the JPEG anti-forensic technique described in this section is based on first-order statistics of transform coefficients, the same principles could be extended to other kinds of forgeries which alter the image histogram.

We start by reviewing in detail how JPEG compression operates and the fingerprints it leaves in the compressed image. We then present the anti-forensic technique of Stamm *et al.* (2010a) to delete the statistical traces of JPEG compression by adding a special anti-forensic noise. In the second part of this section, we move towards the analysis of this anti-forensic techniques, quantifying its side effects in terms of visible noise on the tampered image. Then we propose a strategy to effectively detect the presence of anti-forensic noise.

16.3.1   JPEG Compression and JPEG Compression Footprints

In the JPEG compression standard, the input image is first divided into B non-overlapping pixel blocks of size 8×8. For each block, the two-dimensional discrete cosine transform (DCT) is computed. Let X_i^b, $1 \leq b \leq B$, $1 \leq i \leq 64$ denote the ith transform coefficient of the bth block according to some scanning order (e.g. zig-zag).

---

[1] The material in this section is based on our previous work (Valenzise *et al.* 2013).

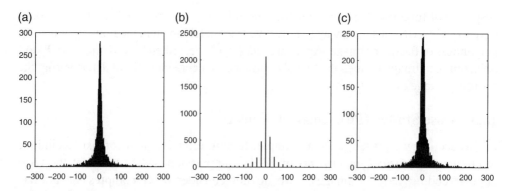

Figure 16.1 (a) Histogram of transform coefficients in the $(2, 1)$ DCT subband for the *Lenna* image. (b) Due to quantization, the coefficients in the JPEG-compressed image have a comb-shaped distribution. (c) The anti-forensic technique in Stamm *et al.* (2010a) enables to approximately restore the original distribution (a), thus removing JPEG-compression footprints.

That is, there is a one-to-one mapping $i \leftrightarrow (r, s)$ between the index i and the position (r, s), $1 \leq r, s \leq 8$, of a coefficient within a DCT block. Let $\mathbf{X}_i = [X_i^1, \ldots, X_i^b]^T$ denote the set of DCT coefficients of the ith subband. Each DCT coefficient X_i^b, $1 \leq i \leq 64$ is quantized with a quantization step size q_i. The set of q_i's forms the quantization table[2] $\mathbf{Q}$. The JPEG standard (ITU-T 1991) does not specify any fixed quantization table in its normative part. However, in the informative part of the standard (Annex K), the quantization table $\mathbf{Q}$ is defined as a scaled version of an image-independent template table, obtained by adjusting a (scalar) quality factor Q. This is the case, for instance, of the quantization tables adopted by the Independent JPEG Group (IJG).[3] The quantization levels W_i^b are obtained from the original coefficients X_i^b as $W_i^b = \text{round}\left(X_i^b / q_i\right)$. The quantization levels are entropy coded and written in the JPEG bitstream. When the bitstream is decoded, the DCT values are reconstructed from the quantization levels as $\tilde{X}_i^b = q_i W_i^b$. Then, the inverse DCT is applied to each block, and the result is rounded and truncated in order to take integer values on $[0, 255]$.

Due to the quantization process, the dequantized coefficients $\tilde{X}_i^b$ can only assume values that are integer multiples of the quantization step size q_i. Therefore, the histogram of dequantized coefficients of the ith DCT subband, that is $\tilde{\mathbf{X}}_i = [\tilde{X}_i^1, \ldots, \tilde{X}_i^B]$, is comb-shaped with peaks spaced apart by q_i. This is depicted in Figure 16.1, which shows the histogram of transform coefficients in the $(2, 1)$ DCT subband before (Figure 16.1a) and after (Figure 16.1b) JPEG compression. The process of rounding and truncating the decompressed pixel values perturbs the comb-shaped distribution

---

[2] In this chapter, we will use interchangeably the terms *quantization table* and *quantization matrix* to indicate the same 8×8 matrix of quantization step sizes used for DCT transform coefficients of each block.

[3] http://www.ijg.org/.

of $\tilde{\mathbf{X}}_i$. However, the DCT coefficient values typically remain tightly clustered around integer multiples of q_i. Hereafter, we refer to this characteristic comb shape of the DCT coefficients' histogram as *JPEG compression footprint*, as it reveals that (i) a quantization process has occurred; and (ii) which was the original quantization step size (Fan and de Queiroz 2003).

16.3.2   JPEG Compression Anti-forensics

Stamm *et al.* (2010a) propose to conceal the statistical traces of JPEG compression by filling the gaps in the comb-shaped distribution of $\tilde{\mathbf{X}}_i$ by adding a dithering, noise-like, signal $\mathbf{N}_i$ in such a way that the distribution of the dithered coefficients $\mathbf{Y}_i = \tilde{\mathbf{X}}_i + \mathbf{N}_i$ approximates the original distribution of $\mathbf{X}_i$. The original AC coefficients ($2 \leq i \leq 64$) are typically assumed to be distributed according to the Laplacian distribution (Lam and Goodman 2000):

$$f_{\mathbf{X}_i}(x) = \frac{\lambda_i}{2}e^{-\lambda_i|x|}, \tag{16.1}$$

where the decay parameter λ_i typically takes values between 10^{-3} and 1 for natural imagery. In practice, only the JPEG-compressed version of the image is available, and the original AC coefficients $\mathbf{X}_i$ are unknown. Therefore, the parameter λ_i in (16.1) must be computed from the quantized coefficients $\tilde{\mathbf{X}}_i$, for example, using the maximum-likelihood method in Price and Rabbani (1999), which will result in an estimated parameter $\hat{\lambda}_i$.

According to Stamm *et al.* (2010a), in order to remove the statistical traces of quantization in $\tilde{\mathbf{X}}_i$, the dithering signal $\mathbf{N}_i$ needs to be designed in such a way that its distribution depends on whether the corresponding quantized coefficients $\tilde{\mathbf{X}}_i$ are equal to zero. That is, for DCT coefficients quantized to zero:

$$f_{\mathbf{N}_i}(n|\tilde{X}_i = 0) = \begin{cases} \frac{1}{c_0}e^{-\hat{\lambda}_i|n|} & \text{if } -\frac{q_i}{2} \leq n < \frac{q_i}{2} \\ 0 & \text{otherwise} \end{cases}, \tag{16.2}$$

where $c_0 = \frac{2}{\hat{\lambda}_i}(1 - e^{-\hat{\lambda}_i q_i/2})$. Conversely, for the other coefficients

$$f_{\mathbf{N}_i}(n|\tilde{X}_i = x) = \begin{cases} \frac{1}{c_1}e^{-\text{sgn}(x)\hat{\lambda}_i(n+q_i/2)} & \text{if } -\frac{q_i}{2} \leq n < \frac{q_i}{2} \\ 0 & \text{otherwise} \end{cases}, \tag{16.3}$$

where $c_1 = \frac{1}{\hat{\lambda}_i}(1 - e^{-\hat{\lambda}_i q_i})$. Note that the value of the DCT coefficient $\tilde{X}_i$ enters the definition of the probability density function (PDF) in (16.3) only through its sign. For some DCT subbands, all the coefficients may be quantized to zero, and $\hat{\lambda}_i$ cannot be determined. In those cases, one can leave the reconstructed coefficients unmodified, that is $\mathbf{Y}_i = \tilde{\mathbf{X}}_i$.

As for the DC coefficients, there is no general model for representing their distribution. Hence, the anti-forensic dithering signal for the DC coefficient ($i = 1$) is sampled from the uniform distribution

$$f_{N_1}(n) = \begin{cases} \frac{1}{q_i} & \text{if } -\frac{q_i}{2} \leq n < \frac{q_i}{2} \\ 0 & \text{otherwise.} \end{cases} \tag{16.4}$$

Figure 16.1c illustrates that the anti-forensic technique enables to approximately restore the original Laplacian distribution, thus removing JPEG-compression footprints. Once the statistical traces of quantization have been removed, the tampered image can be re-saved using a lossless compression scheme and passed off as never compressed. Experiments carried out by Stamm *et al.* (2010a) on 244 images of the UCID database (Schaefer and Stich 2004) demonstrated that the JPEG compression detector in (Fan and de Queiroz 2003) is unable to detect previous JPEG compression in 95.90% of the anti-forensically modied images previously compressed with a quality factor of 90, 92.62% of those previously compressed with a quality factor of 70, and 81.56% of images previously compressed with a quality factor of 50. Notice that an image in their experiment was classified as 'never compressed' only if no entry of the quantization table could be estimated correctly, which significantly biases the decision in favour of declaring an image as JPEG compressed. If this strict condition is relaxed, only two out of the 244 anti-forensically attacked images were detected as previously compressed.

16.3.3   Analysis of Anti-forensic Dithering

As any other block of the image processing chain leaves traces on the modified content, so does the anti-forensic dithering described earlier. When one starts designing a countermeasure for an anti-forensic attack, the first step is to carefully look for possible unnatural and unexpected side effects the anti-forensic algorithm may produce in the tampered images. In the case of JPEG anti-forensics, given the noise-like nature of the dithering in the pixel domain, the dithered picture will be a noisy version of the JPEG-compressed image, and this analysis leads to conclude at least two basic facts:

1. Dithering appears as a noise-like signal in the pixel domain, thus it can break the expectable piece-wise smoothness of natural images;
2. By construction, dithering is suppressed if the image is compressed again using JPEG with the original quantization parameters.

The first fact implies that the anti-forensic dither deteriorates image quality. This deterioration can be objectively measured through proper inter-pixel statistics which follow a different distribution depending on whether the image has been dithered or is an uncompressed original. The second fact suggests a constructive way to detect JPEG compression and its quality factor: only re-compressing the image with the original quantization table the dithering noise will be completely cancelled out from the attacked image, due to the idempotency of quantization.

We describe in detail these properties of anti-forensic dither in the following, in order to show afterwards how they can be used to detect JPEG compression in the presence of an anti-forensic attack.

16.3.3.1 Characterization of Anti-forensic Dithering Energy

The mean-squared-error (MSE) distortion $\hat{D}_i$ between the JPEG-compressed coefficients $\tilde{X}_i$ and the dithered coefficients Y_i in the ith subband can be measured as follows:

$$\hat{D}_i = \frac{1}{B}\sum_{b=1}^{B}(Y_i^b - \tilde{X}_i^b)^2 = \frac{1}{B}\sum_{b=1}^{B}(N_i^b)^2. \tag{16.5}$$

Since the distribution of the dithering signal N_i is known from (16.2) to (16.4), it is possible to obtain an analytical expression of the expected value $D_i = E[\hat{D}_i]$:

$$D_i = \sum_{k=-\infty}^{+\infty}\Pr(\tilde{X}_i = kq_i)\int_{-q_i/2}^{+q_i/2} x^2 f_{N_i}(x|\tilde{X}_i = kq_i)dx, \tag{16.6}$$

where $\Pr(\tilde{X}_i = kq_i)$ represents the probability mass function of the quantized DCT coefficients. For AC coefficients, Equation 16.6 can be rewritten according to the definitions given in (16.2) and (16.3). That is,

$$D_i = m_i^0 D_i^0 + (1 - m_i^0)D_i^1, \quad \text{for } 1 < i \le 64 \tag{16.7}$$

where

$$D_i^0 = \int_{-q_i/2}^{+q_i/2} x^2 f_{N_i}(x|\tilde{X}_i = 0)dx, \tag{16.8}$$

$$D_i^1 = \int_{-q_i/2}^{+q_i/2} x^2 f_{N_i}(x|\tilde{X}_i = kq_i)dx, \tag{16.9}$$

and $m_i^0 = 1 - e^{-\hat{\lambda}q_i/2}$ is the fraction of coefficients quantized to zero.

For DC coefficients, the MSE D_1 is equal to that of a uniform scalar quantizer, that is $D_1 = q_1^2/12$. Instead, for AC coefficients, an expression can be found in closed form by solving the integrals in (16.8) and (16.9), as a function of the quantization step size and the parameter of the Laplacian distribution (Valenzise *et al.* 2011). Figure 16.2a shows the MSE distortion D_i as a function of q_i, for different values of $\hat{\lambda}_i$. As a general consideration, the distortion introduced by the anti-forensic dither gets larger as the quantization step size increases. Indeed, a larger value of q_i implies a wider spacing between the peaks in the comb-shaped distribution of $\tilde{\mathbf{X}}_i$. Thus, a larger amount of noise needs to be added to restore the original coefficient distribution.

The growth of the MSE D_i depends also on the value of $\hat{\lambda}_i$, as illustrated in Figure 16.2b. A larger $\hat{\lambda}_i$ in the Laplacian model (16.1) results in DCT coefficients which are more clustered around zero (i.e. with smaller energy). When $\hat{\lambda}_i$ is sufficiently large, all coefficients fall into the zero bin of the quantizer (i.e. $m_i^0 = 1$ in (16.7)). Therefore, no anti-forensic noise is added, and the distortion is exactly zero.

As an example, Figure 16.3 shows the MSE distortion $\hat{D} = (1/64) \sum_{i=1}^{64} \hat{D}_i$ as a function of JPEG compression quality Q, when the IJG quantization matrices are used. We consider two images with different content characteristics: *Lenna* and *Mandril*. Visual inspection reveals that *Lenna* is smoother than *Mandril*. In the DCT domain, larger values of $\hat{\lambda}_i$ are observed for *Lenna*, especially at high frequency. Therefore, the MSE distortion introduced in *Lenna* is smaller than in *Mandril*. Figure 16.3 also

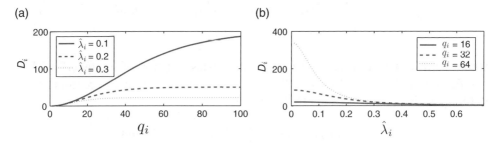

Figure 16.2 The MSE distortion D_i introduced by anti-forensic dithering is a function of the decay parameter $\hat{\lambda}_i$ and of the quantization step size q_i. (a) $D_i(q_i; \hat{\lambda}_i)$ for different values of the decay parameter $\hat{\lambda}_i$ and (b) $D_i(\hat{\lambda}_i)$ for different values of the quantization step size q_i.

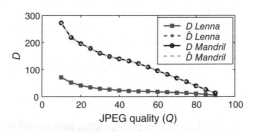

Figure 16.3 The MSE distortion D for two images characterized by different smoothness. The distortion introduced when removing JPEG footprints from images characterized by high-frequency textures is in general higher than for smooth images.

demonstrates that the analytical model describing the expected distortion D introduced by dithering provides an accurate match of the measured MSE $\hat{D}$.

We characterize the distribution of the MSE distortion due to the anti-forensic dither in the DCT domain in order to understand which DCT frequencies are affected more by dithering. This is useful in order to tune the parameters for the anti-forensic detector (with unknown quantization matrix) explained in Section 16.3.4.2. In order to exploit frequency masking and increase coding efficiency, the quantization step sizes q_i's are generally larger at higher frequencies. Hence, we expect larger values of $\hat{D}_i$ in those DCT subbands corresponding to higher frequencies. On the other hand, high-frequency components have lower energy (higher λ_i) due to the piecewise smoothness of natural images. As a result, we typically observe larger values of the MSE distortion at intermediate frequencies. This is illustrated in Figure 16.4a, which shows $\hat{D}_i$ averaged over all the 8×8 blocks of the images in the UCID colour image database (Schaefer and Stich 2004), compressed using JPEG at different quality factors $Q \in [30, 95]$ with the IJG quantization matrices. We notice that $\hat{D}_i$ is not uniformly distributed across the DCT subbands, and it is concentrated at medium frequencies. Indeed, the distribution depends on both the image content and the quantization matrix employed. As a further example, Figure 16.4b and c shows the average MSE distortion when all images in the dataset are compressed at a quality factor equal to, respectively, $Q = 30$ and $Q = 90$. When the quality factor of JPEG compression increases, (i) the overall amount of distortion decreases (see the different scale being used) and (ii) the distribution of the MSE distortion is shifted towards DCT subbands that correspond to higher frequencies.

16.3.3.2   Effect of Re-quantization on the Dithered Image

The distribution of the anti-forensic dither in the ith DCT subband is designed in such a way that it is nonzero only in the interval $\left[-\frac{q_i}{2}, \frac{q_i}{2}\right)$, where q_i is the corresponding

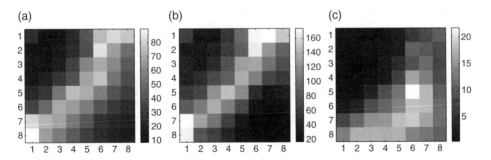

Figure 16.4   MSE distortion $\hat{D}_i$ in the 64 DCT subbands for different JPEG quality factors Q, averaged over the collection of images in the UCID dataset. (a) Average MSE distortion over several quality factors in the range $[30, 95]$, (b) MSE distortion at $Q = 30$ and (c) MSE distortion at $Q = 90$.

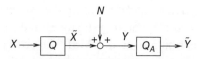

Figure 16.5   Scheme of the re-quantization of a DCT coefficient.

quantization step size (see Equations 16.2 to 16.4). Based on this observation, we show that the anti-forensic noise can be completely cancelled if the dithered image is re-quantized using the same quantization matrix used in the original JPEG compression step.

For clarity of illustration, we start considering a *single* coefficient in one DCT subband, re-quantized according to the scheme illustrated in Figure 16.5. In order to simplify the notation, we drop the subband index, for example $q_i = q$. Let X denote the value of a DCT coefficient in the original (uncompressed) image. During JPEG compression, X is quantized using a uniform quantizer Q with quantization step size q, thus producing $\tilde{X}$. In order to remove the traces of quantization, an adversary adds the dithering signal N, thus producing Y. From the discussion in Section 16.3.2, the net result is that the pdf $f_Y(y)$ is indistinguishable from $f_X(x)$. Then, the dithered coefficient Y is re-quantized with a uniform quantizer Q_A with quantization step size q_A, producing the new coefficient $\tilde{Y}$. We are interested in computing the MSE distortion, $D_A(q_A)$, between $\tilde{X}$ and $\tilde{Y}$. That is,

$$D_A(q_A) = E\left[(\tilde{X} - \tilde{Y})^2\right]$$

$$= \sum_{k=-\infty}^{+\infty} p_k \left[\int_{-\frac{q}{2}}^{+\frac{q}{2}} (\tilde{x}_k - Q_A(\tilde{x}_k + n))^2 f_N(n)\, dn\right], \qquad (16.10)$$

where $\tilde{x}_k = kq$, the expectation is taken with respect to the joint distribution of $\tilde{X}$ and N, $f_N(n)$ is the pdf of the dithering noise as in (16.2)–(16.4), and

$$p_k = \int_{kq-\frac{q}{2}}^{kq+\frac{q}{2}} f_X(x)\, dx \qquad (16.11)$$

is the probability of the original coefficient X falling in the kth quantization bin. Notice that the output of Q_A assumes values at integer multiples of q_A. Therefore, after a change of variables, (16.10) can be written as follows:

$$D_A(q_A) = \sum_{k=-\infty}^{+\infty} p_k \left[\sum_{h=-\infty}^{+\infty} (kq - hq_A)^2 p_{h|k}\right], \qquad (16.12)$$

where

$$
p_{h|k} = \int_{hq_A - \frac{q_A}{2}}^{hq_A + \frac{q_A}{2}} f_N(n - kq)\, dn
\tag{16.13}
$$

is the probability of quantizing a dithered sample to the hth bin of $\mathcal{Q}_A$, given that the original sample was quantized to the kth bin of $\mathcal{Q}$. This is illustrated in Figure 16.6. Notice that $f_N(n - kq)$ is nonzero in the interval $\left[kq - \frac{q}{2}, kq + \frac{q}{2}\right)$.

We observe that when $q_A \to 0$, that is re-quantization is almost lossless, $D_A(q_A) \to \sigma_N^2$, the variance of N. On the other hand, when $q_A \to \infty$, Y is always quantized to zero. Therefore, $D_A(q_A) \to \sigma_{\tilde{X}}^2$, that is the variance of $\tilde{X}$. The dithering noise is cancelled, that is $D_A(q_A) = 0$, when $q_A = \frac{k}{h} q$ *for all the values* h, k for which $p_{h|k} > 0$. This is achieved when $q_A = q$, such that $p_{h|k} = 1$, when $k = h$, and $p_{h|k} = 0$ otherwise. It can be easily seen from Figure 16.6 that this corresponds to the case when all the noise N added to the coefficients in the k-th bin is re-absorbed by the quantized values $kq_A = kq$, resulting in $D_A(q_A) = 0$.

When $q_A \neq q$, re-quantization does not suppress distortion completely, that is $D_A(q_A) > 0$. Specifically, when $q_A > q$, the dithering signal is mostly cancelled.

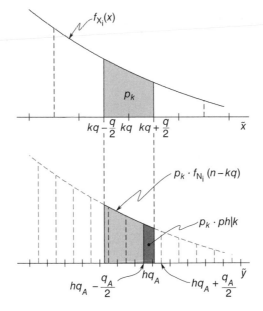

Figure 16.6 Re-quantization of a dithered DCT coefficient (originally quantized with a quantization step size q), with a quantization step q_A. The anti-forensic dither recovers the original distribution of the coefficient $f_X(x)$. When re-quantized, the noise in the kth quantization bin is redistributed into several bins.

Conversely, when $q_A < q$, new non-empty bins in the histogram of $\tilde{Y}$ are created, due to the dithering signal N leaking to neighbouring bins (see Figure 16.6). In other words, the noise N is more accurately reproduced in $\tilde{Y}$. Due to the additive nature of MSE distortion, it is possible to generalize the earlier analysis from individual DCT subbands to the whole image.

As an illustrative example, we measured the MSE distortion between the original JPEG-compressed image and the output of the second JPEG compression for *Lenna*. We considered the widely used JPEG quantization matrices suggested by the Independent JPEG Group. Hence, quantization is adjusted by means of a 100-points quality factor $Q = 1, \ldots, 100$ that scales a template matrix to obtain the quantization steps $q_{A,i}, i = 1, \ldots, 64$ for each DCT subband. Similarly, re-compression is driven by a quality factor Q_A. Figure 16.7 illustrates the mean-square-error distortion between the re-compressed image (at quality factor Q_A) and the JPEG-compressed one, when the latter was originally compressed at $Q = 35, 60, 85$. We notice a trend similar to the one predicted for each DCT coefficient subband, where the distortion is minimized when $Q_A = Q$ (thus $q_{A,i} = q_i$ for all i). The distortion is not exactly zero due to rounding and truncation of pixel values.

Visual inspection of the re-compressed images reveals that, for $Q_A < Q$ the re-compressed image does not contain traces of the dithering signal, which is mostly suppressed together with the high-frequency components of the underlying image. On the other hand, for $Q_A > Q$, the dithering signal is somewhat preserved, and transformed back to the spatial domain, thus resulting in an image affected by grainy noise. This observation triggers the intuition for the detection method illustrated in Section 16.3.4.

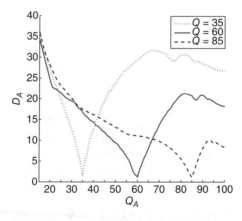

Figure 16.7 MSE distortion between the *Lenna* image, JPEG-compressed at quality Q, and its re-compressed version at quality Q_A. The distortion is amost equal to zero when $Q_A = Q$.

16.3.4 Countering JPEG Compression Anti-forensics

The analysis in Section 16.3.3 suggests that it is possible to identify anti-forensically dithered images by checking whether the noise introduced is annihilated after re-quantization. Unfortunately, in practice we do not have access to the original JPEG-compressed image in order to compute the MSE distortion after re-quantization. Nevertheless, we observe that the presence of the dithering signal in the spatial domain can be detected using a blind noisiness metrics. To this end, any metrics that can robustly measure the amount of noise present in an image could be employed. Thus, the general strategy used in the following consists of two steps: first, the noisiness metric is computed for various levels of quantization of the tested image; and second, according to the response of the metric as a function of the re-quantization, a decision about whether the image was anti-forensically attacked is taken.

We choose to adopt as the noisiness measure the total variation (TV) metric (Rudin *et al.* 1992), which is defined as the ℓ_1 norm of the spatial first-order derivatives of an image. The total variation is more sensitive to small and frequent variations of pixel values due to noise, than to abrupt changes corresponding to edges. We consider two distinct cases. In the first one, we assume that some prior knowledge about the original JPEG coding is available, for example that the original quantization table belongs to a family of quantization matrices corresponding to a certain JPEG implementation (e.g. the IJG implementation). In this setting, the forensic analyst can re-compress the questioned image using the same quantization matrix template as the original. In the second case, we consider the more general setting in which the quantization matrix template is not available. In this case, we only make very loose assumptions about the symmetry properties that characterize typical JPEG quantization tables.

16.3.4.1 Known Quantization Matrix Template

In many JPEG implementations – including the IJG libjpeg software and commercial photo-editing programs such as Adobe Photoshop – it is customary to use pre-determined JPEG quantization tables. The specific quantization table is implicitly identified when the user selects the target quality factor Q. For instance, in the IJG scheme, quantization tables are obtained by properly scaling a template matrix, while in Adobe Photoshop, each JPEG quality factor corresponds to a specific quantization matrix stored in a lookup table.

If the forensic analyst is aware of the specific JPEG implementation that was originally used to encode the image, he can readily generate 8×8 quantization matrices $\mathbf{Q}_A$ given a (scalar) quality factor Q_A. That is, $\mathbf{Q}_A \equiv \mathbf{Q}_A(Q_A)$, where the subscript $_A$ refers to the quality factor used by the analyst. Then, he can re-compress the suspected image using different analysis quality factors Q_A. For each re-compressed image, the total variation $\mathrm{TV}(Q_A) \equiv \mathrm{TV}(\mathbf{Q}_A(Q_A))$ is computed. Figure 16.8 shows the TV as a function of Q_A for two versions of the *Lenna* image, when the IJG scheme

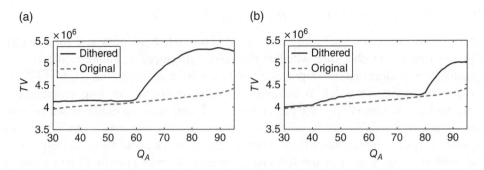

Figure 16.8 Total variation (TV) as a function of the re-compression quality factor Q_A, for two versions of the *Lenna* image. (a) $Q = 60$; (b) $Q = 80$.

is employed. The dashed line corresponds to the genuine, uncompressed image. Not surprisingly, the TV increases smoothly when Q_A increases. Instead, to generate the solid line, the *Lenna* images has been compressed (at quality factor $Q = 60$) and subsequently manipulated to add a dithering signal to restore the original distribution of the DCT coefficients. The apparent slope change at $Q_A = 60$ is due to the fact that noise starts being visible when $Q_A > Q$. We observed that this behaviour is general, and applies also to different kinds of visual content. Therefore, we propose to analyze the $TV(Q_A)$ curve in order to devise a detector that identifies when the traces of JPEG compression have been concealed by an adversary and, in this case, to find the original quality factor Q. We derive a simple but effective threshold detector based on $TV(Q_A)$; more advanced machine learning approaches could be used beyond this simple thresholding scheme.

In order to decide whether an image has been attacked, we consider the first order backward finite difference signal $\Delta TV(Q_A)$, obtained from the total variation curve as follows:

$$\Delta TV(Q_A) = TV(Q_A) - TV(Q_A - 1) \tag{16.14}$$

We deem an image to have been anti-forensically attacked if

$$\left[\max_{Q_A} \Delta TV(Q_A) \right] > \tau, \tag{16.15}$$

where the threshold τ is a parameter that can be adjusted by the detector. In this case, we also estimate the quality factor $\hat{Q}$ of the JPEG-compressed image as follows:

$$\hat{Q} = \left(\arg\max_{Q_A} \Delta TV(Q_A) \right) - 1. \tag{16.16}$$

The -1 term in (16.16) is to compensate for the bias introduced by the approximation of the first order derivative in (16.14).

16.3.4.2 Unknown Quantization Matrix

In the general case, the forensic analyst might not be aware of the JPEG implementation used to originally encode the suspected image. Unlike the case discussed above, $TV(\mathbf{Q}_A)$ cannot be conveniently expressed as a function of only one scalar variable.

Indeed, the most straightforward approach would be to assume a quantization matrix template that contains the same entries for each DCT coefficient. A function $TV(Q_A)$ can be obtained by scaling the values of such constant matrix, based on a scalar quality factor Q_A. However, our experiments showed that this method was ineffective in detecting traces of JPEG compression, as the function $TV(Q_A)$ did not present a distinctive shape as the one illustrated in Figure 16.8. This can be justified by looking at Figure 16.9, which shows the dependency between the quality factor and the quantization step sizes for two pairs of DCT coefficients and two JPEG implementations. The analysis of Figure 16.9 reveals scaling a constant matrix ignores the differences among DCT coefficient subbands. The original quantization step sizes used to JPEG compress the suspected image correspond to different values of Q_A. Hence, the slope discontinuity in Figure 16.8 would not be clearly localized at a single value of Q_A.

The quantity $TV(\mathbf{Q}_A)$ can be expressed as a function of 64 variables $q_{A,i}$, $i = 1, \dots, 64$. However, analyzing the characteristics of $TV(\mathbf{Q}_A)$ in a 64-dimensional space is clearly unfeasible. Interestingly, it is possible to restrict the analysis to a two-dimensional space. Specifically, we consider a re-compression scheme where the quantization table used by the analyst, $\mathbf{Q}_A$, is designed in such a way that quantization affects only two DCT subbands. That is, given a pair of DCT subbands (i_1, j_1) and (i_2, j_2), we vary $q_{A,(i_1,j_1)}$ and $q_{A,(i_2,j_2)}$, whereas we set $q_{A,(i,j)} = 1$ for any $(i,j) \notin \{(i_1,j_1),(i_2,j_2)\}$.

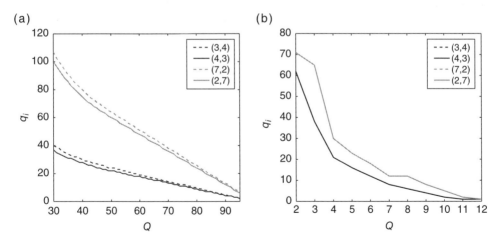

Figure 16.9 Quantization step size q_i as a function of quality factor Q for two pairs of DCT coefficients: (a) IJG implementation and (b) Adobe Photoshop.

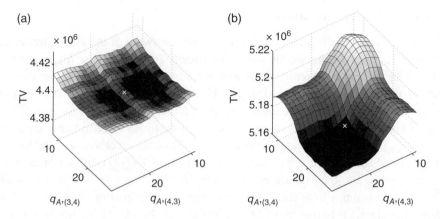

Figure 16.10 Total variation surface for the *Lenna* image, obtained by changing the quantization steps for the DCT coefficients $(3, 4)$ and $(4, 3)$; (a) original and (b) dithered. The white cross corresponds to the true value of $(q_{A,(3,4)}, q_{A,(4,3)})$.

Figure 16.10 illustrates an example of $\text{TV}(\mathbf{Q}_A) = \text{TV}(1, \ldots, q_{A,(3,4)}, q_{A,(4,3)}, \ldots, 1)$ obtained by varying $q_{A,(3,4)}$ and $q_{A,(4,3)}$. Figure 16.10a refers to the uncompressed *Lenna* image. Conversely, Figure 16.10b refers to the same image, but compressed with the IJG libjpeg software at quality factor $Q = 60$, and manipulated by an adversary by adding the anti-forensic dither. We observe that Figure 16.10b exhibits a distinctive behaviour with respect to Figure 16.10a.

In order to further reduce the dimensionality of the problem, we notice that JPEG quantization matrices are designed in such a way that they are approximately symmetric, i.e., $q_{(i,j)} \simeq q_{(j,i)}$. This is illustrated in Figure 16.9 for two pairs of DCT coefficients. We observed this property in most commonly used JPEG quantization matrices, including those employed in IJG libjpeg, in popular photo-editing software such as Adobe Photoshop, and in digital cameras of several brands. As a result, we can further restrict search space by considering symmetric DCT subbands pairs, that is $(i_2, j_2) = (j_1, i_1)$, and by varying both quantization step sizes simultaneously, that is $q_{A,(i_1 j_1)} = q_{A,(i_2 j_2)}$. Intuitively, this corresponds to evaluating the TV function in Figure 16.10 only along the diagonal.

Based on the these observations, the proposed anti-forensic detector works as follows. Each image is recompressed $(q_{\max} - q_{\min} + 1)$ times, by setting, at each round, $q_{A,(i_1 j_1)} = q_{A,(i_2 j_2)} = q_A$, with $q_A = q_{\min}, \ldots, q_{\max}$. Hence, a $(q_{\max} - q_{\min} + 1)$-dimensional vector $\text{TV}_{(i,j)}$ is populated as

$$
\text{TV}_{(i,j)} = \begin{bmatrix} \text{TV}(1, \ldots, q_{\max}, q_{\max}, \ldots, 1) \\ \text{TV}(1, \ldots, q_{\max} - 1, q_{\max} - 1, \ldots, 1) \\ \cdots \\ \text{TV}(1, \ldots, q_{\min}, q_{\min}, \ldots, 1) \end{bmatrix} \tag{16.17}
$$

Note that we sort the elements of the vector $\mathrm{TV}_{(i,j)}$ in decreasing order of q_A, in such a way that the first (last) element correspond to coarser (finer) re-quantization, to retain the same convention adopted for the case of known matrix template.

Figure 16.11 shows $\mathrm{TV}_{(i,j)}$ for two quality factors $Q = \{60, 80\}$ (using IJG implementation) and two pairs of DCT coefficients $(i,j) = \{(4,3), (7,2)\}$. Vectors are normalized to have zero mean for display purposes. We observe that the coefficient $(4,3)$ leads to a vector which is noisier than the one obtained using $(7,2)$, especially at a higher JPEG quality factor. Thus, the latter leads to a more robust detector when it comes to discriminate dithered and original images. This confirms the analysis in Section 16.3.3, where we showed in Figure 16.4 that the MSE introduced by dithering is significantly higher in DCT coefficient $(7,2)$ than in $(4,3)$.

In order to distinguish between uncompressed and dithered images, first, we compute a smoothed vector $\overline{\mathrm{TV}}_{(i,j)}$ by means of local regression using weighted linear least squares and a second degree polynomial model. This operation removes noisy variations of the $\mathrm{TV}_{(i,j)}$ vector. Then, similarly to the case of known matrix template, we compute the first-order derivative

$$\Delta\overline{\mathrm{TV}}_{(i,j)}(q_A) = \overline{\mathrm{TV}}_{(i,j)}(q_A) - \overline{\mathrm{TV}}_{(i,j)}(q_A - 1), \qquad (16.18)$$

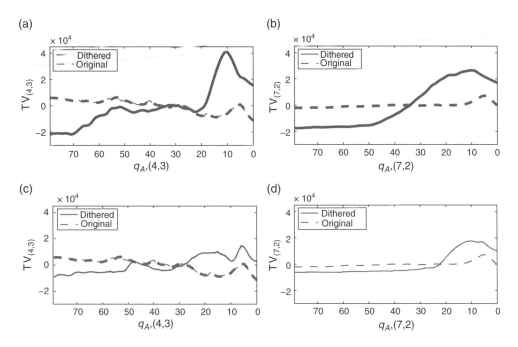

Figure 16.11 Total variation $\mathrm{TV}_{(i,j)}$ as a function of q_A for two JPEG quality factors $Q = \{60, 80\}$ and two pairs of DCT coefficients $(i,j) = \{(4,3), (7,2)\}$: (a) $Q = 60, q_{(4,3)} = 18$, (b) $Q = 60, q_{(7,2)} = 51$, (c) $Q = 80, q_{(4,3)} = 9$ and (d) $Q = 80, q_{(7,2)} = 26$.

and we deem an image to have been anti-forensically attacked if

$$\left[\max_{q_A} \Delta \overline{TV}_{(i,j)}(q_A) \right] > \tau_{(i,j)}. \tag{16.19}$$

The detector in (16.18) and (16.19) exploits only a single pair of DCT coefficient subbands. In order to improve the accuracy of the detector, we merge together the observations gathered from multiple DCT coefficient subbands. Therefore, we consider a set $\mathcal{C} = (i_1, j_1) \ldots, (i_C, j_C)$ of C subbands and we pursue two different information fusion approaches, which correspond to, respectively, pre-classification and post-classification according to the taxonomy proposed by Jain et al. (2005). In the pre-classification approach, we construct a vector $\Delta \overline{TV}_{\mathcal{C}} = [\Delta \overline{TV}_{(i_1, j_1)}^T, \ldots, \Delta \overline{TV}_{(i_C, j_C)}^T]^T$ of size $(q_{max} - q_{min}) \cdot C$ obtained concatenating the vectors corresponding to the individual subbands. The detector is identical to (16.19), where $\Delta \overline{TV}_{(i,j)}$ is replaced by $\Delta \overline{TV}_{\mathcal{C}}$. In the post-classification approach, for each subband we compute a binary decision according to (16.19), and the final decision is obtained by majority voting among the set of C binary decisions.

16.3.4.3 Experimental Results

The effectiveness of JPEG anti-forensic dithering to the forensic detector of Fan and de Queiroz (2003) has been proven by Stamm et al. (2010a) and described in Section 16.3.2. However, as shown before, it is possible to design newer forensic tools to detect JPEG compression even in the presence of anti-forensics. We evaluated the method based on TV described so far, and we compared it with other analysis tools proposed in the steganalysis literature. We show in all the cases that JPEG anti-forensic dithering alone is not sufficient to mask JPEG compression.

Experimental Conditions
We carried out a large-scale test of the algorithms described in Section 16.3.4.1 on 1338 images of the uncompressed colour image database (UCID) (Schaefer and Stich 2004). All the pictures in this dataset have a resolution of 512×384. Without loss of generality, we considered the luma component only. Some of the methods described in the literature require training, for which we adopted a separate dataset. We downloaded 978 uncompressed images (raw scans of film with resolution 2100×1500) of the NRCS dataset.[4] In order to provide a fair comparison, we resampled the images at the same resolution as the UCID dataset used for testing.

We split the dataset in two sets of equal size. The first half contained images that were JPEG-compressed at a random quality factor Q using the IJG implementation. More specifically, the quality factor is uniformly sampled in the set

---

[4] Photo courtesy of USDA Natural Resources Conservation Service. Available at http://photogallery.nrcs.usda.gov/res/sites/photogallery/.

$\{30, 40, 50, 60, 70, 80, 90, 95\}$ with probability 1/8. In order to restore the original statistics of the DCT coefficients, we added an anti-forensic dithering signal according to the method in Stamm *et al.* (2010a). The remaining half contained uncompressed original images.

Detection of JPEG Anti-forensics

To evaluate the detection performance of the TV detector in both the case of known and unknown quantization matrices, we let the threshold τ vary to trace the receiver operating characteristic (ROC) curves shown in Figure 16.12. There, the true positive (TP) rate is the fraction of JPEG-compressed images that were correctly reported to be compressed and the false positive (FP) rate is the fraction of uncompressed images that were reported to be compressed. Typically, the forensic analyst is interested to work at a low target FP rate. Figure 16.12c illustrates a zoomed version of the ROC for FP rates in the interval $[0, 0.2]$. We observe that the detector reaches a TP rate above 0.89 at an FP rate as low as 0.02. Figure 16.12 is obtained by considering all the images in the dataset.

In the case of unknown quantization matrix, we selected a set of DCT coefficient subbands $C = \{(5,3), (5,4), (6,3), (6,4), (6,5), (7,1), (7,2), (7,3), (8,1), (8,2)\}$, based on the analysis in Section 16.3.3. Indeed, they correspond to mid-frequency subbands in which the MSE distortion due to the addition of the dithering noise is largest (see Figure 16.4). We let $\tau_{(i,j)}$ vary to trace the ROC curve for each subband $(i,j) \in C$. To avoid cluttering the figure, results on a set of five coefficients are shown in Figure 16.12a and c and results on the other set in Figure 16.12b and d. In this case, we observe that the performance of the detector depends on the selected DCT subband, with the best results achieved for subbands $(5,4)$, $(6,3)$, $(7,2)$ and $(7,3)$. With respect to the case of known matrix template, a TP rate above 0.89 is reached at an FP rate equal to 0.1, confirming the fact that the knowledge of the matrix template facilitates the work of the forensic analyst.

Figure 16.12 also shows the results obtained with the two information fusion approaches described in Section 16.3.4.2. As for fusion based on pre-classification, that is concatenating the vectors obtained with all subbands, the TP rate for a target FP rate is higher than in the case of a detector based on a single subband. The only exception is at very low FP rates, where a detector based on $(5,4)$ achieves higher TP rate. We argue that more sophisticated fusion methods (e.g. weighting the contribution of each DCT subband differently) might further improve the results. In the case of fusion based on post-classification, majority voting provides a binary decision for each image. As such, instead of the ROC curve we can only report the corresponding TP rate versus FP rate point.

Estimation of Original JPEG Quality Factor

When the matrix template is known, the forensic analyst might be also interested in determining an estimate $\hat{Q}$ of the quality factor Q originally used to compress the image. Table 16.1 reports the performance of the estimator in terms of

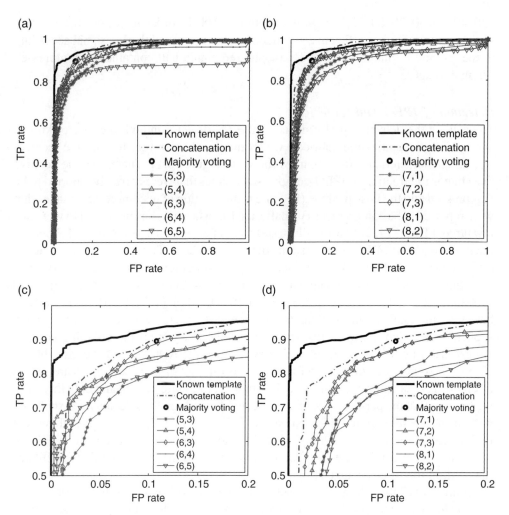

Figure 16.12 ROC curves of the proposed detector. All figures report the curves corresponding to (i) known matrix template; (ii) unknown matrix template, concatenating $C = 10$ DCT subbands and (iii) unknown matrix template, fusing the outputs of the individual detectors by majority voting. In addition, we show one curve for each DCT coefficient subband used by the detector. To avoid cluttering the figure, a set of five coefficients is shown in (a)–(c) and the other set in (b)–(d). (c)–(d): zoom of (a)–(b).

- Bias

$$\mathcal{E}_{\text{bias}} = \frac{1}{\text{TP}} \sum_{p \in \mathcal{S}} (\hat{Q}^{(p)} - Q^{(p)}) \qquad (16.20)$$

Table 16.1   Performance of quality factor estimation.

Q	30	40	50	60	70	80	90	95
$\mathcal{E}_{\text{bias}}$	−12.4	−2.4	−1.8	−1.5	−1.1	−1.0	−0.3	12.3
$\mathcal{E}_{\text{sd}}$	1.1	2.2	1.9	1.6	3.7	0.4	5.0	12.8
$\mathcal{E}_5$	1.00	0.01	0.08	0.04	0.03	0.00	0.01	0.82

- Standard deviation

$$\mathcal{E}_{\text{sd}} = \sqrt{\frac{1}{\text{TP}} \sum_{p \in \mathcal{S}} |\hat{Q}^{(p)} - Q^{(p)}|^2} \qquad (16.21)$$

- Fraction of errors above five units

$$\mathcal{E}_5 = \frac{|\{p \in \mathcal{S}, |\hat{Q}^{(p)} - Q^{(p)}| > 5\}|}{\text{TP}} \qquad (16.22)$$

where $\mathcal{S}$ denotes the set of indexes of TP images classified as positives and matching the ground truth. $Q^{(p)}$ and $\hat{Q}^{(p)}$ denote, respectively, the true and estimated quality factor of the p-th image in the set $\mathcal{S}$. Note that $Q^{(p)}$ is undefined for negative samples (TN + FP) and $\hat{Q}^{(p)}$ is not computed for false negative samples. Both bias and standard deviation are within a few units when $Q \subset [40, 90]$, with larger values at the extremes of the tested range, that is $Q = 30, 95$.

Comparison with Steganalysis
It is possible to observe that methods originally developed for steganalysis can be effectively adopted to reveal the traces of JPEG compression anti-forensics, achieving very good results for a wide range of quality factors.

We compared the TV anti-forensics detector detailed before in this chapter with other methods described in the literature, namely detectors based on the subtractive pixel adjacency matrix (SPAM) features in Pevny *et al.* (2010) and the calibration features in Lai and Böhme (2011). The SPAM features were proposed to detect steganographic methods that embed in the spatial domain by adding a low-amplitude independent stego signal. Although originally designed for a different purpose, these features can be used to build a detector by considering the uncompressed original image as the cover image, and the decompressed JPEG image with added the anti-forensic dithering signal as the stego image. We extracted first order SPAM features using the default parameters suggested by Pevny *et al.* (2010), which leads to a 162-dimensional vector for each image. This feature vector describes the amount of inter-pixel correlation along different directions, and we refer the reader to Pevny *et al.* (2010) for details on how it is computed. In order to design a detector, a support vector machine (SVM) classifier was trained on the NRCS dataset. In order to tune the parameters of the SVM classifier (C, γ), we followed the procedure described in Pevny *et al.* (2010), which suggests five-fold cross-validation with an exhaustive search over a multiplicative grid sampling the parameter space. We did not consider second-order SPAM features as the feature vector has a dimension comparable to the size of the training set, thus being at risk for overfitting.

The calibration features were also originally proposed in the field of steganalysis (Fridrich 2004) and recently adapted to detect JPEG compression anti-forensics (Lai and Böhme 2011). Specifically, a single calibration feature is used as in Lai and Böhme (2011), which measures the ratio of the variance of high frequency subbands. The forensic analyst crops the suspected image $\mathbf{Y}$ in the spatial domain by four pixels in both horizontal and vertical direction to obtain a new image $\mathbf{Z}$. Then the sample variance of 28 high frequency subbands in a set $\mathcal{C}$ is computed for both images. The calibration feature F is calculated as follows:

$$F = \frac{1}{|\mathcal{C}|} \sum_{i \in \mathcal{C}} \left(\frac{\sigma^2_{\mathbf{Z},i} - \sigma^2_{\mathbf{Y},i}}{\sigma^2_{\mathbf{Y},i}} \right) \qquad (16.23)$$

The detector compares the value of F to a threshold τ_F. If $F \leq \tau_F$, the image is considered to be an uncompressed original.

As a metric for comparison, we use the detection *accuracy*, that is the fraction of correct decisions taken by the detector. That is

$$\text{Accuracy} = \frac{TP + TN}{N}, \qquad (16.24)$$

where TN denotes the number of true negatives, that is the uncompressed original images that were detected to be so, and N is the number of images in the dataset. Accuracy is a suitable metric when the dataset is balanced, such as the one adopted in our experiments. For the threshold-based detectors (16.15), (16.19) and (16.23), the accuracy is in fact a function of the threshold τ. We choose the value of τ for each detector by maximizing accuracy(τ) over the NRCS dataset.

Table 16.2 reports the average accuracy both over all tested values of Q and in the $[30, 90]$ interval. For each detector, we also specify if it relies on prior knowledge of the quantization template. The proposed detector achieves the highest accuracy over the range $[30, 90]$, reaching almost 100%, when the quantization matrix template is known. The detector based on calibration features also gives very good results, slightly outperforming the proposed fusion methods as well as the detector based on SPAM features. The latter detector outperforms all the others when $Q = 95$, thus yielding uniform accuracy across the whole range of tested values of Q.

Note that the detector based on SPAM features is prone to give false positives when other kinds of noise are introduced into the image, since it only considers inter-pixel correlations, thus disregarding the underlying processing chain of JPEG

Table 16.2 Accuracy of the detectors vs. Q.

Detector	Known template	30	40	50	60	70	80	90	95	avg.	avg. 30–90
TV(Q_A)	y	**0.99**	**0.99**	**1.00**	**1.00**	**0.99**	**1.00**	**1.00**	0.62	0.93	**1.00**
TV$(q_{A,i})$ – concat.	n	0.97	0.98	0.97	0.97	0.94	0.96	0.93	0.70	0.90	0.96
TV$(q_{A,i})$ – voting	n	0.97	0.98	0.97	0.97	0.94	0.96	0.93	0.70	0.90	0.96
Calibration feat.	n	0.98	**0.99**	0.99	0.98	0.98	0.99	0.96	0.81	**0.97**	0.98
SPAM feat.	n	0.98	0.96	0.98	0.97	0.96	0.98	0.96	**0.94**	0.95	0.97

compression anti-forensics. Using a terminology borrowed from statistics, we could say that SPAM features have a high *sensitivity* to JPEG anti-forensic dithering. That is, they can identify accurately the case in which an image has been compressed and anti-forensically dithered. However, they have a low *specificity*, since they have a higher tendency to classify uncompressed images as anti-forensically dithered.

16.4   Trade-off between Forensics and Anti-forensics

While forensic tools have been widely studied in the past few years, anti-forensics is a rather new field, and only recently a few works have started analyzing the trade-off between forensics and anti-forensics by a more formal standpoint (Barni and Tondi 2013; Böhme and Kirchner 2013; Stamm *et al.* 2012). When confronted with a forensic technique capable of detecting traces of anti-forensics, an intelligent forger will try to modify his anti-forensic technique in such a way to conceal their traces to a forensic analyst, for example by minimizing the anti-forensic fingerprint. For example, in the case of JPEG compression anti-forensics – where anti-forensic dither used to smooth the histogram of coded transform coefficients corresponds to adding noise in the spatial domain – the attacker could reduce the amount of dithering and perform some denoising to fool anti-forensic detectors. This leads to a cat-and-mouse game between a digital forger and a forensic investigator, and poses a number of questions about how they should behave in order to maximize their utility in view of the (rational) actions of the other: How much can a forensic technique be considered reliable given that an opposite anti-forensic method? To which extent should the forger conceal a manipulation, if the analysts has the means to detect his anti-forensic attack? Which is the optimal strategy for a forensic analyst in order to identify digital forgeries?

In this section we describe two anti-forensic performance indicators proposed by Stamm *et al.* (2012), and we show how the interplay between a forger and a forensic analyst can be analyzed using game theory.

16.4.1   Performance Analysis of Anti-forensics

Anti-forensics is generally defined as the ability to reduce the performance of a forensic detector (Böhme and Kirchner 2013). Thus its performance can be evaluated in terms of how much a forensic operator can reduce the detection rate of a given forensic tool. Following Stamm *et al.* (2012), let ψ be a digital multimedia file and $m(\cdot)$ an operator which could manipulate on ψ. Given some test media ψ, a forensic investigator wants to determine if ψ is an original content, or it comes from a manipulation of another file ψ'. That is, he wants to differentiate between two hypotheses:

$$H_0 : \psi \neq m(\psi')$$
$$H_1 : \psi = m(\psi'). \tag{16.25}$$

To identify the correct hypothesis, the forensic analyst will employ a detection algorithm δ_m, which looks for fingerprints left by m in ψ and compares the strength of these traces against some decision threshold. This threshold is generally chosen to maximize the probability of detection, $P_d(\delta_m) = P(\delta_m = H_1 | \psi = m(\psi'))$, given a constraint on the probability of false alarm $P_{fa}(\delta_m) = P(\delta_m = H_1 | \psi \neq m(\psi'))$. The attacker, on the other hand, given the knowledge of δ_m, creates an anti-forensic technique α_m designed to fool the forensic detector.

The performance of anti-forensic α_m with respect to the detector δ_m can be measured through the *anti-forensic effectiveness* $P_{ae}(\alpha_m)$, that is the ability of α_m to cause the missed detection of an altered multimedia file given that the manipulation would be detected if anti-forensics is not used. That is,

$$P_{ae}(\alpha_m) = P(\delta_m(\alpha_m(\psi)) = H_0 | \delta_m(\psi) = H_1, \psi = m(\psi')). \tag{16.26}$$

It is important to notice that an anti-forensic technique does not need to achieve $P_{ae} = 1$ in order to render a forensic detector ineffective. In fact, practical forensic detectors already have some degree of ineffectiveness, which make them miss some true positives even in case no anti-forensic attack is done. Thus, an adversary wants simply to cause a sufficient number of missed detection so that its performance becomes equivalent to that of a random detector, that is $P_d(\delta_m) = P_{fa}(\delta_m)$ for a given constraint on P_{fa}. As a consequence, one can quantify the degree up to which a forensic detector is susceptible to an anti-forensic attack by measuring the *anti-forensic susceptibility* of a detector as follows:

$$S_\alpha(\delta_m, P_{fa}) = \frac{P_d\left(\delta_m^{(P_{fa})}\right) - \max\left(P_d\left(\delta_m^{(P_{fa})}\right)(1 - P_{ae}(\alpha_m)), P_{fa}\right)}{P_d\left(\delta_m^{(P_{fa})}\right) - P_{fa}}, \tag{16.27}$$

where $\delta_m^{(P_{fa})}$ denotes the detection algorithm δ_m operating using the decision threshold associated with the false alarm rate P_{fa}. The numerator of S_α measures the decrease in the performance of δ_m due to the use of α_m. The denominator in (16.27) instead is the difference between the probability of detection achieved by $\delta_m^{(P_{fa})}$ and its corresponding false alarm rate. This is a measure of how well the detector performs with respect to taking random decisions, in the case when *no anti-forensics* is used; therefore, it measures the maximum decrease in performance an attacker may attain by attacking δ_m with anti-forensics. As a result, the anti-forensic susceptibility is a measure between 0 and 1, and reflects how well α_m can decrease the effectiveness of δ_m.

16.4.2 Interplay between Forger and Forensic Analyst Using Game Theory

If the forensic analyst is aware of a possible anti-forensic attack on ψ, he can design a new detector, δ_α which aims at discovering whether the tested content has been manipulated through α_m. In the framework of hypothesis testing, the detector evaluates the two hypotheses:

$$H_{0\alpha} : \psi \neq \alpha_m(m(\psi'))$$

$$H_{1\alpha} : \psi = \alpha_m(m(\psi')). \tag{16.28}$$

Now, the forger can choose the strength with which he wants to conceal the traces of α_m. The stronger these traces, the higher the probability that δ_α detect them. At the same time, if he leaves too few traces, δ_m will detect the manipulation, and the forensic analyst will be again better off. In fact, the forensic investigator will be always achieving his goal whenever he detects one of the two kinds of forgery, that is the original manipulation or the anti-forensic one. Furthermore, since many anti-forensic operations degrade the quality of the digital multimedia file they operate on, a too strong anti-forensic attack may actually fool δ_α, but decrease image quality to such an extent that it will rise suspicion to the forensic analyst or make the image not usable in practice. Therefore, the adversary should take into account the cost of the perceptual distortion $\gamma\left(m(\psi), \alpha_m^{(k)}(m(\psi))\right)$ between the tampered content with and without anti-forensics (where the superscript k is an index of the strength of applied anti-forensics).

Apparently, the optimal anti-forensic strength used by the forger depends on the decision threshold used by δ_m and δ_α, since the attacker has to balance his risk to be detected either through the normal detector or the anti-forensic one. On the other hand, the forensic investigator has to balance his probability of false alarm, assigning more or less importance to false alarms caused by the manipulation or by the anti-forensic attack. Since these are influenced by the choice of the attack strength decided by the attacker, it is clear that both the forger and the forensic investigator's optimal actions depend on the actions of their counterpart.

This situation can be formulated as a game in the following way. The forensic investigator moves first by choosing his probability of false alarm allocation (out of a total false alarm probability budget of ξ) between δ_m and δ_α, as well as the optimal decision thresholds. Thus, the strategy of the forensic investigator is to choose the false alarm level $\eta \in [0, \xi]$ to be allocated to δ_m. The corresponding false alarm level $\tilde{\eta}$ allocated to δ_α is found consequently as the maximum false alarm level such that the resulting total P_{fa} is less than or equal to ξ. The set of strategies of the forger instead is given by the strength $k \in [0, 1]$ he can put on the anti-forensic operation α_m as explained earlier.

For a given pair of strategies (η, k), the utility that the forensic investigator (FI) wants to maximize is the probability that either manipulation or the use of anti-forensics will be detected, that is:

$$U_{FI}(\eta, k) = P\left(\delta_m^{(\eta)}(\psi) = H_{1m} \text{ or } \delta_\alpha^{\tilde{\eta}}\left(\alpha_m^{(k)}(\psi)\right) = H_{1\alpha} | \psi = m(\psi')\right). \tag{16.29}$$

Conversely, the attacker (AT) wants to minimize (16.29), together with the quality loss of the multimedia file, so his utility is

$$U_{AT}(\eta, k) = -U_{FI}(\eta, k) - \gamma\left(m(\psi), \alpha_m^{(k)}(m(\psi))\right). \tag{16.30}$$

By substituting in the appropriate expressions for the probabilistic quantities in each utility function, one can find the Nash equilibrium strategies (η^*, k^*) that neither player has an incentive to deviate from. In practice, however, the analytical evaluation of these utilities is often difficult or impossible. In many forensic scenarios, no known equation exists to express the probabilistic quantities used in each utility function. As a result, the Nash equilibria must often be sought out numerically.

16.5   Conclusions

While the field of forensics has received a great deal of attention in the past decade, the study of anti-forensic techniques is still in its infancy. We have reviewed the principal techniques proposed in the literature, organizing them according to where the forensic detector they fool is positioned in the processing chain which goes from acquisition, to coding and editing. Generally speaking, most of the attacks proposed so far target specific forensic tools, while only a small portion of them are universal, in the sense that the outcome of anti-forensics is to re-establish some natural statistics of the signal which would increase the false negative rate of a generic detector. Targeted attacks are general also more powerful, although they require some prior knowledge by the adversary. Anti-forensic attacks can also be divided into integrated and post-processing attacks. Typically, integrated attacks have access to more information to perform the forgery (i.e. for copy-move forgeries), but in some cases the only feasible solution is post-processing (i.e. for JPEG compression anti-forensics).

In order to show in detail how one can design an anti-forensic technique, and especially, how countermeasures can be found, we studied the case of JPEG compression anti-forensics. We showed how it is possible to detect a JPEG anti-forensic operation through the noisy traces it leaves into the image, for a number of application scenarios (known or unknown quantization matrix). This example also gave us the possibility to show how several tools previously proposed in steganalysis can provide very useful cues for the forensic analyst.

Finally, we concluded by outlining the last research trends in anti-forensics in the directions of formalizing the performance of anti-forensic techniques and modelling the cat-and-mouse game between forensic analyst and forger through a game-theoretic perspective. Although we did not cover other multimedia object than images and video, we want to remark that the audio-forensic field is a very active research area as well. As for the video case, we notice that very little work has been done in terms of anti-forensics, and still there is much to do, for example to consider more realistic video coding and transmission scenarios (e.g. using H.264/AVC or HEVC).

References

Amerini I, Ballan L, Caldelli R, Del Bimbo A and Serra G 2011 A SIFT-based forensic method for copy-move attack detection and transformation recovery. *IEEE Transactions on Information Forensics and Security* **6**(3), 1099–1110.

Amerini I, Barni M, Caldelli R and Costanzo A 2013 Counter-forensics of sift-based copy-move detection by means of keypoint classification. *EURASIP Journal on Image and Video Processing* **2013**(1), 18.

Barni M and Tondi B 2013 The source identification game: An information-theoretic perspective. *IEEE Transactions on Information Forensics and Security* **8**(3), 450–463.

Barni M, Fontani M and Tondi B 2012 A universal technique to hide traces of histogram-based image manipulations. *Proceedings of the ACM International Workshop on Multimedia and Security MM&Sec '12*. ACM, New York, pp. 97–104.

Bayram S, Sencar H and Memon N 2008 A survey of copy-move forgery detection rechniques. *Proceeding of the IEEE Western New York Image Processing Workshop*, Rochester, NY.

Bestagini P, Allam A, Milani S, Tagliasacchi M and Tubaro S 2012 Video codec identification *Proceedings of the* 37th *International Conference on Acoustics, Speech, and Signal Processing (ICASSP 2012)*, Koyoto, Japan, pp. 2257–2260.

Bianchi T and Piva A 2012 Detection of nonaligned double JPEG compression based on integer periodicity maps. *IEEE Transactions on Information Forensics and Security*, **7**(2), 842–848.

Bianchi T., De Rosa A and Piva A 2011 Improved DCT coefficient analysis for forgery localization in JPEG images *Proceedingds of the International Conference on Acoustics, Speech, and Signal Processing*, Prague, Czech Republic.

Böhme R and Kirchner M 2013 Counter-forensics: Attacking image forensics. In *Digital Image Forensics* (eds., Sencar HT and Memon N). Springer, New York, pp. 327–366.

Caldelli R, Amerini I, Ballan L, Serra G, Barni M and Costanzo A 2012 On the effectiveness of local warping against sift-based copy-move detection. *5th International Symposium on Communications Control and Signal Processing*, Rome, Italy, pp. 1–5.

Cao G, Zhao Y, Ni R and Tian H 2010a Anti-forensics of contrast enhancement in digital images *Proceedings of the 12th ACM Workshop on Multimedia and Security, (MM&Sec '10)*. ACM, New York, pp. 25–34.

Cao G, Zhao Y, Ni R, Yu L and Tian H 2010b Forensic detection of median filtering in digital images. *IEEE International Conference on Multimedia and Expo* IEEE, Suntec City, Singapore pp. 89–94 .

Chen C, Shi Y and Su W 2008a A machine learning based scheme for double JPEG compression detection *International Conference on Pattern Recognition*, Jampa, FL, pp. 1–4.

Chen M, Fridrich J, Goljan M and Lukas J 2008b Determining image origin and integrity using sensor noise. *IEEE Transactions on Information Forensics Security* **3**(1), 74–90.

Chen M, Fridrich J, Goljan M and Lukas J 2008c Determining image origin and integrity using sensor noise. *IEEE Transactions on Information Forensics and Security* **3**(1), 74–90.

Do TT, Kijak E, Furon T and Amsaleg L 2010 Deluding image recognition in sift-based cbir systems *Proceedings of the 2nd ACM Workshop on Multimedia in Forensics, Security and Intelligence*, (MiFor '10), Firenze, Italy. pp. 7–12.

Fan Z and de Queiroz RL 2003 Identification of bitmap compression history: JPEG detection and quantizer estimation. *IEEE Transactions on Image Processing* **12**(2), 230–235.

Fan W, K. W, Cayre F and Xiong Z 2013a A variational approach to JPEG anti-forensics *Proceedings of the International Conference on Acoustics, Speech, and Signal Processing*. Vancouver, British Columbia, Canada, pp. 3058–3062.

Fan W, Wang K, Cayre F and Xiong Z 2013b JPEG anti-forensics using non-parametric DCT quantization noise estimation and natural image statistics. *Proceedings of the First ACM Workshop on Information Hiding and Multimedia Security*, ACM, New York, pp. 117–122.

Farid H 2006 Digital image ballistics from JPEG quantization. *Dartmouth College, Department of Computer Science, Tech. Rep. TR2006-583*, Hanoves, NH.

Farid H 2009 Exposing digital forgeries from JPEG ghosts. *IEEE Transactions on Information Forensics and Security* **4**(1), 154–160.

Ferrara P, Bianchi T, De Rosa A and Piva A 2012 Image forgery localization via fine-grained analysis of CFA artifacts. *IEEE Transactions on Information Forensics and Security*, **7**(5), 1566–1577.

Fontani M and Barni M 2012 Hiding traces of median filtering in digital images. *EURASIP European Signal Processing Conference (EUSIPCO12)*, Hanouer, NH.

Fridrich J 2004 Feature-based steganalysis for JPEG images and its implications for future design of steganographic schemes. *Proceeding on Hiding Workshop*, Springer LNCS, Toronto, Ontario, Canada, pp. 67–81.

Fridrich J, Soukal D and Lukás J 2003 Detection of copy-move forgery in digital images *Proceedings of Digital Forensic Research Workshop*, Cleveland, OH.

Gloe T, Kirchner M, Winkler A and Böhme R 2007 Can we trust digital image forensics? *Proceedings of the 15th ACM International Conference on Multimedia*, Augsburg, Germany, pp. 78–86.

Goljan M, Chen M and Fridrich J 2007 Identifying common source digital camera from image pairs. *IEEE International Conference on Image Processing* IEEE, vol. 6, San Antonio, TX, pp. VI–125.

Goljan M, Fridrich J and Chen M 2011 Defending against fingerprint-copy attack in sensor-based camera identification. *IEEE Transactions on Information Forensics and Security* **6**(1), 227–236.

Hsu CY, Lu CS and Pei SC 2009 Secure and robust sift. *Proceedings of the 17th ACM International Conference on Multimedia (MM '09)*, Beijing, China, pp. 637–640.

Huang F, Huang J and Shi Y 2010 Detecting double JPEG compression with the same quantization matrix. *IEEE Transactions on Information Forensics and Security* **5**(4), 848–856.

ITU-T 1991 Recommendation T.81: Digital compression and coding of continuous-tone still images (ISO/IEC 10918-1) International Telecommunication Union – Telecommunication Standardization Bureau.

Jain AK, Nandakumar K and Ross A 2005 Score normalization in multimodal biometric systems. *Pattern Recognition* **38**(12), 2270–2285.

Janesick JR 2001 *Scientific Charge-Coupled Devices*, vol. 117. SPIE Press, Bellingham, WA.

Kirchner M and Böhme R 2007 Tamper hiding: Defeating image forensics. *9th International Workshop on Information Hiding*, Springer, Saint Malo, France, pp. 326–341.

Kirchner M and Böhme R 2008 Hiding traces of resampling in digital images. *IEEE Transactions on Information Forensics and Security* **3**(4), 582–592.

Kirchner M and Böhme R 2009 Synthesis of colour filter array pattern in digital images. *IS&T/SPIE Electronic Imaging*. International Society for Optics and Photonics, pp. 72540K–72540K, Belligham, WA.

Kirchner M and Fridrich J 2010 On detection of median filtering in digital images. *IS&T/SPIE Electronic Imaging*. International Society for Optics and Photonics. pp. 754110–754110, Bellingham, WA.

Kraetzer C, Oermann A, Dittmann J and Lang A 2007 Digital audio forensics: A first practical evaluation on microphone and environment classification *Proceedings of the 9th Workshop on Multimedia & Security (MM&Sec '07)*. ACM, New York, pp. 63–74.

Lai S and Böhme R 2011 Countering counter-forensics: The case of JPEG compression. *13th International workshop on Information Hiding*. Springer, pp. 285–298, Berlin, Germany.

Lam E and Goodman J 2000 A mathematical analysis of the DCT coefficient distributions for images. *IEEE Transactions on Image Processing* **9**(10), 1661–1666.

Lin W, Tjoa S, Zhao H and Liu K 2009 Digital image source coder forensics via intrinsic fingerprints. *IEEE Transactions on Information Forensics and Security* **4**(3), 460–475.

Lowe DG 2004 Distinctive image features from scale-invariant keypoints. *International Journal of Computer Vision* **60**(2), 91–110.

Lukás J and Fridrich J 2003 Estimation of primary quantization matrix in double compressed JPEG images. *Proceedings of Digital Forensic Research Workshop*, Cleveland, OH.

Lukás J, Fridrich J and Goljan M 2006a Detecting digital image forgeries using sensor pattern noise. *Electronic Imaging 2006*. International Society for Optics and Photonics, p. 60720Y, Bellingham, WA.

Lukás J, Fridrich J and Goljan M 2006b Digital camera identification from sensor pattern noise. *IEEE Transactions on Information Forensics and Security* **1**(2), 205–214.

Luo W, Qu Z, Huang J and Qiu G 2007 A novel method for detecting cropped and recompressed image block. *Proceedings of the International Conference on Acoustics, Speech, and Signal Processing*, Honolulu, NH, vol. 2, pp. 217–220.

Luo W, Wu M and Huang J 2008 MPEG recompression detection based on block artifacts. *Proceedings of the SPIE*, San Jose, CA, vol. 6819.

Malik H and Farid H 2010 Audio forensics from acoustic reverberation. *ICASSP*, Dallas, Tx, pp. 1710–1713.

Milani S, Fontani M, Bestagini P, Barni M, Piva A, Tagliasacchi M and Tubaro S 2012 An overview on video forensics. *APSIPA Transactions on Signal and Information Processing*, **1**, e2.

Pevny T, Bas P and Fridrich J 2010 Steganalysis by subtractive pixel adjacency matrix. *IEEE Transactions on Information Forensics and Security* **5**(2), 215–224.

Popescu A and Farid H 2005a Exposing digital forgeries by detecting traces of resampling. *IEEE Transactions on Signal Processing* **53**(2), 758–767.

Popescu A and Farid H 2005b Exposing digital forgeries in colour filter array interpolated images. *IEEE Transactions on Signal Processing* **53**(10), 3948–3959.

Price J and Rabbani M 1999 Biased reconstruction for JPEG decoding. *IEEE Signal Processing Letters* **6**(12), 297–299.

Rudin L, Osher S and Fatemi E 1992 Nonlinear total variation based noise removal algorithms. *Physica D: Nonlinear Phenomena* **60**(1-4), 259–268.

Schaefer G and Stich M 2004 UCID: An uncompressed colour image database *Proceeding of the SPIE: Storage and Retrieval Methods and Applications for Multimedia*, San Jose, CA, vol. 5307, pp. 472–480.

Stamm M and Liu K 2008 Blind forensics of contrast enhancement in digital images *Proceedings of the International Conference on Image Processing*, San Diego, CA, vol. 1, pp. 3112–3115.

Stamm M and Liu K 2010a Forensic detection of image manipulation using statistical intrinsic fingerprints. *IEEE Transactions on Information Forensics and Security* **5**(3), 492–506.

Stamm M and Liu K 2010b Forensic estimation and reconstruction of a contrast enhancement mapping. *IEEE International Conference on Acoustics Speech and Signal Processing*, Dallas, TX, pp. 1698–1701.

Stamm MC and Liu KR 2010c Wavelet-based image compression anti-forensics. *2010 17th IEEE International Conference on Image Processing (ICIP)*, IEEE, Hong Kong, China, pp. 1737–1740.

Stamm M, Tjoa S, Lin W and Liu K 2010a Anti-forensics of JPEG compression *Proceedings of the International Conference on Acoustics, Speech, and Signal Processing*, Dallas, TX.

Stamm M, Tjoa S, Lin W and Liu K 2010b Undetectable image tampering through JPEG compression anti-forensics, China *Proceedings of the International Conference on Image Processing*, Hong Kong, China, pp. 2109–2112.

Stamm M, Lin W and Liu K 2012 Temporal forensics and anti-forensics for motion compensated video. *IEEE Transactions on Information Forensics and Security* **7**(4), 1315–1329.

Steinebach M, Liu H, Fan P and Katzenbeisser S 2010. Multimedia on mobile devices 2010. *Proceedings of SPIE 7542*, pp. 75420B.

Valenzise G, Tagliasacchi M and Tubaro S 2010 Estimating QP and motion vectors in H.264/AVC video from decoded pixels. *Proceedings of the 2nd ACM workshop on Multimedia in Forensics, Security and Intelligence (MiFor '10)*, Firenze, Italy, pp. 89–92.

Valenzise G, Tagliasacchi M and Tubaro S 2011 The cost of JPEG compression anti-forensics *Proceedings of the International Conference on Acoustics, Speech, and Signal Processing*, Prague, Czech Republic.

Valenzise G, Tagliasacchi M and Tubaro S 2013 Revealing the traces of JPEG compression anti-forensics. *IEEE Transactions on Information Forensics and Security* **8**(2), 335–349.

Vazquez-Padin D and Comesana P 2012 ML estimation of the resampling factor. *EEE International Workshop on Information Forensics and Security*, Tenerife, Spain.

Wang W and Farid H 2006 Exposing digital forgeries in video by detecting double mpeg compression. *Proceedings of the 8th Workshop on Multimedia and Security*, Geneva, Switzerland, pp. 37–47 ACM.

Yuan HD 2011 Blind forensics of median filtering in digital images. *IEEE Transactions on Information Forensics and Security* **6**(4), 1335–1345.

Index

Note: Page numbers in *italics* refer to Figures; those in **bold** to Tables.

Handbook of Digital Forensics of Multimedia Data and Devices, First Edition.
Edited by Anthony T.S. Ho and Shujun Li.
© 2015 John Wiley & Sons, Ltd. Published 2015 by John Wiley & Sons, Ltd.
Companion Website: www.wiley.com/go/digitalforensics